Global Perspectives on Health Geography

Series Editor

Valorie Crooks, Department of Geography, Simon Fraser University, Burnaby, BC, Canada

Global Perspectives on Health Geography showcases cutting-edge health geography research that addresses pressing, contemporary aspects of the health-place interface. The bi-directional influence between health and place has been acknowledged for centuries, and understanding traditional and contemporary aspects of this connection is at the core of the discipline of health geography. Health geographers, for example, have: shown the complex ways in which places influence and directly impact our health; documented how and why we seek specific spaces to improve our wellbeing; and revealed how policies and practices across multiple scales affect health care delivery and receipt.

The series publishes a comprehensive portfolio of monographs and edited volumes that document the latest research in this important discipline. Proposals are accepted across a broad and ever-developing swath of topics as diverse as the discipline of health geography itself, including transnational health mobilities, experiential accounts of health and wellbeing, global-local health policies and practices, mHealth, environmental health (in)equity, theoretical approaches, and emerging spatial technologies as they relate to health and health services. Volumes in this series draw forth new methods, ways of thinking, and approaches to examining spatial and place-based aspects of health and health care across scales. They also weave together connections between health geography and other health and social science disciplines, and in doing so highlight the importance of spatial thinking.

Dr. Valorie Crooks (Simon Fraser University, crooks@sfu.ca) is the Series Editor of Global Perspectives on Health Geography. An author/editor questionnaire and book proposal form can be obtained from Publishing Editor Zachary Romano (zachary.romano@springer.com).

Rais Akhtar
Editor

Climate Change and Human Health Scenarios

International Case Studies

Springer

Editor
Rais Akhtar
Formerly Professor of Geography
University of Kashmir
Srinagar, Jammu & Kashmir, India

ISSN 2522-8005 ISSN 2522-8013 (electronic)
Global Perspectives on Health Geography
ISBN 978-3-031-38877-4 ISBN 978-3-031-38878-1 (eBook)
https://doi.org/10.1007/978-3-031-38878-1

This Springer imprint is published by the registered company Springer Nature Switzerland AG
The registered company address is: Gewerbestrasse 11, 6330 Cham, Switzerland

Paper in this product is recyclable.

Foreword

Scientists drew attention to the possibility of future climate changes caused by human activities about four decades ago. At that time, some people welcomed the fact that in a warmer climate there would be less winter-time deaths, especially of old people. Others postulated that warming would have an adverse impact on human health because of the likely spread of diseases such as malaria. Both points of view were correct. However, as time has gone on, it has become increasingly appreciated that climate change will have adverse effects on human health in a multitude of ways. This is partly because we have seen the health effects of an increasing number of serious events, such as wildfires in Australia and California, the droughts that have been afflicting the Horn of Africa, and the catastrophic floods that have inundated large tracts of Pakistan. In addition, we have seen problems of water shortage associated with reductions in snowpack and the wasting and disappearance of glaciers in the "Water Towers" of Asia. Then there is the effect of heatwaves. In the year 2022, we were subjected in Europe to unprecedented summer temperatures, so that heat stress, especially in great cities, was experienced. Another major effect of higher temperatures is that the ranges and durations of certain diseases will expand, as with mosquito-borne malaria and dengue, tick-borne diseases, dust-storm borne allergies and respiratory and cardiovascular complaints, and bacterial and diarrhoeal problems. In addition, it is important to remember that climate change may have an effect on human *mental* health because of the effects of forced migration, flooding and erosion of island states and low-lying areas by sea-level rise, food insecurity, and various environmental disasters. The Intergovernmental Panel on Climate Change, reporting in 2022, (https://www.ipcc.ch/report/ar6/wg2/downloads/report/IPCC_AR6_WGII_Chapter07.pdf), suggests that in coming years, we may be looking at 250,000 excess deaths per year, with the poorest countries and the weakest individuals at the greatest risk.

This book is, therefore, immensely important in understanding one of the greatest threats facing humankind.

Andrew Goudie
University of Oxford
Oxford, UK

Preface

To quote UN Secretary General, humanity is on a "highway to climate hell", who had warned, saying the fight for a liveable planet will be won or lost in this decade. António Guterres told world leaders at the opening of the Cop27 UN climate summit in Sharm El-Sheikh, Egypt: in November 2022 "We are in the fight of our lives and we are losing". These pessimistic words of UN Secretary General reveals the catastrophic scenario of impacts of climate change on human being. On 27th July 2023, he again warned that, "the era of global warming has ended and the era of global boiling has arrived".

The present book entitled *Climate Change and Human Health Scenarios*, which covers countries both from developed and developing regions, with focus growing human health problems, adaptation and mitigation strategies adopted by different governments .In short, Climate change affects the social and environmental determinants of health—clean air, safe drinking water, sufficient food and secure shelter. The assessment between 2030 and 2050, reveals that climate change is expected to cause approximately 250,000 additional deaths per year, from vector borne diseases-, malaria, and diarrhoea, malnutrition and heat stress in developing countries.

The European Environment Agency (EEA) report highlighted "on the impact high temperatures are having on the population, which leads to the largest number of fatalities associated with natural hazards in Europe. Due to climate change, these fatalities are projected to increase substantially unless adaptation measures are taken. Climate-sensitive infectious diseases—another emerging threat—are projected to further spread northwards and cause a higher disease burden in Europe. The report draws on knowledge developed for the European Climate and Health Observatory, which provides access to a wide range of relevant data, tools, publications and other resources informing about climate change impacts for human health." (EEA, 2022).

Such human health scenario has been the source of encouragement to me to edit a book on *Climate Change and Human Health Scenarios* in countries from both developing and developed world.

In Australasia region, with focus on Australia, John Connel asserts that climate change is increasingly influencing Australian lives and livelihoods, with direct and indirect repercussions for health status. Achieving effective climate action has become a key political issue. The most immediate health risks are from extreme weather events related to climate change, including heatwaves, bushfires, cyclones, floods and droughts.

In nearby country Fiji a major challenge is arising from natural hazards are tropical cyclones (TC) and floods. Climate change requires well-functioning social institutions at the local level. These are support systems that help people to better deal with natural hazards. People can reduce the risk that hazards turn out to become disasters best using their own capacities and capabilities.

Mei-Hui Li focuses on an integrated approach in her study on Taiwan, that is relevant to understand impacts of the novel coronavirus disease 2019 (COVID-19) for implications for future public health policy in the face of climate change. Additionally, social inequality and vulnerability of COVID-19 pandemic are discussed by applying syndemic thinking.

Oka Kazutaka in his chapter on Japan highlights that the temperature rise has caused severe heat-related health impacts in Japan, including heatstroke. In Japan, heatstroke causes 65,000 ambulance cases and 1000 deaths annually. Various measures have been implemented by the Japanese government, including the launch of the "Heatstroke Alert", a heat-health warning system to reduce the health impacts caused by heatstroke. This chapter introduces the main measures implemented by the Japanese government. Notably, heatstroke health impacts have been intensively studied in Japan. Based on these studies, this chapter describes the scientific findings and issues related to health impacts, such as heatstroke and heat-related excess mortality.

In Indonesia, Budi Haryanto and his co-authors focus on Nationally Determined Contributions (NDCs) are the necessary non-binding actions plans on climate change targeted by each country as their long-term goals on reducing emissions and combating climate change impacts. Implementation of NDCs includes enhanced ambition on adaptation as elaborated in the programmes, strategies and actions to achieve economic, social and livelihood, and ecosystem and landscape resilience; enhanced clarity on mitigation by adopting the Paris Agreement rule book (Katowice Package) on information to be provided in NDC, as well as updated policies which potentially contribute to additional achievement of NDC target. Indonesia is ongoing to strengthen the NDCs by developing health-inclusive and health-promoting climate targets and policies.

The effects of climate change impact humanity diversely in Thailand, posing a great danger to the fundamental components of health and well-being, including clean air, safe drinking water, wholesome food, secure housing as well as outbreaks of pandemics. This threat to global health could potentially overwhelm health systems, particularly during crisis if not resilient enough to cope with challenges in providing essential services as witnessed during the COVID-19 pandemic which greatly exposed health systems hitherto considered strong. To identify government policies and programs supportive of a climate-resilient health system, a documentary review was carried out between 2012 and 2022 to analyse policy measures that could potentially contribute to developing and sustaining a health system capable of resiliently responding to health needs. A total of 54 articles were identified. After the articles were de-duplicated, the title and abstract were checked, and 42 articles were further screened using full text. Of these, 11 met the eligibility requirements for inclusion in the final analysis. The existing policy measures were broadly categorized at the system and agency level and the level of the population. The climate change adaptation and mitigation policy measures have significant potential for health system resilience worthy of emulation as they aim to strengthen health related systems and agencies and are sensitive to healthcare needs population wide. The policies should be consistently backed by political commitment through sustained advocacy and pragmatic actionable steps to realize the objective of a climate change-resilient health system for the provision of efficient people-centred care through collaboration and partnership.

Malaysia is one of the vulnerable climate change hotspots that ostensibly witness many extreme weather events that amplify the burden of climate-sensitive diseases. Elevating rates of urbanization and population explosion in near future could magnify the implications of climate change in Malaysia, including the exacerbation of warming trends, amplification of environmental disasters and resurgence of infectious diseases. In response to this, environmental and public health institutions at all operational scales need to consciously modify their approaches to articulate evidence-based adaptation and mitigation strategies to build climate resilience across all the sectors in Malaysia. Mainstreaming behaviour change into the adaptation strategies related to the extreme climate events along with infrastructure, technological and policy advancements is another key aspect to strengthening the public health adaptive capacity and responses in Malaysia. The identification and management of constraints and barriers to climate change adaptation in anticipation of existing public health strategies are equally inevitable to address the adverse health impacts and increase the efficiency and sustainability of climate solutions in the country.

It is projected that Bangladesh will be severely impacted by climate change and extreme weather-related events, such as global warming, sea-level rise, catastrophic cyclones, flood, and drought. In addition, the air quality of Bangladesh, especially in the urban areas, has deteriorated in recent years. Dhaka, the country's capital city, is often ranked as one of the worst urban areas in the world for its degraded air quality. The polluted air in and around Dhaka is estimated to affect more than 40 million people. In combination with future climate change issues, air pollution will potentially cause acute public health outcomes in Bangladesh if no action is taken. This chapter examined in situ and satellite-based air quality data; explored the spatio-temporal distribution of air quality measures in urban and rural areas; identified air pollutant hotspots; assessed the connection of air pollutants with climate variables; and then investigated the potential impact of hypothetical climate change-related phenomena on air quality in Bangladesh from a public health perspective.

Chapter "Heatwave Mortality and Adaptation Strategies in India" focuses on heatwave mortality and adaptation strategies in India. The WHO estimates that from 1998–2017, more than 166,000 people died due to heatwaves, including more than 70,000 who died during the last two weeks of August 2003 heatwave in Europe. Indian Meteorological department (IMD) predicts that most parts of India including those in peninsular and coasts of India will experience in increase in duration of heatwave by 12–18 days by 2060.

The study shows that mortality due to heatwaves started declining in the country from 2016. This is because various heat protection plans initiated by the governments both at the Centre and state levels and other organizations helped in the reduction of heatwave related deaths. The introduction of Ahmedabad Heat Action Plans (HAP) in some of the cities in Gujarat, Maharashtra, Odisha and Telangana have demonstrated that such action plans can improve resilience of the citizens and reduce the severe health impacts of Heat Wave.

The Hindu Kush Himalayan (HKH) region occupies areas of eight countries, namely Afghanistan, Bangladesh, Bhutan, China, India, Myanmar, Nepal, and Pakistan. The HKH region is warming at a rate higher than the global average, and precipitation has also increased significantly over the last six decades, along with increased frequency and intensity of some extreme events. Changes in temperature and precipitation have affected and would like to affect the health and well-being of mountain people. This book chapter aims to document how climate change has impacted and will impact the health and well-being of the people in the HKH region and offers adaptation and mitigation measures to reduce the vulnerability and impacts of climate change on the health and well-being of the people.

Examples from Studies in Africa, Include from Kenya, Tanzania, Nigeria and South Africa

In Kenya, Andrew Githeko asserts that Kenya has been impacted by the effects of climate change that include epidemics, geographic range expansion of climate-sensitive diseases, droughts and floods. These diseases cause a high health burden. The public health system has put in place intervention measures. Malaria control relies on insecticides and drugs. Rift Valley Fever is managed by vaccinating livestock while dengue and chikungunya are managed using insecticides and larval source management. Victims of drought and floods depend on humanitarian assistance. Water-borne infections can be reduced using safe drinking water sources. A shift from rain-fed to irrigated crops is expected to reduce food insecurity. These adaptation projects, according to the author, require heavy financial investments from government development budgets.

The United Republic of Tanzania is among the countries that has experienced impacts of climate change including epidemics of climate-sensitive infectious diseases, food and nutrition insecurity. Others include damage to infrastructure caused by flooding and land slides resulting to human injuries, deaths and displacement, and high cost to restore the damaged infrastructure. The climate-sensitive diseases that have occurred in Tanzania include dengue, chikungunya, malaria, Rift Valley fever, leptospirosis, cholera and Human African Trypanosomiasis. The country has recently developed a National Climate Change Strategic Plan to provides a set of interventions on adaptation and mitigation, which are expected to strengthen country's resilience to the impacts of climate change and contribute to the global efforts of reducing greenhouse gas emissions. In addition, the country has developed the Health-National Adaptation Plan to climate change to guide the towards a health system that is more resilient to climate change. However, the efficiency of the operationalization of these strategies are not sufficiently known. Moreover, the vulnerability to climate change and mitigation, adaptation and coping measures at different levels of the health system in different areas of the country have not been well studied. The climate projections in Tanzania indicate that there will be an annual increase of rainfall by 10% by 2100 and temperature is projected to increase by 1.5 to 4.5 °C by 2090. These projections suggest that more impacts from climates change are expected, which call for appropriate mitigation, adaptation, and coping strategies at all levels of health system.

Authors of chapter on South Africa highlight that Southern Africa bears disproportionate consequences of the changing climate. The El Niño Southern Oscillation causes phases of extreme weather events, leading to flooding and extended periods of droughts in different regions of the sub-continent. Our data provides evidence that the extreme El Niño event of 2014–16 increased the risk for perinatal infant mortality in the Northern Cape and North West provinces of South Africa. The maternal health services profile of the Northern Cape and KwaZulu-Natal suggests a compromised health system with limited resilience to respond to the climate crises. The National Department of Health lacks adequate policies and strategies to ensure systems are able to meet the maternal and child health requirements in the context of climate change.

The role of climate change as a trigger factor causing the advancing spread of diseases in Russia has been analyzed. The potential change in ranges due to predicted climate warming was studied according to climate model INM-CM5.0. A series of maps was compiled to identify the territories prone to suitability changes for the infection foci for the period up to 2100. It was determined that regions with temperate and arctic climate may become vulnerable to the emergence of climate-related diseases in the course of environmental changes.

Spain is reaching the forecasts set by the Intergovernmental Panel on Climate Change (IPCC) since 1990–1992. To get a consensus and reach a minimum governmental awareness of the problem, numerous global meetings were necessary in Spain, like in other countries. However, it was clear that there is a need to transfer this reality to society clearly, concisely and forcefully, influencing changes in social norms, political priorities and cultural values. The scientific literature agrees that the most important climate change events affecting human health are: high temperatures, heat waves and ultraviolet radiation, as well as air, soil and water pollution. In addition, torrential rains, droughts, forest fires, diminishing water resources, coastal phenomena and endangered habitats could be also included. The aim of this chapter is to present the state of the art on the effects of climate change on health in Spain. So, methodologically, diseases exacerbated by climate change detected in Spain were organized according to medical specialities and climatic elements, analysing morbidity and mortality. Spain increased its population from 2000 to 2020 by 16.6%

and stabilized its mortality at 9.01% (omitting COVID-19). Other reasons aside, increases in morbidity or mortality above these demographic values may be due to the effects of climate change. Thus, the data consulted indicate that 26.7% of mortality is due to cancers, which increase in women (26% between 2000 and 2020) and stabilize in men. This is followed by heart disease (18.8%), which has fallen since 2000; digestive diseases (11.8%), which have increased by 20.3%; and respiratory and neurological diseases (13.1 and 12%, respectively), which have stabilized since the effect of COVID-19 has been cancelled out; this zoonotic disease, in 2020 alone, increased mortality by 1.37%, and thus increased mortality due to infectious diseases to 18.2% of all deaths. Climate change-related illnesses increase more in women. After reviewing this, we conclude that, in addition to the need to reduce greenhouse gases, mitigation measures should include self-protection against heat, ultraviolet rays and water purification, as well as increased research on the environmental effects of climate on human health. Any of these measures can be understood as "disease sinks".

In France the author insists that after the deadly nature of the 2003 heat wave and the occurrence of numerous brutal, progressive or systemic climatic disasters, the French became aware that the predictions made by the models had become reality, but the issues raised by mitigation and adaptation go beyond the political field to question values such as justice, solidarity and sobriety. However, the insurance system used to repair disasters is at the end of its tether and only prevention, combined with mitigation are necessary for avoiding to fall into the trap of bad-adaptation: injustice, rebound effect. Only a very integrated territorial policy, following the "one health" concept can enable the French to improve their health by fighting against climate change.

The author discusses climate change in the United Kingdom in relation to the health of its people. He begins with a brief discussion of the nature of the existing and predicted impacts of climate change in the UK alongside a history of the politics of climate change in the UK and an overview of government climate change mitigation policy (and particularly its inadequacies). The chapter then concerns itself with the present and future impacts of climate change on human health in the UK. The author examines the impacts of exposure to extreme heat, flooding, and disruption to health services as key examples of the negative impacts of climate change on health. Pertinently the author discusses the uneven distribution of the health risks associated with climate change according to socio-economic status that emerge from climate change mitigation strategies and the associated decarbonization of the economy.

The author of the chapter on Canada suggests that climate change impact is an active area of research, the increasing average temperature and decreasing atmospheric humidity due to climate change is predicted to increase the frequency and intensity of large wildfires in Canada. Wildfire smoke causes immediate respiratory distress, although there is a noted absence of research into prolonged exposure and long-term health outcomes. Further, evacuation from wildfires causes short-term hardships which leads to long-term mental health outcomes. Options to adapt to wildfires are limited and our capacity to prevent ever worsening wildfires in the future may be overwhelmed.

Fire and climate change in the Sierra Nevada of California, USA have a complex interaction with human land management and forest ecology. Fire was an important agent of change for the fire prone forests of this landscape. Many species, such as the Giant Sequoia (*Sequoiadendron giganteum*) evolved to take advantage of frequent fire as this natural process sculpted the environment. Native Americans used fire widely for socioeconomic benefit and fuel reduction with moderate intensity fire encouraged to burn across the land. Euro-American settlement brought about an era of suppression that increased fuels and changed the forest composition and structure. But, suppression was and is the simple seeming solution. Even if suppression is not sustainable it will garner support. Historic suppression has currently brought an extreme fuel problem that has manifested into a greater and greater threat of destructive high intensity fire not typical of this ecosystem. Fire policy was and is slow to change due to risk aversion and lack of urgency. This is not in small part from increased smoke impacts as a result of heavy fuel loads and returning fire to the landscape. These emissions were essentially mortgaged to the current age from previous generations. Fire and land management policies collide with air regulatory policy in California because of already heavily anthropogenically polluted air with little to no capacity for an additional emission source. However, fire and the subsequent smoke are inevitable. Public smoke tolerance is low and a significant deterrent to bringing fire back to California wilderness. This combined with the political polarization of the 21st century and even the simplest science based solutions are unpalatable as can be seen by the myriad of publications outlining a path forward. The need for more frequent low intensity fire over California wild lands is obvious. Climate change is narrowing the options for fire and smoke management as suppression fails. We discuss the challenges and potential solutions to this conundrum that allow fire to act as an agent of change and ecosystem benefit while minimizing the health impacts from smoke and attempt to further the public understanding of tradeoffs in a complex ecosystem process.

Another chapter on USA Focuses on climate change and mental health. Awareness of climate-related mental health issues has important ramifications for the implementation of national healthcare policies. This is more so in the United States (the third most populous country and a leading contributor to global greenhouse gas emission), where over one in ten adults live with severe mental illness and/or a substance use disorder (SUD). In the 2016 *Climate and Health Assessment* report, the U.S. Global Change Research Program (USGCRP) noted (with very high confidence in some instances) that exposure to weather and climate extremes increases the risk of trauma-and-stressor-related disorders, anxiety, depression and substance use, among other psychological issues. Given that mental and SUDs are the leading cause of years lived with disability globally, preventing the exacerbation of this reality by a changing climate is crucial. In this chapter, we introduce the reader to established mental health consequences of climate change with focus on the United States.

In Mexico, Climate change directly affects health through extreme weather events, and indirectly through the effect of these events on the dynamics of pathogens and vector-borne diseases, as well as on the productivity of crops that impact human nutrition. Mexico's geographic location is a relevant factor for exposure to hydrometeorological phenomena, such as cyclones, storms, and floods. This chapter reviews how extreme hydrometeorological events affect ecosystems and therefore morbidity and mortality in Mexico. It also discusses health impacts from the lack of water and food security. This leads to the need of specific public policies for adaptation to climate change. We describe Mexico's institutional framework regarding health and climate change. We discuss whether it has the necessary policies and resources to face this phenomenon. Finally, we make recommendations to make health and general policy climate sensitive.

Climate change, caused by the increase in Greenhouse Gas (GHG) emissions, is inducing significant climatic alterations in Argentina, so it is a priority to deepen the knowledge of its impact, particularly on human health. In this context, the main diseases that cause serious threats to the population are presented. Aspects as relevant as the effects that extreme temperatures expressed as heat waves and cold waves, air or water quality and the possible spread of diseases have on the morbidity and mortality of the population are addressed. It is concluded that the health effects caused by climate change require new strategies to mitigate them, with a multidisciplinary and intersectoral approach, where prevention and health promotion actions are essential in addressing this issue. In addition, the article intends to present a summary of the mitigation and adaptation measures applied in Argentina on the health sector. Both measures are complementary and, although they present different challenges, they converge in the final objective.

In recent years, Brazil, multidisciplinary studies have sought to understand the interaction between environmental factors that modulate people's risks and susceptibilities to climate change. This study aimed to assess the socio-environmental and health vulnerability of the population in the state of Mato Grosso do Sul, Brazil, to climate change through the application of a vulnerability index. The results show sensitivity and adaptive capacity as bottlenecks found in this study. Challenging conditions include unequal access to health services and population occupation concentrated in large urban centers.

In another case study from Brazil on climate change and forced migration, the chapter presents and analyzes the role of public policies in Brazil in the face of the effect of climate change, focusing on displacement and women's health. It describes the role of the Brazilian State and the attention given to the health field of displaced women. In this context, we reflect the social and economic vulnerability of women in Brazil and the fragility the social assistance policies for this group of population who, as a result of the social and political dimension of disasters, lose their homes and become either homeless and/or displaced.

Aligarh, India Rais Akhtar

Acknowledgements

In the process of writing, editing and preparing this book, there have been many people who have encouraged helped and supported me with their skill, thoughtful evaluation of chapters and constructive criticism.

First of all I am indebted to all the contributors of chapters from both developed and developing countries for providing the scholarly and innovative scientific piece of research to make this book a reality. I am also thankful to the reviewers who carefully and timely reviewed the manuscripts.

I am also grateful to Prof. Andrew Goudie, University of Oxford, for writing Foreword, which adds greatly to the book with his thoughtful insights.

I am thankful to my family-wife Dr. Nilofar Izhar, daughter Dr. Shirin Rais, Assistant Professor, AMU, and my son-in law Dr. Wasim Ahmad, Associate Professor IIT Kanpur, who encouraged and sustained me in developing the structure of the book and editing tasks, and I am deeply grateful for their support and indulgence.

Finally and most essentially I am deeply obliged to the Springer and the entire publishing team, without whose patience, immense competence and support this book would not have come to fruition. I specially thank Dr. Robert K. Doe whose energizing leadership ensured that this book would indeed translate to reality. I am also thankful to Prof. A. R. Kidwai, Doris Bleier, Aarthi Padmanabhan, and Jayanthi Narayanaswami for their constant cooperation.

Aligarh, India Rais Akhtar

Contents

About the Editor

Rais Akhtar is Formerly Professor of Geography, University of Kashmir, Srinagar, Jammu & Kashmir, India, National Fellow and Emeritus Scientist (CSIR), CSRD, Jawaharlal Nehru University, New Delhi, and Visiting Professor Department of Geology, AMU, Aligarh.

He has taught at the Jawaharlal Nehru University, New Delhi, University of Zambia, Lusaka and The University of Kashmir, Srinagar. He is recipient of a number of international fellowships including Leverhulme Fellowship (University of Liverpool), Henry Chapman Fellowship (University of London), and Visiting Fellowship, (University of Sussex), Royal Society Fellowship, University of Oxford and, Visiting Professorship, University of Paris-10. Prof. Akhtar was elected Fellow of Royal Geographical Society, London and Royal Academy of Overseas Sciences, Brussels.

Prof. Akhtar delivered invited lectures in about 55 universities geography departments and medical colleges including London School of Hygiene and Tropical Medicine, Liverpool School of Tropical Medicine, Department of Geography, University of Edinburgh, and School of Public Health, Johns Hopkins University, Baltimore.

He was Lead Author (1999–2007), on the Intergovernmental Panel on Climate Change (IPCC), which is the joint winner of Nobel Peace Prize for 2007. Prof. Akhtar is the recipient of Nobel Memento.

Prof. Akhtar has to his credit 91 research papers and 21 Books published from India, United Kingdom, United States, Germany and The Netherlands. His books entitled: *Climate Change and Human Health Scenario in South and Southeast Asia*, was published in 2016 by Springer, and his other books include *Climate Change and Air Pollution* (with C. Palagiano) published by Springer, and *Geographical Aspects of Health and Disease in India* (with Andrew Learmonth) published in 2018 by Concept (New Delhi). His latest books include: *Extreme Weather Events and Human Health*, was published by Springer in early 2020, *Coronavirus (COVID-19) Outbreaks, Environment and Human Behaviour*, and *Coronavirus (COVID-19) Outbreaks, Vaccination, Politics and Society, 2022*, were also published by Springer in 2021 and 2022 respectively.

Prof. Rais Akhtar was Member of Expert Group on Climate Change and Human Health of the Ministry of Health and Family Welfare, Government of India. Prof. Akhtar has been appointed as Corresponding Member of Italian Geographical Society in October 2018.

Introduction: Climate Change and Human Health Scenarios

Rais Akhtar

Abstract It has been scientifically proved that climate change has upset various physical and biological systems globally which in turn, influenced human health in both the developed and developing countries. WHO asserts that "Climate change is impacting human lives and health in a variety of ways. It threatens the essential ingredients of good health—clean air, safe drinking water, nutritious food supply and safe shelter—and has the potential to undermine decades of progress in global health" (WHO, https://www.who.int/health-topics/climate-change#tab=tab_1).

Keywords WHO · PCC · Hospitalisation · Hypothermia · Lyme disease · Malaria · Pollen energy · WHO conferences · Paris climate agreement · Nationally determined contributions

1 Climate Change and Human Health Scenario: Global Case Studies

It has been scientifically proved that climate change has upset various physical and biological systems globally which in turn, influenced human health in both the developed and developing countries. WHO asserts that "Climate change is impacting human lives and health in a variety of ways. It threatens the essential ingredients of good health—clean air, safe drinking water, nutritious food supply and safe shelter—and has the potential to undermine decades of progress in global health" (WHO, https://www.who.int/health-topics/climate-change#tab=tab_1).

The human health impacts of climate change encompadirect and indirect effects and immediate and delayed effects. At the same time, climate change has also been demonstrated to have both a positive and serious negative impact on human health and wellbeing as depicted by Tony McMichael in his pioneering work on climate change and human health (Fig. 1). This has led to the growing recognition mainly

R. Akhtar (✉)
Formerly Professor of Geography, University of Kashmir, Srinagar, Jammu & Kashmir, India
e-mail: raisakhtar@gmail.com

R. Akhtar (ed.), *Climate Change and Human Health Scenarios*, Global Perspectives on Health Geography, https://doi.org/10.1007/978-3-031-38878-1_1

since 1990s that human health is threatened by unsustainable global environmental trends. A calendar entitled *Climate Change is Affecting Your Health*, published by the WHO in 2008, helped in deeper understanding of the seriousness of the problem concerning climate change.

WHO contends that climate change is the single biggest health threat facing humanity, and health professionals worldwide are already responding to the health disaster caused by this emerging crisis. According to an estimate by WHO, "between 2030 and 2050, climate change is expected to cause approximately 250,000 additional deaths per year from malnutrition, malaria, diarrhea and heat stress alone. The direct damage costs to health are estimated to be between US $2 and 4 billion per year by 2030. Areas with weak health infrastructure—mostly in developing countries—will be the least able to cope without assistance to prepare and respond" (WHO, 2021). The IPCC Sixth Synthesis report highlights "a serious climate health scenario, "those include warnings that the world was approaching 'irreversible' levels of global

	MEDIATING PROCESS	HEALTH OLUTCOMES
DIRECT EFFECTS (PRIMARY)	Thermal extremes (e.g. heatwaves)	→ Heat extreme wildfires, floods storms and death.
	Increase in weather disasters	→ Deaths, injuries and psychological disorders
(SECONDARY)	**DISTURBANCES OF ECOLOGICAL SYSTEMS**	
	Range/activity of vectors and infective agents	→ Altered range/pattern of vector borne diseases, Mosquito proliferation
	Water and food borne infective agents.	→ Diarrheal and other infectious diseases
	Altered food (especially crop) and livestock productivity	→ Hunger and under nutrition, impaired child development
INDIRECT EFFECTS (TERTIARY)	Sea-level rise, with population displacement	→ Injuries, infections,crowding, contaminated water, precipitate cholera outbreaks in impoverished communities, psych, disorders.
	Changes in air pollution including pollens and spores	→ Asthma and allergies, other respiratory disorders, acute rhinitis (coryza), Covid-19
	Social demographic dislocations due to assorted climate change impacts	→ Many public health consequences (incl. civil strife?) post traumatic stress long term-depression and survivor guilt

Fig. 1 Climate change and human health. *Source* McMichael (1996) updated by Rais Akhtar (2023)

heating, with catastrophic impacts rapidly becoming inevitable; and that it was 'now or never' to take drastic action to avoid disaster" (Harvey, 2023). The IPCC Sixth Synthesis report also maintains, "in all regions, increases in extreme heat events have resulted in human mortality and morbidity (very high confidence). The occurrence of climate-related food-borne and water-borne diseases (very high confidence) and the incidence of vector-borne diseases (high confidence) has increased." In assessed regions, some mental health challenges are associated with increasing temperatures (high confidence), trauma from extreme events (very high confidence) and loss of livelihoods and culture (high confidence). Climate and weather extremes are increasingly driving displacement in Africa, Asia, North America (high confidence) and Central and South America (medium confidence), with small island states in the Caribbean and South Pacific being disproportionately affected relative to their small population size (high confidence)" (IPCC, 2023).

2 Climate Change: Direct and Indirect Impacts

To elaborate further, the direct health effects of climate change would include altered rates of mortality/morbidity due to heat waves and thermal stresses (Fig. 1) in general, the respiratory consequences of a change in patterns of exposure to aeroallergens and increases in extreme weather events, including storms (as well as snow storms during winter), floods and droughts. The indirect health effects of climate change include (i) alterations in the range and activity of climate-sensitive vector- and virus-borne diseases (such as malaria, dengue, chikungunya and schistosomiasis), (ii) changes in transmission of person-to-person infections (including food poisoning and water-borne pathogens), (iii) the nutritional and health consequences of local and regional changes in agricultural productivity and (iv) the various consequences of rising sea level (McMichael, 2017), and forest fires (Akhtar, 2020) and increased frequencies and intensity of hurricanes. Experts also say that climate change has resulted in the emergence of virus-borne region-based diseases such as Ebola in West Africa and Zika in Latin America (Akhtar, 2020).

Besides the impact on public health, climate change may be far reaching and include death and hospitalizations due to heat waves, as seen during the 2003 heat-wave in the western and central Europe; hypothermia from blizzards as evident in the recent Arctic bomb cyclone which claimed dozens of human lives in Canada and USA (December, 2022); injuries and death from flooding; and potential shifts in the transmission ranges of vector-borne diseases such as hantavirus, West Nile virus, tick-borne encephalitis, lyme disease, malaria and dengue. Most importantly, "the potential population health impacts of environmental changes extend far into future, if environmental conditions deteriorate further, change can be abrupt and unexpected but they can also be protracted and gradual and thus pose considerable challenges to public health." (Semenza & Paz, 2021).

Europeans are not only exposed to direct effects from climate change, but are also vulnerable to indirect effects from infectious disease, many of which are climate sensitive, which is of concern because of their epidemic potential. Climatic conditions have facilitated vector-borne disease outbreaks like chikungunya, dengue and West Nile fever and have contributed to a geographic range expansion of tick vectors that transmit Lyme disease and tick-borne encephalitis in developed countries. Extreme precipitation events have caused water-borne outbreaks and longer summer seasons have contributed to increases in food-borne diseases. Under the Green Deal, The European Union aims to support climate change health policy, in order to be better prepared for the next health security threat, particularly in the aftermath of the traumatic COVID-19 experience (see also Akhtar (ed.) 2021, 2022). To bolster this policy process, there is a need to highlight climate change-related hazards, exposures and vulnerabilities to infectious disease and delineate observed impacts, projected risks, with policy entry points for adaptation to reduce these risks or avoid them altogether (Semenza & Paz, 2021).

Gennaro an Italian scientist and his colleagues (2023), emphasized that, "Climate change affects the quantity, intensity and frequency of precipitation type as well as extreme climate change events such as heat waves, droughts, thunderstorms, floods and hurricanes". Respiratory health can be particularly affected by climate change, which can contribute to the development of asthma and allergic respiratory diseases. Pollen allergens have been shown to trigger the release of immunomodulatory and pro-inflammatory mediators that accelerate the onset of allergy. Allergy to pollen and pollen season at its beginning in duration and intensity is altered by climate change. Studies show that plants exhibit enhanced photosynthesis and reproductive effects and produce more pollen as a response to high atmospheric levels of carbon dioxide (CO_2). Pollen allergy is generally used to evaluate the interrelation between air pollution and allergic respiratory diseases, such as rhinitis and asthma. Lightning storms during pollen seasons can cause exacerbation of respiratory allergy and asthma in patients with pollinosis (Gennaro et al., 2023).

3 WHO Conferences of Health and Climate

The World Health Organization has been publishing reports on climate change and health since 2003, but the three-day conference, hosted by the World Health Organization in Geneva, Switzerland, had brought together leading experts in the fields of health and climate change, and discussed strengthening health system resilience to climate risks; and promoting health while mitigating climate change. Within each of these themes, the conference advanced recommendations on policy options to maximize health benefits and the specific contribution of the health sector in both the developed and developing countries.

The WHO organized a second global conference on health and climate change, in Paris' during July 7–8, 2016 and this was aimed at setting a global health action agenda necessary for the implementation of the Paris Climate Agreement of 2015. The conference emphasized the benefits of switching to cleaner energy sources. It was stressed that the health sector should itself make a greater effort to promote low-carbon healthcare facilities and technologies; to reduce costs as well as climate and environmental impacts.

4 Recent Climatic Hazards

Pakistan declared a state of emergency on August 25, 2022, because of flooding. The flooding was the world's deadliest flood since the 2020 South Asian floods and described as the worst in the country's history. It was also recorded as one of the world's costliest natural disasters of all time.

From December 21 to 26, 2022, a historic extra-tropical cyclone created winter storm conditions, including blizzards, high winds, snowfall or record cold temperatures across the majority of the United States and parts of Canada. Areas which experienced blizzard conditions included parts of Minnesota, Iowa, Wisconsin, Michigan, Ohio, New York and Ontario, with the Buffalo area of New York and the Fort Erie and Kingston areas of Ontario experiencing almost two full days of blizzard/zero-visibility conditions on December 23 and 24. The cold wave affected all U.S. states from Colorado to the eastern seaboard and as far south as Miami (Wikipedia, 2022). The storm and the related cold wave killed at least 100 people.

According to Jared, there are a couple reasons why things were so bad in Buffalo. The first is that Lake Erie was still very warm when the storm hit. This resulted in lake effect snow adding to the snow depth. The other major issue is that this city is very segregated and has communities with entrenched poverty. People living in these areas were disproportionately affected by the storm and most of the deaths have occurred in these areas (Jared, 2022).

In early March 2023, intense storms have slammed both coasts of the US, bringing more rain, flooding and mud slides to California and high winds and heavy snow to the north-east including parts of New York, New Hampshire, Massachusetts and Vermont.

Hannam explained that for most Australians, 2022 will be recalled for the rain and the floods—but also some exceptional heat. In a year marked by flooding, there were at least 10 sites that recorded more than half a meter of rain in a day. Doon Doon, a small rural locality approximately 40 km south-west of Tweed Heads in the Tweed Shire in the New South Wales northern rivers region, topped the lot as the wettest location in 2022 with 758 mm on 28 February (Hannam, 2022).

5 IPCC 2018 Special Report, Global Warming of 1.5 °C

IPCC 2018 Special Report entitled, *Global Warming of 1.5 °C* asserts that emissions of carbon dioxide (CO_2) caused by human activity must reach "net zero" by 2050 to keep the average rise in global temperatures at 1.5 °C above pre-industrial levels to reduce catastrophic climate change risk on populations (IPCC, 2018). However in 2022, UNEP in its Emissions Gap Report contends that "the window is closing! The world is not on track to reach the Paris Climate Agreement goals and global temperatures can reach 2.8 °C by the end of the century" (UNEP, 2022). Such grim scenario would mean significantly more heatwaves, extreme precipitation (as was evident in the recent extreme precipitation and flooding in Pakistan), droughts, and ice melting than if we stayed below that threshold. The 2018 European drought and heatwaves was a period of unusually hot weather that led to record-breaking temperatures and wildfires in many parts of Europe, especially in France, Spain and Portugal during the spring and summer of 2018 (Rubin, 2018). Japan suffered an unprecedented two-week heatwaves in July 2018 that was declared a natural disaster by the government. The number of deaths rose sharply: at least 65 people died of heatstroke during the last week of July, while 22,647 people were hospitalized. In March–April 2022, India and Pakistan experienced a heatwave that was 30 times more likely to have happened because of climate change (D'Souza, 2022).

Most of these impacts will leave an irreversible imprint on ecologies and people, said Jean-Pascal van Ypersele, former vice chair of the Intergovernmental Panel on Climate Change (Hindustan Times, 2023).

6 IPCC (2023) ARI Synthesis Report

More than a century of burning, fossil fuels as well as unequal and unsustainable energy and land use have led to global warming of 1.1 °C above pre-industrial levels. This has resulted in more frequent and more intense extreme weather events that have caused increasingly dangerous impacts on nature and people in every region of the world. But there are multiple, feasible and effective options to reduce greenhouse gas emissions and adapt to human-caused climate change, and they are available now, said scientists in this IPCC report. Taking effective and equitable climate action will not only reduce losses and damages for nature and people, it will also provide wider benefits, the report points out, underscoring the urgency of taking more ambitious action now to secure a livable sustainable future for all.

7 Observation by IPCC Chair on the Synthesis Report

There are multiple, feasible and effective options to reduce greenhouse gas emissions and adapt to human-caused climate change, and they are available now, said scientists in the latest report released by the Intergovernmental Panel on Climate Change (IPCC) on March 20, 2023. "This Synthesis Report underscores the urgency of taking more ambitious action and shows that, if we act now, we can still secure a liveable sustainable future for all," said IPCC Chair Hoesung Lee. It is to be hoped the government in both the developed and developing countries must heed the warning related to worsening climate apocalypse in the world.

The present book encompasses case studies on climate change and human health scenario of 26 countries from both the developed and developing regions.

8 Examples from Selected Countries

It is interesting and relevant to give examples from this book on climate change impacts on health in selected countries.

In the Australasia region, with the focus on Australia, John Connell argues that climate change is increasingly influencing Australian lives and livelihoods, with direct and indirect repercussions for health status. Achieving effective climate action has become a key political issue. The most immediate health risks are from extreme weather events related to climate change, including heatwaves, bushfires, cyclones, floods and droughts. In nearby Fiji, a major challenge is arising from natural hazards which are tropical cyclones (TC) and floods. Climate change requires well-functioning social institutions at the local level. These are support systems that help people to better deal with natural hazards. People can reduce the risk that hazards turn out to become disasters best using their own capacities and capabilities. Mei-Hui Li focuses on an integrated approach in her study on Taiwan, that is relevant to understand impacts of the novel coronavirus disease 2019 (COVID-19) for implications for future public health policy in the face of climate change. Additionally, the social inequality and vulnerability of the COVID-19 pandemic are discussed by applying syndemic thinking.

Oka Kazutaka in his chapter on Japan highlights that the temperature rise has caused severe heat-related health impacts in Japan, including heatstroke. In Japan, heatstroke causes 65,000 ambulance cases and 1000 deaths annually. Various measures have been implemented by the Japanese government, including the launch of the "Heatstroke Alert," a heat-health warning system to reduce the health impacts caused by heatstroke. This chapter introduces the main measures implemented by the Japanese government. Notably, heatstroke health impacts have been intensively studied in Japan. Based on these studies, this chapter describes the scientific findings and issues related to health impacts, such as heatstroke and heat-related excess mortality.

In Indonesia, Budi Haryanto and his co-authors focus on Nationally Determined Contributions (NDCs) which are the necessary non-binding actions plans on climate change targeted by each country as their long-term goals on reducing emissions and combating climate change impacts. Implementation of NDCs includes enhanced ambition on adaptation as elaborated in the programs, strategies and actions to achieve economic, social and livelihood and ecosystem and landscape resilience; enhanced clarity on mitigation by adopting the Paris Climate Agreement rule book (Katowice Package) on information to be provided in NDC, as well as updated policies which potentially contribute to additional achievement of NDC target. Indonesia is ongoing to strengthen the NDCs by developing health-inclusive and health-promoting climate targets and policies.

It is projected that Bangladesh will be severely impacted by climate change and extreme weather-related events, such as global warming, sea-level rise, catastrophic cyclones, flood and drought. In addition, the air quality of Bangladesh, especially in the urban areas, has deteriorated in recent years. Dhaka, the country's capital city, is often ranked as one of the worst urban areas in the world for its degraded air quality. The polluted air in and around Dhaka is estimated to affect more than 40 million people. In combination with future climate change issues, air pollution will potentially cause acute public health outcomes in Bangladesh if no action is taken. This chapter examines in situ and satellite-based air quality data; explores the spatio-temporal distribution of air quality measures in urban and rural areas; identifies air pollutant hotspots; assessed the connection of air pollutants with climate variables; and then investigates the potential impact of hypothetical climate change-relates phenomena on air quality in Bangladesh from a public health perspective.

9 Examples from Studies in Africa, Include from Kenya, Tanzania and South Africa

In Kenya, Andrew Githeko asserts that country has been impacted by the effects of climate change that include epidemics, geographic range expansion of climate-sensitive diseases, droughts and floods. These diseases cause a high health burden. The public health system has put in place intervention measures. Malaria control relies on insecticides and drugs. Rift Valley Fever is managed by vaccinating livestock, while dengue and chikungunya are managed using insecticides and larval source management. Victims of drought and floods depend on humanitarian assistance. Water-borne infections can be reduced using safe drinking water sources. A shift from rain-fed to irrigated crops is expected to reduce food insecurity. These adaptation projects, according to the author, require heavy financial investments from government development budgets.

The United Republic of Tanzania is among the countries that has experienced impacts of climate change including epidemics of climate-sensitive infectious diseases, food and nutrition insecurity. Others include damage to infrastructure caused by flooding and landslides resulting to human injuries, deaths and displacement and high cost to restore the damaged infrastructure. The climate-sensitive diseases that have occurred in Tanzania include dengue, chikungunya, malaria, Rift Valley fever, leptospirosis, cholera and Human African Trypanosomiasis. The country has recently developed a National Climate Change Strategic Plan to provide a set of interventions on adaptation and mitigation, which are expected to strengthen country's resilience to the impacts of climate change and contribute to the global efforts of reducing greenhouse gas emissions. In addition, the country has developed the Health-National Adaptation Plan to climate change to guide the toward a health system that is more resilient to climate change. However, the efficiency of the operationalization of these strategies is not sufficiently known. Moreover, the vulnerability to climate change and mitigation, adaptation and coping measures at different levels of the health system in different areas of the country have not been well studied. The climate projections in Tanzania indicate that there will be an annual increase of rainfall by 10% by 2100, and temperature is projected to increase by 1.5 to 4.5 °C by 2090. These projections suggest that more impacts from climates change are expected, which call for appropriate mitigation, adaptation, and coping strategies at all levels of health system.

Authors of the chapter on South Africa highlight that Southern Africa bears disproportionate consequences of the changing climate. The El Niño Southern Oscillation causes phases of extreme weather events, leading to flooding and extended periods of droughts in different regions of the sub-continent. The data provides evidence that the extreme El Niño event of 2014–16 increased the risk for perinatal infant mortality in the Northern Cape and North West provinces of South Africa. The maternal health services profile of the Northern Cape and KwaZulu-Natal suggests a compromised health system with limited resilience to respond to the climate crises. The National Department of Health lacks adequate policies and strategies to ensure systems are able to meet the maternal and child health requirements in the context of climate change.

The role of climate change as a trigger factor causing the advancing spread of diseases in Russia has been analyzed, by Svetlana Malkhazova and her co-authors. The potential change in ranges due to predicted climate warming was studied according to climate model INM-CM5.0. A series of maps was compiled to identify the territories prone to suitability changes for the infection foci for the period up to 2100. It was determined that regions with temperate and arctic climate may become vulnerable to the emergence of climate-related diseases in the course of environmental changes.

In France, Roussel Isabelle insists that after the deadly nature of the 2003 heatwave and the occurrence of numerous brutal, progressive or systemic climatic disasters, the French became aware that the predictions made by the models had become reality, but the issues raised by mitigation and adaptation go beyond the political field to question values such as justice, solidarity and sobriety. However, the insurance system used to repair disasters is at the end of its tether and only prevention, combined with mitigation are necessary for avoiding to fall into the trap of bad adaptation. Only a very integrated territorial policy, following the "one health" concept can enable the French to improve their health by fighting against climate change.

Tom Douglass discusses climate change in the United Kingdom in relation to the health of its people. He begins with a brief discussion of the nature of the existing and predicted impacts of climate change alongside a history of the politics of climate change in the UK and an overview of government climate change mitigation policy (and particularly its inadequacies). The chapter then concerns itself with the present and future impacts of climate change on human health in the UK. The author examines the impacts of exposure to extreme heat, flooding and disruption to health services as key examples of the negative impacts of climate change on health. Pertinently the author discusses the uneven distribution of the health risks associated with climate change according to socio-economic status that emerge from climate change mitigation strategies and the associated decarbonization of the economy.

Robin Meadows in the chapter on Canada suggests that climate change impact is an active area of research. The increasing average temperature and decreasing atmospheric humidity due to climate change is predicted to increase the frequency and intensity of large wildfires in Canada. Wildfire smoke causes immediate respiratory distress, although there is a noted absence of research into prolonged exposure and long-term health outcomes. Further, evacuation from wildfires causes short-term hardships which leads to long-term mental health outcomes. Options to adapt to wildfires are limited and our capacity to prevent ever worsening wildfires in the future may be overwhelmed.

Donald and his colleagues argue that fire and climate change in the Sierra Nevada of California, USA, have a complex interaction with human land management and forest ecology. Fire was an important agent of change for the fire prone forests of this landscape. Many species, such as the Giant Sequoia (*Sequoiadendron giganteum*) evolved to take advantage of frequent fire as this natural process sculpted the environment. Native Americans used fire widely for socio-economic benefit and fuel reduction with moderate intensity fire encouraged to burn across the land. Euro-American settlement brought about an era of suppression that increased fuels and changed the forest composition and structure. But, suppression was and is the simple seeming solution. Even if suppression is not sustainable, it will garner support. Historic suppression has currently brought an extreme fuel problem that has manifested into a greater and greater threat of destructive high intensity fire not typical of this ecosystem. Fire policy was and is slow to change due to risk aversion and lack of urgency. This is not in small part from increased smoke impacts as a result of heavy fuel loads and returning fire to the landscape. These emissions were essentially mortgaged to the current age from previous generations. Fire and land management policies collide with air regulatory policy in California because of already heavily anthropogenically polluted air with little to no capacity for an additional emission source. However, fire and the subsequent smoke are inevitable. Public smoke tolerance is low and a significant deterrent to bringing fire back to California wilderness. This combined with the political polarization of the twenty-first century and even the simplest science-based solutions are unpalatable as can be seen by the myriad of publications outlining a path forward. The need for more frequent low intensity fire over California wild lands is obvious. Climate change is narrowing the options for fire and smoke management as suppression fails. We discuss the challenges and potential solutions to this conundrum that allow fire to act as an agent of change and ecosystem benefit while minimizing the health impacts from smoke and attempt to further the public understanding of tradeoffs in a complex ecosystem process.

Another chapter on USA by Olayinka and Alemu suggest that there is an accumulating body of evidence showing the human health impacts of climate change. Although research on physical human effects predominates, recent studies show a significant mental health impact of climate- and weather-related events. Awareness of climate-related mental health issues has important ramifications for the implementation of national healthcare policies. This is more so in the United States (the third most populous country and a leading contributor to global greenhouse gas emission), where over one in ten adults live with severe mental illness and/or a substance use disorder (SUD). In the 2016 *Climate and Health Assessment* report, the U.S. Global Change Research Program (USGCRP) noted (with very high confidence in some instances) that exposure to weather and climate extremes increases the risk of trauma-and-stressor-related disorders, anxiety, depression and substance use, among other psychological issues. Given that mental and SUDs are the leading cause of years lived with disability globally, preventing the exacerbation of this reality by a changing climate is crucial. In this chapter, we introduce the reader to established mental health consequences of climate change with a focus on the United States.

In Mexico, Maria and colleagues say climate change directly affects health through extreme weather events, and indirectly through the effect of these events on the dynamics of pathogens and vector-borne diseases, as well as on the productivity of crops that impact human nutrition. Mexico's geographic location is a relevant factor for exposure to hydrometeorological phenomena, such as cyclones, storms and floods. This chapter reviews how extreme hydrometeorological events affect ecosystems and therefore morbidity and mortality in Mexico. It also discusses health impacts from the lack of water and food security. This leads to the need of specific public policies for adaptation to climate change. We describe Mexico's institutional framework regarding health and climate change. The authors discuss whether it has the necessary policies and resources to face this phenomenon and finally make recommendations to put together health and general policy climate sensitive.

Climate change, caused by the increase in Greenhouse Gas (GHG) emissions, is inducing significant climatic alterations in Argentina as discussed by Daniel Oscar, so it is a priority to deepen the knowledge of its impact, particularly on human health. In this context, the main diseases that cause serious threats to the population are presented. Aspects as relevant as the effects that extreme temperatures expressed as heat waves and cold waves, air or water quality and the possible spread of diseases have on the morbidity and mortality of the population are addressed. It is concluded that the health effects caused by climate change require new strategies to mitigate them, with a multidisciplinary and inter-sectoral approach, where prevention and health promotion actions are essential in addressing this issue. In addition, the chapter presents a summary of the mitigation and adaptation measures applied in Argentina on the health sector. Both measures are complementary and, although they present different challenges, they converge in the final objective.

In the chapter by Rhavena and colleagues on Brazil, focus on multidisciplinary studies that have sought to understand the interaction between environmental factors that modulate people's risks and susceptibilities to climate change. This study aims to assess the socio-environmental and health vulnerability of the population in the state of Mato Grosso do Sul, to climate change through the application of a vulnerability index. The results show sensitivity and adaptive capacity as bottlenecks found in this study. Challenging conditions include unequal access to health services and population occupation concentrated in large urban centers.

Another chapter on Brazil by Gislene Santos presents and analyzes the role of public policies in Brazil in the face of the effect of climate change, focusing on displacement and women's health. It describes the role of the Brazilian State and the attention given to the health field of displaced women. In this context, we reflect the social and economic vulnerability of women in Brazil and the fragility the social assistance policies for this population who, as a result of the social and political dimension of disasters, lose their homes and become either homeless and/or displaced.

Thus, it is evident from the above discussion that a greater emphasis is being laid on implementing various adaptation and mitigation strategies in various developmental plans as evident in different regional studies in Australasia, Africa, Europe, North, Central and South America.

References

Akhtar, R. (2020). *Extreme weather events and human health.* Springer.

Akhtar, R. (2021). *Coronavirus (covid-19) outbreaks, environment and human behaviour.* Springer.

Akhtar, R. (2022). *Coronavirus (covid-19) outbreaks, vaccination, politics and society.* Springer.

D'Souza, P. (2022). A recent study has shown that heat related deaths in the country have gone up by more than 50 per cent. *Times of India*, October, 26, New Delhi) C0.

Gennaro, D., Isabella, A.-M., Lorenzo, C., Biagioni, B., María, D. (2023). Climate change, air pollution, pollen allergy and interaction with Sars-Cov-2. In R. Akhtar (Ed.), *Climate change and human health scenario.* Springer (to be published).

Hannam, P. (2022). Australia's record-breaking weather in 2022: A very wet and sometimes very hot year. *TheGuardian*, December 29.

Harvey, F. (2023). What is the IPCC AR6 synthesis report and why does it matter? *The Guardian*, March 19.

Hindustan Times. (2023). Allowing 1.5 °C warming will mean more heatwaves, New Delhi, March 21.

IPCC. (2018). *Global warming of 1.5 C*, Geneva.

IPCC. (2023). *AR6 synthesis report climate change 2023*, Geneva.

Jared, A. (2022). Personal communication, December 28.

McMichael, A. J. (1996). Climate change and human health: The likely impacts and research needs. *Journal of Environment, Disease and Health Care Planning, 1*(2), 1–8.

McMichael, A. J. (2017). *Climate change and health of nations.* Oxford University Press.

Rubin, A. J. (2018). Scorching summer in Europe signals long-term climate changes. *The New York Times*, August 4.

Semenza, C., & Paz, S. (2021). Climate change and infectious disease in Europe: Impact, projection and adaptation. https://www.sciencedirect.com/journal/the-lancet-regional-health-europe, *The Lancet Regional Health—Europe* (Vol. 9) October, 100230.

UNEP. (2022). *Emissions Gap Report*, 2022, Nairobi.

WHO. (2021). *Climate change and health.* Key Facts.

WHO, https://www.who.int/health-topics/climate-change#tab=tab_1

Wikipedia. (2022). December, 23.

Australasia

'Like Shells off the Beach'. Climate Change and Health in Australia

John Connell

Abstract Climate change is increasingly influencing Australian lives and livelihoods, with direct and indirect repercussions for health status. Achieving effective climate action has become a key political issue. The most immediate health risks are from extreme weather events related to climate change, including heatwaves, bushfires, cyclones, floods and droughts. More subtly, some tropical diseases, notably Ross River virus, are moving southwards. Few parts of Australia are hazard free, but floods typify the east coast and cyclones the north. Mental health is worse in inland areas where droughts have been prolonged. While all these hazards have occurred in the past, climate change has exacerbated their impact, and morbidity and mortality have increased in this century. That has been particularly so for First Nations peoples, both Aboriginal and Torres Strait islander populations, where socio-economic disadvantages have long prevailed, and remedial actions and service provision are difficult in remote locations. Present policies and practices, and global trends and forecasts, suggest that all current trends and risks are unlikely to change in the future.

Keywords Climate change · Health · Hazards · Cyclones · Floods · Heat · Drought · Dengue · First nations

1 Introduction

Climate change is increasingly recognised as universally critical to human health and national and international development. The 2018 report of the Lancet Countdown on health and climate change concluded that 'climate change is the biggest global health threat of the twenty-first century (Watts et al., 2018). It has caused a range of impacts on communities, including more frequent extreme weather, air pollution, higher temperatures and a changing distribution of infectious diseases, all with mental health impacts. The Intergovernmental Panel on Climate Change (IPCC) has stated that

---

J. Connell (✉)
The University of Sydney, Sydney, NSW 2006, Australia
e-mail: john.connell@sydney.edu.au

R. Akhtar (ed.), *Climate Change and Human Health Scenarios*, Global Perspectives on Health Geography, https://doi.org/10.1007/978-3-031-38878-1_2

recent decades have involved warming air and ocean temperatures, altered precipitation patterns, rising sea levels and changes in the frequency and intensity of extreme events, such as droughts, floods and storms. The IPCC also asserts, with greater confidence than previously, that current warming is largely attributable to human activity. While climate change was once only a natural phenomenon, human influences have gradually become more significant, and there is increasing certainty that these trends will continue and, in some cases, accelerate, with a still growing population, higher energy and other consumption heights (particularly in high-income countries), land clearance and expanded food production activities (McMichael et al., 2012). These factors are readily apparent in Australia, alongside the production of fossil fuels, with their massive contribution to global warming. Climate change alters human ecological processes, influencing the rates, ranges, seasonality and patterns of injury, disease and death (McMichael & Lindgren, 2011). The most immediate health risks are from extreme weather events and disasters, all of which have posed problems and increased mortality rates in Australia in the past five years. Hazards and disasters will become more intense and frequent through the twenty-first century, and as populations and built environments continue to grow in hazard-prone areas, vulnerability to disasters increases. However, demonstrating the link between climate change, catastrophic events and health status are difficult.

Climate change affects health directly and indirectly, firstly, through primary or direct effects, e.g. injuries and deaths caused by extreme weather events such as cyclones, or increased morbidity and mortality resulting from higher temperatures and heatwaves, particularly amongst vulnerable groups such as elderly people and those with pre-existing cardiovascular and respiratory diseases. Secondly, secondary or indirect effects result from interactions of climate with other systems (e.g. the increasing geographic range of, and population exposed to, vectors that spread disease, and the mental health impacts of both extreme events and slow-onset climate change, or declining agricultural production caused by drought or floods, resulting in poorer nutrition; and changes in the distribution of vectors that spread infectious disease caused by flooding and polluted water sources). Climate change can lead to both physical and mental health problems and stress-related disorders, asthma, respiratory allergies and airway diseases. Thirdly, diffuse, and/or delayed effects (e.g. disruptions to health and social services) follow on from more direct effects (e.g. damage to livelihoods, from floods and drought and coastal erosion and migration). Sea-level rise poses a longer-term threat in many places. These broad impacts have been widely documented (e.g. Rocque et al., 2021). This chapter examines these varied impacts in an Australian context.

The Australian continent has an unusually wide climatic range, from Snowy Mountains to tropical rain forests, and has experienced all the changes and problematic outcomes typically attributable to climate change, increasingly evident in this century—from bushfires to floods—with highly visible consequences, increased mortality (from flood victims to firefighters, and asthma sufferers to dengue victims) and substantial costs of recovery, recuperation and reconstruction. Climate change is amplifying the intensity and frequency of extreme weather events and, in every case, intense events are moving southwards, affecting greater numbers of the population (who, ironically, themselves are moving northwards) and raising concerns over food security. Tropical diseases too are moving southwards. Climate change is also leading to slow-onset changes in climatic and environmental conditions (including sea-level rise, land degradation and loss, declining abundance of fish, contamination of water resources and degradation of coral) that contribute to loss of important environmental amenity and livelihoods, presently most evident in northern Australia, and challenging for Indigenous populations. The number of people facing health problems caused by climate change is likely to rise, and nowhere in Australia is climate change likely to be beneficial to health. Climate change is now literally a hot national issue, especially because of Australia's own contribution to global warming. This chapter first briefly addresses Australian climate policy, then reviews the wide-ranging health impacts from climate change, with particular reference to First Nations people.

2 Australia and Climate Change Policy

Most of the Australian population of 26 million live on the continental periphery, and big cities account for 87% of them, thus the interior is thinly populated. Unusually for a developed country, about 6% of Australians live in the tropics, including about a third of Indigenous First Nation Australians. Australia is the driest inhabited continent, making it particularly vulnerable to the challenges of climate change.

Australia is a prominent exporter of natural resources, energy and food, including coal and iron, with investments continuing in the resources sector. Growth slowed from 2020 with the main market, China, disinclined to import more coal, iron and other products. Australia is one of the larger global coal exporters. Dependence on mining has made the 'lucky country' reluctant to take a strong anti-climate change stance in policy formation, despite pressure from Pacific island states at particular risk from climate change, and despite sedimentation of the Great Barrier Reef.

Conservative coalition (Liberal Party and National Party) governments, following the dismantling of carbon pricing in 2014, took minimal subsequent action on climate policy (Beggs et al., 2021; Crowley, 2021), and lacked the drive to scale back coal mining and provide incentives to transition away from fossil fuels. At the federal level, mining lobby groups exerted influence, resulting in political inaction to formulate and implement policies to address climate change, although, facing a range of climate extremes, farming lobbies were less vocal in opposition. By contrast, positive action existed at individual, local, state and territory levels, with growing uptake of rooftop solar and electric vehicles, and the beginnings of appropriate adaptation planning. This was severely undermined by contrary national policies that increasingly placed Australia out on a limb (Beggs et al., 2021). Public pressure mounted to engage in climate action and six victorious 'teal' independent candidates, whose primary focus was on climate change, made climate action a major factor in the 2022 national election which resulted in a loss for the coalition and victory for the Labour Party.

The loss was in some part because of the image of the previous Prime Minister, Scott Morrison, triumphantly bringing a lump of coal into parliament, being on holiday in Hawaii as 2019 bushfires ravaged the coast of New South Wales and his unwillingness to 'hold a hose' and take climate change seriously.

Climate change action became a necessary priority for the new Labour Government and climate legislation was endorsed in mid-2022. The climate bill, the first climate change policy for a decade, enshrines into law two national greenhouse gas emissions targets: a 43% cut below 2005 levels by 2030, and a reduction to 'net zero' by 2050. Within weeks of the bill being passed, doubts were raised over the genuine will of the government to achieve this outcome and over the existence, development and implementation of policies that would achieve this, especially as the economy moved through a downturn. While critically important, it is not action by the Australian government alone that has the primary influence on climate change in Australia, but it did allow for some measure of optimism.

3 Heat and Heat Islands

Seemingly, the most obvious of climate change influences is increased temperatures. Australia had its third hottest year on record in 2017, including the highest historical maximum temperatures during winter and very low rainfall, providing a context for heat-related extreme weather events such as bushfires and heatwaves. Eight of Australia's ten warmest years on record have occurred since 2005. Onslow (Western Australia) experienced a temperature of 50.7° as recently as January 2022, the highest temperature ever recorded on the continent. The heat was on.

Heatwaves are the 'silent killer', the deadliest natural hazard in Australia, causing 55% of all-natural hazard-related deaths and costing the Australian workforce around US $6.2 billion every year. The intensity, frequency and duration of heatwaves have all increased in this century, and climate projections show that they will probably further intensify throughout the century. The steepest rise is projected for eastern coastal regions where most of the Australian population live (Nishant et al., 2022). The number of hot days and warm nights is likewise expected to increase during this century and more frequent and prolonged heatwaves have already intensified urban heat islands in such cities as Sydney and Brisbane, and increased inland temperatures by as much as 10 °C more than in coastal zones. Higher urban population densities, with increased migration, and the expansion of high-density, low socio-economic status neighbourhoods in vulnerable low-lying flood-prone areas, as in Cairns, in Far North Queensland, places stress on facilities (including health care services) and on residents. As the 2019 Living Melbourne Strategy summarises:

> In Melbourne, deaths begin to rise when the mean daily temperature reaches 28°C, with hospital admissions for heart attack increasing by 10.8 per cent when the mean daily temperature reaches 30°C. When the average temperature is higher than 27°C for three consecutive days, hospital admissions increase by 37.7 per cent (quoted in Walls, 2022)

More intense urban heat islands and densely populated urban areas, including the expanding suburbs of cities like Sydney and Melbourne, are likely to contribute to asthma, and other respiratory allergies. Asthma is already the second main cause of other chronic diseases affecting children. Heat exposure is associated with increased hospitalisations for urologic diseases, as in Queensland, where a 1 °C increase in temperature is associated with a 3% increase in hospitalisation. Males and the elderly ($\geq$ 60 years old) were more affected by heat exposure than females and younger groups. Overall, nearly one-fifth of hospitalisations for urologic diseases were attributable to heat exposure (Lu et al., 2022). Heatwaves contribute to acute cerebrovascular accidents, and aggravate chronic respiratory, cardiac and kidney conditions and psychiatric illness.

Higher temperatures and urban densities have also worsened urban pollution. Less directly, greater heat will increase morbidity and mortality (linked to heat stroke and dehydration), increase the risks from outdoor activities (including sunbathing) and from bushfires, so increasing the stress on people, animals and plants, damaging crops and vegetation, increasing demand on water and other resources, with a negative impact on economic development (Borchers Arriagada et al., 2019). They are also costly. The heatwaves in 2009 and 2014 in Victoria contributed to 374 and 167 excess deaths, respectively, and the 2016 thunderstorm asthma event in Victoria—the outcome of an unusual conjuncture of meteorological conditions—was reported to have contributed to the death of nine people and a 3000% increase in asthma-related intensive care unit admissions, placing significant demands on the health system (Johnston et al., 2021). Heat waves can also bring mental health problems and stress-related disorders, especially in regional areas, where livestock (and wild animals) has perished in extreme conditions. There is minimal indication that national and urban planning are taking any note of designing cities and transport systems to cope with increasingly hot futures.

3.1 Tropical Diseases

Climate change is influencing the geographic range, seasonality and incidence of various infectious diseases, such as malaria, dengue and diarrheal diseases. The geographic and seasonal patterns of several infectious diseases, endemic in adjoining Papua New Guinea (PNG) are moving southwards, requiring Australian response to close its borders to infections. Several of the diseases linked to climate change occur in association with hazards such as floods and cyclones.

Malaria was previously endemic in the Torres Strait, but Australia was declared free of malaria in 1981, however infections are regularly diagnosed in the Torres Strait with most cases imported from PNG, where it is endemic, and local transmission is rare (Preston-Thomas et al., 2012). Otherwise, malaria is yet to enter Australia. Mosquito elimination in the Torres Strait islands is not feasible because of their geography. Malaria poses no current public health threat in northern Australia, but is a constant concern.

Floods have the potential to produce a surge of mosquito populations by stimulating egg hatching or providing breeding opportunities for adult mosquitoes. High numbers of mosquitoes thus increase the opportunity for disease transmission. In February 2019, a major flood in Townsville (Queensland) typically brought a greater occurrence of mosquito-borne diseases (Adekunle et al., 2019). Warmer weather may see an annual extension of the 'mosquito season', which brings risks to humans such as malaria and dengue fever.

The spread southwards of other tropical diseases, such as dengue, is occurring, with an increase in disease burden, a wider spatial distribution of dengue cases and more people exposed to climatically suitable areas of dengue as climate change proceeds. Dengue fever is a mosquito-borne infection endemic in most tropical and subtropical countries and is caused by one or more of the four different dengue viruses. Both species of mosquitoes that spread dengue viruses are present in Australia. Although their distribution has receded over the last five decades from multiple states to Queensland alone, several outbreaks have been recorded in Queensland and its future re-expansion is a constant threat (Beebe et al., 2009; Pyke, 2018) especially since it has been recorded as far south as Rockhampton (Walker et al., 2021). Japanese encephalitis, similarly spread by mosquitoes, has also moved southwards from the Torres Strait in the last decade.

With malaria and dengue effectively under control, Ross River Fever (RRF), named after Townsville's river, is the most common and widespread mosquito-borne disease in Australia, resulting in considerable health and economic cost to communities (Woodruff and Bambrick 2008). There is a strong correlation between contracting RRF when living in close proximity to wetlands, as in Western Australia, where suburbs are still being developed, enhancing the risk (Jardine et al., 2015). Although a positive association between flooding and RRF outbreaks is largely circumstantial (Tall et al., 2014), it is likely that floods do contribute to its incidence and that climate warming increases the risk and spread of RRF.

Evidence shows that the reach and risk of such infectious 'tropical' diseases has extended and increased as the tropics have extended southwards, exacerbated by poorer nutrition (weakening immune systems), inadequate healthcare systems, lack of clean water and poor sanitation, posing the greatest problems in Far North Queensland and the Northern Territory that has necessitated a constant vigilance against their arrival and spread in Australia.

3.2 *Bushfire*

In this century, bushfires have become more numerous and more dangerous in Australia and in most parts of the world. Numerous bushfire have been recorded in Australia, but especially in southern Australia rather than in the wetter tropics. Based on current trends, they will probably increase in the future, with major environmental, social and economic consequences, as perhaps the most vivid and destructive markers of climate change.

Climatic trends, especially increased temperatures, have given rise to conditions that are more favourable to bushfires, and where they burn more intensely and frequently (Clarke et al., 2013). Years of drought, low relative humidity, high temperatures and high forest fuel loads led to catastrophic fire conditions across Australia during 2019 and 2020. Bushfires, over wider areas, have become seemingly an annual occurrence, usually matched by rural drought.

Most parts of Victoria, South Australia, New South Wales (NSW) and southern Western Australia have been affected, with fires destroying houses in the suburbs of metropolitan Sydney and Canberra, in this century, alongside many houses and property in much smaller towns. Seemingly regular bushfires have been punctuated by horrendous events, such as Ash Wednesday 1983 (where 75 people died in Victoria and South Australia), Black Saturday in Victoria in 2001 (when 171 people died) and on the NSW coast 2019–20 (when 34 died). Bushfires were no new phenomenon, hence the extent to which climate changes has contributed to their greater impact, is uncertain.

Bushfires have obvious direct effects including the increasing risks of burns, smoke inhalation, heat stress and dehydration (Johnston et al., 2009). Indirectly, they aggravate lung and heart conditions through particle pollution associated with smoke (McManus, 2021; Ranse et al., 2022). These may precede trauma and long-term mental health impacts, sometimes compounded by and directly related to the loss of homes, pets, livestock and businesses. Some were despondent enough to abandon their homes and move to reduce associations of fire with place and minimise further stress, as after the January 2005 Black Tuesday bushfires in the Eyre Peninsula (South Australia) that resulted in nine deaths, extensive injury, and the loss of livestock, property, infrastructure and livelihoods (Watts et al., 2023). During the so-called Black Summer, from November 2019 to February 2020, a significant deterioration in air quality was experienced in Canberra, and in Australia's largest cities, Melbourne and Sydney, as a result of fires, unpleasant as much as injurious to health, preventing some outdoor activities (including sport), limiting exercise and worsening well-being. Over the same period, bushfire smoke was responsible for 417 excess deaths, more than 1100 hospitalisations for cardiovascular problems and over 2000 for respiratory problems, many with asthma, despite their remaining indoors (Beyene et al., 2022; Borchers Arriagada et al., 2020). Smoke inhalation and its impacts are not evenly shared. Indigenous people may be affected more strongly than the rest of the population, appearing to have a higher risk of being admitted for cardio-respiratory effects.

Very substantial human costs, in injury and deaths amongst rural residents and fire fighters (on the ground and in the air) occurred alongside very high costs of reconstruction. Destruction of businesses affected the tourist industry. While some Australians deliberately sought to take holidays in the recovering regions, to support reconstruction, as they previously did in drought-stricken areas (Schweinsberg et al., 2019) it took time before the businesses (especially shops and accommodation) were restored. Moreover wildlife, especially koalas, was killed both making the regions where they normally lived less attractive to tourists and prompted realisations that their status was becoming increasingly endangered, with long-term ecological and touristic consequences. Overall, the 2019–20 Black Summer bushfires resulted in more than 450 deaths—two thirds of whom were 'excess deaths' from the heatwaves—and more than 4000 hospitalisations and $1.95 bn in smoke-related healthcare costs (Johnston et al., 2021): the most costly fires in human lives, health and incomes in Australian history.

4 Cyclones

Climate change is increasing the destructive power of tropical cyclones, with increased wind speeds and greater related rainfall. Rising sea levels, and cleared coastal regions, make accompanying storm surges more damaging. While climate change may mean fewer tropical cyclones overall, those that do form are likely to be more intense, destructive and costly, as predicted for Australia and more specifically the Torres Strait (Hall et al., 2021; Steffen & Bradshaw, 2021). Australia averages eleven tropical cyclones in a season, all in northern Australia, with the northwest of Western Australia being the most cyclone prone part of Australia's coastline. Between 1970–71 and 2003–04, seventy-two of the total of 146 coastal cyclone crossings in Australia were over this coastline between Broome and Exmouth. However that area is relatively thinly populated so that some of the more devastating cyclones in Australia have been in Far North Queensland and the Northern Territory (famously in the destruction of Darwin in 1975). As with other hazards, the impact of cyclones is moving southwards. Australia experienced a stormy 2010s, with some of the nation's most damaging storm events occurring within this period.

The most direct impacts of cyclones are some combination of mainly coastal flooding and damage to property and infrastructure. Adequate forecasting usually prevents immediate casualties, but they cause mental stress which can undermine the resilience of individuals and communities, placing further physical, emotional and financial burdens on recovery efforts. Natural ecosystems, and agricultural systems, can suffer serious impacts from wind damage. Tropical Cyclone Yasi in 2011 was one of the most powerful cyclones to have affected Australia since records began, and proved one of Australia's costliest natural disasters, causing problems across various sectors, with widespread damage and flooding across much of Queensland. More than 10,000 people were displaced from their homes, and it destroyed 30% of the houses in coastal town of Tully. Damage to power lines left 150,000 homes without electricity. At least 75% of the banana crop was destroyed, and damage to sugar cane farms amounted to about AU $500 million. Coral damage was reported across much of the Great Barrier Reef Marine Park. Indirect impacts on health status were considerable if immeasurable.

5 Floods

Often consequent on cyclones and storm surges, floods are the most expensive type of natural disaster in Australia, increasing risks of injury, communicable disease transmission, distress and acute and chronic anxiety disorders. Flash floods follow more intense heavy rainfall events, with instantaneous and unpredictable damage, including drownings. A propensity to establish suburbs and other economic activities on flood plains, as in western Sydney, has contributed to the increased extent of economic damage and human distress following particularly severe twenty-first century floods.

Between 1852 and 2011 more than 900 people died and over 1300 were injured in floods in Australia. The estimated cost of property damage during that period was nearly A$5 billion, including damage to houses, other buildings, infrastructure (such as power and transportation systems) and loss of crops and livestock. Floods pose public health risks by spreading water-borne diseases and carrying pollutants from land into waterways. Surviving, flood-affected livestock can suffer long-term health conditions, including parasites and bacterial infections, with implications for animal welfare and farm productivity.

The most economically devastating floods followed unusually heavy and prolonged rains over eastern Australia in late 2010 and early 2011, owing partly to Cyclone Tasha. These costs were probably exceeded by the east coast floods of 2022 that devastated the Lismore region particularly, with subsequent floods in northern Victoria, submerging the most productive agricultural region in the state. Record annual rainfall and recurrent floods ruined agricultural crops, resulting in significant price increases, reducing the range and quantity of foods that could be purchased by the relatively poor (in the wake of the COVID-19 panic and the global downturn) so worsening diets and leading to inadequate nutrition.

During these events various people died while trying to cross flooded rivers, bridges and causeways, many homes and businesses were flooded and people displaced. So devastating was the Lismore floods, despite massive levees being constructed after earlier floods, that many people and businesses left the town and serious consideration given to relocating it elsewhere (Rugendyke and Renouf, 2022).

6 Drought

Floods and droughts often seem to complement each other in regional Australia, with few years escaping either. Drought is very familiar in the inland, virtually determining the extent of settlement and feasible forms of livelihood. Climate change is making drought conditions in southwest and southeast Australia worse, as shifts in weather systems have decreased rainfall in both regions. The duration of drought is projected to increase in future across southern and especially south-western Australia. The increased intensity and frequency of hot days and heatwaves has exacerbated drought conditions.

The gradual onset of droughts that are more frequent, prolonged and widespread, if less obviously dramatic than other hazards, is a significant cause of adverse mental health amongst rural Australians, especially when drawn out over time and when finance and insurance is unavailable (Ng et al., 2015) that is particularly so for First Nations Australians for whom the land has a powerful spiritual and livelihood value (Rigby et al., 2011; see below). Declines in physical health are particularly prevalent amongst the elderly (Horton et al., 2010). The debilitating mental health consequences of drought have been well documented in Australia with suicide rates as much as 40% higher than in urban areas, and correlated with ongoing drought. In NSW, the relative risk of suicide can increase by up to 15% for rural males aged 30–49 as the severity of drought increases (Hanigan et al., 2012). Rural Australians face barriers to engagement with mental health services, including increased physical distance, reduced service availability, a culture of self-reliance and reluctance to discuss mental health issues partly due to perceived stigma: overall a 'harsher' physical and social environment (Brown, 2017; Morrissey & Reser, 2007; Zidersch et al., 2009). Children too are stressed by drought and by the loss of wildlife.

Drought has obvious consequences for agricultural production, from the death of livestock (and inability to achieve market prices for weakened animals) to the disappearance of topsoil in strong winds. Drought conditions affect the security of water supply, for domestic and agricultural use, and have required the installation of mobile desalination plants in the Torres Strait islands. Changes in flooding and drying cycles, and hotter summers, are likely to further reduce agricultural productivity.

Some crops are particularly affected. In the northern Murray-Darling Basin, which produces 93% of Australia's cotton, rainfall has decreased significantly since the 1990s, coinciding with accelerated climate change. The impact on reduced yields threatens the future of cotton farming, while, competing with cotton, a sustainable water supply is vital for the ecological health of wetlands, waterholes and floodplains (Speer et al., 2022). By contrast, grape farmers in South Australia are yet to be constrained by climate change, but projected future climate change means socio-economic stress in the farming community is likely to worsen (Fleming et al., 2015). Wheat farmers in Western Australia are affected by climate change, exacerbating their worries about the weather, undermining notions of self-identity and contributing to cumulative and chronic forms of place-based distress, culminating in heightened risks of depression and suicide (Ellis & Albrecht, 2017). While cotton and wheat are more dependent on a secure and regular supply of water, other crops and livestock are affected, making rural production (and employment) increasingly uncertain.

Droughts are stressful, with wide-ranging effects on health, including impacts on nutrition, infectious diseases, on forest fires (causing air pollution), and on livestock and wildlife. After 2020 that was intensified by the arrival of the COVID-19 pandemic (multiplying the stress burden) especially alongside labour shortages and reduced income levels occurred. In rural NSW, the combined incidence of the two brought higher distress levels to many, affecting sleep patterns, exercise, nutrition, alcohol and cigarette consumption (Chan et al., 2022) that was accentuated a year later by plagues of mice that destroyed crops and overwhelmed farmers, coming after years of drought, and a period of heavy rain that boosted plant growth, creating ideal conditions for the hungry rodents to reproduce exponentially. Farms and fields were overrun. Not surprisingly, as droughts worsened, and climate change promised only more of the same, despairing farmers saw the combination of problems and pestilence as being of biblical proportions.

7 Indigenous Australia

Vulnerability to the impacts of climate change, both in Australia and globally, depends on geographic, social, economic and biological factors. At every scale, climate change adversely affects health and well-being and disproportionately impacts socially, culturally and economically vulnerable groups and individuals, in the nation as a whole, and as it does in NSW (Boylan et al., 2018). Two Indigenous groups, or First Nations people, are present in Australia: the Melanesian Torres Strait islanders and the socially diverse Aboriginal population. In mid-2016, the 70,880 Torres Strait Islander people made up 9% of the total 798,400 Aboriginal and Torres Strait Islander peoples, which was 3.3% of the total Australian population. Many Torres Strait islanders have moved south from their home islands between continental Australia and PNG, to urban centres, notably Cairns, and the Aboriginal population, who have a presence in all states but are proportionately of greater numerical significance in Western Australia, the Northern Territory and Queensland, where many live in remote settlements with limited access to health care and other services. Australia remains the only developed country in the world where trachoma still exists in endemic proportions. Difficult social circumstances prevail in several remote Aboriginal townships, such as Wadeye and Aurukun; health is poor (with both infections and NCDs), often exacerbated by alcohol problems, and life expectancy is significantly shorter (about 10 years) than that of the overall Australian population (Australian Indigenous Health Infonet, 2022). Despite diverse strategies, government efforts at 'closing the gap' have been unable to reduce or remedy these disadvantages and inequalities.

Overall, climate change increases problems and inequality, as people with fewer material, social and health resources are more vulnerable to the adverse impacts of climate change. In the most extreme instances, some coastal and inland areas may become uninhabitable. These health impacts are greater for Torres Strait Islanders due to their remote location with limited healthcare services, lower economic resources and a higher burden of pre-existing health conditions. The islands are the only Australian location with the invasive mosquito, *Aedes albopictus*, associated with the spread of dengue. More than anywhere else in Australia, the Torres Strait islands have already experienced some of the more damaging features of climate change, including enhanced coastal erosion from sea-level rise (Green et al., 2010) and the need for continued vigilance over disease transmission from PNG. The northernmost island, Saibai, is only 1.7 m above sea level, and several other islands are little different. Several have been flooded in this century.

The Bramble Cay melomys, the first Australian (and Torres Strait) mammal to become extinct as a result of climate change, as sea-level rise destroyed the flora on its home island, may be a metaphor for future changes in the islands. The United Nations Human Rights Committee (UNHRC) stated in September 2022 that the Australian government had violated Torres Strait Islanders' rights to enjoy culture and family life by failing to adequately protect them against the adverse effects of climate change and ordered it to pay for the harm caused. Torres Strait Islanders appealed to the UNHRC after erosion of their islands, and in one case of a graveyard, where one claimant discovered what he claimed to have been his great-grandmother's remains eroded by rising seas: 'We were picking her up like shells off the beach' (quoted in Kampmark, 2022; 20). Seawalls have not prevented inundation of parts of the more low-lying islands, prompting thoughts of migration. Many Torres Street Islanders have moved to Bamaga, on the adjoining Cape York Peninsula, or to Queensland towns, initially for economic reasons but in this century because of concern over flooding (Green et al., 2010). Nonetheless, Torres Street Islanders may become some of the first Australian citizens to be forced to move because of climate change, raising issues for identity and the stress from forced displacement. As one Indigenous woman has said: 'without action to stop climate change, people will be forced to leave their country and leave behind much of what makes them Aboriginal' (quoted in Godden et al., 2022).

Despite much migration to urban centres, living on 'country' (traditional land) is of great importance to both First Nations people to maintain cultural responsibilities, identity and kinship connections, which also brings health benefits (Creamer & Hall, 2019). At Aurukun (North Queensland), Aboriginal Australians who returned to country intermittently were able to counter feelings of disempowerment and despondency arising from living solely in the township of Aurukun. Engaging with country (if not actually returning to live there) built cultural resilience in the face of multiple economic, environmental and social challenges, thereby benefiting physical and psychosocial health and well-being (Green & Martin, 2017). However, as temperatures increase, houses become overheated, electricity unreliable (so that refrigeration is inadequate for storing medicines and food) remote settlements are vulnerable due to their isolated location and limited infrastructure, including transport (Lansbury-Hall & Crosby, 2022; Quilty et al., 2022). Heat waves may even cause adverse pregnancy outcomes and trigger stress and interpersonal violence (HEAL Network & CRE-STRIDE, 2021), while climate change (notably through bushfires) is destroying valuable rock art.

Five climate-sensitive diseases in the Torres Strait and Cape York region are of some concern: tuberculosis, dengue, Ross River Fever, melioidosis and nontuberculous mycobacterial infection. The region constitutes only 0.5% of Queensland's population but has a disproportionately high proportion of the state's cases: 20.4% of melioidosis, 2.4% of tuberculosis and 2.1% of dengue (Hall et al., 2021). Dengue is largely confined to the north, but with an increase in the disease burden, a wider spatial distribution of dengue cases and more people living in climatically suitable areas. Ross River disease has similarly been prevented from spreading much further. Melioidosis is predominant amongst Aboriginal populations as in the remote Katherine region of northern Australia. In a population with high rates of chronic disease, social inequities and extreme remoteness, its impact is exacerbated by severe weather events, disproportionately affecting First Nations Australians, where mortality from the disease is significant. Most First Nations Australians with melioidosis also had diabetes and a history of hazardous alcohol consumption. Several cases were the result of flooding sustained by wading through flood waters or cleaning up after flooding, and other cases were exacerbated by it. Diabetes management, public health intervention for hazardous alcohol consumption, provision of housing to address homelessness and patient education on melioidosis prevention in First Nations languages would diminish the incidence (Hodgetts et al., 2022). As with several other diseases, meloidosis is exceptional in otherwise healthy people but prevalent where co-morbidity exists. Tuberculosis (TB) is of particular concern in the Torres Strait because of proximity to PNG, where multi-drug resistant TB is prevalent, especially as the conditions that enhance the diffusion of TB (higher humidity and rainfall and greater temperatures) are all increasing. Since both First Nations peoples mostly intend to remain living on their traditional country, where delivering services is particularly difficult, climate change brings risks of both direct and indirect human health impacts (Hall et al., 2021), will intensify inequalities and do nothing to close the gap.

8 Conclusion: 'A Sunburnt Country ... A Land of Drought and Flooding Rains'

Australia is already experiencing numerous impacts of climate change on physical and mental health, primarily in northern Australia, where cyclones dominate, but its effects are spreading southwards and intensifying (most marked through drought and flood cycles, and most visible in bushfires). Half the electorates most vulnerable to climate change are in Queensland (Hutley et al., 2022), but 'natural' disasters are becoming ubiquitous in Australia with the 2010s decade suggesting a 'new normal' of more substantial disasters. It provides an example of climate injustice: impacts are particularly pronounced amongst First Nations peoples, who experience a higher burden of co-morbidities than the general Australian population, notably in the Torres Strait where some islands are already being intermittently inundated. By contrast, urban dwellers arguably have the advantages of substantial disaster mitigation infrastructures, and superior healthcare systems. In regional Australia, pressures on an overstretched healthcare system are well-known, and adequate human resources are rare, despite there being significant mental health care issues.

There are no obvious advantages to climate change in Australia, while all projections of changing weather patterns and hazard events indicate increased risks of cyclones, intense rainfall, flooding and droughts. Less perceptible increases in heat will occur everywhere, with particular effect in urban heat islands. Worsened weather-related events, such as bushfires, floods and cyclones, place additional burdens on the health system where they already challenge existing capacity. Disasters have important and long-term psychosocial, mental health and community impacts, especially where they seem never-ending. The ever-present possibility of hazards in Australia, and the inherent uncertainty and anxiety, that are a part of living with such phenomena, constitute powerful stresses exacerbated by climate change.

Belatedly national policies are shifting towards more effective climate action, but only slowly. It needs extreme optimism to assume that achieving net zero carbon emissions by 2050 will occur while, in the meantime, warming continues. Action is more evident at the local scale. Even then, to reduce the incidence and severity of bushfires, or ensure water resources are available, more effective climate change mitigation policies are crucial but elusive (Jalaludin & Morgan, 2021; Nursey-Bray et al., 2020) while responding to the resilience, agency and needs of Indigenous populations requires appropriate 'strategic localism' (Nursey-Bray et al., 2019). Prevention is preferable to cure, as injuries, diseases, mental stresses and reconstruction, are costly. Even more costly, in stress and finance, is forced migration because of floods, drought or excess heat.

Expanding the health system to cope with climate change, so that it is more flexible, requires investment in personnel, infrastructure and service coordination, alongside close collaboration with the non-health sectors. Like virtually all environmental issues and challenges, hazards require informed, sustained and interdisciplinary community preparedness and response. Resilience to urban heat requires cooperation and discipline across multiple physical scales (Blashki et al. 2011; Bambrick et al., 2011). Health systems face the challenge of managing the increased burden of treating those whose health is affected by climate change while remaining resilient to extreme weather events and changing disease patterns that requires expanded and diversified approaches to bushfire mitigation and adaptation to living in an increasingly hot and fire-prone country. The health costs of climate change will not be reduced without changes in other sectors of the economy—involving, e.g. urban planning, that reduces transport congestion and vehicular emissions, favours public transport, involves tree planting and provision of open space, while focussing on the sectors of society that are more vulnerable due to their physical locations, social and economic disadvantage. In urban areas, green infrastructure and green spaces are slowly becoming more significant; for the first time in October 2022, heat officers were appointed in Melbourne to develop policies to reduce heat islands, yet elsewhere building in wetlands and urban sprawl proceeds without public transport.

Health problems from climate change are currently more mental than physical but, unless trends are reversed, both will be increasingly significant and demand that health adaptation plans take increasing account of more tropical diseases and worsened mental health. Australia faces a difficult future, where the natural and physical environment seemed at particular risk in 2022: the COVID-19 pandemic remained in place, floods were swamping large parts of the country, droughts simultaneously existed island and a plague of mice were devouring what was left of rural grain production. Five metre high walls were being constructed on vulnerable northern Sydney beaches as coastal surges became more threatening. While these changes, and their outcomes in mental stress, were not all directly the result of climate change, they suggested the vulnerability of Australia to climate change and to a suite of hazards with negative effects on health that hopefully reminded politicians and planners that multi-scalar action was crucial.

References

Adekunle, A., Oyelola, A., & Rahman, K. (2019). Flooding in Townsville, North Queensland, Australia, in February 2019 and its effects on mosquito-borne diseases. *International Journal of Environmental Research and Public Health, 16*(8), 1393.

Australian Indigenous Health Infonet. (2022). *Overview of Aboriginal and Torres Strait Islander health status, 2021*. Australian Indigenous Health Infonet.

Bambrick, H., Capon, G., Barnett, G., Beaty, R., et al. (2011). Climate change and health in the urban environment: Adaptation opportunities in Australian cities. *Asia-Pacific Journal of Public Health Supplement, 23*(2), 67S-79S.

Beebe, N., Cooper, R., Mottram, P., & Sweeney, A. (2009). Australia's dengue risk driven by human adaptation to climate change. *PLoS Neglected Tropical Diseases, 3*(5), e429.

Beggs, P., Zhang, Y., McGushin, A., Trueck, S., et al. (2021). The 2021 report of the *MJA–Lancet* countdown on health and climate change: Australia increasingly out on a limb. *Medical Journal of Australia, 215*(9), 390–392.e2.

Beyene, T., Harvey, E., Van Buskirk, J., McDonald, V., et al. (2022). 'Breathing fire': Impact of prolonged bushfire smoke exposure in people with severe asthma. *International Journal of Environmental Research and Public Health, 19*(12), 7419.

Blashki, G., Armstrong, G., Berry, H., Weaver, H., et al. (2011). Preparing health services for climate change in Australia. *Asia-Pacific Journal of Public Health Supplement, 23*(2), 133S-143S.

Borchers Arriagada, N., Bowman, D., Palmer, A., & Johnston, F. (2019). Climate change, wildfires, heatwaves and health impacts in australia. In R. Akhtar (Ed.), *Extreme weather events and human health* (pp. 99–116). Springer.

Borchers Arriagada, N., Palmer, A., Bowman, D., Morgan, G., et al. (2020). Unprecedented smoke-related health burden associated with the 2019–20 bushfires in eastern Australia. *Medical Journal of Australia, 213*(6), 282–283.

Boylan, S., Beyer, K., Schlosberg, D., Mortimer, A., et al. (2018). A conceptual framework for climate change, health and wellbeing in NSW, Australia. *Public Health Research and Practice, 28*(4), e2841826.

Brown, P. (2017). Mental health in rural Australia. *Australian Journal of Rural Health, 25*, 258–259.

Chan, K., Xue, H., Carlson, J., Gray, J., et al. (2022). Impact of COVID-19 on lifestyle and mental wellbeing in a drought-affected rural Australian population. *Rural and Remote Health, 22*, 7231.

Clarke, H., Lucas, C., & Smith, P. (2013). Changes in Australian fire weather between 1973 and 2010. *International Journal of Climatology, 33*, 931–944.

Creamer, S., & Hall, N. (2019). Receiving essential health services on country: Indigenous Australians, native title and the United Nations declaration. *Public Health, 176*, 15–20.

Crowley, K. (2021). Fighting the future: The politics of climate policy failure in Australia (2015–2020). *Wires Climate Change, 12*(5), e725.

Ellis, N., & Albrecht, G. (2017). Climate change threats to family farmers' sense of place and mental wellbeing: A case study from the Western Australian Wheatbelt. *Social Science and Medicine, 175*, 161–168.

Fleming, A., Dowd, A. -M., Gaillard, E., Park, S., & Howden, M. (2015). "Climate change is the least of my worries": Stress limitations on adaptive capacity. *Rural Society, 24*(1), 24–41.

Godden, N., Naidu, K., & Boom, K. (2022). Climate change hits some of us much harder than others—But affected groups are fighting back. *The Conversation,* 14 October.

Green, D., Alexander, L., McInnes, K., Church, J., et al. (2010). An assessment of climate change impacts and adaptation for the Torres Strait Islands, Australia. *Climatic Change, 102*, 405–433.

Green, D., & Martin, D. (2017). Maintaining the healthy country-healthy people Nexus through sociocultural and environmental transformations: Challenges for the Wik aboriginal people of Aurukun, Australia. *Australian Geographer, 48*(3), 285–309.

Hall, N., Barnes, S., Canuto, C., Nona, F., & Redmond, A. (2021). Climate change and infectious diseases in Australia's Torres Strait Islands. *Australian and New Zealand Journal of Public Health, 45*(2), 122–128.

Hanigan, I., Butler, C., Kokic, P., & Hutchinson, M. (2012). Suicide and drought in New South Wales, Australia, 1970–2007. *Proceedings of the National Academy of Sciences, 109*, 13950–13955.

HEAL Network and CRE-Stride. (2021). *Climate change and aboriginal and Torres Strait islander health.* Lowitja Institute.

Hodgetts, K., Kleinecke, M., Woerle, C., Kaestli, M., et al. (2022). Melioidosis in the remote Katherine region of northern Australia. *PLoS Neglected Tropical Diseases, 16*(6), e0010486.

Horton, G., Hanna, L., & Kelly, B. (2010). Drought, drying and climate change: Emerging health issues for ageing Australians in rural areas. *Australasian Journal on Ageing, 29*, 2–7.

Hutley, N., Dean, A., Hart, N., & Daley, J. (2022). *Uninsurable Nation: Australia's most climate-vulnerable places.* Climate Council.

Jalaludin, B., & Morgan, G. (2021). What does climate change have to do with bushfires? *Australian Health Review, 45*(1), 4–6.

Jardine, A., Neville, P., & Lindsay, M. (2015). Proximity to mosquito breeding habitat and Ross River virus risk in the peel region of Western Australia. *Vector-Borne and Zoonotic Diseases, 15*(2), 141–146.

Johnston, F. (2009). Bushfires and human health in a changing environment. *Australian Family Physician, 38*(9), 720–724.

Johnston, F., Borchers-Arriagada, N., Morgan, G., Jalaludin, B., et al. (2021). Unprecedented health costs of smoke-related PM2. 5 from the 2019–20 Australian megafires. *Nature Sustainability, 4*(1), 42–47.

Kampmark, B. (2022). Australia trashes Torres Strait Islander Rights. *Green Left*, 1362, 4 October, 20.

Lansbury-Hal, N., & Crosby, L. (2022). Climate change impacts on health in remote indigenous communities in Australia. *International Journal of Environmental Health Research, 32*(3), 487–502.

Lu, P., Xia, G., Zhao, Q., Green, D., et al. (2022). Attributable risks of hospitalizations for urologic diseases due to heat exposure in Queensland, Australia, 1995–2016. *International Journal of Epidemiology, 51*, 144–154.

McManus, P. (2021). A more-than-urban political ecology of bushfire smoke in eastern Australia, 2019–20. *Australian Geographer, 52*(3), 243–256.

McMichael, A., & Lindgren, E. (2011). Climate change: Present and future risks to health, and necessary responses. *Journal of Internal Medicine, 270*(5), 401–413.

McMichael, C., Barnett, J., & McMichael, A. (2012). An Ill Wind? Climate change, migration, and health. *Environmental Health Perspectives, 120*(5), 646–654.

Morrissey, S., & Reser, J. (2007). Natural disasters, climate change and mental health considerations for rural Australia. *Australian Journal of Rural Health, 15*(2), 120–125.

Ng, F., Wilson, L., & Veitch, C. (2015). Climate adversity and resilience: The voice of rural Australia. *Rural and Remote Health, 15*(4), 3071.

Nishant, N. J., F, Guo Y, Herold N., et al. (2022). Future population exposure to Australian heatwaves. *Environmental Research Letters, 17*(6), 064030.

Nursey-Bray, M., Palmer, R., Smith, T., & Rist, P. (2019). Old ways for new days: Australian Indigenous peoples and climate change. *Local Environment, 24*(5), 473–486.

Nursey-Bray, M., Palmer, R., Stuart, A., Arbon, V., & Rigney, L. (2020). Scale, Colonisation and adapting to climate change: Insights from the Arabana People, South Australia. *Geoforum, 114*, 138–150.

Preston-Thomas, A., Gair, P., Hosking, K., Devine, G., & Donohue, S. (2012). An outbreak of plasmodium falciparum malaria in the torres strait. *Community Diseases Intelligence, 36*(2), E180-5.

Pyke, A. (2018). The origins of dengue outbreaks in northern Queensland, Australia, 1990–2017. *Microbiology Australia, 39*(2), 93–95.

Quilty, S., Jupurrurla, N., Baillie, R., & Gruen, R. (2022). Climate, housing, energy and Indigenous health: A call to action. *Medical Journal of Australia, 217*(1), 9–12.

Ranse, J., Luther, M., Hertelendy, A., & Skinner, R. (2022). Impact of fine particulate matter (PM2.5) smoke during the 2019/2020 Australian bushfire disaster on emergency department patient presentations. *Journal of Climate Change and Health, 6*, 100113.

Rigby, C., Rosen, A., Berry, H., & Hart, C. (2011). If the land's sick, we're sick: The impact of prolonged drought on the social and emotional well-being of Aboriginal communities in rural New South Wales. *Australian Journal of Rural Health, 19*(5), 249–254.

Rocque, R., Beaudoin, C., Ndjaboue, R., Cameron, L., et al. (2021). Health effects of climate change: An overview of systematic reviews. *British Medical Journal Open, 11*, e046333.

Rugendyke, B., & Renouf, J. (2022). 'I simply haven't got it in me to do it again': Imagining a new heart for flood-stricken Lismore. *The Conversation,* 18 March.

Schweinsberg, S., McManus, P., Darcy, S., & Wearing, S. (2019). 'Drought tourism' as compassion. *Annals of Tourism Research, 83*, 102843.

Speer, M., Hartigan, J., & Leslie, L. (2022). Cotton on: One of Australia's most lucrative farming industries is in the firing line as climate change worsens. *The Conversation*, 14 October.

Steffen, W., & Bradshaw, S. (2021). *Hitting home: The compounding costs of climate inaction.* Climate Council.

Tall, J., Gatton, M., & Tong, S. (2014). Ross River virus disease activity associated with naturally occurring non-tidal flood events in Australia: A systematic review. *Journal of Medical Entomology, 51*, 1097–108.

Walker, J., Pyke, A., Florian, P., Moore, F., et al. (2021). Re-emergence of dengue virus in regional Queensland: 2019 dengue virus outbreak inRockhampton, Central Queensland, Australia. *Community Diseases Intelligence, 45.*

Walls, W. (2022). Melbourne now has chief heat officers. Here's why we need them and what they can do. *The Conversation,* 17 October.

Watts, N., Amann, M., Ayeb-Karlsson, S. B., K, et al. (2018). The Lancet Countdown on health and climate change: From 25 years of inaction to a global transformation for public health. *The Lancet, 391*(10120), 581–630.

Watts, R., Liu, W., Van Hooff, M., & McFarlane, D. (2023). Incidence and factors impacting PTSD following the 2005 Eyre Peninsula bushfires in South Australia—A 7 year follow up study. *Australian Journal of Rural Health, 31*(1), 132–137.

Woodruff, R., & Bambrick, H. (2008). Climate change impacts on the burden of Ross River virus disease. In R. Garnaut (Ed.), *Garnaut climate change review* (pp. 1–15). Cambridge University Press.

Ziersch, A., Baum, F., Darmawan, I., Kavanagh, A., & Bentley, R. (2009). Social capital and health in rural and urban communities in South Australia. *Australian and New Zealand Journal of Public Health, 33*, 7–16.

Climate Change and Human Health in Fiji: Policies and Equity

Eberhard H. Weber

Abstract Climate Change policies have made good progress on national and regional levels in the Pacific Island Region. Link between climate change and health challenges have been insufficiently addressed to date. Here additional efforts are required to reduce climate change impacts and enhance effective protection of people's health. This paper discusses major health challenges in Fiji that can come along with climate change. There is time and scope to achieve that health institutions and networks can make a difference and create climate change safer societies. Governments and civil society would to do a disservice to the country by waiting until resources flow and/or essential changes are happening from alone. Changes need to particularly address challenges and incapacities poorer sections of Fiji's society have to adequately respond to threats arising from climate and other challenges.

Keywords Fiji · Vector and water borne diseases · Equity · Multiple hazards · Food security · Disasters

1 Introduction

Scientists agree that climate change is not a serious threat for the global future, it has already started. Since 1990, the Intergovernmental Panel on Climate Change (IPCC) has published six Assessment Reports (AR) on a huge number of aspects about the causes of, impacts from and adaptation to climate change (IPCC, 1990, 1996b, 2001, 2007, 2014b, 2023). In addition, IPCC published a big number of reports on special topics of climate change (e.g., IPCC, 2012a, 2012b on extreme events; 2011 on renewable energy, 1998 on regional impacts; and many more). What is missing is a special report on health challenges that arise from climate change. Even the first two ARs were rather short and general on climate change and related health outcomes. This changed with AR3 (2001). All later reports give health great importance as well, in the meanwhile supplemented by a chapter on well-being.

E. H. Weber (✉)
The University of the South Pacific, Suva, Fiji Islands
e-mail: eberhard.weber@usp.ac.fj

R. Akhtar (ed.), *Climate Change and Human Health Scenarios*, Global Perspectives on Health Geography, https://doi.org/10.1007/978-3-031-38878-1_3

In 2014, the World Health Organization (WHO) published models indicating that in 2030, climate change impacts cause an annual mortality of about globally 250,000. Major health risks arise from diarrheal diseases, malaria, malnutrition and heat-related illness (WHO, 2014). Climate change can exacerbate existing health risks (already existing heat waves, endemic diseases, hazards that intensify) or introduce entirely new health risks to countries (Doelle & Seck, 2021).

When looking at the relationship between human health and climate change three major aspects stand out: (1) the impacts climate change has on human health (2) possible strategies/policies countries take up to mitigate climate change impacts on human health, and (3) differences in health impacts within societies that arise from inequalities. In this chapter, the focus of these three perspectives is on Fiji, a Republic of just under 1 million inhabitants in the Pacific Islands region.

2 Vector- and Water-Borne Diseases as Climate Change Impacts

Fiji has a humid tropical maritime climate. There are no extremely high or low temperatures. Maximum day temperatures rarely exceed 35 °C (95 °F). Heat waves likely experienced in India, the Middle East, and in Europe, with days exceeding maximum day temperatures of 40 °C (104 °F) (Adélaïde et al., 2022; Marx et al., 2021; Miller et al., 2021), are unknown in Fiji. The heat disasters as experienced 2023 in many part of the world as well as resulting forest fires are unknown to Fiji.

Predictions from an assessment in 2005 indicted that climate change causes an increase in the incidence of dengue fever, diarrhea and nutrition related illness in Fiji. The assessment sees a considerate risk in dengue fever outbreaks until 2100 (Government of Fiji, 2005).

Among vector-borne diseases, dengue fever is the most serious. It is endemic to Fiji, which is free from endemic malaria. The Fiji government fears that an increase in average temperature will lead to more dengue fever outbreaks and bring malaria to the country. The argument appears logical: Climate change extents the range of mosquitoes that carry malaria and dengue fever, as climate allegedly influences the distribution of these insects (Barnett & Campbell, 2015). It appears, however, unclear, if climate is the (only) factor that prevents a wider distribution of these vectors. Species of the Anopheles mosquito that can transmit malaria are not restricted to tropical and sub-tropical regions. Indeed, they are almost ubiquitous. Malaria was endemic in Europe until very recently (Gelabert et al., 2017). Malaria outbreaks have been recorded in Northern Canada (Fallis, 1984), Scandinavia (Chen et al., 2021), and Russia. Here, malaria occurred as far North as Arkhangelsk in the arctic (Brabin, 2014; Bruce-Chwatt & De Zulueta, 1980; see also the map of global distribution of malaria vectors, CDC, 2012). The reason that endemic malaria did disappear from most places in Europe was not climate change (this time a decline in temperature), but a widespread, systematic approach to eradicate the vector of this dangerous disease.

Annually, there are between 20 and 100 cases of leptospirosis in Fiji. When epidemic outbreaks happen, numbers can be much higher (Togami et al., 2018). The disease is caused by various strains of Leptospira bacteria, which is very sensitive to higher temperatures and rainfall patters as they occur in tropical countries. The disease can affect liver, kidney and meninges, and other organs. Transmission to humans occurs through contact with urine, blood, or tissues of infected animals or contaminated water, especially from streams, swamps, ponds, and sewers. People exposed to the feces of infected animals are at highest risk to get infected. During floods and cyclones, the risk of transmission increases (PCCAPHH, 2012). Very similar to typhoid fever, leptospirosis outbreaks often accompany flood disasters (Mayfield et al., 2018). Following the TC/floods in January and March 2012 in the Western Division, some 576 cases and 40 deaths from leptospirosis have been recorded (Syakbanah & Fuad, 2021).

Typhoid fever is endemic in Fiji. It happens also in epidemic outbreaks (de Alwis et al., 2018; Watson et al., 2017). Since the past 20 years, an increase of typhoid in Fiji has been observed and frequent outbreaks recorded (Singh et al., 2022; Strobel et al., 2022). Typhoid fever is caused by the bacteria Salmonella enterica Serovar Typhi. Symptoms can be mild, but the fever can also be live threatening, especially when no medical treatment is available. Symptoms start six to 30 days after contracting the disease. Increasing fever over several days is usually accompanied by weakness, abdominal pain, constipation, headaches, and vomiting. Transmission is through fecal–oral route. Infections happen through contaminated water or insects (usually flies) which contaminate food. Frequent ways to spread the disease are fruits and vegetables exposed to human manure, unboiled, contaminated milk, kava prepared using unsafe water or unclean hands, flies carrying bacteria from open latrines to food that is uncovered. Rivers and streams contaminated by sewage during flooding are important methods of infections during disasters. Significant socio-economic risk factors are poverty, poor sanitation, and hygiene.

3 Food Security, Non-communicable Disease, and Climate Change

In Fiji, conditions for a highly productive agriculture are rather good. Important parameters are soil quality, topography, and others. Physical parameters enable or restrict plant growth and influence plant survival. Extremes, e.g., very low or very high temperatures, the lack of water (drought), too much water (flood, water logging) are factors that can devastate plant growth (Langan et al., 2022; Orimoloye et al., 2022; Rhaman et al., 2022). Especially under high temperatures, pathogens and plant pests/diseases thrive (Arya, 2021).

A changing climate can affect crops and livestock in various ways. This is not always negative. An increase in the number of warmer days, longer periods in temperate climate zone that are frost-free, prolong the vegetation period and extend areas that can be under cultivation. Higher temperatures and higher CO_2 concentration is not only a liability to agriculture; they can support plant growth and output. CO_2 is the most important requirement for photosynthesis. If crop yields increase as a result of higher temperatures and CO_2 concentration depends on the type of photosynthesis of a particular plant (C3 and C4 photosynthesis pathways, Molotoks et al., 2021; Sonmez et al., 2022), but predictions are complex and difficult to make.

In many developing countries, agriculture and related activities provide livelihood to more than 70% of the population, usually the poorer sections of society (Daramola & James, 2021; Konuma, 2021). In Pacific Island countries, the importance of coastal fisheries is very high (Leal Filho et al., 2021; Roscher et al., 2022; Veitayaki, 2021). Additional stress on ecological fragile food production systems in societies with many socially vulnerable populations can create challenges of hunger and starvation (Kemmerling et al., 2022; Nelson et al., 2010; Quak, 2021).

More intense (and possibly more frequent) natural hazards have the potential to affect agricultural production in Fiji severely. IPCC (2014a) expects that extreme events exacerbate existing vulnerabilities (IPCC, 2014a). Agriculture is a crucial component of food security. People cannot eat food that has not been produced. People with land can produce their own food, but people without access to land have to buy it. People cannot eat food that is available, when they don't have money to buy it. When much food is imported, people become highly dependent on what is happening elsewhere, how commodity prices are changing. We right now witness how a war in Ukraine severely compromises food security of people living in Africa (Behnassi & El Haiba, 2022; Berkhout et al., 2022; Glauben et al., 2022).

People can establish rights over food in various ways: access to land allows them to grow own food; people's income allow them to buy food in the (super)market. During disasters, people can get free or subsidized food from governments and NGOs. Social welfare programs exist for very poor, destitute people, who have nobody to care for them. Such systems support people, who otherwise would be unable to satisfy their food and other basic needs. Formal social security, the institutionalized protection against sickness, accidents, old age, pregnancy, and other risks of the daily life are open mainly for people working in the formal part of Fiji's economy.

Despite the crucial value of such support systems, most important protective measures against the impacts of climate change are those that make people independent from government intervention. To be able to earn living wages that help to escape poverty. In addition, one needs to fine-tune instruments to identify most vulnerable sections in society. Only this can lead to policies and measures aiming to support the most vulnerable. Vulnerability goes beyond poverty. Social vulnerability has a distinct structural dimension created over long time. Coping can protect people in the short run. Over a longer time, coping can be disastrous as it increases what Watts and Bohle (1993) call people's baseline vulnerability. When people have to deal with multiple threats at the same time, or in close succession to each other, coping is not enough. People can deal with a drought or a cyclone. When another disaster strikes before they have fully recovered from a previous one, then it often happens that people are no longer able to get out of such situations without losses. Development achievement of years is then collapsing like card houses that have become instable.

Poor people spend a considerable part of their disposable income on food. This can be up to 95%. During 'normal' situations, they just have enough to afford as much food as they require. Increases in food prices can bring them into situations, when they do not possess enough money required for a diet that safeguards a health life.

Other challenges happen as well: when maximum temperatures increase because of climate change people respond with a decline in activities. This can lead to a rise in obesity. Non-communicable diseases (NDC) like diabetes, cardio-vascular illnesses, and musculoskeletal disorders follow.

To relate such diseases to climate change (alone) appears artificial. Interventions to reduce morbidity need to happen anyway. In western societies, ideals are increasingly about low(er) weight indicating a 'perfect body'. When humans made first representation of their bodies some 35,000 years ago, they depicted obese females. The Venus figurines are corpulent, shown as obese figures, in an era, when the global climate was changing considerably: it became colder and the last ice age followed (Johnson et al., 2021). Today some people particularly in non-western societies still associate obesity with wealth, fertility, and well-being (Bishwajit, 2017; Sumińska et al., 2022). Other studies, however, suggest that often obesity is a sign of low(er) social status, like Addo et al. (2009) have shown for women in Ghana, or of lower educational level (Pérez-Ferrer et al., 2018).

Obesity is a special form of malnutrition. In Fiji (and most other Pacific Island societies), it has become a serious cause for morbidity and mortality. Other forms of food insecurity relate to endocrine (hormones), nutritional and metabolic diseases (e.g., diabetes), which are the second most common cause of morbidity and mortality in Fiji, next to diseases of the circulatory system (Ministry of Health, 2011, Table 51).

4 Health Impacts of Flooding

Scientists assume that climate change will increase intensities of meteorological and hydrological hazards. With such hazards, severe health challenges come along. A major challenge in Fiji arising from natural hazards are tropical cyclones (TC) and floods. What the latter are concerned the Western Division of the country is a hotspot for floods. Towns like Nadi and Ba are frequently affected. Floods often cause displacements of people and at times even death. They put thousands at risk of sliding into poverty, when they lose valuable assets, which had taken them decades to build and which they cannot replace easily.

Ba district is known for its severe problems of flooding. Many people suffered from the floods of January 2009 and January 2012. In 2012, the Fiji government declared 'a state of natural disaster emergency' for the Ba region. Looking at people affected, the most vulnerable groups have been women and children, but farmers were affected what livelihood generation is concerned.

Research on flooding in five villages around Ba covered 97 households. Both floods, in 2009 and January 2012, affected all these households. Some 30% of households had to evacuate their homes twice. Around 85% found refuge with extended families, neighbors, and friends. The remaining 15% went to evacuation centers. More than 75% of households participating in the study reported a considerable loss of livelihood to the two floods. The short time between both floods as well as their magnitude made it difficult to many households to recover fully from the 2009 flood before then the next flood hazard came in 2012.

The damage to houses was the most serious impact. Erosion to agricultural land, however, had severe consequences to livelihood generation. Farmland situated along the Ba River was lost during the floods. One quarter of households included in the survey lost agricultural land due flooding. A few households even reported that they had lost their entirely land.

After the floods in 2012, cases of dengue fever increased in Ba. Dengue case numbers peaked a month following the floods. The floods compromised water supply and safety leading to a high incidence of diarrhea. The situation could be mitigated in areas to where trucks supplied water and where WASH kits (containing water purification tablets, water containers, and soap) were distributed to people. Some 9600 sachets of Oral Rehydration Salts (ORS) were also distributed (UNOCHA Pacific, 2012).

5 Multiple Hazards, Disasters, and Health

Although recent TCs have been the strongest ever recorded in Fiji (TC Winston, 2016, TC Herold 2020 and TC Yasa 2020), it is not fully understood, if TC has become more frequent and more intense. Prior to the past severe TCs indications suggested that frequency and intensity of tropical cyclones affecting Fiji between 1969/70 and 2009/10 were on a downward trend (Government of Fiji, 2012).

From the perspective of health, TCs can cause injuries and death to people, and increase the dangers of vector- and water-borne diseases. In extreme cases, TCs can have severe impacts on people's food security, when no outside support is available.

Strongest impacts, however, relate to multiple hazards that happen at the same time, or in short succession to each other. Between 2020 and 2022, there been five tropical cyclones (TC) in Fiji and a major flooding event. The worst were two category 5 tropical cyclones (TC Harold and TC Yasa) and two category 3 tropical cyclones (TC Cody and TC Ana). These hazards added challenges and hardship to the medical and economic crisis caused by the COVID-19 pandemic. The SARS-CoV-2-virus made rescue, relief, and rehabilitation activities difficult when tens of thousands of people were seeking refuge in emergency centers.

A month after COVID-19 had reached Fiji TC Harold (April 2020) happened, a category 5 cyclone, one of the strongest that ever made landfall in Fiji. TC Harold caused considerable damage all over Fiji, but particularly in Kadavu. Relief efforts could not start from the port of Suva as Fiji's capital was under lockdown. Instead,

Natovi jetty, some 60 km north of Suva, became the starting point for relief and rehabilitation efforts. At the end of 2020, TC Yasa (December 2020) was the second category 5 TC that hit Fiji in 2020. Maximum wind speeds (260 km/h) were even higher than during TC Harold. Around 23,000 people spent days in 456 evacuation centers. More than 2000 homes have been completely destroyed and around 6200 were partially damaged (Talei, 2021).

Just a month after TC Yasa, TC Ana (end of January, early February 2021) caused severe damages. The last major TC happening subsequently with the COVID-19 pandemic was TC Cody, which struck Fiji in early January 2022. Widespread flooding caused much damage in Fiji particularly in the western and central divisions with 4630 people at 162 evacuation centers (RNZ, 13.01.2022). Around the same time, the downfall of volcanic ash after a volcanic eruption in neighboring Tonga affected Fiji.

The most severe challenges that come with such double/multiple exposure to hazards is to provide people protection against cyclones and floods in evacuation centers, and at the same time assure that people are not exposed to the highly contagious Delta, and later Omicron variant of SARS-CoV-2-virus, which dominated COVID-19 infections in 2021 and 2022. Another logistical challenge has been to provide food for a big number of people in emergency centers and COVID-19 lockdown areas.

Under a changing climate such double/multiple exposure to hazards, natural, and human-made, the like can become more frequent. Once Fiji had overcome the most severe repercussion of COVID-19, another 'hazard' affected global society with the Russian invasion of Ukraine. Implications outside Eastern Europe, e.g., impacts on global inflation and food security in developing countries are considerate (Weber, 2022).

Fiji, like many Pacific Island Countries, experiences transition that weaken their societies. Social capital can provide essential benefits when people are at risk from disasters that arise from 'natural' hazards, but also from adverse events that are human-made. Weakening social capital deprives people from support structures in societies where often social security exists only for fractions of people, this working in the formal sectors of the economy.

Food security in the context of disasters plays a major role in Pacific Islands (Campbell, 2006; Thaman, 1982a, 1982b). Currey (1980) highlights that historically hazards were the major events that have compromised food security in Pacific islands to an extend that famines occurred. This was in the eighteenth and nineteenth century, when societies were far more isolated compared to the twenty-first century.

Research conducted in communities of Fiji (Yila et al., 2014), the Solomon Islands, and Samoa suggests that food insecurity is *not* unavoidable, neither when climate change intensifies nor when other hazards strike. Societies can master such challenges. Resilience helps people to deal with the impacts of climate change. Governments can make their health systems ready for such changes. Societies eradicated erstwhile deadly diseases. This can be achieved for health challenges arising from climate change. Improvements of food and livelihood security seem to be possible as well as improvements to the overall conditions and levels of people's lives. Although two research activities look at impacts of tsunamis and do not directly relate to climate change, the studies provide insights into people's resilience to severe disasters.

Research on immediate tsunami impacts in Samoa and flooding in Fiji were conducted right after the hazard events. The third field study on the 2007 tsunami in the Solomon Islands was conducted some five years after the actual hazard. A second round of fieldwork was undertaken some 4.5 years after the tsunami in Samoa. In all cases, people's food security was at risk, but nowhere serious food crises happened. Mechanisms in place helped to prevent this. None of the communities visited reported widespread lack of food immediate after the disaster events, nor some considerable time later.

6 Climate Change Policies as They Relate to Health

Often governments took up climate change related issues from the perspective of getting more funds from the international community. Demands were addressed to countries that have high present and historical emissions in GHG to compensate losses and damages caused by climate change. The Fiji Government along with many other governments in SIDS lobby strongly to include community relocation in such provisions about potential loss and damage (Lund, 2021). McNamara et al. (2021) highlight the health dimension in the loss and damages debate, including direct as well as indicted health impacts arising from climate change. Some are of economic nature, but many are non-economic, at least at first sight, such as mental health impacts.

Such requests are justified beyond any doubt. SIDS, including Fiji, has little contributed to the accumulation of GHG in the Earth's atmosphere, but SIDS bears the brunt of damages in comparison with the resources they have available. It would be, however, a disastrous mistake to wait until international compensation mechanisms are in place. Governments and civil society have to take up urgent steps at home to prepare countries for an uncertain future. This includes the preparation of their health systems.

At home, Fiji could and should have done more to revamp its health sector. Not only in international comparison, but also within the Pacific Island region, Fiji is situated in the lower ranks concerning health indicators. Among 11 Pacific Island countries listed by the World Bank, Fiji ranks second last in budget provisions for health. In 2019, just 3.8% of the country's budget went toward health (Fig. 1). Tuvalu's budget dedicates a share that is seven times as big for expenses in the health sector. Other PICs give much bigger importance to the health sectors as well. From an international perspective, Fiji stood at 160th place among the 188 countries the World Bank provided data.

Comparing budgetary allocations at times can be challenging. Different countries allocate health expense in varying ways. Looking at some health outcomes, Fiji equally is among the lower ranked countries in the Pacific (as well as internationally). Among ten PIC listed by the World Bank, six (Vanuatu, Tuvalu, Solomon Islands,

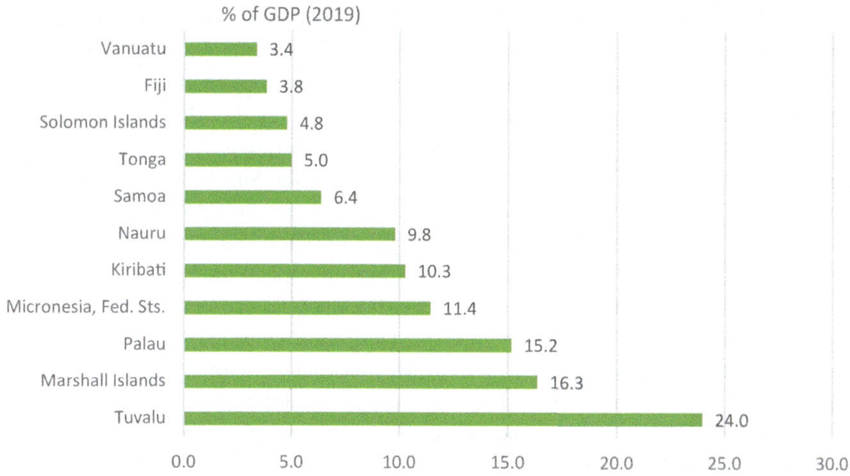

Fig. 1 Government expenditure on health in selected Pacific Island Countries (2019)

Palau, Samoa, and Tonga) have lower infant mortality rates than Fiji. Tonga with the lowest rate (11.4 per 1000 live births) is considerably below Fiji's rate with 27.4 deaths per 1000 live births. Another health indicator is life expectancy. For Fiji, it stood at 67.6 years in 2020. Only PNG (64.7 years) had a lower life expectancy among eight PICs listed by the World Bank. Samoa was highest (73.5 years), followed by the Solomon Islands (73.1 years). In these two countries, people on average live more than 5.5 years longer than in Fiji (all data: https://data.worldbank.org/).

Fiji's rather weak health system has little to do with climate change. Considerable improvements could make the country and the health of its citizens more resilient to climate change. Efforts to advance toward the Sustainable Development Goals (SDG) are crucial steps. Five of the 17 SDGs appear to be crucial from a climate change and health perspective: SDG 1 (end poverty), SDG 3 (Good health and well-being), SDG 6 (clean water and sanitation), SDG 10 (reduce inequality), and SDG 13 (climate action) (Pérez-Peña et al., 2021).

7 Climate Change, Health and Equity

Climate change can have severe impacts on physical and social environments in which people live. Vulnerabilities abound, but not everybody is equally and in the same way vulnerable to climate change. Inequities that exist in health issues generally do also matter in health challenges that arise from a changing climate. This means, among others, that a changing climate is not universally harmful. 'Poor people face a double burden of inequality—from uneven development and climate change' (Pelling & Garschagen, 2019). Health impacts due to climate change are unevenly and inequitably distributed. Women and children; people living in poverty, particularly in developing countries are most severely affected (Balasubramanya et al., 2020). The imperative to see climate change as a severe, or even the biggest development challenge is expressed by many (Kelman, 2014; Leal Filho, et al., 2020; Mathew et al., 2021; Owusu & Nursey-Bray, 2019). Morton et al. (2019) suggest that health sectors, particular public health systems, can benefit a lot from the Sustainable Development Goals (SDGs) addressing climate change and other threats to future health.

Climate change adversely affects some groups in society more than others. People already in poverty suffer more from of climate change impacts than others. This is true also for health impacts. Climate change makes poor people poorer as they have lesser capacities to adequately respond to impacts of climate change. Poor people are often excluded from decision making processes for relevant climate action (Obringer & Nateghi, 2021). There is even convincing indication that climate change becomes a multiplier of poverty pushing people into poverty, people, who are just slightly above the poverty line, but cannot cope with new challenges climate change brings (Dugassa, 2021; Ehsan et al., 2022). Severe impacts on food security as well as increase in diseases affect poor people more than non-poor.

A study in the USA suggests that climate change is an important driver worsening poverty and health in decades ahead. Indeed climate change and its health impacts can generate a huge number of new poor, particularly elderly, people who cannot afford increasing costs for housing and medical services, especially when they have to deal with disasters that are more ferocious and new diseases (Tonn et al., 2021).

Improving public health and well-being, and protecting people from the impacts of climate change is an urgent policy requirement.

There is much empirical evidence from outside Fiji that poor people disproportionally suffer from natural hazards (Hallegatte & Walsh, 2021). Spatial considerations play a big role: poor people more often than others live in places where hazards are most damaging: in wetlands, mangrove forests, on sub-standard coastal land and along rivers (Hallegatte et al., 2020; Kawata, 2022; Koto, 2020). Poor people's housing structures struggle to withstand forces of TC. Their socio-economic situation makes it difficult and lengthy to recover from disasters. Impacts of climate change not only affect poor severely, climate change has the potential to create new poverty and exacerbate existing inequalities. Climate change has the potential to undo positive results of development efforts and repel people in their efforts to secure a decent standard of living (Defe & Matsa, 2021; Dugassa, 2021; Sibiya et al., 2022).

Evidence from Fiji shows that climate change and extreme events can have severe impacts on rural and urban poor, who anyway face many challenges and are less able to cope or adapt to additional pressures (Brown et al., 2017). In rural areas, but also in informal settlements in Suva, poor people often live close to nature (French et al., 2021; Weber & Koto, 2021; Weber et al., 2022). They derive bigger parts of their livelihoods from the resource surrounding them. When resources degrade in quality and decline in productivity, poor people are socially and economically least mobile, not able to move to other place to making a living.

8 Conclusion

It is very important to have well functioning social institutions at the local level. People then can better deal with natural hazards. People can reduce the risk that hazards turn out to become disasters best using their capacities and capabilities. Such social capital helps also in long-term recovery and rehabilitation after disasters. Although social capital can become essential, the formal establishment of safety nets is crucial to protect particular poor groups and minority group, where social capital is often insufficiently developed. Informal systems can easily erode when damages are far beyond of what such systems can handle.

Damaging impacts of climate change might reach dimensions that people's capacities and capabilities are insufficient to protect them. Support from outside will be more often required than earlier. Dangers that people 'trust too much' on such external support and diminish internarial capacities to respond to threats are very relevant. Already Thaman (1982a) highlights that external interventions during and after disasters in Pacific Island Countries have to pay more attention to local contexts, capacities, and constraints. External support structures should strengthens local populations' agency and resilience rather than compromise them. External support must not compromise internal capabilities. It needs to build upon local support and knowledge systems. Whenever relocation—temporal are permanent—becomes essential all stakeholders involved need to assess land tenure and other local systems carefully.

References

Addo, J., Smeeth, L., & Leon, D. A. (2009). Obesity in urban civil servants in Ghana: Association with pre-adult wealth and adult socio-economic status. *Public Health, 123*(5), 365–370.

Adélaïde, L., Chanel, O., & Pascal, M. (2022). Health effects from heat waves in France: An economic evaluation. *The European Journal of Health Economics, 23*(1), 119–131.

Arya, N. (2021). A review on effects of climate change on plant diseases. *ACADEMICIA: An International Multidisciplinary Research Journal, 11*(11), 896–902.

Balasubramanya, S., & Stifel, D. (2020). Water, agriculture & poverty in an era of climate change: Why do we know so little? *Food Policy, 93*, 101905.

Barnett, J., & Campbell, J. (2015). *Climate change and small island states: Power, knowledge and the South Pacific*. Routledge.

Behnassi, M., & El Haiba, M. (2022). Implications of the Russia–Ukraine war for global food security. *Nature Human Behaviour*, 1–2.

Berkhout, P., Bergevoet, R., & van Berkum, S. (2022). A brief analysis of the impact of the war in Ukraine on food security (No. 2022–033). Wageningen Economic Research.

Bishwajit, G. (2017). Household wealth status and overweight and obesity among adult women in Bangladesh and Nepal. *Obesity Science & Practice, 3*(2), 185–192.

Brabin, B. J. (2014). Malaria's contribution to World War One–the unexpected adversary. *Malaria Journal, 13*(1), 1–22.

Brown, P., Daigneault, A., & Gawith, D. (2017). Climate change and the economic impacts of flooding on Fiji. *Climate and Development, 9*(6), 493–504.

Bruce-Chwatt, L. J., & De Zulueta, J. (1980). *The rise and fall of Malaria in Europe*. Oxford University Press.

Campbell, J. R. (2006). *Traditional disaster reduction in Pacific Island communities*. GNS Science Report 38.
CDC. (2012). Global Distribution (Robinson Projection) of Dominant or Potentially Important Malaria Vectors. https://www.cdc.gov/malaria/about/biology/mosquitoes/map.html
Chen, T. T., Ljungqvist, F. C., Castenbrandt, H., Hildebrandt, F., Ingholt, M. M., Hesson, J. C., Ankarklev, J., Seftigen, K., & Linderholm, H. W. (2021). The spatiotemporal distribution of historical malaria cases in Sweden: a climatic perspective. *Malaria Journal, 20*(1), 1–14.
Currey, B. (1980). Famine in the Pacific losing the chances for change. *GeoJournal, 4*, 447–466.
Daramola, D. G., & James Jr, H. S. (2021). The social protection function of agriculture. In *Handbook on the Human Impact of Agriculture* (pp. 233–255). Edward Elgar Publishing.
de Alwis, R., Watson, C., Nikolay, B., Lowry, J. H., Thieu, N. T. V., Van, T. T., Ngoc, D. T. T., Rawalai, K., Taufa, M., Coriakula, J., Lau, C. L., Nilles, E. J., John Edmunds, W., Kama, M., Baker, S., & Cano, J. (2018). Role of environmental factors in shaping spatial distribution of *Salmonella enterica* serovar Typhi, Fiji. *Emerging Infectious Diseases, 24*(2), 284.
Defe, R., & Matsa, M. (2021). The contribution of climate smart interventions to enhance sustainable livelihoods in Chiredzi District. *Climate Risk Management, 33*, 100338.
Doelle, M., & Seck, S. L. (2021). *Introducing loss and damage*. Edward Elgar Publishing.
Dugassa, B. (2021). Climate change and public health challenges in the horn of Africa: The need for sustainable leadership and institutions. *American Journal of Public Health, 9*(1), 5–17.
Ehsan, S., Begum, R. A., & Maulud, K. N. A. (2022). Household external vulnerability due to climate change in Selangor coast of Malaysia. *Climate Risk Management, 35*, 100408.
Fallis, A. M. (1984). Malaria in the 18th and 19th centuries in Ontario. *Canadian Bulletin of Medical History, 1*(2), 25–38.
French, M., Trundle, A., Korte, I., & Koto, C. (2021). Climate resilience in urban informal settlements: Towards a transformative upgrading agenda. In Climate Resilient Urban Areas (pp. 129–153). Palgrave Macmillan, Cham.
Gelabert, P., Olalde, I., De-DIos, T., Civit, S., & Lalueza-Fox, C. (2017). Malaria was a weak selective force in ancient Europeans. *Scientific Reports, 7*(1), 1–10.
Glauben, T., Svanidze, M., Götz, L., Prehn, S., Jamali Jaghdani, T., Đurić, I., & Kuhn, L. (2022). The war in Ukraine, agricultural trade and risks to global food security. *Intereconomics, 57*(3), 157–163.
Government of Fiji. (2005). *Climate change—The Fiji Islands response. Fiji's first national communication under the framework convention on climate change*. https://unfccc.int/resource/docs/natc/fjinc1.pdf
Government of Fiji. (2012). *Republic of Fiji national climate change policy*. Secretariat of the Pacific Community, Suva, Fiji.
Hallegatte, S., & Walsh, B. (2021). Natural disasters, poverty and inequality: New metrics for fairer policies. In *The Routledge handbook of the political economy of the environment* (pp. 111–131). Routledge.
Hallegatte, S., Vogt-Schilb, A., Rozenberg, J., Bangalore, M., & Beaudet, C. (2020). From poverty to disaster and back: A review of the literature. *Economics of Disasters and Climate Change, 4*(1), 223–247.
IPPC. (1990). *Climate change: The IPCC impacts assessment; report prepared for IPCC by Working Group II*. Australian Gov. Publ. Service.
IPCC. (1996a). *Climate change, 1995: Impacts, adaptations, and mitigation of climate change: Scientific-technical analyses: Contribution of working group II to the second assessment report of the Intergovernmental Panel on Climate Change*. Cambridge University Press.
IPCC. (1996b). *Climate change 1995: The science of climate change*. Cambridge University Press.
IPCC. (1998). *The regional impacts of climate change: An assessment of vulnerability: A special report of IPCC Working Group II*. Cambridge Univ. Press.
IPCC. (2001). *Impacts, adaptation, and vulnerability*. Cambridge Univ. Press.
IPCC. (2007). *Climate change 2007: Impacts, adaptation and vulnerability: Working Group II contribution to the fourth assessment report of the IPCC*. Cambridge Univ. Press.

IPCC. (2012a). IPCC special report on renewable energy sources and climate change mitigation. Intergovernmental Panel on Climate Change.

IPCC. (2012b). *Managing the risks of extreme events and disasters to advance climate change adaptation: Special report of the Intergovernmental Panel on Climate Change*. IPCC.

IPCC. (2014a). *Fifth Assessment Report (AR5): Climate Change 2013/2014: Climate change 2014: Impacts, adaptation, and vulnerability*. Cambridge University Press.

IPCC. (2014b). *Climate change 2013: The physical science basis: Working group I contribution to the fifth assessment report of the intergovernmental panel on climate change*. Cambridge University Press.

IPCC. (2022). *Climate change 2022: Impacts, adaptation and vulnerability*. Switzerland, World Meteorological Organization, IPCC Secretariat.

IPCC. (2023). *Climate change 2021: The physical science basis: Working group I contribution to the sixth assessment report of the intergovernmental panel on climate change*. Cambridge University Press. https://doi.org/10.1017/9781009157889

Johnson, R. J., Lanaspa, M. A., & Fox, J. W. (2021). Upper paleolithic figurines showing women with obesity may represent survival symbols of climatic change. *Obesity, 29*(1), 11–15.

Kawata, Y. (2022). Green infrastructures in megacity Jakarta: Current status and possibilities of Mangroves for flood damage mitigation. In *Green infrastructure and climate change adaptation* (pp. 335–351). Springer.

Kelman, I. (2014). No change from climate change: Vulnerability and small island developing states. *The Geographical Journal, 180*(2), 120–129.

Kemmerling, B., Schetter, C., & Wirkus, L. (2022). The logics of war and food (in) security. *Global Food Security, 33*, 100634.

Konuma, H. (2021). Agriculture, food security and nutrition for most vulnerable people and our new generation. In *The basics of human civilization* (pp. 27–36). CRC Press.

Koto, C. (2020). *Coping strategies of Fiji's urban poor*. GRIN Verlag.

Langan, P., Bernád, V., Walsh, J., Henchy, J., Khodaeiaminjan, M., Mangina, E., & Negrão, S. (2022). Phenotyping for waterlogging tolerance in crops: current trends and future prospects. *Journal of Experimental Botany*, erac243, https://doi.org/10.1093/jxb/erac243

Leal Filho, W., Ha'apio, M. O., Lütz, J. M., & Li, C. (2020). Climate change adaptation as a development challenge to small Island states: A case study from the Solomon Islands. *Environmental Science & Policy, 107*, 179–187.

Leal Filho, W., Krishnapillai, M., Sidsaph, H., Nagy, G. J., Luetz, J. M., Dyer, J., Otoara Ha'apio, M., Havea, P. H., Raj, K., Singh, P. and Rogers, T., & Azadi, H. (2021). Climate change adaptation on small island states: An assessment of limits and constraints. *Journal of Marine Science and Engineering, 9*(6), 602.

Lund, D. (2021). Navigating slow-onset risks through foresight and flexibility in Fiji: Emerging recommendations for the planned relocation of climate-vulnerable communities. *Current Opinion in Environmental Sustainability, 50*, 12–20.

Marx, W., Haunschild, R., & Bornmann, L. (2021). Heat waves: A hot topic in climate change research. *Theoretical and Applied Climatology, 146*(1), 781–800.

Mathew, T. H., Agnes, R. S., & Louis, N. (2021). Climate change impacts and effects on health system: A challenge towards achievement of sustainable development goal 3 in South Africa. *Technium Social Sciences Journal, 26*, 950.

Mayfield, H. J., Smith, C. S., Lowry, J. H., Watson, C. H., Baker, M. G., Kama, M., Nilles, E.J., & Lau, C. L. (2018). Predictive risk mapping of an environmentally-driven infectious disease using spatial Bayesian networks: A case study of leptospirosis in Fiji. *PLoS Neglected Tropical Diseases, 12*(10), e0006857.

McNamara, K. E., Westoby, R., & Chandra, A. (2021). Exploring climate-driven non-economic loss and damage in the Pacific Islands. *Current Opinion in Environmental Sustainability, 50*, 1–11.

Miller, S., Chua, K., Coggins, J., & Mohtadi, H. (2021). Heat waves, climate change, and economic output. *Journal of the European Economic Association, 19*(5), 2658–2694.

Ministry of Health. (2011). *Annual Report 2011*. Ministry of Health of the Republic of Fiji.

Molotoks, A., Smith, P., & Dawson, T. P. (2021). Impacts of land use, population, and climate change on global food security. *Food and Energy Security, 10*(1), e261.

Morton, S., Pencheon, D., & Bickler, G. (2019). The sustainable development goals provide an important framework for addressing dangerous climate change and achieving wider public health benefits. *Public Health, 174*, 65–68.

Nelson, G. C., Rosegrant, M. W., Palazzo, A., Gray, I., Ingersoll, C., Robertson, R., ... & You, L. (2010). *Food security, farming, and climate change to 2050: scenarios, results, policy options*. International Food Policy Reserach Institute.

Obringer, R., & Nateghi, R. (2021). What makes a city 'smart' in the Anthropocene? A critical review of smart cities under climate change. *Sustainable Cities and Society, 75*, 103278.

Orimoloye, I. R., Belle, J. A., Orimoloye, Y. M., Olusola, A. O., & Ololade, O. O. (2022). Drought: A common environmental disaster. *Atmosphere, 13*(1), 111. https://doi.org/10.3390/atmos13010111

Owusu, M., & Nursey-Bray, M. (2019). Socio-economic and institutional drivers of vulnerability to climate change in urban slums: The case of Accra, Ghana. *Climate and Development, 11*(8), 687–698.

PCCAPHH. (2012). *Fiji: Protecting health from climate change working paper: Climate-sensitive infectious diseases in Fiji*. 2011 Summary report from Fiji's piloting climate change adaptation to protect human health project. WHO-SP and Fiji Ministry of Health, Suva.

Pelling, M., & Garschagen, M. (2019). Put equity first in climate adaptation. *Nature, 569*(7756), 327–329. https://doi.org/10.1038/d41586-019-01497-9

Pérez-Ferrer, C., McMunn, A., Zaninotto, P., & Brunner, E. J. (2018). The nutrition transition in Mexico 1988–2016: The role of wealth in the social patterning of obesity by education. *Public Health Nutrition, 21*(13), 2394–2401.

Pérez-Peña, M. D. C., Jiménez-García, M., Ruiz-Chico, J., & Peña-Sánchez, A. R. (2021). Analysis of research on the SDGs: The relationship between climate change. *Poverty and Inequality. Applied Sciences, 11*(19), 8947.

Quak, E. J. (2021). The drivers of acute food insecurity and the risk of famine. K4D Helpdesk Report no.1057. Institute of Development Studies. https://doi.org/10.19088/K4D.2021.132

Rhaman, M. S., Kibria, M. G., & Hoque, A. (2022). Climate change and its adverse impacts on plant growth in South Asia: Current status and upcoming challenges. *Phyton, 91*(4), 695.

Roscher, M. B., Eriksson, H., Harohau, D., Mauli, S., Kaltavara, J., Boonstra, W. J., & van der Ploeg, J. (2022). Unpacking pathways to diversified livelihoods from projects in Pacific Island coastal fisheries. *Ambio*, 1–11.

Sibiya, N., Sithole, M., Mudau, L., & Simatele, M. D. (2022). Empowering the voiceless: Securing the participation of marginalised groups in climate change governance in South Africa. *Sustainability, 14*(12), 7111.

Singh, R., Lal, S., Khan, M., Patel, A., Chand, R., & Jain, D. K. (2022). The COVID-19 experience in the Fiji Islands: Some lessons for crisis management for small island developing states of the Pacific region and beyond. *New Zealand Economic Papers, 56*(1), 67–72.

Sonmez, M. C., Ozgur, R., Uzilday, B., Turkan, I., & Ganie, S. A. (2022). Redox regulation in C3 and C4 plants during climate change and its implications on food security. *Food and Energy Security*, e387.

Strobel, A. G., Airs, S., Nguyen, C., Vadei, T. R., Matanitobua, S., Kama, M., Watson, C. H., Crump, J. A., Mulholland, E. K., Strugnell, R. A., & Parry, C. M. (2022). Assessment of rapid diagnostic tests for typhoid diagnosis and assessment of febrile illness outbreaks in Fiji. *The American Journal of Tropical Medicine and Hygiene, 106*(2), 543.

Sumińska, M., Podgórski, R., Bogusz-Górna, K., Skowrońska, B., Mazur, A., & Fichna, M. (2022). Historical and cultural aspects of obesity: From a symbol of wealth and prosperity to the epidemic of the 21st century. *Obesity Reviews*, e13440.

Syakbanah, N. L., & Fuad, A. (2021). Human leptospirosis outbreak: A year after the 'Cempaka' tropical cyclone. *Jurnal Kesehatan Lingkungan, 13*(4), 211–218.

Talei, F. (2021). Cyclone yasa damage costs hit $500m: Seruiratu 6184 houses were partially damaged and 2141 completely destroyed. Fiji Sun, February 13

Thaman, R. R. (1982a). Deterioration of traditional food systems, increasing malnutrition and food dependency in the Pacific Islands. *Journal of Food and Nutrition, 39*, 109–121.

Thaman, R.R. (1982b). Hurricane Isaac and Tonga: A natural or cultural disaster? *Review, 3*(8), 22–35. School of Social and Economic Development, USP.

Togami, E., Kama, M., Goarant, C., Craig, S. B., Lau, C., Ritter, J. M., Imrie, A., Ko, A. I., & Nilles, E. J. (2018). A large Leptospirosis outbreak following successive severe floods in Fiji, 2012. *American Journal of Tropical Medicine and Hygiene, 99*(4), 849–851.

Tonn, B., Hawkins, B., Rose, E., & Marincic, M. (2021). A futures perspective of health, climate change and poverty in the United States. *Futures, 131*, 102759.

UNOCHA Pacific. (2012). *Fiji. Flooding tropical depression TD17F-2012.* Situation report No. 4. 12th April 2012. UNOCHA Pacific, Suva.

Veitayaki, J. (2021). Securing coastal fisheries in the Pacific: Critical resources for food, livelihood and community security. *Development Bulletin, 82*, 56–60.

Watts, M. J., & Bohle, H. G. (1993). Hunger, famine and the space of vulnerability.*GeoJournal, 30*(2), 117–125.

Watson, C. H., Baker, S., Lau, C. L., Rawalai, K., Taufa, M., Coriakula, J., Thieu, N. T. V., Van, T. T., Ngoc, D. T. T., Hens, N., Lowry, J. H., & Edmunds, W. J. (2017). A cross-sectional seroepidemiological survey of typhoid fever in Fiji. *PLoS Neglected Tropical Diseases, 11*(7), e0005786.

Weber, E. (2022). People's management of risks from extreme weather events in the Pacific Island region. In V. Ongoma & H. Tabari (Eds.), *Climate impacts on extreme weather: Current to future changes on a local to global scale.* Elsevier.

Weber, E., & Koto, C. (2021). Disturbing the creation of a spatial system: Outside intervention and urbanization in the Republic of the Marshall Islands. In *The Routledge handbook of small towns* (pp. 386–403). Routledge.

Weber, E., Koto, C., Kopf, A. Bhagwan, M. Nawaqalevu, A., & Halter, N. (2022). History and recent processes of displacement in Suva, Fiji's capital. In A. Neef, T. Moreda, S. Mollett & C. Ngin (Eds.), *Routledge handbook of global land and resource grabbing.*

World Health Organization. (2014). *Quantitative risk assessment of the effects of climate change on selected causes of death, 2030s and 2050s.* WHO.

Yila, O., Weber, E., & Neef, A. (2014). The role of social capital in post-flood response and recovery among downstream communities of the Ba River, Western Viti Levu, Fiji Islands. In *Risks and conflicts: Local responses to natural disasters* (pp. 79–107). Emerald Group Publishing Limited.

Double Exposure Framework of COVID-19 Pandemic and Climate Change

Mei-Hui Li

Abstract Social and ecological factors for infectious diseases have received much attention in public health science. This chapter adopted an integrated approach to understand the effects of COVID-19 on public health policy amid climate change. First, a brief history of the one health approach, syndemic theory, and a socio-ecological model were analyzed. Then, case reports on wildlife, domesticated animals, and pets with severe acute respiratory syndrome coronavirus 2 (SARS-CoV-2) were described. Next, prevention measures and economic crises during the COVID-19 pandemic were organized in terms of their environmental effects. Social inequality and vulnerability during the COVID-19 pandemic were examined by using syndemic theory. Finally, the double exposure framework was used as an integrated tool comprising the one health approach, syndemic theory, and social-ecological model. The framework also illustrates the importance of the social and ecological dimensions of the COVID-19 pandemic and climate change.

Keywords Social inequality · Social-ecological model · One health · Syndemic · SARS-CoV-2

1 Introduction

Research on the interaction of COVID-19 pandemic, human health, and environmental change is still lacking. The pandemic cannot be understood through public health or medical science-based approaches alone. Both ecosystems and human society strongly affect the risk of infection and disease outcomes. This interrelationship is essential to understand how pandemics and climate change can be addressed. In this chapter, the literature on the ecological, environmental, and social effects of COVID-19 and climate change is analyzed through a one health approach, the syndemic theory, and socio-ecological model.

M.-H. Li (✉)
Department of Geography, National Taiwan University, Taipei, Taiwan
e-mail: meihuili@ntu.edu.tw

R. Akhtar (ed.), *Climate Change and Human Health Scenarios*, Global Perspectives on Health Geography, https://doi.org/10.1007/978-3-031-38878-1_4

During the 1990s, researchers developed the one health approach, syndemic theory, and a socio-ecological model to understand and promote public and environmental health. These tools were used in many scientific publications across disciplines during the COVID-19 pandemic, including veterinary medicine, medical and public health, and the social sciences. However, few studies have used all three tools to explore the interaction between climate change and the COVID-19 pandemic.

This chapter is organized in five sections. First, a brief history of the one health approach, syndemic theory, and socio-ecological model is introduced. Second, case reports on disease transmission among wildlife, domesticated animals, and pets with severe acute respiratory syndrome coronavirus 2 (SARS-CoV-2) are discussed. Third, prevention measures and economic crises due to COVID-19 are examined to establish an overview of its environmental effects. Social inequality and vulnerability caused by the COVID-19 pandemic are then discussed using syndemic theory. Finally, the double exposure framework is proposed as an integrated tool to illustrate the importance of the social and ecological dimensions of COVID-19 and climate change.

2 Background

2.1 One Health Approach

The one health approach has received attention in veterinary medicine, animal science, wildlife biology, and environmental sciences (Evans & Leighton, 2014). The purpose of the one health approach is to understand the interconnection among humans, animals, and environmental health, prevent infectious disease, and protect ecosystems. Many infectious diseases have emerged or reemerged worldwide since 1990 and are major public health concerns. Researchers and medical practitioners have applied the one health approach to analyze infectious zoonotic diseases. In addition, many researchers have suggested that the one health approach can be applied to the social sciences to analyze the interaction among social, ecological, and environmental factors that influence the risk of transmission at the individual and community levels (Lapinski et al., 2015; Woldehanna & Zimicki, 2015).

2.2 Socio-ecological Model

The social sciences are crucial for understanding diseases and public health (Badgley et al., 1963; Waitzkin, 1981). Urie Bronfenbrenner developed a socio-ecological model to understand human development (Bronfenbrenner, 1977). The model provides a conceptual framework to examine health and environmental problems and understand the interaction among individuals, communities, and the environment (Kilanowski, 2017). For example, the socio-ecological model has been used to assess the mental health and social well-being of health care workers (Hennein, et al., 2021; Magruder et al., 2022) and community members (Chachar et al., 2021; Lin et al., 2021) during the COVID-19 pandemic and resulting lockdowns. The model has also been used to describe individuals in their physical and social environments who have been affected by COVID-19.

2.3 Syndemic Theory

The term "syndemic" was first coined by Merrill Singer in a study on substance abuse, violence, and HIV in the United States (Singer, 1996). Syndemic theory involves identifying the concentration of a disease in a particular environment and social context, the movement of a disease through biological, social, and psychological pathways, and clusters due to various factors in a sociobiological context and environment (Singer & Mendenhall, 2022). More than 100 studies have been published on social inequality, premedical conditions, and social contexts during the COVID-19 pandemic. Mendenhall (2020) recognized the importance of syndemics in understanding the effects of COVID-19 but reminded readers that "the COVID-19 syndemic is not global: context matters." Many researchers have emphasized the need for a syndemic approach to understand the effects of COVID-19 more broadly and at the local level (Caron & Adegboye, 2021; Di Ciaula et al., 2022; Gravlee, 2020; Irons, 2020; McGowan & Bambra, 2022; Mendenhall, 2020).

The one health approach, syndemic theory, and socio-ecological model have been used to analyze climate change and the COVID-19 pandemic (Fronteira et al., 2021; Persad-Clem et al., 2022; Zhang et al., 2022). Integrating the one health approach into syndemic theory can provide solutions for medical scientists and epidemiologists addressing complex public health problems (Fronteira et al., 2021; Rock et al., 2009). Figure 1 illustrates the conceptual framework of the one health approach, syndemic theory, and socio-ecological model across the physical, ecological, and social dimensions of the COVID-19 pandemic and climate change.

3 Disease Ecology and COVID-19

3.1 Animals and COVID-19 Transmission

Many infectious diseases in humans are of zoonotic origin. When COVID-19 was first reported in Wuhan, China, in December 2019, bats were suggested as the natural host of SARS-CoV-2 (Zhou et al., 2020). However, the intermediate host

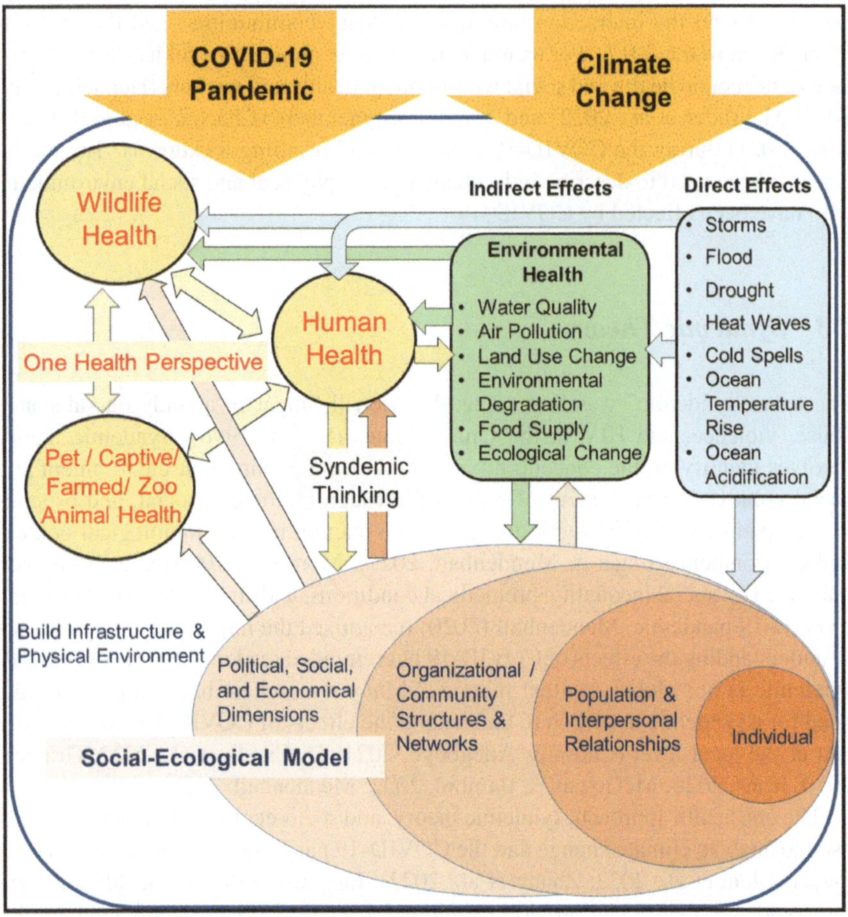

Fig. 1 Conceptual framework of the interaction between the COVID-19 pandemic and climate change in the one health approach, syndemic theory, and socio-ecological model

remains unidentified (Mahdy et al., 2020; Shi et al., 2020; Zhao et al., 2020). Scientific evidence on the zoonotic origins of SARS-CoV-2 is still lacking. One possible pathway of SARS-CoV-2 transmission is direct viral transmission from horseshoe bats to humans and indirect transmission through pangolins or other animals, in addition to interspecies contamination from either wild or captive animals to humans (Domingo, 2021). Frutos et al. (2022) hypothesized that SARS-CoV-2 was not of zoonotic origin and may have been present in human populations before the COVID-19 pandemic and evolved into novel variants. Three possible natural and zoonotic origins of SARS-CoV-2 are depicted in Fig. 2.

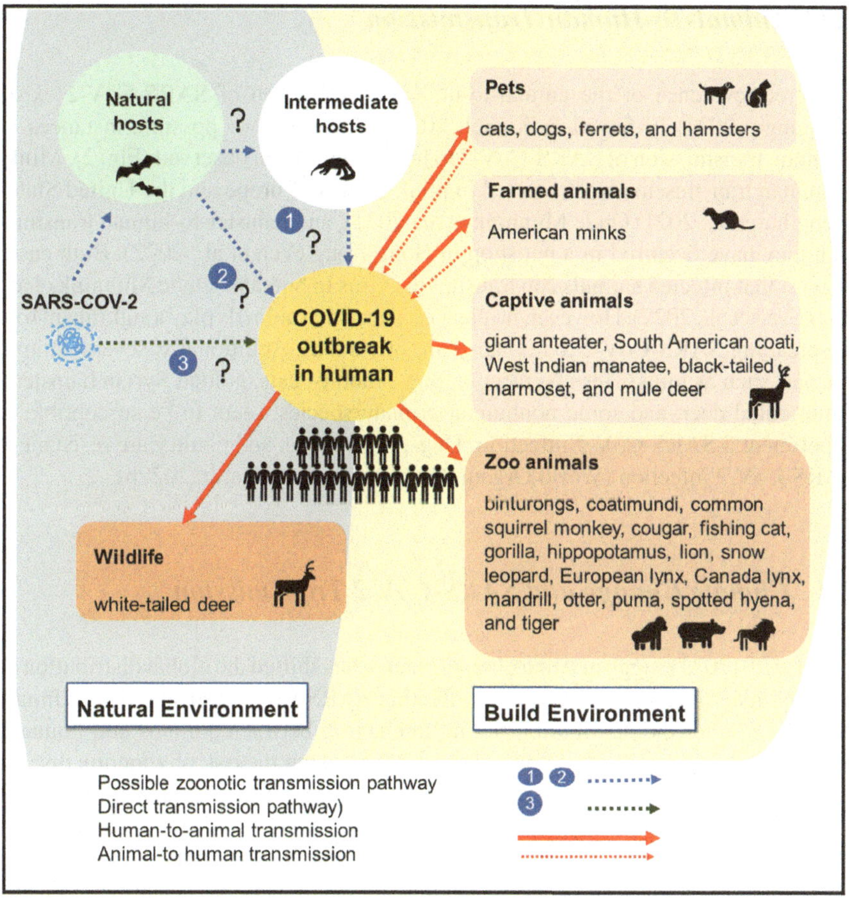

Fig. 2 Natural infections in animals through human-to-animal transmission and three hypotheses of the natural and zoonotic origins of SARS-CoV-2 (*Data Source* World Organization for Animal Health, 2022a)

3.2 Human-To-Animal Transmission

As of September 30, 2022, SARS-CoV-2 infection has been reported in 26 animal species in 36 countries in the Americas, Africa, Asia, and Europe (World Organization for Animal Health, 2022a). Human-to-animal transmission has been reported in pets, domesticated animals, captive animals, zoo animals, and wild animals (Fig. 2). SARS-CoV-2 infection of wild white-tailed deer (*Odocoileus virginianus*) due to spillovers from humans has also been documented, as has deer-to-deer transmission (Chandler et al., 2021; Kuchipudi et al., 2022; Willgert et al., 2022). However, no deer-to-human transmission has been observed (Hale et al., 2022).

3.3 Animal-To-Human Transmission

No direct evidence of the animal-to-human transmission of SARS-CoV-2 exists (Domingo, 2021; El-Sayed & Kamel, 2021). However, two possible instances of zoonotic transmission of SARS-CoV-2 to human have been observed (Fig. 2). Mink-to-human transmission was reported in mink farms in Europe and the United States (Fenollar et al., 2021; Oude Munnink et al., 2021), and hamster-to-human transmission may have occurred in a pet shop in Hong Kong (Yen et al., 2022). Both cases suggest that infected animals can transmit the virus to humans (Oude Munnink et al., 2021; Yen et al., 2022). However, no clear evidence that animals play a significant role in spreading SARS-CoV-2 to humans currently exists. Additionally, several animal species, such as minks, ferrets, raccoon dogs, rabbits, cats, golden Syrian hamsters, white-tailed deer, and some nonhuman primate species, seem to be susceptible to experimental SARS-CoV-2 infection. Dogs, pigs, cattle, and poultry are resistant to SARS-CoV-2 infection (World Organization for Animal Health, 2022b).

3.4 Climate Change and SARS-CoV-2 Transmission

Beyer et al. (2021) suggested that climate change has shifted the global distribution of bats and may have been a key factor affecting SARS-CoV-2 transmission. Climate change can alter the distribution of and interaction between animals and humans, leading to changes in vectors and agents and increasing the risk of zoonotic disease (Carlson et al., 2022). Carlson et al. (2022) created a phylogeographical model of the mammal—virus network to simulate potential hotspots for viral transmission in climate-change and land-use scenarios for 2070. They predicted that high-elevation areas, biodiverse areas, and areas with a high human population density in Asia and Africa would become hotspots and that interspecies transmission would increase by 4000 fold (Carlson et al., 2022).

4 Environmental Effects of COVID-19

Many strategies to control the spread of SARS-CoV-2 have greatly affected human activity, the economy, and health-care systems. COVID-19 had several positive and negative environmental effects. Numerous studies have investigated the environmental effects of the COVID-19 pandemic (Loh et al., 2021; Rume & Islam, 2020). Figure 3 presents how COVID-19 prevention measures altered human and economic activity and affected the environment and ecology.

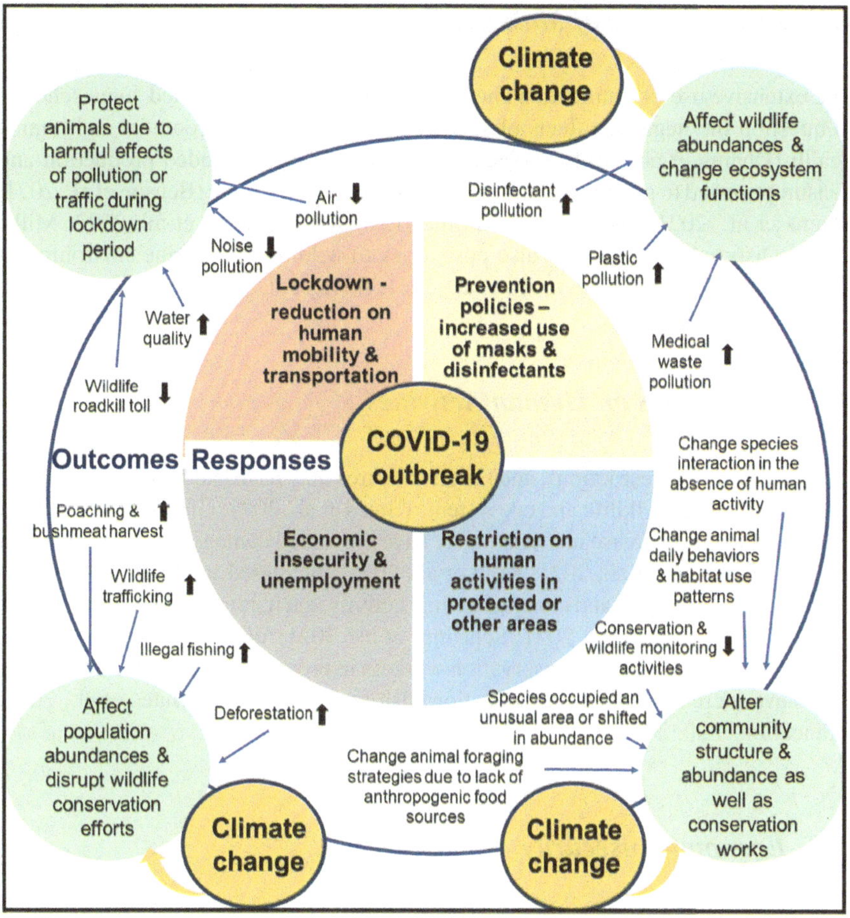

Fig. 3 Environmental effects of COVID-19 prevention measures and economic crises

4.1 Reductions in Human Mobility and Transportation

The effects of COVID-19 lockdowns on air quality worldwide have been widely researched (Marquès & Domingo, 2022; Zang et al., 2022). By contrast, assessments of COVID-19 lockdowns on water quality have been limited to Brazil, China, India, Mexico, Nepal, and Turkey (de Oliveira et al., 2022; Kutralam-Muniasamy et al., 2022; Liu et al., 2022; Muduli et al., 2021; Pant et al., 2021; Tokatlı & Varol, 2021; Yunus et al., 2020). Preventive measures to control the COVID-19 pandemic led to a reduction in human mobility and economic activity and temporarily decreased water pollution (Fig. 3). Additionally, the reduced traffic decreased the amount of wildlife roadkill (Bíl et al., 2021; Driessen, 2021; LeClair et al., 2021).

4.2 Increased Use of Masks and Disinfectants

The extensive use of sanitization chemicals and disinfectants has led some scholars to question the negative effects of such chemicals on the environment and human health (Dhama et al., 2021; Parveen et al., 2022). Their extended production and consumption led to greater pollution from bottles and packaging (Benson et al., 2021; Shams et al., 2021) and affected wildlife (Eisfeld-Pierantonio et al., 2022; Mills, 2022). Disposable masks may also pose a risk to wildlife and marine environments (Dharmaraj et al., 2021; Hiemstra et al., 2021; Silva et al., 2021).

4.3 Restrictions on Human Activities

Lockdowns, travel restrictions, and border closures may have had both beneficial and harmful effects on wildlife and ecosystems (Bates et al., 2021; Gibbons et al., 2022). For example, the decrease in tourism may have prevented damage to the environment (Hentati-Sundberg et al., 2021; Koju et al., 2021), increased urban wildlife (Silva-Rodríguez et al., 2021), and changed animal activity and habitats in urban and natural environments (Koju et al., 2021; Schrimpf et al., 2021; Wilmers et al., 2021). The reduction in the number of conservation workers in protected areas due to lockdowns may have increased illegal deforestation, fishing, and hunting (Bates et al., 2021; Brancalion et al., 2020).

4.4 Economic Insecurity

Because COVID-19 reduced local industrial activity and increased economic insecurity across South Asia and Africa, illegal wildlife poaching increased (Koju et al., 2021), and wildlife conservation efforts were disrupted (Manenti et al., 2020; Rahman et al., 2021; Stenhouse et al., 2022). For example, a higher rate of deforestation and poaching was reported in Bangladesh in 2020 (Rahman et al., 2021).

5 Syndemic Theory of COVID-19 and Climate Change

Disease concentration refers to two or more epidemics occurring in a particular temporal or geographical context due to the harmful physical and social environment at the structural level (Singer & Mendenhall, 2022). A clear link exists between social inequality, vulnerability to climate change (Adger, 1999; Thomas et al., 2019), and the COVID-19 outbreak in both developing and developed countries (Chakraborty,

2021; de Souza et al., 2020; Dennison, 2021; Vandentorren et al., 2022; Yadav et al., 2020; Yun et al., 2022; Zhang et al., 2021).

5.1 Air Pollution and COVID-19

Air pollution can increase vulnerability to SASR-CoV-19 infection among disadvantaged groups (Marquès & Domingo, 2022). Di Ciaula et al. (2022) proposed three possible mechanisms of this increase: (1) that air pollutants affect the respiratory and immune systems and increase susceptibility to SAS-CoV-2 infection, (2) that particulate matter carries SAS-CoV-2 deeper into the airways, and (3) that air pollutants lead to inflammation, oxidative stress on the airways, and severe disease outcomes. Individuals who are the most affected by air pollution and climate change have the greatest risk of developing severe illness from COVID-19.

5.2 Climate Change and COVID-19

Climate change can facilitate infectious disease transmission and may increase the prevalence and incidence of human helminth infection (Blum & Hotez, 2018). Concurrent helminth and COVID-19 infection, which can worsen health outcomes, have been reported (Cepon-Robins & Gildner, 2020; Fonte et al., 2022; Naidoo et al., 2021) and exemplifies the interaction among climate change, COVID-19, and human health. Temperature, humidity, ultraviolet radiation, and extreme weather events can increase the risk of SARS-CoV-2 transmission by altering viral persistence in the environment, the functioning of the immune system, and human behavior (Weaver et al., 2022).

5.3 COVID-19 Outcomes and Noncommunicable Diseases

The burden of noncommunicable diseases attributable to high temperatures increased worldwide from 1990 to 2019 (Song et al., 2021). Patients with noncommunicable diseases (e.g., diabetes, hypertension, chronic respiratory illnesses, and chronic kidney and liver conditions) have a higher risk of developing severe COVID-19 symptoms and higher mortality (Nikoloski et al., 2021). Two biological mechanisms may explain the link between chronic illnesses and COVID-19 infection and its outcomes: (1) the higher expression of enzymes that convert angiotensin, which facilitates the entry of the virus into the host body, and (2) a stronger hyperinflammatory reaction from cytokines, which triggers multiorgan failure (Cole et al., 2020; Nikoloski et al., 2021). COVID-19 disrupted essential health services and resources for patients with noncommunicable diseases (World Health Organization, 2020).

5.4 Social Inequality

Epidemics that occur at the population level often traverse through biological or sociobiological pathways (Singer & Mendenhall, 2022). Three pandemics, namely obesity, undernutrition, and climate change, have been described as globally syndemic in the Lancet Commission report because they usually occur at the same time and in places with complex interactions and similar underlying societal drivers (Swinburn et al., 2019). Syndemic theory indicates that concurrent disease clusters are shaped and amplified by social, biological, and political factors. These factors affect the dynamics and interaction between diseases (Boes et al., 2021). Structural racism and socioeconomic inequality worsened COVID-19 outcomes in the United States, Europe (Abedi et al., 2021; Gauthier et al., 2021; McGowan & Bambra, 2022; Meurisse et al., 2022; Munford et al., 2022), and Asia (Chung et al., 2021; Yoo et al., 2022; Yoshikawa & Kawachi, 2021). Gender inequality due to the gendered division of labor, unemployment, and economic instability led to a greater risk of SARS-CoV-2 exposure among women (Flor et al., 2022). Certain social groups that are most affected by climate change were predisposed to higher environmental and social vulnerability, which affected their health and resilience during the COVID-19 pandemic.

6 Double Exposure Framework of COVID-19 and Climate Change

Higher exposure to climate hazards and SARS-CoV-2 infection can result in higher vulnerability to the adverse effects of climate change and COVID-19 among disadvantaged groups. Both COVID-19 and climate change negatively affect the health and coping mechanisms of such groups. "Double exposure" refers to exposure to climate change and economic globalization in a particular region, sector, ecosystem, or social group (O'Brien & Leichenko, 2000). The main features of the double exposure framework are the dynamic interactions among processes, viral exposure, environments, responses, and outcomes (Leichenko & O'Brien, 2008). Processes can alter the environment and viral exposure and produce responses that result in health outcomes; such outcomes can affect the environment, viral exposure, and responses. Similar environmental and social contexts can influence exposure pathways and responses from both processes and produce positive or negative outcomes at certain levels of viral exposure (Leichenko & O'Brien, 2008). The double exposure framework can elucidate how climate change and COVID-19 interact and shape social inequality and resilience.

Both climate change and COVID-19 are global processes that affect each region, country, social group, and ecosystem differently (Khojasteh et al., 2022). COVID-19 is a biological and ecological process that resulted in a global economic crisis, and climate change is a biophysiochemical process due to human economic activity. Both processes exacerbate economic crises and social inequality worldwide (Khojasteh et al., 2022). Both processes have distinct features and vary in their effects but can increase the vulnerability of marginalized groups. For example, individuals who are negatively affected by climate change (e.g., extreme weather events or food insecurity) have a higher risk of SARS-CoV-2 infection and poorer health outcomes. Figure 4 presents the one health approach, syndemic theory, and socio-ecological model incorporated into the double exposure framework to model the complex interaction between climate change and the COVID-19 pandemic.

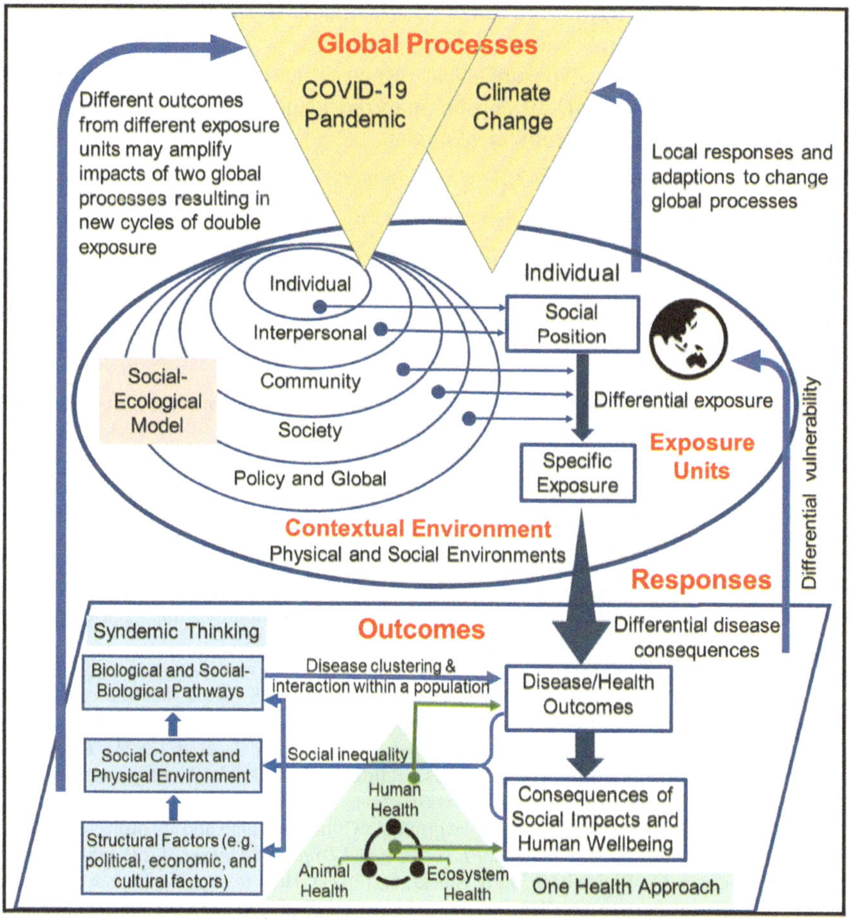

Fig. 4 Double exposure framework of COVID-19 and climate change with one health approach, syndemic theory, and socio-ecological model (adapted from Diderichsen et al. 2001; Gravlee 2020)

7 Conclusion

Human activity, climate change, and the COVID-19 pandemic interact through complex pathways. Economic activity exacerbates the adverse effects of climate change worldwide. Human health and well-being are affected by the direct and indirect effects of climate change. Human mobility and activity were reduced or restricted to control SARS-CoV-2 transmission globally, and such strategies affected people and the environment differently in each region. The effects of these two processes can overlap and produce similar outcomes in certain social groups. This study revealed that a double exposure framework can provide a broader and more structured conceptual model to identify connections between climate change and COVID-19. By using the framework, processes, exposure levels, responses, and outcomes associated with climate change and COVID-19 can be analyzed using the one health approach, syndemic theory, and socio-ecological model.

References

Abedi, V., Olulana, O., Avula, V., Chaudhary, D., Khan, A., Shahjouei, S., et al. (2021). Racial, economic, and health inequality and COVID-19 infection in the United States. *Journal of Racial and Ethnic Health Disparities, 8*(3), 732–742.

Adger, W. N. (1999). Social vulnerability to climate change and extremes in coastal Vietnam. *World Development, 27*(2), 249–269.

Badgley, R. F., Mitton, G. T., Carpenter, H., Robinson, C. E. A., Robertson, E. C., Goodacre, R. H., Jones, T. L., Doughty, J. H., & Osborne, J. E. (1963). Social sciences and public health. *Canadian Journal of Public Health, 54*(4), 147–163.

Bates, A. E., Primack, R. B., Biggar, B. S., Bird, T. J., Clinton, M. E., Command, R. J., et al. (2021). Global COVID-19 lockdown highlights humans as both threats and custodians of the environment. *Biological Conservation, 263*, 109175.

Benson, N. U., Bassey, D. E., & Palanisami, T. (2021). COVID pollution: Impact of COVID-19 pandemic on global plastic waste footprint. *Heliyon, 7*(2), e06343.

Beyer, R. M., Manica, A., & Mora, C. (2021). Shifts in global bat diversity suggest a possible role of climate change in the emergence of SARS-CoV-1 and SARS-CoV-2. *Science of the Total Environment, 767*, 145413.

Bíl, M., Andrášik, R., Cícha, V., Arnon, A., Kruuse, M., Langbein, J., et al. (2021). COVID-19 related travel restrictions prevented numerous wildlife deaths on roads: A comparative analysis of results from 11 countries. *Biological Conservation, 256*, 109076.

Blum, A. J., & Hotez, P. J. (2018). Global "worming": Climate change and its projected general impact on human helminth infections. *PLoS Neglected Tropical Diseases, 12*(7), e0006370.

Boes, S., Sabariego, C., Bickenbach, J., & Stucki, G. (2021). How to capture the individual and societal impacts of syndemics: The lived experience of COVID-19. *BMJ Global Health, 6*(10), e006735.

Brancalion, P. H., Broadbent, E. N., De-Miguel, S., Cardil, A., Rosa, M. R., Almeida, C. T., et al. (2020). Emerging threats linking tropical deforestation and the COVID-19 pandemic. *Perspectives in Ecology and Conservation, 18*(4), 243–246.

Bronfenbrenner, U. (1977). Toward an experimental ecology of human development. *American Psychologis, 32*, 513–531.

Carlson, C. J., Albery, G. F., Merow, C., Trisos, C. H., Zipfel, C. M., Eskew, E. A., et al. (2022). Climate change increases cross-species viral transmission risk. *Nature, 607*, 555–562.

Caron, R. M., & Adegboye, A. R. A. (2021). COVID-19: A syndemic requiring an integrated approach for marginalized populations. *Frontiers in Public Health, 9*, 675280.

Cepon-Robins, T. J., & Gildner, T. E. (2020). Old friends meet a new foe: A potential role for immune-priming parasites in mitigating COVID-19 morbidity and mortality. *Evolution, Medicine, and Public Health, 2020*(1), 234–248.

Chachar, A. S., Younus, S., & Ali, W. (2021). Developmental understanding of death and grief among children during COVID-19 pandemic: Application of Bronfenbrenner's bioecological model. *Frontiers in Psychiatry, 12*, 654584.

Chakraborty, J. (2021). Convergence of COVID-19 and chronic air pollution risks: Racial/ethnic and socioeconomic inequities in the US. *Environmental Research, 193*, 110586.

Chandler, J. C., Bevins, S. N., Ellis, J. W., Linder, T. J., Tell, R. M., Jenkins-Moore, M., et al. (2021). SARS-CoV-2 exposure in wild white-tailed deer (*Odocoileus virginianus*). *Proceedings of the National Academy of Sciences, 118*(47), e2114828118.

Chung, G. K. K., Chan, S. M., Chan, Y. H., Yip, T. C. F., Ma, H. M., Wong, G. L. H., et al. (2021). Differential impacts of multimorbidity on COVID-19 severity across the socioeconomic ladder in Hong Kong: A syndemic perspective. *International Journal of Environmental Research and Public Health, 18*(15), 8168.

Cole, S. A., Laviada-Molina, H. A., Serres-Perales, J. M., Rodriguez-Ayala, E., & Bastarrachea, R. A. (2020). The COVID-19 pandemic during the time of the diabetes pandemic: Likely fraternal twins? *Pathogens, 9*(5), 389.

de Oliveira, M. E. G., da Silva, M. V., de Almeida, G. L. P., Pandorfi, H., Lopes, P. M. O., Manrique, D. R. C., et al. (2022). Investigation of pre and post environmental impact of the lockdown (COVID-19) on the water quality of the Capibaribe and Tejipió rivers, Recife metropolitan region, Brazil. *Journal of South American Earth Sciences, 118*, 103965.

de Souza, C. D. F., do Carmo, R. F., & Machado, M. F. (2020). The burden of COVID-19 in Brazil is greater in areas with high social deprivation. *Journal of Travel Medicine, 27*(7), 1–3.

Dennison, M. (2021). The great leveller? COVID-19's dynamic interaction with social inequalities in the UK. *History and Philosophy of the Life Sciences, 43*(4), 1–4.

Dhama, K., Patel, S. K., Kumar, R., Masand, R., Rana, J., Yatoo, M., et al. (2021). The role of disinfectants and sanitizers during COVID-19 pandemic: Advantages and deleterious effects on humans and the environment. *Environmental Science and Pollution Research, 28*(26), 34211–34228.

Dharmaraj, S., Ashokkumar, V., Hariharan, S., Manibharathi, A., Show, P. L., Chong, C. T., & Ngamcharussrivichai, C. (2021). The COVID-19 pandemic face mask waste: A blooming threat to the marine environment. *Chemosphere, 272*, 129601.

Di Ciaula, A., Moshammer, H., Lauriola, P., & Portincasa, P. (2022). Environmental health, COVID-19, and the syndemic: Internal medicine facing the challenge. *Internal and Emergency Medicine, 17*(8), 2187–2198.

Diderichsen, F., Evans, T., & Whitehead, M. (2001). The social basis of disparities in health. In T. Evans (Ed), *Challenging inequities in health: From ethics to action* (pp. 12–23).

Domingo, J. L. (2021). What we know and what we need to know about the origin of SARS-CoV-2. *Environmental Research, 200*, 111785.

Driessen, M. M. (2021). COVID-19 restrictions provide a brief respite from the wildlife roadkill toll. *Biological Conservation, 256*, 109012.

Eisfeld-Pierantonio, S. M., Pierantonio, N., & Simmonds, M. P. (2022). The impact of marine debris on cetaceans with consideration of plastics generated by the COVID-19 pandemic. *Environmental Pollution, 300*, 118967.

El-Sayed, A., & Kamel, M. (2021). Coronaviruses in humans and animals: The role of bats in viral evolution. *Environmental Science and Pollution Research, 28*, 19589–19600.

Evans, B. R., & Leighton, F. A. (2014). A history of one health. *Revue Scientifique Et Technique, 33*(2), 413–420.

Fenollar, F., Mediannikov, O., Maurin, M., Devaux, C., Colson, P., Levasseur, A., et al. (2021). Mink, SARS-CoV-2, and the human-animal interface. *Frontiers in Microbiology, 12*, 663815.

Flor, L. S., Friedman, J., Spencer, C. N., Cagney, J., Arrieta, A., Herbert, M. E., et al. (2022). Quantifying the effects of the COVID-19 pandemic on gender equality on health, social, and economic indicators: A comprehensive review of data from March, 2020, to September, 2021. *The Lancet, 399*, 2381–2397.

Fonte, L., Acosta, A., Sarmiento, M. E., Norazmi, M. N., Ginori, M., de Armas, Y., & Calderón, E. J. (2022). Overlapping of pulmonary fibrosis of postacute COVID-19 syndrome and tuberculosis in the helminth coinfection setting in sub-Saharan Africa. *Tropical Medicine and Infectious Disease, 7*(8), 157.

Fronteira, I., Sidat, M., Magalhães, J. P., de Barros, F. P. C., Delgado, A. P., Correia, T., Daniel-Ribeiro, C. T., & Ferrinho, P. (2021). The SARS-CoV-2 pandemic: A syndemic perspective. *One Health, 12*, 100228.

Frutos, R., Pliez, O., Gavotte, L., & Devaux, C. A. (2022). There is no "origin" to SARS-CoV-2. *Environmental Research, 207*, 112173.

Gauthier, G. R., Smith, J. A., García, C., Garcia, M. A., & Thomas, P. A. (2021). Exacerbating inequalities: Social networks, racial/ethnic disparities, and the COVID-19 pandemic in the United States. *The Journals of Gerontology B, 76*(3), e88–e92.

Gibbons, D. W., Sandbrook, C., Sutherland, W. J., Akter, R., Bradbury, R., Broad, S., et al. (2022). The relative importance of COVID-19 pandemic impacts on biodiversity conservation globally. *Conservation Biology, 36*(1), e13781.

Gravlee, C. C. (2020). Systemic racism, chronic health inequities, and COVID-19: A syndemic in the making? *American Journal of Human Biology, 32*(5), e23482.

Hale, V. L., Dennis, P. M., McBride, D. S., Nolting, J. M., Madden, C., Huey, D., et al. (2022). SARS-CoV-2 infection in free-ranging white-tailed deer. *Nature, 602*(7897), 481–486.

Hennein, R., Mew, E. J., & Lowe, S. R. (2021). Socio-ecological predictors of mental health outcomes among healthcare workers during the COVID-19 pandemic in the United States. *PLoS ONE, 16*(2), e0246602.

Hentati-Sundberg, J., Berglund, P. A., Hejdström, A., & Olsson, O. (2021). COVID-19 lockdown reveals tourists as seabird guardians. *Biological Conservation, 254*, 108950.

Hiemstra, A. F., Rambonnet, L., Gravendeel, B., & Schilthuizen, M. (2021). The effects of COVID-19 litter on animal life. *Animal Biology, 71*(2), 215–231.

Irons, R. (2020). Pandemic… or syndemic? Re-framing COVID-19 disease burden and 'underlying health conditions.' *Social Anthropology, 28*(2), 286–287.

Khojasteh, D., Davani, E., Shamsipour, A., Haghani, M., & Glamore, W. (2022). Climate change and COVID-19: Interdisciplinary perspectives from two global crises. *Science of the Total Environment, 844*, 157142.

Kilanowski, J. F. (2017). Breadth of the socio-ecological model. *Journal of Agromedicine, 22*(4), 295–297.

Koju, N. P., Kandel, R. C., Acharya, H. B., Dhakal, B. K., & Bhuju, D. R. (2021). COVID-19 lockdown frees wildlife to roam but increases poaching threats in Nepal. *Ecology and Evolution, 11*(14), 9198–9205.

Kuchipudi, S. V., Surendran-Nair, M., Ruden, R. M., Yon, M., Nissly, R. H., Vandegrift, K. J., et al. (2022). Multiple spillovers from humans and onward transmission of SARS-CoV-2 in white-tailed deer. *Proceedings of the National Academy of Sciences, 119*(6), e2121644119.

Kutralam-Muniasamy, G., Pérez-Guevara, F., Roy, P. D., Elizalde-Martínez, I., & Chari, S. V. (2022). Surface water quality in the upstream of the highly contaminated Santiago River (Mexico) during the COVID-19 lockdown. *Environmental Earth Sciences, 81*(11), 1–14.

Lapinski, M. K., Funk, J. A., & Moccia, L. T. (2015). Recommendations for the role of social science research in one health. *Social Science & Medicine, 129*, 51–60.

LeClair, G., Chatfield, M. W., Wood, Z., Parmelee, J., & Frederick, C. A. (2021). Influence of the COVID-19 pandemic on amphibian road mortality. *Conservation Science and Practice, 3*(11), e535.

Leichenko, R., & O'Brien, K. (2008). *Environmental change and globalization: Double exposures.* Oxford University Press.

Lin, P. I., Srivastava, G., Beckman, L., Kim, Y., Hallerbäck, M., Barzman, D., et al. (2021). A framework-based approach to assessing mental health impacts of the COVID-19 pandemic on children and adolescents. *Frontiers in Psychiatry, 12*, 655481.

Liu, D., Yang, H., Thompson, J. R., Li, J., Loiselle, S., & Duan, H. (2022). COVID-19 lockdown improved river water quality in China. *Science of the Total Environment, 802*, 149585.

Loh, H. C., Looi, I., Ch'ng, A. S. H., Goh, K. W., Ming, L. C., & Ang, K. H. (2021). Positive global environmental impacts of the COVID-19 pandemic lockdown: A review. *GeoJournal, 87*, 4425–4437.

Magruder, L., Wilke, D., Radey, M., Cain, M., & Yelick, A. (2022). COVID-19's social ecological impacts on health and human services worker well-being. *Social Work in Public Health, 37*(3), 233–243.

Mahdy, M. A., Younis, W., & Ewaida, Z. (2020). An overview of SARS-CoV-2 and animal infection. *Frontiers in Veterinary Science, 7*, 596391.

Manenti, R., Mori, E., Di Canio, V., Mercurio, S., Picone, M., Caffi, M., et al. (2020). The good, the bad and the ugly of COVID-19 lockdown effects on wildlife conservation: Insights from the first European locked down country. *Biological Conservation, 249*, 108728.

Marquès, M., & Domingo, J. L. (2022). Positive association between outdoor air pollution and the incidence and severity of COVID-19. A review of the recent scientific evidences. *Environmental Research, 203*, 111930.

McGowan, V. J., & Bambra, C. (2022). COVID-19 mortality and deprivation: Pandemic, syndemic, and endemic health inequalities. *The Lancet Public Health, 7*(11), e966–e975.

Mendenhall, E. (2020). The COVID-19 syndemic is not global: Context matters. *The Lancet, 396*(10264), 1731.

Meurisse, M., Lajot, A., Devleesschauwer, B., Van Cauteren, D., Van Oyen, H., Van den Borre, L., & Brondeel, R. (2022). The association between area deprivation and COVID-19 incidence: A municipality-level spatio-temporal study in Belgium, 2020–2021. *Archives of Public Health, 80*(1), 109.

Mills, G. (2022). Effect of pandemic-related debris on wildlife. *The Veterinary Record, 191*(4), 144.

Muduli, P. R., Kumar, A., Kanuri, V. V., Mishra, D. R., Acharya, P., Saha, R., et al. (2021). Water quality assessment of the Ganges River during COVID-19 lockdown. *International Journal of Environmental Science and Technology, 18*(6), 1645–1652.

Munford, L., Khavandi, S., & Bambra, C. (2022). COVID-19 and deprivation amplification: An ecological study of geographical inequalities in mortality in England. *Health & Place*, 102933.

Naidoo, P., Ghazi, T., Chuturgoon, A. A., Naidoo, R. N., Ramsuran, V., Mpaka-Mbatha, M. N., et al. (2021). SARS-CoV-2 and helminth co-infections, and environmental pollution exposure: An epidemiological and immunological perspective. *Environment International, 156*, 106695.

Nikoloski, Z., Alqunaibet, A. M., Alfawaz, R. A., Almudarra, S. S., Herbst, C. H., El-Saharty, S., et al. (2021). Covid-19 and non-communicable diseases: Evidence from a systematic literature review. *BMC Public Health, 21*(1), 1068.

O'Brien, K. L., & Leichenko, R. M. (2000). Double exposure: Assessing the impacts of climate change within the context of economic globalization. *Global Environmental Change, 10*(3), 221–232.

Oude Munnink, B. B., Sikkema, R. S., Nieuwenhuijse, D. F., Molenaar, R. J., Munger, E., Molenkamp, R., et al. (2021). Transmission of SARS-CoV-2 on mink farms between humans and mink and back to humans. *Science, 371*, 172–177.

Pant, R. R., Bishwakarma, K., Qaiser, F. U. R., Pathak, L., Jayaswal, G., Sapkota, B., et al. (2021). Imprints of COVID-19 lockdown on the surface water quality of Bagmati river basin, Nepal. *Journal of Environmental Management, 289*, 112522.

Parveen, N., Chowdhury, S., & Goel, S. (2022). Environmental impacts of the widespread use of chlorine-based disinfectants during the COVID-19 pandemic. *Environmental Science and Pollution Research, 29*, 85742–85760.

Persad-Clem, R., Hoerster, K. D., Romano, E. F. T., Huizar, N., & Maier, K. J. (2022). Climate to COVID, global to local, policies to people: A biopsychosocial ecological framework for syndemic prevention and response in behavioral medicine. *Translational Behavioral Medicine, 12*(4), 516–525.

Rahman, M. S., Alam, M. A., Salekin, S., Belal, M. A. H., & Rahman, M. S. (2021). The COVID-19 pandemic: A threat to forest and wildlife conservation in Bangladesh? *Trees, Forests and People, 5*, 100119.

Rock, M., Buntain, B. J., Hatfield, J. M., & Hallgrímsson, B. (2009). Animal–human connections, "one health", and the syndemic approach to prevention. *Social Science & Medicine, 68*(6), 991–995.

Rume, T., & Islam, S. D. U. (2020). Environmental effects of COVID-19 pandemic and potential strategies of sustainability. *Heliyon, 6*(9), e04965.

Schrimpf, M. B., Des Brisay, P. G., Johnston, A., Smith, A. C., Sánchez-Jasso, J., Robinson, B. G., et al. (2021). Reduced human activity during COVID-19 alters avian land use across North America. *Science Advances, 7*(39), eabf5073.

Shams, M., Alam, I., & Mahbub, M. S. (2021). Plastic pollution during COVID-19: Plastic waste directives and its long-term impact on the environment. *Environmental Advances, 5*, 100119.

Shi, J., Wen, Z., Zhong, G., Yang, H., Wang, C., Huang, B., et al. (2020). Susceptibility of ferrets, cats, dogs, and other domesticated animals to SARS–coronavirus 2. *Science, 368*(6494), 1016–1020.

Silva, A. L. P., Prata, J. C., Mouneyrac, C., Barcelò, D., Duarte, A. C., & Rocha-Santos, T. (2021). Risks of Covid-19 face masks to wildlife: Present and future research needs. *Science of the Total Environment, 792*, 148505.

Silva-Rodríguez, E. A., Gálvez, N., Swan, G. J., Cusack, J. J., & Moreira-Arce, D. (2021). Urban wildlife in times of COVID-19: What can we infer from novel carnivore records in urban areas? *Science of the Total Environment, 765*, 142713.

Singer, M. (1996). A dose of drugs, a touch of violence, a case of AIDS: Conceptualizing the SAVA syndemic. *Free Inquiry in Creative Sociology, 24*(2), 99–110.

Singer, M., & Mendenhall, E. (2022). Syndemics in global health. In M. Singer, P. I. Erickson & C. E. Abadía-Barrero (Eds.), *A Companion to Medical Anthropology* (2nd edn., pp. 126–144).

Song, J., Pan, R., Yi, W., Wei, Q., Qin, W., Song, S., et al. (2021). Ambient high temperature exposure and global disease burden during 1990–2019: An analysis of the global burden of disease study 2019. *Science of the Total Environment, 787*, 147540.

Stenhouse, A., Perry, T., Grützner, F., Rismiller, P., Koh, L. P., & Lewis, M. (2022). COVID restrictions impact wildlife monitoring in Australia. *Biological Conservation, 267*, 109470.

Swinburn, B. A., Kraak, V. I., Allender, S., Atkins, V. J., Baker, P. I., Bogard, J. R., et al. (2019). The global syndemic of obesity, undernutrition, and climate change: The lancet commission report. *The Lancet, 393*(10173), 791–846.

Thomas, K., Hardy, R. D., Lazrus, H., Mendez, M., Orlove, B., Rivera-Collazo, I., et al. (2019). Explaining differential vulnerability to climate change: A social science review. *Wiley Interdisciplinary Reviews: Climate Change, 10*(2), e565.

Tokatlı, C., & Varol, M. (2021). Impact of the COVID-19 lockdown period on surface water quality in the Meriç-Ergene River Basin. *Northwest Turkey. Environmental Research, 197*, 111051.

Vandentorren, S., Smaïli, S., Chatignoux, E., Maurel, M., Alleaume, C., Neufcourt, L., et al. (2022). The effect of social deprivation on the dynamic of SARS-CoV-2 infection in France: A population-based analysis. *The Lancet Public Health, 7*(3), e240–e249.

Waitzkin, H. (1981). The social origins of illness: A neglected history. *International Journal of Health Services, 11*(1), 77–103.

Weaver, A. K., Head, J. R., Gould, C. F., Carlton, E. J., & Remais, J. V. (2022). Environmental factors influencing COVID-19 incidence and severity. *Annual Review of Public Health, 43*, 271–291.

Willgert, K., Didelot, X., Surendran-Nair, M., Kuchipudi, S. V., Ruden, R. M., Yon, M., et al. (2022). Transmission history of SARS-CoV-2 in humans and white-tailed deer. *Scientific Reports, 12*(1), 12094.

Wilmers, C. C., Nisi, A. C., & Ranc, N. (2021). COVID-19 suppression of human mobility releases mountain lions from a landscape of fear. *Current Biology, 31*(17), 3952–3955.

Woldehanna, S., & Zimicki, S. (2015). An expanded one health model: Integrating social science and one health to inform study of the human-animal interface. *Social Science & Medicine, 129*, 87–95.

World Health Organization. (2020). *The impact of the COVID-19 pandemic on noncommunicable disease resources and services: Results of a rapid assessment*. World Health Organization.

World Organisation for Animal Health. (2022a). SARS-CoV-2 in Animal-Situation Report 17. Accessed November 15, 2022a.

World Organisation for Animal Health. (2022b). OIE Technical Factsheet: Infection with SARS-COV-2 in Animals.

Yadav, U. N., Rayamajhee, B., Mistry, S. K., Parsekar, S. S., & Mishra, S. K. (2020). A syndemic perspective on the management of non-communicable diseases amid the COVID-19 pandemic in low-and middle-income countries. *Frontiers in Public Health, 8*, 508.

Yen, H. L., Sit, T. H., Brackman, C. J., Chuk, S. S., Gu, H., Tam, K. W., et al. (2022). Transmission of SARS-CoV-2 delta variant (AY. 127) from pet hamsters to humans, leading to onward human-to-human transmission: a case study. *The Lancet, 399*(10329), 1070–1078.

Yoo, D. S., Hwang, M., Chun, B. C., Kim, S. J., Son, M., Seo, N. K., & Ki, M. (2022). Socioeconomic inequalities in COVID-19 incidence during different epidemic phases in South Korea. *Frontiers in Medicine, 9*, 840685.

Yoshikawa, Y., & Kawachi, I. (2021). Association of socioeconomic characteristics with disparities in COVID-19 outcomes in Japan. *JAMA Network Open, 4*(7), e2117060.

Yun, J. Y., Sim, J. A., Lee, S., & Yun, Y. H. (2022). Stronger association of perceived health with socio-economic inequality during COVID-19 pandemic than pre-pandemic era. *BMC Public Health, 22*(1), 1757.

Yunus, A. P., Masago, Y., & Hijioka, Y. (2020). COVID-19 and surface water quality: Improved lake water quality during the lockdown. *Science of the Total Environment, 731*, 139012.

Zang, S. T., Luan, J., Li, L., Yu, H. X., Wu, Q. J., Chang, Q., & Zhao, Y. H. (2022). Ambient air pollution and COVID-19 risk: Evidence from 35 observational studies. *Environmental Research, 204*, 112065.

Zhang, R., Tang, X., Liu, J., Visbeck, M., Guo, H., Murray, V., et al. (2022). From concept to action: A united, holistic and one health approach to respond to the climate change crisis. *Infectious Diseases of Poverty, 11*(1), 1–6.

Zhang, Y., Khullar, D., Wang, F., Steel, P., Wu, Y., Orlander, D., et al. (2021). Socioeconomic variation in characteristics, outcomes, and healthcare utilization of COVID-19 patients in New York City. *PLoS ONE, 16*(7), e0255171.

Zhao, J., Cui, W., & Tian, B. P. (2020). The potential intermediate hosts for SARS-CoV-2. *Frontiers in Microbiology, 11*, 580137.

Zhou, P., Yang, X. L., Wang, X. G., Hu, B., Zhang, L., Zhang, W., et al. (2020). A pneumonia outbreak associated with a new coronavirus of probable bat origin. *Nature, 579*(7798), 270–273.

Heat-Related Health Impacts of Climate Change and Adaptation Strategies in Japan

Kazutaka Oka

Abstract The temperatures in Japan have risen by 1.28 °C per 100 years owing to climate change. Accordingly, the number of extremely hot days (days with a daily maximum temperature of ≥ 35 °C) has increased in summer since the 1990s. This temperature rise has caused severe heat-related health impacts in Japan, including heatstroke. In Japan, heatstroke causes 65,000 ambulance cases and 1,000 deaths annually. Various measures have been implemented by the Japanese government, including the launch of the "Heatstroke Alert," a heat-health warning system to reduce the health impacts caused by heatstroke. This chapter introduces the main measures implemented by the Japanese government and local governments. Notably, heat-relate health impacts have been intensively studied in Japan. Based on these studies, this chapter describes the scientific findings and issues related to health impacts, such as heatstroke and heat-related excess mortality. Furthermore, this chapter highlights the collaborative study project with local governments on heat environments and heatstroke conducted by the National Institute for Environmental Studies.

Keywords Climate change · Urban heat island · Heatstroke · Excess mortality · Adaptation · Heat-health warning system · Wet bulb globe temperature · Local government

1 Introduction

Climate change has caused temperatures to rise worldwide (IPCC, 2021). In Japan, temperatures have risen by 1.28 °C per 100 years (JMA, 2022). Moreover, the average annual temperature in 2021 was the third highest on record, after that in 2020 and 2019 (JMA, 2022). With the temperature rise, the number of extremely hot days (days with a daily maximum temperature of ≥ 35 °C) has increased during summer in Japan since the 1990s (JMA, 2022) (Fig. 1).

K. Oka (✉)
Center for Climate Change Adaptation, National Institute for Environmental Studies, Tsukuba, Japan
e-mail: oka.kazutaka@nies.go.jp

R. Akhtar (ed.), *Climate Change and Human Health Scenarios*, Global Perspectives on Health Geography, https://doi.org/10.1007/978-3-031-38878-1_5

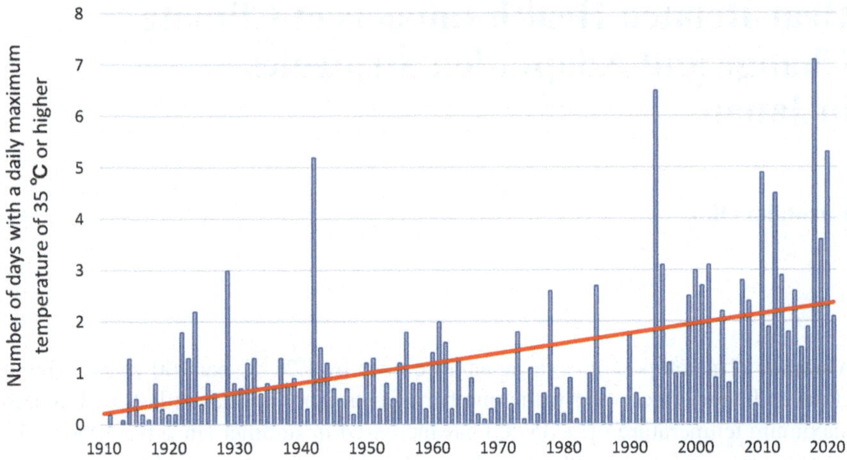

Fig. 1 Annual number of days with a daily maximum temperature of ≥ 35 °C (extremely hot days) nationwide (average of 13 locations) between 1910 and 2021 (data provided by the Japan Meteorological Agency). The red line shows the regression line of extremely hot day over this period

The health impacts of rising temperatures, such as heatstroke, have become severe in Japan. Recently, heatstroke has caused 65,000 ambulance cases and 1000 deaths annually in Japan, and these high levels have remained unchanged (Government of Japan, 2022). Furthermore, the elderly accounted for approximately 50% of patients with heatstroke transported by ambulance, followed by adults (> 30%) and juveniles (approximately 10%) (FDMA, 2022). Additionally, approximately 40% of these cases occurred at residences (FDMA, 2022). Notably, the factors that cause heatstroke among the elderly in residences include the unavailability of air conditioners or their disuse despite being available (Government of Japan, 2022). Additionally, the elderly account for approximately 80% of deaths by heatstroke (MHLW, 2021). Unfortunately, the health impacts of heatstroke among the elderly will increase in the future with the increasingly aging population in Japan and will be accelerated by the rising temperatures due to climate change.

Besides the health impacts of gradually increasing temperature, those associated with extreme heat events (where temperatures can be substantially higher than "usual" hot days) have become increasingly severe recently. In June 2021, the heat wave hit British Columbia, Canada, and caused a daily maximum temperature of 49.6 °C, resulting in > 600 heat-related deaths (Government of Canada, 2022). According to the Intergovernmental Panel on Climate Change (IPCC) Sixth Assessment Report (IPCC, 2021), the frequency and intensity of extreme heat events are projected to increase with the progress of climate change. Notably, extreme heat events will have significant health impacts and possibly cause medical crises owing to a drastic increase in the number of patients with heatstroke and the accompanying increase in the burden on medical facilities.

Mitigation measures are needed to reduce future temperature increases, while adaptation measures are essential to reduce the health impacts and prevent medical crises in extreme heat. This chapter introduces the main measures taken by the Japanese government and local governments to reduce health impacts. Next, studies on heat-related health impacts in Japan are highlighted, and their results are summarized.

2 Adaptation Strategies and Measures by the Japanese Government

In Japan, heatstroke has already become a severe health problem, and its prevention is an integral part of the government's efforts for climate change adaptation that directly affects the lives and health of Japanese people. This section introduces some of the primary measures implemented by the Japanese government to reduce the risk of heatstroke in Japan.

2.1 Action Plan for Heatstroke Prevention

Since 2007, the Japanese government has held a "Liaison Conference of Ministries and Agencies concerned with Heatstroke" and promoted various measures to reduce health impacts. Nevertheless, heatstroke has remained high in Japan, and concerns about exacerbating the risk due to the ongoing climate change remain. In response to this situation, in March 2021, the government replaced the existing conference with the "Heatstroke Prevention Council" to promote measures against heatstroke (Government of Japan, 2021a).

In March 2021, the "Heatstroke Prevention Council" formulated the "Action Plan for Heatstroke Prevention," referred to as the "2021 HP plan" in this chapter (Government of Japan, 2021b), which outlines priority measures to prevent heatstroke, including measures for the elderly with an exceptionally high number of deaths, and strengthening cooperation with local communities and industries. In March 2022, the 2021 HP plan was revised for reviewing the measures implemented by the government, which is referred to as the "2022 HP plan" in this chapter. These measures address new issues, such as heatstroke prevention in different regions and preparing for extreme heat events, similar to the heat waves of Canada in June–July 2021 (Government of Japan, 2022). Figure 2 shows an overview of the 2022 HP plan.

In the 2022 HP plan, the government set a mid-term goal for heatstroke prevention to reduce death cases to < 1000 per year and a goal for the summer of 2022. Following the measures outlined in the 2022 HP plan, the national government, local governments, industry, various organizations, and citizens aim to work together to promote heatstroke prevention.

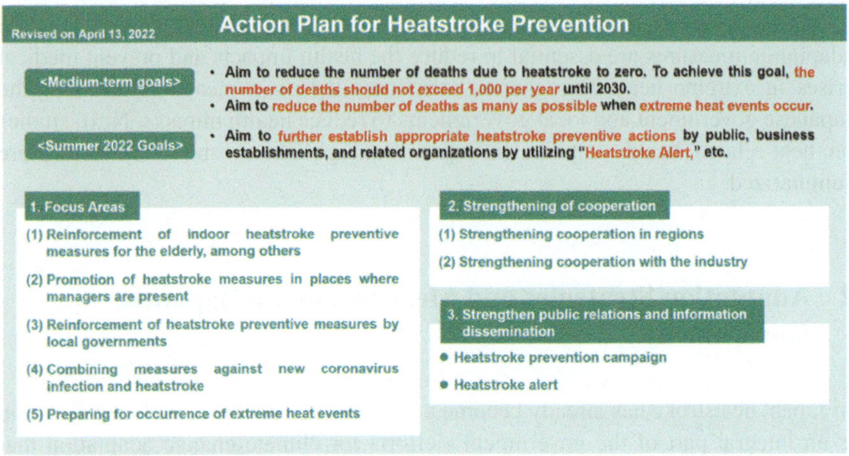

Fig. 2 Outline of the "Action Plan for Heatstroke Prevention 2022." This figure is modified based on the material prepared by the Japanese government

2.2 *Heatstroke Alert*

The public is alerted to heatstroke through the "High-Temperature Warning" issued by the Japan Meteorological Agency (JMA) and the information on heat index through the wet bulb globe temperature (WBGT) issued by the Ministry of the Environment, Japan (MOEJ). Since 2020, the MOEJ and the JMA have collaborated to launch a new heat-health warning system, "Heatstroke Alert," to provide information that improves heatstroke preventive actions (MOEJ, 2022a).

The "Heatstroke Alert" broadly disseminates necessary information to encourage preventive actions when weather conditions are predicted to pose a high risk of heatstroke. Moreover, WBGT, which has high relation with heatstroke, is used as the standard index for announcing alerts; when WBGT is predicted to be $\geq$ 33 °C, an alert is issued at the prefecture level. The disseminated information includes the predicted WBGT value and specific heatstroke preventive actions. The "Heatstroke Alert" was first implemented in the KantoKoshin region in July 2020 and was rolled out nationwide in 2021. Alerts are distributed to relevant ministries and agencies, local governments, media organizations, and private businesses in the same manner as other disaster prevention and weather information released by the JMA (MOEJ, 2022a). Figure 3 shows the flow of the alert transmission. The total number of alerts issued nationwide was 613 in 2021 and 889 in 2022 (MOEJ, 2022b).

For the "Heatstroke Alert," a uniform alert threshold (WBGT of 33 °C) has been adopted. However, the heatstroke incidence varies regionally because heat tolerance differs from region to region owing to various climatic characteristics in Japan. Additionally, concerns regarding the severe health impacts of extreme heat events above the alert threshold remain. To address these issues, a review of alert thresholds is currently underway (MOEJ, 2022c).

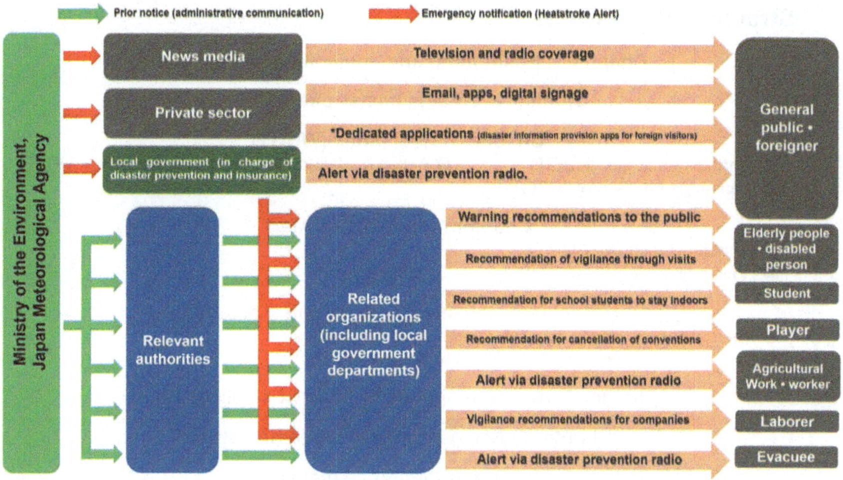

Fig. 3 Transmission of "Heatstroke Alert." The figure is modified from a material prepared by the Ministry of the Environment, Japan

2.3 *Guide for Preparing Heatstroke Prevention Guidelines for Schools*

Recently, approximately 5000 heatstroke cases have occurred in schools annually (MOEJ and the Ministry of Education, Culture, Sports, Science, and Technology (MEXT), 2021a). MEXT, through prefectural boards of education, has issued notices to elementary, junior high, and high schools regarding the prevention of heatstroke accidents. MEXT has been promoting measures to prevent heatstroke in schools, such as issuing reminders for the health management of children and students (MOEJ and MEXT, 2021b).

Some local boards of education have already prepared manuals and guidelines for heatstroke prevention and have been promoting various measures to prevent heatstroke accidents. However, the quality of the content of these manuals and guidelines varies. The MOEJ and MEXT conducted surveys and interviews on examples of heatstroke preventive measures taken in schools and other existing issues. Based on these surveys, the MOEJ and MEXT prepared the "Guide for preparing heatstroke prevention guidelines for schools" in October 2000 (MOEJ and MEXT, 2021b).

This guide outlines the items to be described in the school guidelines and points to be considered when preparing the guidelines. Figure 4 shows the structure of the guide and describes heat index (WBGT), "Heatstroke Alert," preventive measures, and how to respond to heatstroke occurrence.

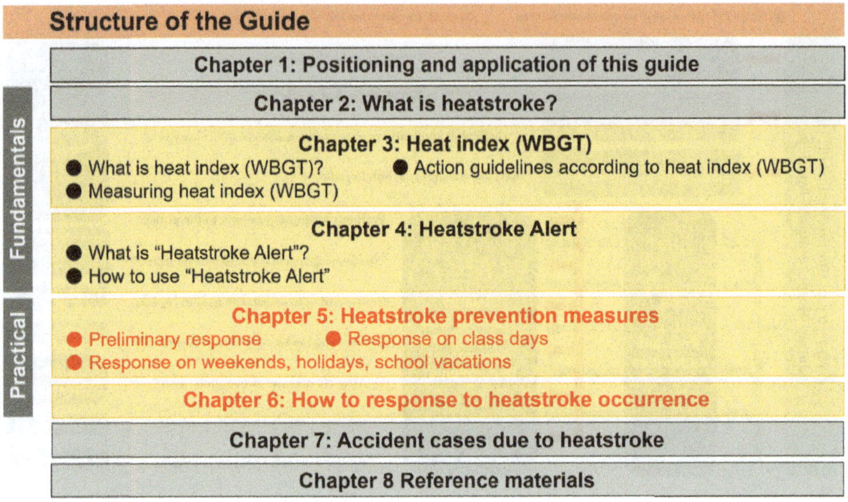

Fig. 4 Structure of the "Guide for preparing heatstroke prevention guidelines for schools." The figure is modified based on the material prepared by the Ministry of the Environment, Japan, and the Ministry of Education, Culture, Sports, Science, and Technology

2.4 Model Project to Promote Air Conditioning Through Subscriptions

In Japan, approximately 80% of heatstroke deaths occur among the elderly aged ≥ 65 years (MHLW, 2021). Similarly, the elderly accounted for > 90% of the heatstroke deaths in the 23 wards of Tokyo in 2020 (Tokyo Medical Examiner's Office, 2020). Of the total number of heatstroke deaths in Tokyo, approximately 93% occurred indoors due to not using air conditioning. Among those who did not use air conditioning, approximately 35% did not have air conditioning installed (Tokyo Medical Examiner's Office, 2020). To prevent heatstroke, installing and promoting the appropriate use of air conditioning are essential.

However, one of the issues in introducing air conditioning is the initial cost of its installation. In response, the MOEJ implemented a model project in 2022 to establish a business that promotes air conditioning using a subscription service. By implementing this business, which reduces the initial cost of installing air conditioning, this project aims to promote the spread of air conditioning installation, thereby promoting heatstroke prevention (MOEJ, 2021).

In this project, local governments and private companies are collaborating to conduct a trial subscription business for residential and commercial air conditioners in public facilities, such as gymnasiums, which are used as evacuation centers in the event of a natural disaster (Fig. 5). Besides verifying the economics and effectiveness of this business, associated data are being collected to investigate improved heatstroke preventive measures (MOEJ, 2021).

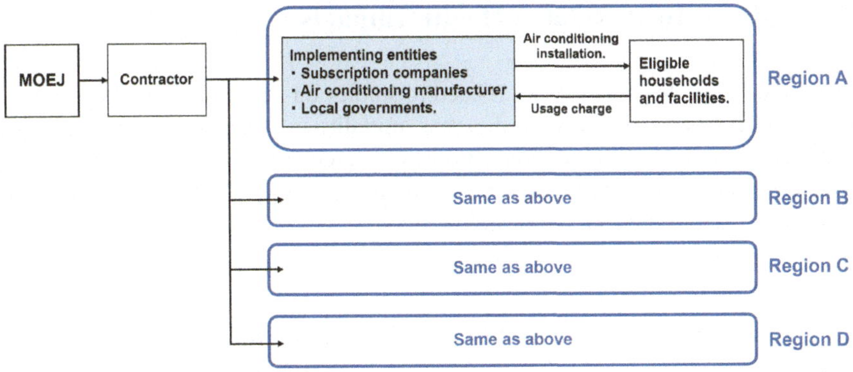

Fig. 5 Example scheme promoting air conditioning by subscription service adopted in a local government (Region A). This scheme will be applied to other local governments (Regions B, C, and D). This figure is modified from the material prepared by the Ministry of the Environment, Japan

3 Adaptation Strategies and Measures by Local Governments

In urban areas, temperature increases owing to urban heat island are observed, in addition to climate change (JMA, 2018). The occurrence of urban heat island is attributed to increasing anthropogenic heat, modification of land surface, and increasing complexity of urban structure (JMA, 2018). Local governments where urban heat island occurs have implemented various measures to mitigate the above factors. Furthermore, local governments have implemented efforts to prevent the occurrence of heatstroke associated with the rising temperatures, focusing on awareness-raising activities through websites and pamphlets (CEIS, 2019).

In Japan, the Climate Change Adaptation Act was enacted in 2018. This Act established a legal framework for the Japanese government, local governments, business entities, and citizens to work together and cooperate in promoting adaptation measures (MOEJ, 2018). It states that local governments are obligated to make efforts to prepare local climate change adaptation plan. At the prefectural level, regional climate change adaptation plans have been prepared in 46 of all 47 prefectures as of February 2023 (NIES, 2023). In the health sector, heat-related health impact (e.g., heatstroke and heat-related excess mortality) and infectious diseases are generally included. Regarding the heat-related health impacts, the main adaptation measure is to spread awareness of the prevention of heatstroke occurrence through websites and pamphlets. Adaptation measures such as the installation and appropriate use of air conditioning are also often included. Note that in the northern regions (the cooler region) of Japan, air conditioner penetration rates are lower than those in other regions (Bureau of Statistics Japan, 2015). Installation of air conditioning in the northern regions will be one of the crucial adaptation measures for further temperature increases.

4 Study on Heat-Related Health Impacts

The scientific findings and issues related to studies on heat-related health impacts, such as heatstroke and heat-related excess mortality in Japan, are described in this section. Furthermore, this section introduces the collaborative study project with local governments on heat environments and heatstroke conducted by the National Institute for Environmental Studies (NIES).

4.1 Heatstroke and Heat-Related Excess Mortality

Heatstroke

In Japan, heatstroke has been studied vigorously. These studies include the development of prediction models for heatstroke incidence (Ikeda & Kusaka, 2021; Kodera et al., 2019; Ogata et al., 2021; Oka & Hijioka, 2021; Tamura et al., 1995), analysis of regional and seasonal differences in heatstroke incidence (Fujibe et al., 2020; Iwamoto & Ohashi, 2021; Oka et al., 2023a, 2023b; Ueno et al., 2021; Yokoyama & Fukuoka, 2006), analysis of differences in heatstroke incidence by age (Fujibe et al., 2020; Oka et al., 2023a; Ueno et al., 2021), and degree of severity (Fujibe et al., 2020; Oka et al., 2023a). Some important scientific findings from these studies are summarized below:

① Heatstroke incidence risk at the end of the rainy season and early summer is higher than in late summer.
② Heatstroke incidence is the highest among the elderly, followed by juveniles and adults.
③ Heatstroke incidence risk in cooler areas is higher than that in warmer regions.

The first finding can be attributed to the fact that people are still unaccustomed to hot temperatures at the end of the rainy season and in early summer (short-term heat acclimatization has not yet progressed) (Fuse et al., 2014; Ono, 2013). Regarding the second finding, physiological characteristics, such as the elderly being particularly susceptible to heat, can be addressed as a primary factor (Kenney & Munce, 2003; Miyake et al., 2010; Nakai et al., 1999; Semenza et al., 1996). The third finding is due to factors such as acclimatization to heat (Anderson & Bell, 2011; Folkerts et al., 2020; Sanderson et al., 2017) and regional differences in the implementation level of heatstroke preventive measures (Anderson & Bell, 2011; Folkerts et al., 2020; Ostro et al., 2010; Sanderson et al., 2017; Sera et al., 2020; Tobías et al., 2021).

Despite vigorous studies on heatstroke, there are still issues to be addressed; these issues require consideration in developing a prediction model for heatstroke incidence. First, how to account for the effect of short-term heat acclimatization (where the body gradually becomes accustomed to hot temperatures as the hot days continue) in the prediction model. Some studies have addressed this issue (Nakamura et al., 2022; Oka & Hijioka, 2021). Oka and Hijioka (2021) developed a prediction model that incorporated the effect of short-term heat acclimatization by introducing a meteorological variable considering the temperature information experienced in the past. The developed prediction model is expected to be supported by physiological knowledge related to short-term heat acclimation. Second, few studies have measured the effects of heatstroke preventive measures (for example, "Heatstroke Alert") on reducing heatstroke. To understand the effects of the currently introduced measures and to consider the measures that will be necessary for the future, it is necessary to conduct studies evaluating the effects of the measures and developing prediction models from the results obtained. Third, issues related to future projections should be addressed. Heat adaptation is expected to progress in cooler regions via the acquisition of physiological heat acclimatization and the implementation of measures. Oka et al. (2023b) showed a correlation between the summer average daily maximum WBGT (maxWBGTms) and the daily maximum WBGT at which heatstroke incidence starts to increase (WBGT threshold) (Fig. 6). Based on this relationship, Oka et al. (2023c) conducted a future projection of heatstroke, taking into account the effect of heat adaptation by shifting the WBGT threshold to a higher WBGT when the maxWBGTms increase in the future. Nevertheless, future model development should consider the physiological mechanisms of long-term heat acclimatization and the effects of implementing heatstroke preventive measures.

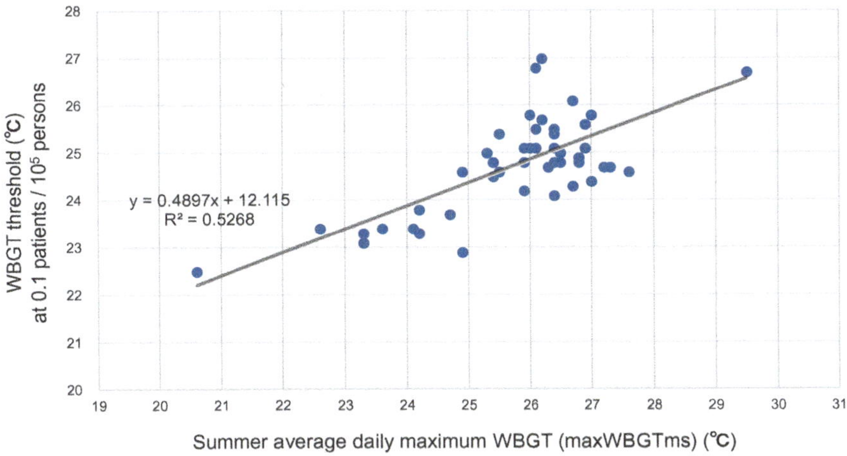

Fig. 6 Association between the summer average daily maximum wet bulb globe temperature (maxWBGTms) and the maximum WBGT at which heatstroke incidence starts to increase (WBGT threshold; defined as the value at which heatstroke incidence is 0.1 persons per 10^5 people). The time period is from May 2015 to September 2019. Each plot shows the values for each of the 47 prefectures in Japan. Based on this relationship, the effect of heat adaptation can be considered by shifting the WBGT threshold to the higher WBGT side when the maxWBGTms increases in the future. The WBGT data from the Heat Illness Prevention System developed by MOEJ (2020) were used

Heat-Related Excess Mortality

Besides heatstroke, the direct health impact of hot temperatures is heat-related excess mortality. When total mortality is examined with temperature, it shows a V- or U-shaped relationship with the minimum mortality temperature (MMT), where mortality is the lowest (El-Zein et al., 2004; Kunst et al., 1993). Mortality increases on the high-temperature side of the MMT. This type of mortality is called heat-related excess mortality. MMT can be estimated as the 80th to 85th percentile value of the daily maximum temperature distribution in the region (Honda et al., 2007), and later, they used the point estimate of the 84th percentile for the future projection study (Honda et al., 2014).

An interesting question is the association between total mortality and temperature changes through heat adaptation considering rising temperatures owing to climate change. Several studies have analyzed how MMT shifts (whether there is a time lag before the MMT catches up with the new (higher) 84th percentile of the daily maximum temperature distribution) and how the risk for a certain level of heat changes with the progress of climate change (Chung et al., 2017; Gasparrini et al., 2015; Petkova et al., 2014). Based on the results of these studies, the changes in the association between total mortality and temperature under the changing climate vary by region. Before such studies, Honda et al. (2014) conducted a future projection of heat-related excess mortality that considered heat adaptation, assuming 100% adaptation when MMT has reached the 84th percentile value at the time of evaluation, 0% adaptation when MMT remained the same as current, and 50% adaptation when MMT was intermediate between both cases. Further studies on the effects of heat adaptation on heat-related excess mortality are expected to progress, and with the results, future projections of heat-related excess mortality will be performed.

4.2 Collaborative Study Project with Local Government

Implementing heatstroke preventive measures has become an urgent issue for local governments. However, the health impacts of heatstroke differ from region to region. Therefore, it is necessary to understand and analyze the current situation in each region to plan and implement appropriate measures. Notably, the NIES initiated a collaborative study project with the local government to address this challenge in 2020. As of December 2022, ten local governments have joined this project, including nine prefectures and one city. This project involves the following items:

① Establishment of a forum

This project provides a forum to exchange opinions on issues and studies related to the heat environment and heatstroke. The forum is generally held thrice per year.

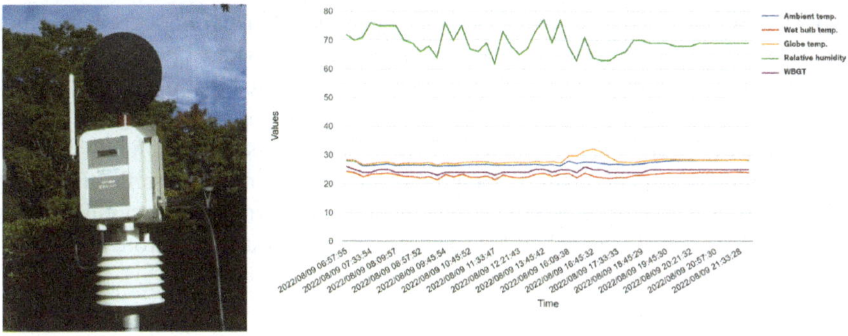

Fig. 7 WBGT observation instrument (left) and viewing screen of observed data uploaded to a cloud system (right)

② WBGT observation of the hot environment

NIES prepares WBGT observation instruments and lends them to the participating local governments. Next, the local governments conduct WBGT observations at locations where the risk of heatstroke is expected to be high (for example, schools and farmlands). To conduct WBGT observations efficiently in remote locations, NIES developed a system for viewing observed data on a cloud system with the cooperation of the private sector (Fig. 7).

The observations showed the extent to which WBGT varies depending on the geographical location of schools within the same prefecture and the direction and number of floors of school buildings within the same school.

③ Collection and analysis of data on the number of patients with heatstroke

The participating local governments collect data on the number of patients with heatstroke transported by ambulance recorded at each fire department. Next, they analyze the data considering regional characteristics, for example, the age structure of heatstroke cases and the location of occurrence. Additionally, NIES develops a tool to observe time-series trends in the number of patients with heatstroke and WBGT and analyze the association between them. The data analysis showed that some areas had a higher heatstroke incidence than others within a prefecture.

④ Prediction of the future number of patients

Using the latest prediction model of heatstroke incidence developed by NIES, NIES and local governments predict the number of individuals with heatstroke in the future with the progress of climate change.

⑤ Implementation of public awareness campaigns to reduce heatstroke

The participating local governments conduct public awareness campaigns to disseminate information on heatstroke and promote preventive measures. These include

WBGT observations by students at schools, the establishment of booths at sporting events (for example, professional soccer games), and amusement parks.

Through the studies in items ② through ④, it is expected that the regional characteristics of the hot environment and health impacts (for example, geographical differences in WBGT and the number of patients with heatstroke and future changes in the number of patients) will be identified, and detailed measures reflecting these characteristics will be developed and implemented. Furthermore, behavioral changes are expected to reduce the risk of heatstroke by raising awareness through activities in item ⑤.

5 Concluding Remarks

The Japanese government and local governments have implemented various heatstroke preventive measures to reduce heat-related health impacts such as heatstroke. However, the increase in the elderly population in Japan and rising temperatures due to climate change will further increase the health impacts. Additionally, medical crises resulting from significant health impacts would be severe in extreme heat events. To secure the lives and health of Japanese people, mitigation measures to reduce the temperature increase are vital, and adaptation measures to reduce the risk of heatstroke and prevent medical crises. The findings from scientific studies will contribute to planning and implementing adaptation plans and measures.

Acknowledgements This chapter was written using the study results from the Climate Change Adaptation Research Program and the Collaborative Study Project on Adaptation with Local Government, both from the National Institute for Environmental Studies, and the Environment Research and Technology Development Fund (JPMEERF20231007) of the Environmental Restoration and Conservation Agency provided by Ministry of the Environment, Japan, as well as relevant information collected while conducting the studies. I thank Professor Y. Honda for his valuable comments on writing this chapter.

References

Anderson, G. B., & Bell, M. L. (2011). Heat waves in the United States: Mortality risk during heat waves and effect modification by heat wave characteristics in 43 U.S. communities. *Environmental Health Perspectives, 119*(2), 210–218. https://doi.org/10.1289/ehp.1002313

Bureau of Statistics Japan. (2015). *2014 national consumption survey*. https://www.stat.go.jp/data/zensho/2014/kekka.html

Center for Environmental Information Science (CEIS). (2019). *Report on adaptation strategy survey for hot environment in FY 2018*. https://www.env.go.jp/air/report/h30-01/30.html

Chung, Y., Noh, H., Honda, Y., Hashizume, M., Bell, M. L., Guo, Y. L. L., & Kim, H. (2017). Temporal changes in mortality related to extreme temperatures for 15 cities in Northeast Asia: Adaptation to heat and maladaptation to cold. *American Journal of Epidemiology, 185*(10), 907–913. https://doi.org/10.1093/aje/kww199

El-Zein, A., Tewtel-Salem, M., & Nehme, G. (2004). A time-series analysis of mortality and air temperature in greater Beirut. *Science of the Total Environment, 330*, 71–80. https://doi.org/10.1016/j.scitotenv.2004.02.02

Fire and Disaster Management Agency (FDMA). (2022). *Transport by ambulance due to heatstroke in 2022*. https://www.fdma.go.jp/disaster/heatstroke/items/r4/heatstroke_geppou_202205-09.pdf

Folkerts, M. A., Bröde, P., Botzen, W. J. W., Martinius, M. L., Gerrett, N., Harmsen, C. N., & Daanen, H. A. M. (2020). Long term adaptation to heat stress: Shifts in the minimum mortality temperature in the Netherlands. *Frontiers in Physiology, 11*, 225. https://doi.org/10.3389/fphys.2020.00225

Fujibe, F., Matsumoto, J., & Suzuki, H. (2020). Regional features and temporal variations of heat-stroke ambulance transport rates in Japan—Comparison with mortality. *J. Heat Isl. Inst. Int., 15*, 1–13. https://www.heat-island.jp/web_journal/download/20A001.pdf

Fuse, A., Saka, S., Fuse, R., Araki, T., Kin, S., Miyauchi, M., & Yokota, H. (2014). Weather data can predict the number of heat stroke patient. *Japanese Association for Acute Medicine, 25*, 757–765. https://doi.org/10.3893/jjaam.25.757

Gasparrini, A., Guo, Y., Hashizume, M., Kinney, P. L., Petkova, E. P., Lavigne, E., Zanobetti, A., Schwartz, J. D., Tobias, A., Leone, M., Tong, S., Honda, Y., Kim, H., & Armstrong, B. G. (2015). Temporal variation in heat–mortality associations: A multicountry study. *Environmental Health Perspectives, 123*, 1200–1207. https://doi.org/10.1289/ehp.1409070

Government of Canada. (2022). *Surviving the heat: The impacts of the 2021 western heat domed in Canada*. https://science.gc.ca/site/science/en/blogs/science-health/surviving-heat-impacts-2021-western-heat-dome-canada. Accessed on Dec 26, 2022.

Government of Japan. (2021a). *Holding of "heatstroke prevention council"*. https://www.wbgt.env.go.jp/pdf/ic_rma/R0402/ref01.pdf

Government of Japan. (2021b). *Action plan for heatstroke prevention 2021*. https://www.env.go.jp/content/900517263.pdf

Government of Japan. (2022). *Action plan for heatstroke prevention 2022*. https://www.env.go.jp/content/900502212.pdf

Honda, Y., Kabuto, M., Ono, M., & Uchiyama, I. (2007). Determination of optimum daily maximum temperature using climate data. *Environmental Health and Preventive Medicine, 12*(5), 209–216. https://doi.org/10.1265/ehpm.12.209

Honda, Y., Kondo, M., McGregor, G., Kim, H., Guo, Y. -L., Hales, S., & Kovats, S. (2014). In S. Hales, et al. (Eds.), *Heat-related mortality, in quantitative risk assessment of the effects of climate change on selected causes of death, the 2030s and 2050s* (pp. 17–24). World Health Organization. https://www.who.int/publications/i/item/9789241507691

Ikeda, T., & Kusaka, H. (2021). Development of models for predicting the number of patients with heatstroke on the next day considering heat acclimatization. *Journal of the Meteorological Society of Japan, 99*(6), 1395–1412.https://doi.org/10.2151/jmsj.2021-067

IPCC. (2021). Climate Change 2021: The physical science basis. In V. Masson-Delmotte, P. Zhai, A. Pirani, S. L. Connors, C. Péan, S. Berger, N. Caud, Y. Chen, L. Goldfarb, M. I. Gomis, M. Huang, K. Leitzell, E. Lonnoy, J. B. R. Matthews, T. K. Maycock, T. Waterfield, O. Yelekçi, R. Yu & B. Zhou (Eds.), *Contribution of working group I to the sixth assessment report of the intergovernmental panel on climate change*. Cambridge University Press, Cambridge, United Kingdom and New York, NY, USA. https://doi.org/10.1017/9781009157896

Iwamoto, Y., & Ohashi, Y. (2021). Assessing the climatological relationship between heatstroke risk and heat stress indices in 47 prefectures in Japan. *GeoHazards, 2*(4), 321–331. https://doi.org/10.3390/geohazards2040017

Japan Meteorological Agency (JMA). (2018). *Heat Island monitoring report 2017*. https://www.env.go.jp/air/report/h30-01/30.html

Japan Meteorological Agency (JMA). (2022). *Climate change monitoring report 2021*. https://www.data.jma.go.jp/cpdinfo/monitor/2021/pdf/ccmr2021_all.pdf

Kenney, W. L., & Munce, T. A. (2003). Invited review: Aging and human temperature regulation. *Journal of Applied Physiology, 95*, 2598–2603. https://doi.org/10.1152/japplphysiol.00202.2003

Kodera, S., Nishimura, T., Rashed, E. A., Hasegawa, K., Takeuchi, I., Egawa, R., & Hirata, A. (2019). Estimation of heat-related morbidity from weather data: A computational study in three prefectures of Japan over 2013–2018. *Environment International, 130*, 104907. https://doi.org/10.1016/j.envint.2019.104907

Kunst, A. E., Looman, C. W., & Mackenbach, J. P. (1993). Outdoor air temperature and mortality in the Netherlands: A time-series analysis. *American Journal of Epidemiology, 137*, 331–341. https://doi.org/10.1093/oxfordjournals.aje.a116680

Ministry of Environment, Japan (MOEJ). (2018). *Cabinet approval of the climate change adaptation act*. https://www.env.go.jp/press/105165.html. Accessed on Mar 2, 2023.

Ministry of Environment, Japan (MOEJ). (2020). *Heat illness prevention information*. https://wbgt.env.go.jp/en/. Accessed on Jul 30, 2021.

Ministry of Environment, Japan (MOEJ). (2021). *Model project to promote air conditioning by subscriptions*. https://www.env.go.jp/content/900470344.pdf

Ministry of Environment, Japan (MOEJ). (2022a). *What is "heatstroke alert"*. https://www.wbgt.env.go.jp/about_alert.php. Accessed on Dec 26, 2022.

Ministry of Environment, Japan (MOEJ). (2022b). *Status of "heatstroke alert" announcement*. https://www.wbgt.env.go.jp/alert_record.php. Accessed on Dec 26, 2022.

Ministry of Environment, Japan (MOEJ). (2022c). *Introduction of higher criteria of issuing "heatstroke alert" (draft)*. https://www.wbgt.env.go.jp/pdf/sg_pcm/R0401/doc04.pdf

Ministry of Environment, Japan (MOEJ) and Ministry of Education, Culture, Sports, Science and Technology (MEXT). (2021a). *Background of the establishment of the "study group on the promotion of heatstroke prevention at school"*. https://www.wbgt.env.go.jp/pdf/sg_sef/R0201/doc01-1.pdf

Ministry of Environment, Japan (MOEJ) and Ministry of Education, Culture, Sports, Science and Technology (MEXT). (2021b). *Guide for preparation of heatstroke prevention guidelines for schools*. https://www.mext.go.jp/content/210528-mxt_kyousei01-000015427_02.pdf

Ministry of Health, Labour and Welfare (MHLW). (2021). *Annual heatstroke deaths by age group in 5-year age group during 1995–2020*. https://www.mhlw.go.jp/toukei/saikin/hw/jinkou/tokusyu/necchusho20/dl/nenrei.pdf

Miyake, Y., Aruga, T., Inoue, K., Okudera, H., Kitahara, T., Shimazaki, S., Tsuruta, R., & Yokota, H. (2010). Characteristics of heatstroke patients in Japan; Heatstroke STUDY2008. *Japanese Association for Acute Medicine, 21*, 230–244. https://doi.org/10.3893/jjaam.21.230

Nakai, S., Itoh, T., & Morimoto, T. (1999). Deaths from heat-stroke in Japan: 1968–1994. *International Journal of Biometeorology, 43*, 124–127. https://doi.org/10.1007/s004840050127

Nakamura, S., Kusaka, H., Sato, R., & Sato, T. (2022). Heatstroke risk projection in Japan under current and near future climates. *Journal of the Meteorological Society of Japan, 100*(4), 597–661. https://doi.org/10.2151/jmsj.2022-030

National Institute for Environmental Studies, Japan. (2023). *List of local climate change adaptation plans*. https://adaptation-platform.nies.go.jp/local/plan/list.html. Accessed on Mar 2, 2023.

Ogata, S., Takegami, M., Ozaki, T., Nakashima, T., Onozuka, D., Murata, S., Nakaoku, Y., Suzuki, K., Hagihara, A., Noguchi, T., Iihara, K., Kitazume, K., Morioka, T., Yamazaki, S., Yoshida, T., Yamagata, Y., & Nishimura, K. (2021). Heatstroke predictions by machine learning, weather information, and an all-population registry for 12-hour heatstroke alerts. *Nature Communications, 12*, 4575. https://doi.org/10.1038/s41467-021-24823-0

Oka, K., & Hijioka, Y. (2021). Prediction of the number of heatstroke patients transported by ambulance in Japan's 47 prefectures: Proposal of heat acclimatization consideration. *Environmental Research Communications, 3*, 125002. https://doi.org/10.1088/2515-7620/ac3d21

Oka, K., Honda, Y., Phung, V. L. H., & Hijioka, Y. (2023a). Potential effect of heat adaptation on association between number of heatstroke patients transported by ambulance and wet bulb globe temperature in Japan. *Environmental Research, 216*(3), 114666. https://doi.org/10.1016/j.envres.2022.114666

Oka, K., Honda, Y., & Hijioka, Y. (2023b). Launching criteria of "heatstroke alert" in Japan according to regionality and age group. *Environmental Research Communications*. Accepted https://doi.org/10.1088/2515-7620/acac03

Oka, K., Honda, Y., Phung, V. L. H., & Hijioka, Y. (2023c). Prediction of climate change impacts on heatstroke cases in Japan's 47 prefectures with the effect of long-term heat adaptation. *Environmental Research, 232*(1), 116390. https://doi.org/10.1016/j.envres.2023.116390

Ono, M. (2013). Heat stroke and the thermal environment. *Japan Medical Association Journal, 56*(3), 199–205. https://www.med.or.jp/english/journal/pdf/2013_03/199_205.pdf

Ostro, B., Rauch, S., Green, R., Malig, B., & Basu, R. (2010). The effects of temperature and use of air conditioning on hospitalizations. *American Journal of Epidemiology, 172*(9), 1053–1061. https://doi.org/10.1093/aje/kwq231

Petkova, E. P., Gasparrini, A., & Kinney, P. L. (2014). Heat and mortality in New York City since the beginning of the 20th century. *Epidemiology, 25*(4), 554–560. https://doi.org/10.1097/EDE.0000000000000123

Sanderson, M., Arbuthnott, K., Kovats, S., Hajat, S., & Falloon, P. (2017). The use of climate information to estimate future mortality from high ambient temperature: A systematic literature review. *PLoS ONE, 12*(7), e0180369. https://doi.org/10.1371/journal.pone.0180369

Semenza, J. C., Rubin, C. H., Falter, K. H., Selanikio, J. D., Flanders, W. D., Howe, H. L., & Wilhelm, J. L. (1996). Heat-related deaths during the July 1995 heat wave in Chicago. *New England Journal of Medicine, 335*, 84–90. https://doi.org/10.1056/nejm199607113350203

Sera, F., Hashizume, M., Honda, Y., Lavigne, E., Schwartz, J., Zanobetti, A., Tobías, A., Iñiguez, C., Vicedo-Cabrera, A. M., Blangiardo, M., Armstrong, B., & Gasparrini, A. (2020). Air conditioning and heat-related mortality: A multi-country longitudinal study. *Epidemiology, 31*(6), 779–787. https://doi.org/10.1097/ede.0000000000001241

Tamura, K., Ono, M., Ando, M., & Murakami, M. (1995). Heatstroke incidence and temperature based on emergency transport data. *Jpn. J. Biometeor., 32*(2), 111–114. https://www.jstage.jst.go.jp/article/seikisho1966/32/2/32_2_111/_pdf/-char/ja

Tobías, A., Hashizume, M., Honda, Y., Sera, F., Ng, C. F. S., Kim, Y., Roye, D., Chung, Y., Dang, T. N., Kim, H., Lee, W., Íñiguez, C., Vicedo-Cabrera, A., Abrutzky, R., Guo, Y., Tong, S., Coelho, M. d. S. Z. S., Saldiva, P. H. N., Lavigne, E., et al. (2021). Geographical variations of the minimum mortality temperature at a global scale: A multicountry study. *Environmental Epidemiology, 5*(5), e169. https://doi.org/10.1097/EE9.0000000000000169

Tokyo Medical Examiner's Office. (2020). *Heat stroke fatalities in the summer of 2020*. https://www.fukushihoken.metro.tokyo.lg.jp/kansatsu/oshirase/R02-heatstroke-sokuhou.html. Accessed on Dec 26, 2022.

Ueno, D., Hayano, D., Noguchi, E., & Aruga, T. (2021). Investigating age and regional effects on the relation between the incidence of heat-related ambulance transport and daily maximum temperature or WBGT. *Environmental Health and Preventive Medicine, 26*, 116. https://doi.org/10.1186/s12199-021-01034-z

Yokoyama, T., & Fukuoka, Y. (2006). Regional tendency of heat disorders in Japan. *Jpn. J. Biometeor., 43*(4), 141–151. https://doi.org/10.11227/seikisho.43.145

Climate-Resilient and Health System in Thailand

Uma Langkulsen and Augustine Lambonmung

Abstract The effects of climate change impact humanity diversely, posing a great danger to the fundamental components of health and well-being, including clean air, safe drinking water, wholesome food, secure housing as well as outbreaks of pandemics. This threat to global health could potentially overwhelm health systems, particularly during crisis if not resilient enough to cope with challenges in providing essential services as witnessed during the COVID-19 pandemic which greatly exposed health systems hitherto considered strong. To identify government policies and programs supportive of a climate-resilient health system, a documentary review was carried out between 2012 and 2022 to analyse policy measures that could potentially contribute to developing and sustaining a health system capable of resiliently responding to health needs. A total of 54 articles were identified. After the articles were deduplicated, the title and abstract were checked, and 42 articles were further screened using full text. Of these, 11 met the eligibility requirements for inclusion in the final analysis. The existing policy measures were broadly categorized at the system and agency level and the level of the population. The climate change adaptation and mitigation policy measures have significant potential for health system resilience worthy of emulation as they aim to strengthen health related systems and agencies and are sensitive to healthcare needs population wide.

Keywords Climate change · Resilient · Health system · Policy · Well-being · Thailand

U. Langkulsen (✉) · A. Lambonmung
Faculty of Public Health, Thammasat University, Bangkok 12120, Pathum Thani, Thailand
e-mail: u.langkulsen@gmail.com; augustine.l@fph.tu.ac.th

A. Lambonmung
Tamale Teaching Hospital, TL 16, Tamale, Ghana

R. Akhtar (ed.), *Climate Change and Human Health Scenarios*, Global Perspectives on Health Geography, https://doi.org/10.1007/978-3-031-38878-1_6

1 Introduction

Deserts are growing larger because of climate change, and heat waves and wildfires are occurring more frequently. The permafrost is melting, the glaciers are retreating, and the sea ice is disappearing due to increased warmth in the Arctic (Oppenheimer et al., 2019). These threaten the fundamental components of health and well-being, including clean air, safe drinking water, wholesome food, and secure housing, impacting human lives and health variously (WHO, 2021). Conflicts and migration of people potentially could be the biggest impact (International Organization for Migration, 2021; Kaczan & Orgill-Meyer, 2020). These consequences emanating from the rapid environmental changes are regarded by the WHO as the greatest threat to global health in the twenty-first century. Globally, climate change is estimated to result in an extra 250,000 fatalities annually from hunger, malaria, diarrhea, and heat stress between 2030 and 2050. By 2030, it is anticipated that direct health harm would cost an estimated US $2–4 billion annually (excluding expenditures in health-determining industries like agriculture, water, and sanitation) (WHO, 2021). Some of the impacts will last for millennia, even if attempts to reduce future warming are successful (Ebi et al., 2021).

Areas in the mid-latitudes are anticipated to warm disproportionately, regions around the Pacific and Indian Oceans, which currently experience high precipitation variability and exposed cities, where urban heat islands' effect can exacerbate extreme climate events, thereby exposing these zones to more vulnerable health effects due to climate change (Kiguchi et al., 2021; Toimil et al., 2020). Thailand's geography and exposure to climate-induced calamities date back centuries ago, and it ranks high in terms of its vulnerability to flooding, including coastal flash, and river floods (Limsakul, 2020; World Bank Group & Asian Development Bank, 2021). The majority of its yearly losses attributable to hazards are caused by floods (World Bank Group & Asian Development Bank, 2021). There have been a number of climate-related disasters of various magnitudes and intensities across the length and breadth of the country in recent past decades (Limsakul, 2020). The nation's susceptibility to climatically connected natural hazards was evidenced by massive havoc wrought by record-breaking flooding in 2011, which was caused by a very powerful monsoon and tropical storm Nock-ten causing a landfall. Over eight hundred fatalities were recorded in the floods which affected 13.6 million people and devastated 20,000 km^2 of agricultural land (Gale & Saunders, 2013). Economic and insurance losses from the storm were estimated to have cost several billions of dollars (World Bank Group & Asian Development Bank, 2021). More recently in 2019, Tropical storm Pabuk caused a similar devastation particularly in the southern coastal region of Nakhon Si Thammarat (Thanathanphon et al., 2020). A previous study forecasted that the country is rapidly being exposed to higher risks of disasters in the ensuing decades as a result of the effects of climate change and rising human activities (Gale & Saunders, 2013). To safeguard life and livelihoods, pragmatic actions by actors to build climate change-resilient health system to capably predict, be ready for, and appropriately respond to climate-related health hazards, events, trends, or disruptions are required (Ebi et al., 2021; Puntub & Greiving, 2022; WHO, 2022). It is essential to develop a health system resilient to the impact of climate change by adopting workable policy measures and interventions seeking to enhance healthcare service to meet the health needs at all levels (Drewry & Oura, 2022; WHO, 2022). Although the country currently has had an effective universal health coverage (UHC) since 2002 (WHO, 2015a), there is a need to continuously organize people, institutions and facilities, and resources for the health system to be resilient in the provision of healthcare services. Making sure the physical structures that make up the healthcare infrastructure are resistant to severe weather and preparations for the effects on key utilities would also ensure the safety of workers, patients, and visitors (ASPR TRACIE, 2022), especially during crisis. It is essential to incorporate climate-related health risk policy measures to effectively and efficiently determine how climate change may alter or produce new climate-related health hazards and then take steps to better manage those dangers (Piankusol, 2021). This chapter discusses how government policies, mitigation strategies, and adaptation measures for climate change can contribute to developing a climate-resilient and health system in Thailand to help alleviate the impacts of climate change on human health and well-being.

2 Methods

Search engines SCOPUS and Google Scholar were used to search through literature for scholarly articles on impacts of climate change and health system resilience that had been published between 2012 and 2022. The full text of articles' links were carefully searched. After determining which articles were appropriate, the full-text articles were examined, and data were retrieved for assessment. Only original and full-text articles were included. Publications on government policies and programs, climate change mitigation measures, adaptation strategies, and health system resilience deemed relevant were also included. In this analysis, gray literature was considered if they were directly related to the objectives.

Articles addressing climate change, non-health concerns, protocol papers, experimental research articles, letters, editorials, and comments have been excluded. The steps for documentary review recommendations were followed in the presentation of the study's findings. Based on the title and abstract, initial screening of the publications was undertaken. Suitability of the publications was simultaneously examined for inclusion; discussion was used to settle any dispute. Information considered important was extracted for analysis.

Figure 1 illustrates the selection process for the documentary review. A total of 54 articles were identified. After articles were de-duplicated, the title and abstract were checked; this yielded 42 articles. Of these, 11 met the eligibility requirements for inclusion in the final synthesis. They are broadly divided into systems and agencies and population levels.

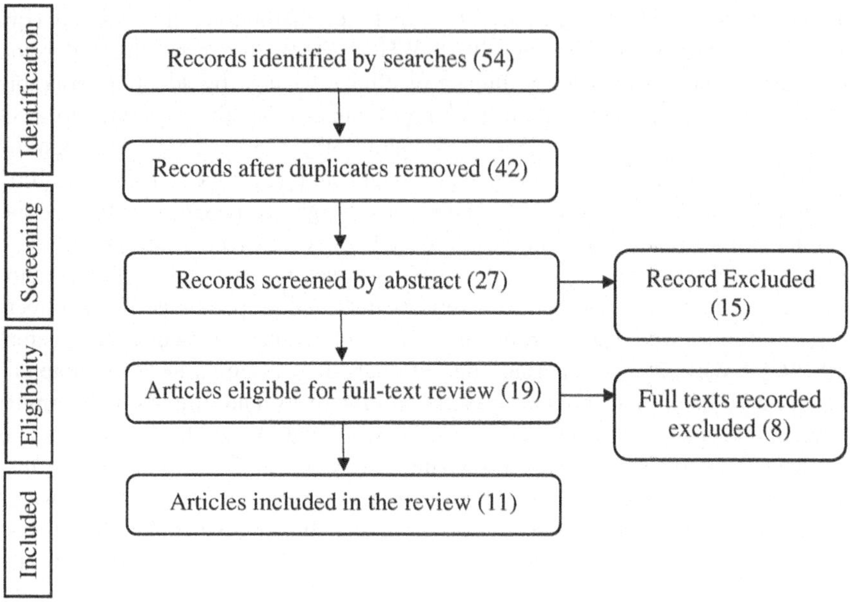

Fig. 1 Study selection for review

The included studies on government policies, mitigation strategies, and adaptation measures for climate-resilient and health system are put into two main categories as relating to: systems and agencies and population levels. The categorization is based on the aim of the study in determining how these factors could contribute to a climate-resilient and health system in Thailand. Publications on climate change-related policies and programs which were examined to contain relevant information and data that could potentially help to create health system resilience were used. Various policies and programs, adaptations and mitigation measures and strategies on the impacts of climate change at system and agency and population levels can potentially make healthcare system function smoothly in times of climate-related health catastrophes to make them more resilient in the discharge of their mandate and are aligned with the objectives of the study which were included in the latter category.

3 Results

The characteristics of the included studies are summarized in Table 1 that lists search terms associated with Thailand and climate change and health system.

Table 1 Government policies and measure and their potential for health system resilience

Policy/measure	Potential for health system resilience
Health National Adaptation Plan (HNAP) Phase 1 (2021–2030) (Health Impact Assessment Division, 2021)	To lower morbidity, lessen the effects on health, and make controlling the hazards associated with climate change on Asia's health a priority
Thailand's Third National Communication (Office of Natural Resources and Environmental Policy and Planning: ONEP, 2020)	Collaborate with response partners, organization stakeholders, and healthcare professionals to communicate risks posed by extreme weather events to susceptible patients and ensure people who live/work in settings that put them at increased/higher risk of becoming infected or exposed to hazards stay informed
Thailand's 2nd Updated Nationally Determined Contribution (ONEP, 2022)	This initiative aims to increase the ability of public health systems to manage health risks and mitigate health impacts from climate change by creating health impact surveillance and preventive mechanisms and improving access to high-quality public health services
The Twelfth National Economic and Social Development Plan (2017–2021) (Office of the National Economic and Social Development Board: NESDB, 2017)	Evaluation and re-evaluation of how the health system provides services to vulnerable populations, considering the potential for these groups to face health risks that are disproportionate, many, or complicated because of climate change. Minority groups, low-income neighborhoods, immigrant groups, the elderly, children, and people with disabilities may be among the populations of concern
Climate Change Master Plan 2015–2050 (ONEP, 2015)	Sectors which policies relate to public health, they may assist in the development of significant policy changes, focusing on the prevention of diseases and threats to health in response to climate change by disseminating information and knowledge to the target groups and developing the effectiveness of basic health services in managing health risks resulting from climate change
Thailand: National Disaster Risk Management Plan (Department of Disaster Prevention and Mitigation: DDPM, 2015)	Implement comprehensive and inclusive economic, structural, legal, social, health, cultural, educational, environmental, and institutional changes to avoid and decrease risk, exposure, and susceptibility to disasters, improve readiness for response and recovery, and boost resilience
Thailand Power Development Plan 2015–2036 (Energy Policy and Planning Office: EPPO, 2015)	Seeks to explore alternative utility sources by establishing more solar and wind power
Thailand Smart Grid Policy Plan and Roadmaps (Hoonchareon, 2015)	Ensure high-power quality and dependability (voltage regulation), system recovery and restoration are swift during disasters

(continued)

Table 1 (continued)

Policy/measure	Potential for health system resilience
The National Water Resources Management Strategies 2015–2026 (Office of the National Water Resources: ONWR, 2015)	Help prepare for secondary impacts that will accompany climate impacts such as water shortages by drilling of wells
Thailand's Transport Infrastructure Development Strategy, 2015–2022 (Kunadhamraks, 2015)	Be aware of the likelihood of adverse events that may not directly threaten facility but may have profound effects on patients and their ability to access and utilize healthcare services
The Kingdom of Thailand Health System Review (WHO, 2015a)	With assistance from development partners, some significant health reforms have been adopted, incorporating climate change-resilient responses into complex policy processes and contexts

3.1 Health System in Thailand

Thailand's health system has undergone major reforms dating several decades back. The significant changes have occurred in the nineteenth century and more recently in 2002 (fourth reorganization) which birthed the current universal health coverage. The UHC replaced the third reorganization in 1992 of the Ministry of Public Health. Operating in a five-tier system, from central, regional, provincial, district, and sub-district levels, the UHC has significantly improved access to services and ensured universal access to essential health services since its inception. Health infrastructure has massively been expanded across the nation increasing access to medical facilities for everybody. Appropriate and equitable workforce allocation in the health sector has been improved under the UHC. Thailand's healthcare workforce has been distributed rather evenly throughout the range of economic conditions in each province improving upon the initiatives for rural development and retention over the previous four decades (Witthayapipopsakul et al., 2019).

The UHC has also contributed to increased financial risk management, increasing population coverage by providing financial security, promoting social solidarity, participatory governance, and health literacy. It has laid grounds for a coalition involving the public and the private health sectors. The partnerships have given rise to group leadership, organizational coordination, and robust infrastructure. Though the UHC considers the entire society, the wide range of abilities, solid relationships, and social support can improve people's health and community resilience (WHO, 2015a). The adoption of feasible climate-related policies and programs, adaptation measures, and mitigation strategies by the government that could contribute to determining largely how the effects of climate change may alter or produce new climate-related health sensitive hazards and then taking steps to better manage those risks to avert and/or alleviate and manage catastrophic effects on human health and well-being. To further enhance and sustain the UHC, the need to consider the most promising and practical policy options well integrated, coordinated, community-based, and person-focused has been suggested (Sumriddetchkajorn et al., 2019).

3.2 *Emergency Response in Thailand*

Thailand's disaster management system has evolved over four decades. Since the Civil Threat Prevention Act was enacted in 1979, the nation's first comprehensive disaster management law, The Royal Thai Government (RTG) created the DDPM under the Ministry of Interior in 2002. The DDPM functions as the primary organization for disaster risk management (DRM) and response. The Disaster Prevention and Mitigation Act (DPMA) which serves as the primary legislative framework for DRM and disaster response was enacted in 2007 in response to the tsunami in the Indian Ocean in 2004 which wreaked havoc on the people living along Thailand's southern coast (CFE-DM, 2022). Recognizing Thailand's vulnerabilities to epidemics, drought, heat waves, earthquakes, tsunamis, floods, and landslides including as food and water insecurity as well as technology-related risks like chemical mishaps, the government keeps improving its disaster management strategy. The Emergency Medical Institute of Thailand (EMIT) was founded under the Emergency Medical Act B.E. 2551 (2008). The EMIT is tasked with system development, including the master plan for emergency medical services, as well as the formulation of policies and the provision of services (WHO, 2015a). However, calls have been made on the need for making the emergency medical service (EMS) system more resilient by improving its effectiveness and efficiency. The need for more medical personnel, an ambulance with better equipment, and organization-wide coordination was observed (Sittichanbuncha et al., 2014).

A previous study reported a significant increase in the number of EMS patients for both traumatic and non-traumatic patients and a lengthened period of EMS surgery for both groups during the COVID-19 pandemic period and thus recommended that emergency workers should be well-trained and well-resourced ambulances in order to reduce the time that ambulances are required to provide care during pandemics in the future (Huabbangyang et al., 2022). Phase 1 (2021–2030) of the HNAP also offers crucial direction for multi-sectoral partnerships that could build and expand health system resilience to climate change (Health Impact Assessment Division, 2021). Essentially, fortifying the resilience of individuals and their localities against the adverse health consequences of climate change should be desired to improve access and utilization of health care delivery (Langkulsen et al., 2022a, 2022b). Favorable governmental policies and programs committed to reducing pollution and pragmatic efforts in strengthening emergency health response systems to efficiently respond to climate change-related health hazards nationwide are essential and should be prioritized.

3.3 System and Agency Levels

Political stability and functional state institutions could be beneficial to a country's ability to combat climate change-related health risks now and in the future since the prospects of health system and agencies are more likely to be positive in a stable democratic regime. In respect of this, the following policies and programs included in this category are regarded as having potentials to strengthen health system and agencies to be more resilient in helping to combat and respond adequately to climate change health-related calamities in a stable political atmosphere. The comprehensive health systems reforms in the past might have encouraged partenrship and collaborations and perhaps directly boosted the efforts for resilience. With the help of foreign development partners, some significant health changes adopted in the 2000s were locally proposed, effectively implemented, and funded. In shaping each reform, various state and non-state actors exerted varying levels of influence, incorporating climate change-resilient responses into complex policy processes and contexts (WHO, 2015a). As aimed by the Climate Change Master Plan 2015–2050, in sectors where policies relate to public health, they may assist in the development of significant policy changes, focusing on the prevention of diseases and threats to health in response to climate change by disseminating information and knowledge to the target groups and developing the effectiveness of basic health services in managing health risks resulting from climate change (ONEP, 2015). By implementing integrated and inclusive socio-economic, structural, inequities, health, educational, environmental, technological, and institutional measures, this program can prevent new disaster risks and reduce existing ones. It may prevent and reduce hazards' exposure and vulnerability to disasters by increasing preparedness for response and recovery and therefore strengthening resilience. It seeks to safeguard against the possibility of unfavorable events, such as road hazards or closures that could delay or prevent emergency medical services from arriving in time to help. Such dangers may not pose a direct threat to the facility but may have a significant impediment on patients and their ability to access care (Kunadhamraks, 2015). Thailand by this policy, anticipates and prepares for secondary impacts that could accompany climate change-induced emergencies like supply chain difficulties, water shortages utility outages, and effects on temperature-sensitive medications and equipment. Plans in seeking to explore alternative utility sources like water, establishing solar or wind power and ensure higher power quality and dependability through a self-managed microgrid system and swiftly recover and fault localization during a natural disaster could contribute greatly in helping the health system in the performance of essential functions (EPPO, 2015; Hoonchareon, 2015; ONWR, 2015). These aim to forestall possible disruption in supply of utility vital for the effective functioning of health system.

3.4 Population Level

Included in this category are publications considered to assist the health system in helping individuals and groups to bounce back from setbacks arising from climate change health-related hazards. It has created health impact surveillance and preventive mechanisms and improved access to high-quality public health services by the generality of the population and an increased in the ability of the health system to manage health risks and mitigate health impacts from climate change. It seeks to lower morbidity, lessen the effects on health, and make controlling of the health hazards associated with climate change a priority (Health Impact Assessment Division, 2021). It would inform communities and individuals who live or work in environments where they are more likely to contract diseases or be exposed to specific hazards posed by harsh weather conditions and other extreme climate events are kept informed. Work together with response partners, organizations, stakeholders, and medical professionals to provide clear guidance on mitigation strategies and adaptation practices (ONEP, 2020). Evaluation and re-evaluation of how the health system provide services to vulnerable populations, considering the potential for these groups to face health risks that are disproportionate, many, or complicated because of climate change. Minority groups, low-income neighborhoods, immigrant groups, the elderly, children, and people with disabilities may be among the populations of concern. Implement comprehensive and inclusive economic, structural, legal, social, health, cultural, educational, environmental, and institutional changes to avoid and decrease risk, exposure, and susceptibility to disasters, improve readiness for response and recovery, and boost resilience (DDPM, 2015).

3.5 Implementation of Climate Change-Resilient Policies

In order to build climate-resilient and environmentally sustainable health system, policies and programs to improve capacity to safeguard and enhance the health of target communities in an unstable and changing climate, and enable them to optimize resource use and minimize the inefficiencies of health system, especially in times, catastrophic events are essential (Fox et al., 2019). The emphasis is on the organization and management of policy process and decision-making. These procedures can aid in diligent public problem-solving in a way that improves policy efficacy by employing deliberative processes that adopts implementation science and encourage efforts to sure that health systems are continuously and progressively enhanced, remain effective, capable of assisting in reducing inequities and vulnerabilities (Corvalan et al., 2020). The implementation of these interventions through policy process is receiving the necessary political support and commitment by implementers to realizing the desire policy outcomes. This is evidenced by the pragmatic practical approach as discussed below, in implementing the various formulated policy programs as summarized in Table 1.

I. Political Commitment

To achieve the universally agreed-upon objectives of decarbonizing economies and constructing resilience to warmer and harsher climates, policy actions by governments at various levels over the coming decades are essential (Fox et al., 2019). Political costs and gains are often well calculated by the political class in matters of public policy, and hence, political conditions necessary for a successful policy are crucial (Compton et al., 2019). The RTG has demonstrated political will and leadership at national and international stages of its commitment to the reduction of pollutants and wastes into the environment through policy instruments and implementation of same. The Royal Thai Government is not only a party to the Paris Agreement, but it also has through policy initiatives instituted various measures to contribute to the achievement of the objectives of the agreement. As stated by the Minister for Natural Resource and Environment at the COP27, 'Thailand has walk the talk' in actions toward climate resilience. The Minister who led the RTG Delegation at the COP conference highlighted some progress achieved under Article 6.2 of the Paris Agreement and further outlined plans for future actions by the country through collaboration and called for a more global concerted efforts for adaptation measures to mitigate the impacts of climate change as no one is 'protected' from climate change-induced disasters (The Nation Thailand, 2022). Therefore, planning and implementing solutions to the climate change crisis must consider health systems seriously (Fox et al., 2019). Although policies cannot succeed on their own in building climate change-resilient health systems; rather, their advancement to that end depends on the political climate in the implementation process, which is becoming increasingly clear and that the political environment is far more complicated than previously believed (Hudson et al., 2019). Hence, the level of political will demonstrated by the RTG at the global and national stages could be significant contribution to building a resilient health system and overall benefit climate change harm reduction activities.

II. Practice

Some of the policies aim to appropriately educate about present and probable future climatic conditions for emergency and catastrophe impacts. Enhance the health system's capability for risk management to handle remaining risks and uncertainties while reducing overall susceptibility and exposure to hazards. Giving communities the tools, they need to avoid and address the health concerns brought on by extreme weather occurrences that enhance the care delivered by the health system in specific contexts. The analysis shows strong public health commitments to attaining climate-resilient goals through sectoral integration with sectors that influence health and well-being (Puntub & Greiving, 2022). Actionable steps in communities and institutions have deliberately been established for increased access to health services at all levels across the populations. In Thailand, all Thais households have three health promotion officers, including (a) village health volunteer (VHV), responsible for 8–15 households. The VHV observes signs and symptoms of illness and refers the treatment-related information and educates the community about risk reduction factors, (b) public health officer or nurse at sub-district level. This health officer is put in charge of 1–3 villages of approximately 1200 people for their medical and health services and help to facilitate participation in community-based health care for preventive diseases. Lastly, (c) Family Physician who oversees about 10,000 people in 1–3 sub-districts providing consultation, referrals in physical, mental, social health, and conducts home visit. The family physician also supports the development and empowerment of VHV and public health officer or nurse. The three health promotion officers collaborate and work together closely to strengthen health security (Thai Health Promotion Foundation: ThaiHealth, 2022). The ThaiHealth supports the work of these three health promotion officers to enhance their knowledge and empower them to improve their performance at the sub-district, district, and provincial levels. Although it is connected to the larger environmental and social determinants of health, this approach is based on the fundamental duties of the health sector because of this, the policies should be applied in a flexible manner to account for various national contexts and iteratively. Taking advantage of fresh information, experience, and lessons from similar contexts as well as considering other climate change dynamics would be beneficial (WHO, 2015b). Periodic assessment, engagement, and participation at all levels would improve health behavior for individual, family and enhance health response and management thereby reducing inequalities in access to healthcare services. A sustainable approach to the implementation of these policy interventions in a supportive political environment could positively affect the resilience of health system in assisting to guarantee that services are accessible and of a high standard, as well as that institutions are less expensive and more readily available. A resilient climate change health system would also allow for a systematic and effective way of responding to and addressing the increasingly health risks induced by climate variability and change.

4 Discussion

Due to the impact of climate on several social and environmental aspects that affect health and well-being, including clean air, safe drinking water, food safety secure housing and a stable socio-economic environment, to build and maintain the resilience of health system, special attention must be paid to management at several levels, which could politically be determined through sound government policy measures as was determined in the Caribbean (Quintana et al., 2021). The government policies could strengthen systems and agencies and the population for building a resilient health system. They were identified as factors that could significantly be of importance to building a climate-resilient health system for a smooth care delivery at all levels in Thailand and beyond if well implemented. Global health security (GHS) index indicates that despite considerable responses provided by nations to the COVID-19 pandemic, all nations are still gravely ill-equipped to deal with potential epidemic and pandemic threats (Bell & Nuzzo, 2021). The findings as stated above could present significant policy options to a resilient health system as the creation of fresh plans, programs, and ideologies through policies and mitigation and adaptation strategies could enhance stakeholder capacity, collaboration, and participation as has been recommended for a robust health system (Health Care Without Harm, 2022; Puntub & Greiving, 2022). Improving public understanding of the connection between climate change and health and encouraging active community involvement in climate and health-related actions and decisions are essential for climate change effect ready healthcare delivery system (Drewry & Oura, 2022). As revealed through the study on how climate change adaptations and mitigation measures are crucial in developing a resilient health system, the need to understand a community's capacity to adapt to, respond to, and recover from climate-related risks and accurate measurement of that community's circumstances prior to a negative event are a necessity due to the possibility of a significant inequalities at specific geographic scales (Langkulsen et al., 2022a, 2022b). Similarly, successful risk reduction, such as disaster risk management and building more robust infrastructure, can be prioritized by conducting vulnerability and adaptation assessments and help in developing health system adaptation plans in containing climate change health hazards by adopting suitable modern technologies locally (Ebi et al., 2021). The policy measures as analyzed are worthy of emulation as they could contribute greatly to building health systems resilient to climate change if well supported politically through sustained advocacy and pragmatic actionable steps towards the provision of efficient people centered healthcare system.

This review could be limited by the dependent on secondary data and could have miss out on some unpublished important practical actionable oriented steps that are essentially important to climate-resilient and health system.

Though the RTG and its development partners need to be commended for their leadership roles by the institution of practical measures toward the global efforts on mitigating the impacts of climate change, the following recommendations are proposed for consideration by policymakers and implementers at all levels. Continuous assessment of climate-related health risks at national and community levels should be analyzed by government agencies in collaboration and in partnership with academia and civil society organizations (CSOs) with emphasis on specific and peculiar health vulnerabilities. Well-trained and motivated health workforce in sufficient numbers for healthcare delivery should be ensured at all times particulalry in under served regions. Premium should be placed on having adequately sized, well-trained teams that can respond to demands anywhere in the country and beyond. A variety of personnel including public health, mental health, operations, clinicians, program administration, development, and communications who promote gender diversity, equality, and inclusion are always ready for deployment, ensuring that tools and resources are adequate. Consumable and non-consumable ranges from intravenous fluids and oxygen supplies to appropriate and sufficient immunizations and essential drugs, such as antibiotics, are readily available. Supply chain groups chosen by employees to simplify purchase orders, standardize requests from healthcare institutions to warehouses, and adhere to various protocols required to manage and deliver healthcare across the nation.

Crucially, create sustainable health system through leadership, governance, knowledge, and funding. To deliver high-quality health care consistently and effectively to the population. Several systems must be in place, including supply chain management for safety, operational record keeping, equipment, medical informatics (electronic health record) expertise, and financial and accounting systems to track revenue and expenses as well as leadership and management structures for prudent decision-making. Ensure safe and conducive environment for attending to patients. To treat patients, there is the need to create, expand, and equip facilities with electricity, clean water, oxygen, isolation, and good sanitation facility to meet the needs of healthcare providers and provide a therapeutic environment for patients. Sometimes, such premises already exist, but in most cases, the facilities need to be reconstructed or built from scratch during an emergency or crisis. The establishment of youth-friendly reproductive health centers and the opening of a specialized unit for patients with existing health challenges such as the disabled and non-communicable diseases (NCDs) like cancers are also recommended.

Finally, social support is recommended to assist vulnerable communities, households, and individuals. Social support can be a critical component of patient treatment and could have a significant impact on how well patients recover from sickness and socio-economic difficulties. This assistance may take different forms, such as food, shelter, transportation, and financial support, to guarantee optimal and holistic healthcare delivery.

5 Conclusion

The effects of climate change are a worrisome situation for nations, affecting economic development, livelihood, and sustainable development. Climate change might present challenges that could overwhelm health system as was witnessed during the COVID-19 pandemic. More resilient health system can be developed by utilizing technology to advance healthcare infrastructure, expand and improve medical services, and develop an inviting business environment for investment and innovation in medicine. The government climate mitigation and adaptation strategy policy measures as analyzed are worthy of emulation. Continuous political will with actionable steps and advocacy through partnership and collaboration could ensure a resilient health system capable of responding effectively and efficiently to deliver people-centered care, particularly during a crisis.

References

ASPR TRACIE. (2022). *Climate change resilience and healthcare system considerations*. Retrieved from https://asprtracie.hhs.gov/technical-resources/resource/10650/climate-change-resilience-and-healthcare-system-considerations

Bell, J. A., & Nuzzo, J. B. (2021). *Global health security index: Advancing collective action and accountability amid global crisis*. Retrieved from https://www.ghsindex.org/wp-content/uploads/2021/12/2021_GHSindexFullReport_Final.pdf

CFE-DM (Center for Excellence in Disaster Management and Humanitarian Assistance). (2022). *Disaster management reference handbook—Thailand (January 2022)*. Retrieved from https://reliefweb.int/report/thailand/disaster-management-reference-handbook-thailand-january-2022

Compton, M. E., Luetjens, J., & Hart, P. (2019). Designing for policy success. *International Review of Public Policy, 1*(2), 119–146.

Corvalan, C., Villalobos Prats, E., Sena, A., Campbell-Lendrum, D., Karliner, J., Risso, A., Wilburn, S., Slotterback, S., Rathi, M., Stringer, R., & Berry, P. (2020). Towards climate resilient and environmentally sustainable health care facilities. *International Journal of Environmental Research and Public Health, 17*(23), 8849. https://doi.org/10.3390/ijerph17238849

Department of Disaster Prevention and Mitigation (DDPM), Ministry of Interior. (2015). *Thailand: National disaster risk management plan*. Retrieved from https://www.preventionweb.net/publication/thailand-national-disaster-risk-management-plan

Drewry, J., & Oura, C. A. L. (2022). Strengthening climate resilient health systems in the Caribbean. *The Journal of Climate Change and Health, 6*, 100135. https://doi.org/10.1016/j.joclim.2022.100135

Ebi, K. L., Vanos, J., Baldwin, J. W., Bell, J. E., Hondula, D. M., Errett, N. A., Hayes, K., Reid, C. E., Saha, S., Spector, J., & Berry, P. (2021). Extreme weather and climate change: Population health and health system implications. *Annual Review of Public Health, 42*, 293–315.

Energy Policy and Planning Office (EPPO), Ministry of Energy. (2015). Thailand Power Development Plan 2015–2036 (PDP2015). Retrieved from https://www.eppo.go.th/images/POLICY/ENG/PDP2015_Eng.pdf

Fox, M., Zuidema, C., Bauman, B., Burke, T., & Sheehan, M. (2019). Integrating public health into climate change policy and planning: State of practice update. *International Journal of Environmental Research and Public Health, 16*(18), 3232. https://doi.org/10.3390/ijerph16183232

Gale, E. L., & Saunders, M. A. (2013). The 2011 Thailand flood: Climate causes and return periods. *Weather, 68*, 233–237.

Health Care Without Harm. (2022). *Climate resilience for health care and communities: Strategies and case studies (January 2022)*. Retrieved from https://noharm-uscanada.org/climate-resilience-for-health-care-and-communities

Health Impact Assessment Division, Department of Health, Ministry of Public Health. (2021). *Health national adaptation plan (HNAP) phase 1 (2021–2030)*. Retrieved from https://hia.anamai.moph.go.th/th/hnap/205088

Hoonchareon, N. (2015). Thailand smart grid policy plan and roadmaps. Retrieved from https://www.appp.or.th/imgadmins/document/11335609.pdf

Huabbangyang, T., Trakulsrichai, S., Yuksen, C., & Sricharoen, P. (2022). The impact of the coronavirus disease 2019 (Covid-19) pandemic on the use of emergency medical services system in Bangkok, Thailand. *Open Access Emergency Medicine, 14*, 429–440.

Hudson, B., Hunter, D., & Peckham, S. (2019). Policy failure and the policy-implementation gap: Can policy support programs help? *Policy Design and Practice, 2*(1), 1–14. https://doi.org/10.1080/25741292.2018.1540378

International Organization for Migration. (2021). Migration and migrants: A global overview. In M. McAuliffe & A. Triandafyllidou (Eds.), *World migration report 2022*. IOM. Retrieved from https://publications.iom.int/books/world-migration-report-2022-chapter-2

Kaczan, D. J., & Orgill-Meyer, J. (2020). The impact of climate change on migration: A synthesis of recent empirical insights. *Climatic Change, 158*, 281–300.

Kiguchi, M., Takata, K., Hanasaki, N., Archevarahuprok, B., Champathong, A., Ikoma, E., Jaikaeo, C., Kaewrueng, S., Kanae, S., Kazama, S., & Kuraji, K. (2021). A Review of climate-change impact and adaptation studies for the water sector in Thailand. *Environmental Research Letters, 16*(2), 023004. https://doi.org/10.1088/1748-9326/abce80

Kunadhamraks, P. (2015). *Thailand's transport infrastructure development strategy 2015–2022*. Retrieved from http://www.thaitransport.org/news/2558.07.01%20proceeding/presentation%20TSTS%2015%20.......02%2004%202015/panel%2001.pdf

Langkulsen, U., Cheewinsiriwat, P., Rwodzi, D. T., Lambonmung, A., Poompongthai, W., Chamchan, C., Boonmanunt, S., Nakhapakorn, K., & Moses, C. (2022a). Coastal erosion and flood coping mechanisms in Southern Thailand: A qualitative study. *International Journal of Environmental Research and Public Health, 19*(19), 12326. https://doi.org/10.3390/ijerph191912326

Langkulsen, U., Rwodzi, D. T., Cheewinsiriwat, P., Nakhapakorn, K., & Moses, C. (2022b). Socio-economic resilience to floods in coastal areas of Thailand. *International Journal of Environmental Research and Public Health, 19*(12), 7316. https://doi.org/10.3390/ijerph19127316

Limsakul, A. (2020). Trends in Thailand's extreme temperature indices during 1955–2018 and their relationship with global mean temperature change. *Applied Environmental Research, 42*(2), 94–107.

Office of Natural Resources and Environmental Policy and Planning (ONEP), Ministry of Natural Resources and Environment. (2015). *Climate change master plan 2015–2050*. Retrieved from https://climate.onep.go.th/wp-content/uploads/2019/07/CCMP_english.pdf

Office of Natural Resources and Environmental Policy and Planning (ONEP), Ministry of Natural Resources and Environment. (2020). *Thailand's third national communication*. Retrieved from https://unfccc.int/documents/181765

Office of Natural Resources and Environmental Policy and Planning (ONEP), Ministry of Natural Resources and Environment. (2022). *Thailand's 2nd updated nationally determined contribution*. Retrieved from https://unfccc.int/documents/620602

Office of the National Economic and Social Development Board (NESDB), Office of the Prime Minister. (2017). *The twelfth national economic and social development plan (2017–2021)*. Retrieved from https://www.nesdc.go.th/nesdb_en/ewt_w3c/main.php?filename=develop_issue

Office of the National Water Resources (ONWR), Office of the Prime Minister. (2015). *The national water resources management strategies 2015–2026*. Retrieved from http://www.onwr.go.th/en/?page_id=4207

Oppenheimer, M., Glavovic, B., Hinkel, J., Van de Wal, R., Magnan, A.K., Abd-Elgawad, A., Cai, R., Cifuentes-Jara, M., Deconto, R.M., Ghosh, T., & Hay, J. (2019). Sea level rise and implications for low-lying islands, coasts and communities. In H. -O. Pörtner, D. C. Roberts, V. Masson-Delmotte, P. Zhai, M. Tignor, E. Poloczanska, K. Mintenbeck, A. Alegría, M. Nicolai, A. Okem, J. Petzold, B. Rama & N. M. Weyer (Eds.) *IPCC special report on the ocean and cryosphere in a changing climate* (pp. 321–445). Cambridge University Press, Cambridge, UK and New York, NY, USA. https://doi.org/10.1017/9781009157964.006

Piankusol, C. (2021). *Health care system and health policy*. Retrieved from https://w1.med.cmu.ac.th/commed/wp-content/uploads/2020/12/L4.-Health-care-system-and-health-policy-ICH1-2020_compressed.pdf?x92916

Puntub, W., & Greiving, S. (2022). Advanced operationalization framework for climate-resilient urban public health care services: Composite indicators-based scenario assessment of Khon

Kaen City, Thailand. *International Journal of Environmental Research and Public Health, 19*(3), 1283. https://doi.org/10.3390/ijerph19031283

Quintana, A. V., Venkatraman, R., Coleman, S. B., Martins, D., & Mayhew, S. H. (2021). COP26: An opportunity to shape climate-resilient health systems and research. *The Lancet Planetary Health, 5*(12), e852–e853. https://doi.org/10.1016/S2542-5196(21)00289-8

Sittichanbuncha, Y., Prachanukool, T., Sarathep, P., & Sawanyawisuth, K. (2014). An emergency medical service system in Thailand: Providers' perspectives. *Journal of the Medical Association of Thailand, 97*(10), 1016–1021.

Sumriddetchkajorn, K., Shimazaki, K., Ono, T., Kusaba, T., Sato, K., & Kobayashi, N. (2019). Universal health coverage and primary care, Thailand. *Bulletin of the World Health Organization, 97*(6), 415–422.

Thai Health Promotion Foundation (ThaiHealth). (2022). *Three health promotion officers*. Retrieved from https://www.thaihealth.or.th/รู้จัก-3-หมอ-นักสร้างเสริ/ (in Thai)

Thanathanphon, W., Luangdilok, N., Chettanawanit, K., & Sisomphon, P. (2020). Storm surge forecasting and inundation estimation during tropical Strom Pabuk 2019 in the Southern Coast of Thailand. In *Proceedings of the 22nd IAHR-APD Congress 2020*, Sapporo, Japan. Retrieved from https://iahr-apd2020.eng.hokudai.ac.jp/htdocs/static/mirror/proceedings/pdf/6-9-1.pdf

The Nation Thailand. (2022). *Tackling climate change a United Battle*. Retrieved from https://www.nationthailand.com/thailand/40022208

Toimil, A., Losada, I. J., Nicholls, R. J., Dalrymple, R. A., & Stive, M. J. F. (2020). Addressing the challenges of climate change risks and adaptation in Coastal Areas: A review. *Coastal Engineering, 156*, 103611. https://doi.org/10.1016/j.coastaleng.2019.103611

Witthayapipopsakul, W., Cetthakrikul, N., Suphanchaimat, R., Noree, T., & Sawaengdee, K. (2019). Equity of health workforce distribution in Thailand: An implication of concentration index. *Risk Management and Healthcare Policy, 12*, 13–22.

World Bank Group and Asian Development Bank. (2021). *Climate risk country profile: Thailand*. Retrieved from https://www.adb.org/publications/climate-risk-country-profile-thailand

World Health Organization, Regional Office for the Western Pacific. (2015a). *The Kingdom of Thailand health system review*. Health Systems in Transition, *5*(5). Retrieved from https://apps.who.int/iris/handle/10665/208216

World Health Organization. (2015b). *Operational framework for building climate resilient health systems*. Retrieved from https://www.who.int/publications/i/item/9789241565073

World Health Organization. (2021). *Climate change and health*. Retrieved from https://www.who.int/news-room/fact-sheets/detail/climate-change-and-health

World Health Organization. (2022). *Measuring the climate resilience of health systems*. Retrieved from https://apps.who.int/iris/handle/10665/354542

Climate Change Adaptation and Public Health Strategies in Malaysia

Nasrin Aghamohammadi, Logaraj Ramakreshnan, Rama Krishna Supramanian, and Yin Cheng Lim

Abstract Human activities have placed immense pressure on the climate stability of the tropical anthroposcapes that put the sustainability of environmental resilience and public health at stake. Malaysia is one of the vulnerable climate change hotspots that ostensibly witness many extreme weather events that amplify the burden of climate-sensitive diseases. Elevating rates of urbanization and population explosion in near future could magnify the implications of climate change in Malaysia, including the exacerbation of warming trends, amplification of environmental disasters and resurgence of infectious diseases. In response to this, environmental and public health institutions at all operational scales need to consciously modify their approaches to articulate evidence-based adaptation and mitigation strategies to build climate resilience across all the sectors in Malaysia. Mainstreaming behaviour change into the adaptation strategies related to the extreme climate events along with infrastructure, technological and policy advancements is another key aspect to strengthening the public health adaptive capacity and responses in Malaysia. The identification and management of constraints and barriers to climate change adaptation in anticipation of existing public health strategies are equally inevitable to address the adverse health impacts and increase the efficiency and sustainability of climate solutions in the country.

Keywords Climate adaptation · Climate change · Climate mitigation · Public health · Tropical country

N. Aghamohammadi (✉) · R. K. Supramanian · Y. C. Lim
Department of Social and Preventive Medicine, Faculty of Medicine, University of Malaya, 50603 Kuala Lumpur, Malaysia
e-mail: nasrin@ummc.edu.my; nasrin@curtin.edu.au

L. Ramakreshnan
Institute for Advanced Studies, University of Malaya, 50603 Kuala Lumpur, Malaysia

N. Aghamohammadi
School of Design and the Built Environment, Curtin University, Perth 6102, Australia

R. Akhtar (ed.), *Climate Change and Human Health Scenarios*, Global Perspectives on Health Geography, https://doi.org/10.1007/978-3-031-38878-1_7

1 Introduction

Unprecedented changes in urbanization, population and resource demands across Malaysia affected its climate system and ecosystem resilience at the regional level. The expansion of urban areas has drastically increased the anthropogenic activities that contribute to the hike of urban temperature and greenhouse gas (GHG) emissions, which in turn accelerated climate change. As a consequence, the country is adversely affected by extreme weather events that are likely to cause deleterious effects on various sectors, the economy and labour productivity by severely impacting the public health sector of Malaysia. Being one of the significant contributors of GHG emissions in the world, especially from the energy sector (Fig. 1), the temperature of Malaysia is projected to rise by 0.3–4.5 °C, which will cause the sea level to rise by about 95 cm over hundred years (Alam et al., 2012). However, avoiding the worst climate impacts on the country will require reversing the upwards trend in all sectors and rapidly decreasing emissions to net zero by 2050 as highlighted in the 27th Conference of Parties (COP27).

The country experienced about 51 major disasters between 1998 and 2018, affecting more than three million people with 281 mortality and total damages estimated at Ringgit Malaysia (RM) 8 billion (approximately USD 2 billion) (Bhuiyan et al., 2018). In fact, the December 2021 flood event in Malaysia is one of the worst disasters that the country has experienced with losses estimated at RM 6.3 billion (USD 1.4 billion) within the space of one week (Rahman, 2022). Such disasters lead to the emergence and rise of climate-sensitive diseases due to environmental degradation and contamination as well as the creation of breeding grounds for disease-carrying vectors. The impact of climate change on public health is an area of substantial concern in Malaysia. Malaysia's aspiration to reduce its carbon dioxide emissions as announced during COP27 is not only crucial to tackle the climate change that the country is experiencing, but also to reduce its burden on the public health sector. With regard to this, the present chapter will discuss the interplay of climate

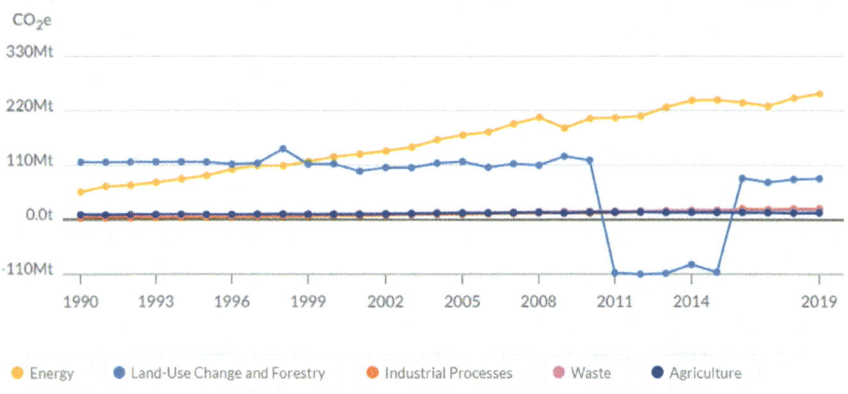

Fig. 1 Historical GHG emissions according to sectors in Malaysia (Friedrich et al., 2020)

change scenarios and public health adaptation strategies in Malaysia, followed by the constraints and barriers of those adaptation strategies to address the health issues driven by climate change in Malaysia.

2 Climate Change Scenario in Malaysia

The potential for climate change in Malaysia is driven by a combination of natural and anthropogenic factors. For instance, unsystematic urbanization, urban sprawl into floodplains and poor drainage when combined with natural factors such as heavy monsoon rains and El Niño-Southern Oscillation (ENSO) cause flash floods, which have become a common scene in the lives of many Malaysians every year (Zulkarnain et al., 2020). The recent World Bank's report asserted that both Peninsular and East Malaysia experienced surface mean temperature increase of 0.14–0.25 °C per decade between 1970 and 2013 (World Bank Group, 2021). By the 2090s, the average temperatures are projected to accelerate by 3.11 °C under the highest emissions pathway of Representative Concentration Pathway (RCP) 8.5, which is just 0.6 °C lesser than the global average. On an important note, the warming climate will further increase the frequency and intensity of heat waves that bring cascading implications to environmental and public health. Based on the temperature data acquired over nine meteorological stations in Peninsular Malaysia between 2001 and 2010, the longest heatwave with an amplitude of 29.4–33.0 °C has occurred for a period of 24 days in Ipoh, Perak (Suparta & Yatim, 2019). Moreover, the highest Excess Heat Factor (EHF) Index of 9.1 $°C^2$ is recorded in the well-urbanized areas of Kuala Lumpur compared to the less urbanized areas in 2002 (Suparta & Yatim, 2019). The accumulation of anomalous heat in the city centres exacerbates the Urban Heat Island (UHI) phenomena that compound the temperature rise and heat stress in the urban areas of Malaysia. UHI is a result of urban surface modifications that have converted the natural green areas into non-evaporating and non-transpiring surfaces by metallic, asphaltic and concrete materials with different thermal conductivities (Nwakaire et al., 2020). High intensities of UHIs have been recorded at night-time in many parts of Malaysia such as in Kuala Lumpur (6.0 °C) (Harun et al., 2020), Putrajaya (1.9–3.1 °C) (Morris et al., 2015), Petaling Jaya (1.71 °C) (Ramakreshnan et al., 2019) and Cameron Highlands (1.4 °C) (Ibrahim et al., 2018). The findings of Rahaman et al. (2022) revealed forest decimation and urbanization in Penang city as the main culprits for the acceleration of land surface temperature by 8 °C and a 38% increase in UHI hotspots between 1996 and 2021. Transboundary emissions of smoke-haze from peatland fires and slash-and-burn of biomass are recurrent disasters that record hazardous air quality in Malaysia every year. In addition to high Particulate Matter ($PM_{2.5}$ and PM_{10}) concentrations, the flux and concentrations of black carbon, carbon monoxide (CO) and ozone (O_3) are strongly aligned with haze, which has lethal consequences on human beings (Khan et al., 2020; Shith et al., 2021; Sundram et al., 2022).

Copious rainfall due to global warming has magnified the frequency and extremity of flood events in Malaysia. Recently, short-duration flash floods have grown more unpredictable with rainfall events in Malaysia that have a detrimental cumulative impact on the environment, public mobility, property damages, massive clean-up costs as well as the livelihood and socioeconomic activities of the affected community (Along et al., 2022). Between 2010 and 2016, about 204 flash floods have occurred in Kuala Lumpur with total damage as high as RM48.7 million, equal to 0.04% of the gross domestic product (GDP) of Kuala Lumpur in 2016 (Bhuiyan et al., 2022). Contrarily, deficits in the precipitation (meteorological) and surface and sub-surface water flow (hydrological) affect the country in terms of extended drought conditions, often during El Niño events (Tan et al., 2022). Currently, Malaysia reports a standardized precipitation evaporation index (SPEI) of less than − 2, indicating a high probability of severe meteorological drought conditions (World Bank Group, 2021). The findings outlined that severe droughts can lead to a reduction in agricultural and aquaculture productivity, freshwater supply and amplification of vector-borne diseases. For instance, a severe drought in 2014 affected 8,000 paddy farmers in Kelantan, causing crop losses worth USD 22 million (Tan et al., 2017). Temperature increases also bring negative implications to the coastal areas by rising sea water levels that enhance the damage caused by cyclone-induced storm surges when combined with the increase in precipitation intensity and wind speed (Walsh et al., 2016). Based on the retrospective data over 20 years (1993–2015), east and west Malaysia are experiencing a sea-level rise of 3.3 and 5.0 mm per year (Hamid et al., 2018). Such sea-level rise is predicted to impose significant impacts on Malaysia's coastal zones, mangrove sites, landmass, plantation productivity and socioeconomic activities. The saltwater inundation and tidal surges at the coastal plains and river deltas can also damage the productive food basins in this region and cause lower crop yields (Binns et al., 2021). Since temperature is one of the critical determinants affecting life cycles and transmission of most infectious agents, climate change, global warming and UHI might be able to cause a spike in the incidence of vector-borne diseases such as cholera, malaria, dengue fever, hand, foot and mouth disease (HFMD) as well as the recent coronavirus disease (COVID-19) (Wahid et al., 2021). Besides humans, climate change can also lead to concomitant increases in basal stem rot disease by *Ganoderma boninense* among oil palms, one of the leading plantation commodities that contribute to the GDP of Malaysia (Paterson, 2019).

3 Climate Change and Public Health in Malaysia

Accelerated climate change and environmental degradation remain incontrovertible threats to public health in Malaysia. Besides the immediacy and gravity of this threat, the rapid impact will be on the vulnerable (i.e. children, older adults and underprivileged people) and marginalized (i.e. indigenous people) populations (Mahmood & Guinto, 2022). According to the United Nations Children's Fund's

(UNICEF) Climate Change Risk Index, Malaysia was reported as one of the least-performing countries where children are at a high exposure to climate and environmental shocks (UNICEF, 2021). The findings from the National Youth Climate Change Survey emphasized that 91.3% of Malaysian youth were found to have experienced climate and environment-related effects (UNDP, UNICEF & EcoKnights, 2020). Importantly, the indigenous youth were more likely to have experienced these effects, registering higher risks to this group compared with the others.

Extreme heat manifestation in the form of heatwaves and UHI imposes severe threats to the health (heat-related mortality and morbidity) and thermal comfort of urban communities in Malaysia (Aghamohammadi et al., 2022). The heat stress level of Kuala Lumpur is moderate, in the range of 30.1–38.6 °C physiological equivalent temperature (PET), and positively associated with psychosomatic pain, psychological anxiety and somatization-related symptoms with females at higher risk (Aghamohammadi et al., 2021). The findings of the survey conducted among the urbanites of Mont Kiara, Jalan Raja Chulan and Setia Alam revealed heat exhaustion (89.4%) and respiratory problems (87.3%) as commonly reported physical health impacts of urban heat, followed by some psychological impacts such as anxiety (79.0%), depression (74.4%) and aggressive behaviour (74.0%) (Wong et al., 2018). The recurrent haze, when coupled with urban heat, exacerbates bronchial asthma and chronic obstructive pulmonary disease for both inpatient and outpatient cases compared with non-haze episodes (Jaafar et al., 2021). Indeed, daily inpatient data for 14 haze-related illnesses from 2005, 2006, 2008 and 2009 indicated an increase of 31% in inpatient cases compared with normal days, representing an average annual economic loss of RM 273,000 (USD 91,000) (Othman et al., 2014). A cumulative direct medical cost difference of haze-related respiratory illnesses from 2012 until 2015 between haze and non-haze episodes in Selangor was approximately RM 13.1 (USD 4.1) million for inpatient cases, RM 1.4 (USD 440) million for outpatient cases and RM 14.6 (USD 4.5) millions for total cases (Jaafar et al., 2018). The assessment of exposure to concentrated heavy metals during major floods in the flood-prone Pahang River basin disclosed possible carcinogenic risks associated with Lead (Pb) and Cadmium (Cd) (Alam et al., 2021). In a survey conducted among 602 flood victims from three districts (i.e. Pekan, Kuantan and Temerloh) of Pahang, Puteh et al. (2018) identified that the incidence of cough, flu, post-traumatic stress disorder (PTSD) and depression was higher after the flood, which also corresponds to a rise in the purchase of prescription medications from RM 24.40 to RM 31.02.

The coastal flooding and storms due to sea-level rise could elevate the risk of indoor mould growth from excess dampness, with severe impacts on respiratory disease among Malaysian coastal communities. Besides, such phenomena can also increase risks of drowning, injury and displacement. In fact, coastal disasters such as tsunamis can predispose the affected coastal communities to mental health problems (Krishnaswamy et al., 2012). Since the lifecycle of most of the pathogens is influenced by weather conditions, the risk of acquiring malarial infection caused by *Plasmodium knowlesi* in the coastal areas of Sabah during extreme rainy days is reported to be high among the coastal communities (Azzeri et al., 2020). On another important note, diseases previously heading for elimination such as leprosy, tuberculosis, malaria, leptospirosis and Zika may be reinvigorated by climate change and the accompanying environmental, economic and social disruptions (Binns et al., 2021).

4 Public Health Adaptation Strategies to Climate Change in Malaysia

Over the years, the Malaysian government has implemented numerous policies and roadmaps to decrease the public health implications of climate change associated with heat waves, storms and floods, droughts, the disruption of food systems, increases in zoonoses and vector-borne diseases and mental health issues (Alhoot et al., 2016). This is because many potentially catastrophic risks to human health are related to changes in temperature and precipitation, which have had significant socioeconomic impacts. Since climate change is already contributing to the Malaysian illness burden, there is an immense need to address this rising issue, which is likely to increase in the near future. The following section describes the main public health adaptation strategies including policies and action plans that promote their implementation in an integrated and balanced manner.

4.1 National Policy on Climate Change 2009

In 2009, the Malaysian government recognizes the impacts of climate change and framed the National Policy on Climate Change to promote the efficient use of resources and environmental conservation (MNRE, 2009). The policy serves as the main blueprint to guide various stakeholders to address the challenges of climate change based on five principles, which underpin ten strategic thrusts to set the national direction in responding to those challenges:

(i) Development on a sustainable path: to integrate climate change responses into national development plans to fulfil the country's aspiration for sustainable development;
(ii) Conservation of environment and natural resources: to strengthen implementation of climate change actions that contribute to environmental conservation and sustainable use of natural resources;
(iii) Coordinated implementation: to incorporate climate change considerations into the implementation of development programmes at all levels;
(iv) Effective participation: to improve participation of stakeholders and major groups for effective implementation of climate change responses;
(v) Common but differentiated responsibilities and respective capabilities: international involvement on climate change will be based on the principle of common but differentiated responsibilities and respective capabilities.

In line with the aim of the National Policy on Climate Change to establish and implement a national Research and Development agenda on climate change considering public health services and delivery, the Ministry of Health in 2019 established 11 Thematic Working Groups (TWG) to identify the main threats to the environment and public health as well as to formulate strategies to overcome the issues (MOH, 2020). The 11 TWGs are working on various issues such as:

- Air quality;
- Water and sewage system;
- Solid waste;
- Hazardous waste management;
- Climate change;
- Contingency readiness and environmental emergency plans;
- Health impact assessment;
- Information technology;
- Urban drainage;
- Environmental health experts;
- Vector-borne diseases.

4.2 Cross-Sectoral Policy—National Green Technology Policy and Master Plan

Along with the National Climate Change Policy 2009, the National Green Technology Policy was introduced in 2009 to help lead the growth of the nation's green technology industry. The programme also intends to promote the use and spread of green technologies in the nation to aid in the reduction of carbon emissions and promote good-quality living and a healthy environment. The five key goals of the policy are as follows:

- Decreasing growth of energy consumption while enhancing economic development;
- Facilitating growth of the green technology industry and enhancing its contribution to the national economy;
- Increasing national capabilities and capacity for innovation in green technology development and enhancing Malaysia's green technology competitiveness in the global arena;
- Ensuring sustainable development and conserving the environment for future generations;
- Enhancing public education and awareness of green technology and encouraging its widespread use.

The Green Technology Master Plan 2017–2030, on the other hand, has earmarked green growth as one of six priorities altering the trajectory of the nation's growth

(MEGTW, 2017). The GTMP creates a framework which facilitates the mainstreaming of green technology into the planned developments of Malaysia while encompassing the four pillars set in the National Green Technology Policy (NGTP) such as energy, environment, economy and society. The development of green cities is emphasized under the Master Plan to lower carbon emissions and improve urbanites' quality of life. In line with the Low-Carbon Cities programme spearheaded by the Ministry of Energy, Green Technology and Water (MEGTW) and Malaysian Green Technology and Climate Change Corporation (MGTC), the establishment of cycling or bikeways infrastructure in the cities is emphasized to promote a healthy and green lifestyle leading to a healthy and cleaner environment.

4.3 National Environmental Health Action Plans (NEHAPs)

Many countries have agreed to develop and implement National Environmental Health Action Plans (NEHAPs). The National Environmental Health Action Plan (NEHAP) outlines ideas for improving environmental health in the country and identifies the roles and responsibilities of various stakeholders. The Ministry of Health (MOH) issued NEHAP in 2013, to improve the environmental and public health quality while achieving the sustainable development goals (MOH, 2013). It was implemented in response to rising rates of food, water and vector-borne diseases throughout the years. Climate change mitigation measures can help to prevent major communicable and non-communicable diseases caused by key economic sectors, lowering healthcare expenditures through fewer deaths and diseases. Harnessing climate activities for health benefits has the potential to revolutionize the climate discussion by increasing public and policymaker willingness to act, which promotes coordinated stakeholder involvement and partnerships.

4.4 Existing Gaps in the Public Health Adaptation Strategies to Address Climate Change in Malaysia

Numerous adaptation strategies have been proposed to deal with climate change. However, the major shortcoming is the lack of connection between the strategies that have already been implemented or in the process of being implemented as well as the lack of coordination between agencies and ministries. According to the NEHAP 2016–2020, the majority of the suggested action plans were not implemented, and there was no clear implementation strategy (MOH, 2020). Under institutional components and critical support functions for environmental health, the strategy was largely focused on generating recommendations and research papers, but no mitigation steps were outlined to address the pollution sources.

On mitigation assessment, there is a lack of adequate funding, a consistent regulatory environment and a strong institutional structure for mitigation efforts. Actions have been taken to improve the institutional framework for mitigation implementation and to rationalize the regulatory framework's uniformity. However, funding remains a barrier to implementation. With international assistance, national mitigation action implementation would be accelerated and improved.

With the heightening concern towards climate change, there is a concern that any strategies taken by the government to cope with climate change impacts could be viewed as adaptation without really considering the projected climate trends. One such example is the MOH-initiated Vector-Borne Disease Control Programme in 1986 which replaced the Malaria Eradication Programme in 1967. The programme does not focus only on the control of Malaria infections but also on other vector-borne diseases such as dengue, filariasis, typhus and yellow fever. Despite the success rate in reducing vector-borne diseases through such programmes, it is also essential to include their forecasts in near future with respect to climate projections for early preparedness, adaptation and mitigation work.

5 Constraints and Barriers of Public Health Adaptation Strategies to Address Climate Change in Malaysia

Public health adaptation can be defined as key strategies taken to reduce adverse health impacts or enhance resilience in response to observed or expected changes in the climate system and associated extremes (Pörtner et al., 2022). These strategies can be in the form of short- or long-term actions that can be implemented in a proactive manner. However, it can also be carried out in a reactive manner due to unforeseen circumstances, particularly in response to immediate threats to public health (Weber & Stern, 2011). Public health adaptation is increasingly being recognized as an important determinant and inevitable part of the success of strategies addressing climate change. Adaptation of public health strategies is vital to address anticipated current and future climate change threats in Malaysia. In order to encourage communities to adapt and implement these strategies effectively, it is pivotal to garner a proper understanding of the challenges and barriers associated with the implementation of those adaptation strategies (Huang et al., 2011).

Despite having well-constructed and sound policies on paper to tackle climate change, implementation of strategies laid out may take a different course due to uncontrollable factors resulting in the inability to achieve the targeted goals. For example, in the year 2009, Australia faced a series of bushfires such as the Melbourne Bushfires which had claimed 172 lives. The bushfire intensified and spread uncontrollably due to weather factors, including rising ambient temperatures and decreased rainfalls (Yu et al., 2020). Despite having a comprehensive bushfire prevention policy, extreme weathers had made it hard to prevent or control bushfire in the country.

Climate change reports by the Malaysian Meteorological Department (MetMalaysia) showed an increase in the surface mean temperature of 0.13–0.24 °C per decade. Meantime, the surface minimum and maximum temperature increases are around 0.19–0.30 °C and 0.17–0.23 °C per decade, respectively. From the year 1990 onwards, increasing trends in rainfall have been observed in Peninsular Malaysia, Sabah and Sarawak. In light of changing weather patterns, Malaysia adopted the National Policy on Climate Change in 2009 to ensure climate-resilient development. This policy serves as a framework to mobilize and guide government agencies, industry, community as well as other stakeholders and major groups in addressing the challenges of climate change in a holistic manner.

Flooding has been an annual issue encountered in certain flood-prone basins around the country, mainly around the East Coast area and some areas in the state of Selangor despite the alert systems already in place. One of the main issues faced in addressing the flood issues in Malaysia is the lack of projected flood maps covering all the flood-prone basins that will assist with risk assessments as well as the development of an accurate and timely early warning system (MESTECC, 2018). The key barrier is mainly due to the technological limitations. Evidence has shown that access to and the use of technology is regarded as an important determinant of adaptive capacity (Huang et al., 2011). For instance, the use of geographic information system (GIS), a system that creates, manages, analyzes and maps all types of data, can serve as a useful assessment tool to assist health professionals with the reallocation of resources and risk mitigation (Khashoggi & Murad, 2020). However, the development and implementation of new technologies require expertise, knowledge, a supportive environment and funding. There is a substantial disparity in the adoption of technologies within Malaysia. There are still limitations in terms of accessibility to internet connectivity in certain remote and rural areas within the country. In the agricultural sector, there is a need for technological adaptations for soil conservation and to increase the irrigation efficiency. In addition, there is also a need for advancement in the area of renewable energies and the implementation of energy-efficient technologies at the residential and commercial premises. The lack of integration between the monitoring systems will lead to a lack of understanding to evaluate the combined impacts of sea-level rise, storm surges, abnormally high tides and rainfall which could lead to severe flooding.

Another key barrier to public health adaptation strategies is the lack of awareness and understanding of the full chain of implications of climate change impacts by the key stakeholders in all relevant sectors. Public health adaptation strategies to address climate change in the Malaysian context are required in various sectors including economy, energy, transportation, environment and solid waste management. This is also similar for other countries globally. However, different sectors are often found to be working in silos resulting in inefficient use of resources and ultimately inability to achieve desired outcomes. In addition to the lack of inter-sectoral collaboration, there are various non-governmental organizations (NGOs) such as Global Environment Centre (GEC), Malaysian Climate Change Group (MCCG), Environmental Protection Society Malaysia (EPSM) and many others that are working towards the common goal. However, there is a communication gap between these NGOs and the public sectors that could bring all parties under the same umbrella to create an ecosystem that allows for better sharing of knowledge, information, strategies and even technological inputs.

Another important gap observed is the institutional and organizational arrangements within Malaysia governing human interactions at various levels from policymakers all the way to grass root levels of the communities. The government at times can be said to be typically organized in specialized policy domains when it comes to addressing climate change, making implementation and adaptation of these strategies difficult. Fragmentation and policy contradictions are often serious problems at all levels of the government. For instance, the Malaysian government currently provides chemical fertilizers to farmers to increase their productivity. However, this government policy should be more environmentally friendly to enable more subsidies to be provided on organic fertilizers than chemical fertilizers (Alam et al., 2012). The disparities of available resources across the different states in Malaysia also play an important role and may influence local government policies in terms of prioritization of other public needs that are deemed more critical. Given the complexity of the health impacts of climate change, it is difficult to record all adaptations which are being implemented. Hence, there is a need for a platform to share and discuss public health adaptation strategies at all levels within the government.

Social capital is also an important factor when it comes to supporting collective initiatives which enable residents to coordinate community-based public health adaptation measures aimed at climate change adaptation. In Malaysia, there are many vulnerable communities mainly dependent on the agricultural sector that are at greater risk of climate change impact. Failure to adapt could lead to social disruption and population displacement that could result in morbidity and mortality among these vulnerable communities. The most critical problem is to identify the appropriate adaptation policies that favour the most vulnerable groups. Targeted policies and adaptation strategies are needed to be evaluated for suitability and sustainability for these vulnerable communities. The lack of proper engagement with these vulnerable communities is one of the lacking elements in proper adaptation and implementation of translated scientific evidence of climate change impacts on human health. Direct engagement between various key stakeholders is needed to improve regional adaptation policy to tackle climate change (Tonmoy et al., 2020). Access to information regarding climate change and its impacts also needs to be improved among these communities as most of these communities live in remote areas away from the general population. There is a gap in knowledge on climate change and its impact on human health among vulnerable communities. Besides, they may not be aware of the country's policies and adaptation strategies for addressing climate change in Malaysia. As we know, an individual's knowledge of climate change is necessary but does not suffice for adaptation. There needs to be additional key information such as perceptions of risk, vulnerability and adaptive capacity to be shared with these communities. This will in turn affect their decision-making or behavioural change, especially in implementing local public health adaptation strategies. Individual interpretation of information can be influenced by personal experience, values, priorities and other contextual factors that also need to be considered when framing such adaptation strategies.

Climate change is no longer a distant possibility in the future but is rather a rapidly ongoing reality that requires immediate and effective actions from policymakers and communities. Public health adaptation strategies to address climate change require multisectoral and multidisciplinary responses in which individuals, communities, governments, international organizations and the research community collaborate closely to address the adverse health impacts of climate change. The highest priority should be given to collaborative multidisciplinary research on the assessment of potential health effects of climate change, projections of health effects under various climate and socioeconomic scenarios, identification of health benefits of greenhouse gas mitigation strategies and assessment of affordable adaptation options. Further research is also needed to study the effectiveness of existing public health adaptation strategies and provide evidence-based recommendations to ensure a successful implementation of strategies laid to address climate change. The focus of such research areas has the potential to make a significant contribution to mitigating the effects of climate change on human health as well as fortifying partnerships for sustainable development.

6 Conclusion

Both natural and man-made factors have put an immense pressure on this Earth's climate system, which in turn has become one of the biggest menaces to the public health. Being one of the climate change hotspots in the tropics, pragmatic actions and strategies are indispensable to alleviate the burden of climate-sensitive diseases in Malaysia. While the topmost priority is given to the mitigation approaches, strategies pertaining to the climate adaptation are equally important to reduce the health risks posed by extreme weathers, reduce climate-altering pollution levels and lower risks associated with climate change impacts. To materialize this, the health sector should participate in developing and executing relevant adaptation policies. Climate adaptation measures should be incorporated into relevant national health systems' policies and plans in areas where the health sector has primary decision-making over greenhouse gases emissions' sources from healthcare activities. Effective engagement with youth groups, non-governmental and civil society organizations and the private sector while providing them with a platform to advocate for climate justice will also help raise public awareness about opportunities for climate and health. Meantime, the identification and management of constraints and barriers associated with the country's climate change adaptation strategies are also crucial to recognize the existing gaps for improvements in the public health sector.

Acknowledgements The authors would like to extend their gratitude to Universiti Malaya for providing the space and resources to undertake this work.

References

Aghamohammadi, N., Fong, C. S., Idrus, M. H. M., Ramakreshnan, L., & Sulaiman, N. M. (2021). Environmental heat-related health symptoms among community in a tropical city. *Science of the Total Environment, 782*, 146611.

Aghamohammadi, N., Ramakreshnan, L., Fong, C. S., Noor, R. M., Hanif, N. R., & Sulaiman, N. M. (2022). Perceived impacts of Urban Heat Island phenomenon in a tropical metropolitan city: Perspectives from stakeholder dialogue sessions. *Science of the Total Environment, 806*, 150331.

Alam, L., Rahman, L. F., Ahmed, M. F., Bari, M., Masud, M. M., & Mokhtar, M. B. (2021). Assessment of health risks and individuals' willingness to participate in drinking water management at flood-prone Pahang River Basin, Malaysia. *Environmental Geochemistry and Health, 43*(5), 2049–2063.

Alam, M., Siwar, C., Talib, B., Mokhtar, M., & Toriman, M. (2012). Climate change adaptation policy in Malaysia: Issues for agricultural sector. *African Journal of Agricultural Research, 7*(9), 1368–1373.

Alhoot, M. A., Tong, W. T., Low, W. Y., & Sekaran, S. D. (2016). Climate change and health: The Malaysia scenario. In *Climate change and human health scenario in South and Southeast Asia* (pp. 243–268). Springer.

Along, N. Z. B., Ahmed, I., & MacKee, J. (2022). Flood knowledge management by multiple stakeholders: An example from Malaysia. *International Journal of Disaster Resilience in the Built Environment*, (ahead-of-print).

Azzeri, A., Ching, G. H., Jaafar, H., Mohd Noor, M. I., Razi, N. A., Then, A. Y. H., Suhaimi, J., Kari, F., & Dahlui, M. (2020). A review of published literature regarding health issues of coastal communities in Sabah, Malaysia. *International Journal of Environmental Research and Public Health, 17*(5), 1533.

Bhuiyan, T. R., Reza, M. I. H., Choy, E. A., & Pereira, J. J. (2018). Facts and trends of urban exposure to flash flood: a case of Kuala Lumpur City. In *Improving flood management, prediction and monitoring* (Vol. 20, pp. 79–90). Emerald Publishing Limited.

Bhuiyan, T. R., Er, A. C., Muhamad, N., & Pereira, J. J. (2022). Evaluating the cumulative costs of small-scale flash floods in Kuala Lumpur, Malaysia. *Journal of Hydrology, 612*, 128181.

Binns, C., Low, W. Y., & Wai Hoe, V. C. (2021). Red alert: Climate change and public health in the Asia Pacific Region. *Asia Pacific Journal of Public Health, 33*(8), 810–811.

Friedrich, J., Ge, M., & Pickens, A. (2020). *This interactive chart shows changes in the world's top 10 emitters.* Accessed at https://www.wri.org/insights/interactive-chart-shows-changes-worlds-top-10-emitters

Hamid, A. I. A., Din, A. H. M., Hwang, C., Khalid, N. F., Tugi, A., & Omar, K. M. (2018). Contemporary sea level rise rates around Malaysia: Altimeter data optimization for assessing coastal impact. *Journal of Asian Earth Sciences, 166*, 247–259.

Harun, Z., Reda, E., Abdulrazzaq, A., Abbas, A. A., Yusup, Y., & Zaki, S. A. (2020). Urban heat island in the modern tropical Kuala Lumpur: Comparative weight of the different parameters. *Alexandria Engineering Journal, 59*(6), 4475–4489.

Huang, C., Vaneckova, P., Wang, X., FitzGerald, G., Guo, Y., & Tong, S. (2011). Constraints and barriers to public health adaptation to climate change: A review of the literature. *American Journal of Preventive Medicine, 40*(2), 183–190.

Ibrahim, M. H., Latiff, N. A. A., Ismail, K., & Isa, N. K. M. (2018, April). Effect of urbanization activities towards the formation of urban heat island in Cameron Highlands, Malaysia. In *IOP conference series: Earth and environmental science* (Vol. 148, No. 1, p. 012022). IOP Publishing.

Jaafar, H., Azzeri, A., Isahak, M., & Dahlui, M. (2021). The impact of haze on healthcare utilizations for acute respiratory diseases: Evidence from Malaysia. *Frontiers in Ecology and Evolution, 860.*

Jaafar, H., Isahak, M., & Dahlui, M. (2018). Economic burden of haze-related respiratory illnesses in Selangor, Malaysia. *Value in Health, 21*, S104.

Khan, M. F., Hamid, A. H., Ab Rahim, H., Maulud, K. N. A., Latif, M. T., Nadzir, M. S. M., … Yusoff, S. (2020). El Niño driven haze over the Southern Malaysian Peninsula and Borneo. *Science of the Total Environment, 730*, 139091.

Khashoggi, B. F., & Murad, A. (2020). Issues of healthcare planning and GIS: A review. *ISPRS International Journal of Geo-Information, 9*(6), 352.

Krishnaswamy, S., Subramaniam, K., Indran, T., & Low, W. Y. (2012). The 2004 tsunami in Penang, Malaysia: Early mental health intervention. *Asia Pacific Journal of Public Health, 24*(4), 710–718.

Mahmood, J., & Guinto, R. R. (2022). Lessons from climate reports for the Malaysian medical community. *The Malaysian Journal of Medical Sciences: MJMS, 29*(3), 1.

Ministry of Energy, Green Technology and Water (MEGTW). (2017). *Green technology master plan Malaysia.* Retrieved from https://www.pmo.gov.my/wp-content/uploads/2019/07/Green-Technology-Master-Plan-Malaysia-2017-2030.pdf

Ministry of Energy Science Technology Environment & Climate Change (MESTECC). (2018). Malaysia third national communication and second biennial update report to the united nations framework convention on climate change. *Ministry of Energy, Science, Technology, Environment And Climate Change.* https://www.mestecc.gov.my/web/documents/jqhRxaP3ff20SNzCaBfiyqiOdsLhuBa9Jyz5D0C_gQ0

Ministry of Health (MOH). (2013). *Guidance document: National Environmental Health Action Plan (NEHAP) Malaysia.* Retrieved from http://nehapmalaysia.moh.gov.my/wpcontent/uploads/2015/11/guidance_doc_malaysia.pdf

Ministry of Health (MOH). (2020). *National Environmental Health Action Plan (NEHAP) Malaysia: Action plan for environmental health 2016–2020.* Retrieved from http://nehapmalaysia.moh.gov.my/wp-content/uploads/2021/04/Action-Plan-TWG-NEHAP-2016_2020-Final-Dis-2020.pdf

Ministry of Natural Resources and Environment Malaysia (MNRE). (2009). *National policy on climate change 2009.* Retrieved from https://policy.asiapacificenergy.org/sites/default/files/National%20Policy%20on%20Climate%20Change.pdf

Morris, K. I., Salleh, S. A., Chan, A., Ooi, M. C. G., Abakr, Y. A., Oozeer, M. Y., & Duda, M. (2015). Computational study of urban heat island of Putrajaya, Malaysia. *Sustainable Cities and Society, 19*, 359–372.

Nwakaire, C. M., Onn, C. C., Yap, S. P., Yuen, C. W., & Onodagu, P. D. (2020). Urban Heat Island Studies with emphasis on urban pavements: A review. *Sustainable Cities and Society, 63*, 102476.

Othman, J., Sahani, M., Mahmud, M., & Ahmad, M. K. S. (2014). Transboundary smoke haze pollution in Malaysia: Inpatient health impacts and economic valuation. *Environmental Pollution, 189*, 194–201.

Paterson, R. R. M. (2019). *Ganoderma boninense* disease deduced from simulation modelling with large data sets of future Malaysian oil palm climate. *Phytoparasitica, 47*(2), 255–262.

Pörtner, H. O., Roberts, D. C., Adams, H., Adler, C., Aldunce, P., Ali, E., … Birkmann, J. (2022). *Climate change 2022: Impacts, adaptation and vulnerability.* IPCC Sixth Assessment Report.

Puteh, S. E. W., Siwar, C., Hod, R., Nawi, A. M., Idris, I. B., Ahmad, I. S., Idris, N. D. M., Alias, N. A., & Taha, M. R. (2018). Burden of health-related issues and community empowerment in Malaysia's east coast flood. In *Improving flood management, prediction and monitoring.* Emerald Publishing Limited.

Rahaman, Z. A., Kafy, A. A., Saha, M., Rahim, A. A., Almulhim, A. I., Rahaman, S. N., … Al Rakib, A. (2022). Assessing the impacts of vegetation cover loss on surface temperature, urban heat island and carbon emission in Penang city, Malaysia. *Building and Environment, 222*, 109335.

Rahman, S. (2022). *Malaysia's floods of december 2021: Can future disasters be avoided?* ISEAS—Yusof Ishak Institute, 26. Retrieved from https://www.iseas.edu.sg/wp-content/uploads/2022/03/ISEAS_Perspective_2022_26.pdf

Ramakreshnan, L., Aghamohammadi, N., Fong, C. S., Ghaffarianhoseini, A., Wong, L. P., & Sulaiman, N. M. (2019). Empirical study on temporal variations of canopy-level Urban Heat

Island effect in the tropical city of Greater Kuala Lumpur. *Sustainable Cities and Society, 44*, 748–762.

Shith, S., Awang, N. R., Latif, M. T., & Ramli, N. A. (2021). Fluctuations in nighttime ground-level ozone concentrations during haze events in Malaysia. *Air Quality, Atmosphere & Health, 14*(1), 19–26.

Sundram, T. K. M., Tan, E. S. S., Cheah, S. C., Lim, H. S., Seghayat, M. S., Bustami, N. A., & Tan, C. K. (2022). Impacts of particulate matter (PM2. 5) on the health status of outdoor workers: observational evidence from Malaysia. *Environmental Science and Pollution Research*, 1–11.

Suparta, W., & Yatim, A. N. M. (2019). Characterization of heat waves: A case study for Peninsular Malaysia. *Geographia Technica, 14*(1), 146–155.

Tan, M. L., Yusop, Z., Chua, V. P., & Chan, N. W. (2017). Climate change impacts under CMIP5 RCP scenarios on water resources of the Kelantan River Basin, Malaysia. *Atmospheric Research, 189*, 1–10.

Tan, M. L., Zhang, F., Juinn, C., Derek, C., Yu, K. H., Shaharudin, S. M., Chan, N. W., & Asyirah, A. R. (2022). Spatio-temporal analysis of precipitation, temperature and drought from 1985 to 2020 in Penang, Malaysia. *Water Supply, 22*(5), 4757–4768.

Tonmoy, F. N., Cooke, S. M., Armstrong, F., & Rissik, D. (2020). From science to policy: Development of a climate change adaptation plan for the health and wellbeing sector in Queensland, Australia. *Environmental Science & Policy, 108*, 1–13.

UNDP (United Nations Development Programme), UNICEF (United Nations Children's Fund) & EcoKnights. (2020). *Change for climate: Findings from the national youth climate change survey Malaysia*. Putrajaya.

UNICEF (United Nations Children's Fund). (2021). *The climate crisis is a child rights crisis: Introducing the children's climate risk index.*

Wahid, N. A. A., Suhaila, J., & Rahman, H. A. (2021). Effect of climate factors on the incidence of hand, foot, and mouth disease in Malaysia: A generalized additive mixed model. *Infectious Disease Modelling, 6*, 997–1008.

Walsh, K. J., McBride, J. L., Klotzbach, P. J., Balachandran, S., Camargo, S. J., Holland, G., … Sugi, M. (2016). Tropical cyclones and climate change. *Wiley Interdisciplinary Reviews: Climate Change, 7*(1), 65–89.

Weber, E. U., & Stern, P. C. (2011). Public understanding of climate change in the United States. *American Psychologist, 66*(4), 315.

Wong, L. P., Haridah, A., Nasrin, A., Sima, A., & Nik, M. N. S. (2018). Physical, psychological, and social health impact of temperature rise due to urban heat island phenomenon and its associated factors. *Biomedical and Environmental Sciences, 31*(7), 545–550.

World Bank Group. (2021). *Climate risk country profile: Malaysia*. The World Bank Group and the Asian Development Bank.

Yu, P., Xu, R., Abramson, M. J., Li, S., & Guo, Y. (2020). Bushfires in Australia: A serious health emergency under climate change. *The Lancet Planetary Health, 4*(1), e7–e8.

Zulkarnain, S. H., Tsujimura, M., Yuzir, M. A., Razali, M. N., & Tarmidi, Z. (2020). A review of climate change (floods) and economic attributes response to residential property value in Malaysia. *Journal of Water and Climate Change, 11*(4), 1084–1094.

Status of Nationally Determined Contributions in Indonesia: A Review on Climate Change Health Impacts

Budi Haryanto, Jatna Supriatna, Triarko Nurlambang, and Marsum

Abstract Nationally Determined Contributions (NDCs) are the necessary non-binding actions plans on climate change targeted by each country as their long-term goals on reducing emissions and combating climate change impacts. In Indonesia, implementation of NDCs includes enhanced ambition on adaptation as elaborated in the programmes, strategies and actions to achieve economic, social and livelihood, and ecosystem and landscape resilience; enhanced clarity on mitigation by adopting the Paris Agreement rule book (Katowice Package) on information to be provided in NDC, as well as updated policies which potentially contribute to additional achievement of NDC target; National context that relates the existing condition, milestones along with national development for the period of 2020–2024, and indicative pathways towards long-term vision (Vision of Indonesia 2045 and the Long-Term Strategy on Low Carbon and Climate Resilient Development 2050); Translating the Paris Agreement Rule Book (Katowice Package) into Indonesia's context with a view to enhance effectiveness and efficiency in implementing the agreement and in communicating its progress and achievement as part of the responsibility of the party to the agreement.

Keywords Indonesia · NDCs on health · Climate change · Health resilience

B. Haryanto (✉)
Department of Environmental Health, Faculty of Public Health, Universitas Indonesia, Depok City, Indonesia
e-mail: bharyant@cbn.net.id

B. Haryanto · J. Supriatna · T. Nurlambang
Research Center for Climate Change, I-SER, Universitas Indonesia, Depok City, Indonesia

J. Supriatna
Department of Biology, Faculty of Mathematics and Basic Sciences, Universitas Indonesia, Depok City, Indonesia

T. Nurlambang
Department of Geography, Faculty of Mathematics and Basic Sciences, Universitas Indonesia, Depok City, Indonesia

Marsum
Polytechnic of Health Semarang, Ministry of Health, Semarang, Indonesia

R. Akhtar (ed.), *Climate Change and Human Health Scenarios*, Global Perspectives on Health Geography, https://doi.org/10.1007/978-3-031-38878-1_8

1 Introduction

The Paris Agreement is an international agreement on climate change that aims to hold the global average temperature rise below 2 °C above pre-industrial levels and continue efforts to reduce temperature increases to 1.5 °C above pre-industrial levels. In addition, the Paris Agreement aims to increase adaptation capacity to the negative impacts of climate change, toward climate resilience and low emission development, without threatening food production and prepare funding schemes for low emission development and climate resilience. The Paris Agreement, which is legally binding and applied to all countries (legally binding and applicable to all) with the principle of differentiated shared responsibility based on their respective capabilities (common but differentiated responsibilities and respective capabilities), assigns responsibilities to developed countries to provide funds, capacity building, and transfer of technology to developing countries. In addition, the Paris Agreement mandates a more effective and efficient increase in bilateral and multilateral cooperation to carry out climate change mitigation and adaptation actions with funding support, technology transfer, capacity building supported by transparency mechanisms, and sustainable governance.

Indonesia, together with members of the international community through the 21st UNFCCC Conference of Parties (COP) in Paris, has adopted the Paris Agreement to the United Nations Framework Convention on Climate Change, which was followed up with the signing of the Agreement on April 22, 2016, in New York, United States of America. The Paris Climate Change Conference resulted in a new agreement called the Paris Agreement, which resulted in a deal regarding the NDC, which regulates and projects the potential for reducing GHG emissions to be carried out by States Parties in the post-2020 time frame. Indonesia submitted the NDC to the UNFCCC Secretariat ahead of COP-22 Marrakech in 2016 as an elaboration of the NDC and, at the same time, replaced the Intended Nationally Determined Contribution (INDC) submitted to the UNFCCC Secretariat before COP-21 Paris. The first NDC Indonesia document was integral to the NDC Implementation Strategy which is intended as a guide for the synergy of each component of the nation, starting from Ministries/Agencies, Local Governments, Academics, Business Sector, Non-Governmental Organizations, and the General Community. The strategy is to achieve national commitments to reducing GHG emissions and achieving low emission and climate-resilient development goals, as in the NDC document. The NDC targets were 29% (unconditional) and 41% (conditional or with international assistance) against 2030 business as usual. Through the NDC Implementation Strategy, inter-sectoral synergies can be strengthened to fulfill national commitments to the international world that align with national goals and aspirations (Masripatin et al., 2017).

The enhanced NDC (E-NDC) was submitted on September 23, 2022 to UNFCCC. Indonesia is committed to reducing carbon emissions to maintain global temperature rise by increasing the Enhanced Nationally Determined Contribution target to 31.89% or the equivalent of 912 million tons of CO_2 in 2030 and up to 43.2% with international assistance. Previously, Indonesia targeted a reduction in carbon emissions of 29% or the equivalent of 835 million tons of CO_2. If Indonesia only does business as usual and makes no changes to reduce the use of fossil fuels, Indonesia will produce 1.5 Giga tons of CO_2 in 2060. The enhanced NDC is the transition toward Indonesia second NDC, which will be aligned with the Long-Term Strategy of Low Carbon and Climate Resilience (LTS-LCCR) (KLHK, 2022). In addition to increasing the E-NDC target, Indonesia's efforts to achieve Net-Zero Emissions (NZE) in 2060 or sooner, including through the conversion of fossil fuels to Liquefied Natural Gas (LNG), the use of electric stoves, the use of biofuels to replace and accelerate the installation of rooftop solar panels as well as converting motorized vehicles to electricity program. In addition, the forestry and land use sectors (FOLU) aim to commit to net-zero emissions by 2060 or sooner. Likewise, the NDC strategy seeks to reduce the potential loss of the country's GDP by 3.45% due to climate change by 2050 by increasing resilience in the four basic socio-economic development needs: food, water, energy, and environmental health (Masripatin et al., 2017).

Globally, worry about the effects of climate change on health and potential harm to communities is growing (European Commission, 2010). Addressing climate change requires numerous transformational areas, including primary prevention, cross-sectoral action, strengthened health sector, building support, increased evidence and communication, and monitoring (WHO, 2020). Less than 1 in 5 nations have evaluated the health co-benefits of their national climate, according to the WHO Health and Climate Change Global Survey (2021). Just 13% of current NDCs, according to the WHO evaluation of Nationally Determined Contributions (NDCs) (2021), commit to measuring or keeping track of the health benefits of climate policies or targets (WHO, 2021). The health sector would be given priority in the climate resilience targets, according to the Indonesia NDC. Given that the National Priority Program No. 6 is focused on climate change and disasters, the promises to combat the impacts of climate change have also been incorporated into the mid-term development plan for 2020–2024. In the area of health, Indonesia also has policy tools for assessing climate risk as accordance with Ministry of Health (MoH) Regulation No. 35/2012 and 1018/2011, the health sector must analyze risks and adapt to climate change. The implementation of Carbon Economic Value for Achieving Nationally Determined Contribution Targets and Control of GHG Emissions in National Development under Presidential Regulation No. 98/2021 has enhanced this policy. The health sector is one of the priority development areas that this strategy firmly specifies must adapt to climate change. Indonesia has implemented health co-benefits for climate policies and targets, which are uncommon in the present global NDCs, based on the Roadmap NDC on Adaptation (WHO, 2021). Additionally, a technical team for climate change adaptation has been constituted by the Ministry of Health. The necessary tools directed to compile data and information have also been initiated. Unfortunately, the tools are still at the national level and have not been operationalized regularly.

2 Implementation of NDCs in Indonesia: Updates, Issues and Options

The Indonesian Ministry of Environment and Forestry (MoEF) provided a quick overview of Nationally Determined Contributions (NDCs) which are a nation's commitment to combating climate change. Forestry and energy are the two most important sectors for achieving emission reduction goals out of the five (energy, waste, Industrial Process and Product Uses (IPPU), agriculture, and forestry). The eight strategies to implement the roadmap of NDCs are: (1) Policy instruments for climate change adaptation and disaster risk; (2) Integration into development planning and financial mechanisms; (3) Increasing climate literacy on vulnerability and risk; (4) Landscape-based approach for comprehensive understanding; (5) Strengthening local capacity on best practices; (6) Improved knowledge management; (7) Stakeholder participation; and (8) Application of adaptive technology.

Indonesia's climate change adaptation and mitigation aims to reduce risks, increase adaptive capacity, strengthen resilience, and reduce vulnerability to climate change in all sectors. The goal of Indonesia's climate change adaptation is to reduce risks, enhance adaptive capacity, strengthen resilience, and reduce vulnerability to climate change on all development sectors. In achieving the adaptation goal, Indonesia focuses on three areas of resilience: economic resilience, social and livelihood resilience, and ecosystem and landscape resilience. The NDC Adaptation Road Map, which operationally prioritizes various disciplines, including food, water, energy, health, and ecosystems, has detailed these three resilience categories. To slow down global warming, Indonesia has been implementing the *Proklim*, or Climate Village Program, since 2012. In 2016, this program was made a national initiative. In 2021, a total of 3270 locations launched *Proklim* programs; it is anticipated that by 2024, there will be 20,000 *Proklim* locations in Indonesia.

In the priority development areas, such as the health sector, this strategy is a hard mandate that requires climate change adaptation. Indonesia established health co-benefits for climate policies and targets based on the Roadmap NDC on Adaptation, which are uncommon in the present global NDCs (WHO, 2021). A technical team for climate change adaptation has also been established by the Ministry of Health. The scorecard covers updated and enhanced NDCs, which the UNFCCC requested to be submitted in the lead up to COP-26. NDCs were assessed based on their attention to five health categories: health impacts, health in adaptation measures, health co-benefits, economics and finance, and bonus points available for overall prominence and integration of health. Three points were available for each category, with a total possible 'health score' of 15 (Fig. 1).

The NDC for Indonesia discusses the health and social effects of climate change, including how drought, floods, and other natural catastrophes affect the most vulnerable communities' access to food and water. Additionally, it lists critical actions and mentions that the health sector is a focus area for its adaptation planning. There is a general mention of the idea that reduced emissions have positive health effects. The NDC points out that eliminating subsidies for fossil fuels has freed up money for

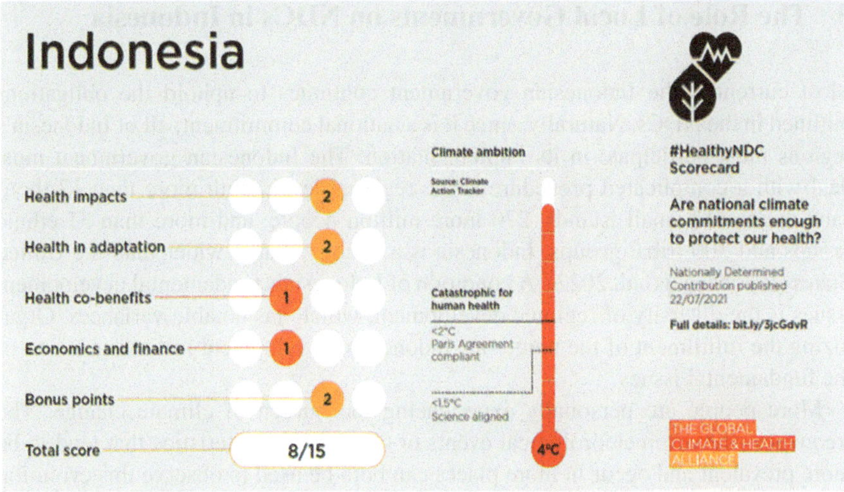

Fig. 1 NDCs' health of Indonesia status reported on Healthy NDC scorecard

other priorities, including as health. The addition of paragraphs containing in-depth health content and language referencing the right to health and a healthy environment received bonus marks. In its NDC, Indonesia speaks about gender and intergenerational equity. Despite these references to health, it is determined that Indonesia's climate ambition is consistent with 2 °C of warming.

The risks of climate change on health are already acknowledged in Indonesia. In the health sector, many studies focused on dengue and malaria. The location of vulnerability studies is spread across areas in Indonesia. Many studies have also conducted sector-based vulnerability and risk assessments on specific locations as updated in the Climate Outlook 2019 and the Roadmap NDC on Adaptation released by the MoEF as the National Focal Point (NFP) of Climate Change in Indonesia. This compilation highlighted the climate change impacts that are evident in Indonesia. As for the health sector, the Ministry of Health conducted a health risk assessment in the process of developing the draft Health National Action Plan (HNAP). The vulnerability and risk assessment was completed following the IPCC climate risk assessment framework (2014), the concept of risk analysis regulated by Ministerial Decree of MoEF No.33/2016, and the disaster risk analysis regulated by the Head Decree of National Disaster Management Agency No. 02/2012. The vulnerability and risk assessment utilized climate information for assessing the risks of the disease incidences, namely DHF, diarrhea, pneumonia, and malaria, in Indonesia. Understanding the risks of climate change to the health sector, the government of Indonesia has included climate change and disaster in the mid-term development plan. This commitment is stated in Presidential Regulation No.18/2020 concerning the National Medium-Term Development Plan for 2020–2024.

3 The Role of Local Governments on NDCs in Indonesia

Most currently, the Indonesian government continues to uphold the obligations outlined in the NDCs. Naturally, since it is a national commitment, all of Indonesia's regions must participate in its implementation. The Indonesian government must deal with a complicated procedure at the regional level. With more than 17 thousand major and small islands, 270 more million people, and more than 31 ethnic groups and 300 tribal groups, Indonesia is a country that is wider than the United States (Indonesia.go.id, 2022). A condition of Indonesia's fundamental development issues is the diversity of regional development, which has notable variances. Organizing the fulfillment of the numerous Indonesian development initiatives is one of the fundamental issues.

More people are personally experiencing the burden of climate change. The frequency of hydrometeorological events or climatic characteristics that tend to be more prevalent and occur in more places can both be used to observe this evolution in Indonesia (see Table 1). The size and severity of the hazard or catastrophe event region can both be directly impacted by climate change. There are no districts or cities in Indonesia that are classed as having a low or no disaster risk, according to the Indonesian Disaster Risk Index (IRBI) (BNPB, 2022a, 2022b). The majority of Indonesia's districts and cities are categorized as having a high disaster risk, while the remaining 50% are categorized as having a medium disaster risk.

Other risks, such as landslides and tornadoes, are also systemically caused by these particular geographical and climatological circumstances, hydrometeorological events, and weather patterns. The Working Group II AR-6-IPCC Report and pages 5–9 of the IPCC Chair's Vision Paper AR-6 both discuss the causes of the climate change process and how they are linked systemically (2017). It demonstrates that the occurrence has been linked with anthropogenic factors in order to anticipate adaptation and mitigation of hazards or disasters, as specified in the

Table 1 Consequences of hazards and disasters in Indonesia 2022

Types of hazard	Hazard area (Ha)	Number of human casualties (injured, sick and dead)	Value of loss (economic and physical)—billion rupiah
By climatic factors (floods, flash floods, landslides, droughts, forest fires, tornadoes)	380,896.858	636,443	6,579,226.983
By non-climatic factors (earthquakes, eruptions, liquefaction)	74,850.633	214,504	1,518,221.413
COVID-19	4615.650	173,830	N/A
Total	460,363.141	1,024,777	

Source Downloaded and processed from *Inarisk* December 2022 (http://inarisk.bnpb.go.id/infografis/new)

Sendai Framework on global disaster management. After the COP agreement in Paris in 2015 and the SDGs aim by 2030, this 'school of thought' may be profoundly applied. Therefore, the interaction between this hydrometeorological event and the socio-cultural-economic processes on the surface of the earth cannot be separated.

Local governments play a strategic and significant role in accomplishing NDC aims within the framework of this systemic linkage. Even if it is acknowledged that the real state of the NDC national policy's implementation depends greatly on the roles and responsibilities of local governments and stakeholders, including those from private businesses, social organizations, mass organizations, and community organizations. So how can local governments and all stakeholders effectively apply the procedures of national policies handled by the federal government? Of course, the execution of this NDC follows the same principles as the SDGs, which define the "No One Left Behind" principle and can be accomplished by incorporating the development sector or other Goals. Additionally, Goal 13 of the SDGs is conceptually important to this NDC. According to the 2020–2024 National Mid-Term Development Plan, the SDGs themselves have been identified as the mainstay of Indonesia's national development (RPJMN). Additionally, Presidential Regulation No. 111 of 2022 has been published to further underline the dedication to reaching the SDGs development target. In general, the following illustration can be used to represent significant policies for the implementation of NDCs in the context of achieving the SDGs (Fig. 2).

The national development policy on achieving the NDC target must be elaborated into an implementation regulation at the regional level that is integrated into the Regional Development Plan. From the picture above, a unified policy is designed that has the same goal, namely achieving sustainable development (through the achievement of Indonesia's SDGs targets) (Nurlambang and Tambunan in Jatna (ed.), 2021).

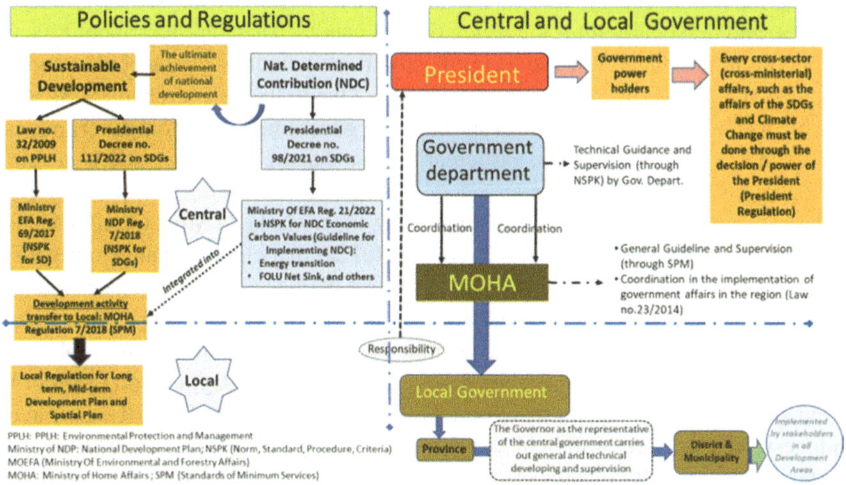

Fig. 2 Applying NDC within SDGs policies by Central and Local Government in Indonesia. *Source* Processed products

Synchronization between policies and relevant laws and regulations is the first step. Among them include energy transition policies (President Regulation No. 112 of 2022) and FOLU (Forestry and Other Land Uses) net sink 2030 (as stated in President Regulation No. 98 of 2021 and MoEF regulation No. 198 of 2022) as the largest contributor to greenhouse gases in Indonesia (DG Climate Change Control, MoEF, 2022). Furthermore, implementation at the regional level can be integrated into one medium-term development plan policy and also regional spatial plans at the local government level. This integration is realized through SPM (Minimum Service Standards) which is regulated in the Domestic Regulation, as stipulated in the Ministry of Home Affairs (*Permendagri*) No. 7 of 2018. This *Permendagri* has integrated the KLHS or Strategic Environmental Assessment (SEA) which includes a study of climate change affairs including the results of the energy transition process and a net sink FOLU in its study content. Moreover, this KLHS policy has been determined to be mandatory for all regional planning policies based on Law No. 32/2009 on Environmental Management and Protection (Nurlambang and Tambunan, 2021).

Additionally, it has been established that SPM through *Permendagri* No. 7 of 2018 can be implemented to each region by releasing the appropriate regional regulations. By the end of 2019, it is now predicted that the greenhouse gas reduction objective will only have been achieved to a degree of about 24.4%. By 2030, it is intended to handle FOLU up to 17.2% and energy transition up to 11%, for a target of 29% (Business as Usual scenario based on funding from the Government of the Republic of Indonesia itself). The remainder, less than 1%, is split among numerous industries, including trash and agriculture (MoEF, 2019). According to the national strategy plan, it is required to lessen deforestation, lessen degradation, and regulate forest and land fires in order to attain FOLU net sink.

Regarding the energy transition, the goal is to have 23% of energy sources from renewable resources available by 2030 (Simamora, 2019). Additionally, it must develop renewable energy sources, such as electricity from hydrogen and ammonia sources, and drastically limit the usage of coal energy. There are just about 8 years remaining to get there 2030, so a significant effort including all stakeholders must be made and carried out according to schedule. Due to this, it is crucial for SPM to be able to mobilize stakeholders from both the government and non-governmental sectors in a variety of locations, particularly those that are the hub of FOLU net sink and energy transition initiatives.

The declaration of Brazilian President Luiz Inacio Lula da Silva (Tempo Magazine, pp. 80–82, November 27, 2022 edition 2022a) that climate change mitigation must go hand in hand with social justice and also include economic justice was just confirmed at the COP 27 in Egypt. Obviously, there is a practical aspect to this remark that involves the actual circumstances in the location. One of them is to implement a strategy that aims to achieve zero deforestation or forest degradation in order to avoid an average temperature increase of 1.5 °C. The president of Brazil encouraged Indonesia and the Congo to help safeguard tropical forests around the world in the context of the global concern of climate change. The governments of Congo and Indonesia both accepted this request. This initiative is a crucial negotiating chip to reaffirm developed nations' commitment to providing funds for initiatives to recover from climate change-related losses and damages.

Local governments in Indonesia will manage the FOLU net sink program to enable implementation if the new promise made as a result of COP 27 is fulfilled. Of course, getting cooperation from outside the Indonesian government is a possibility in order to fulfill the NDC's pledge to cut greenhouse gas emissions by 41%. In the meantime, the Ministry of Energy and Mineral Resources no longer functions in isolation in order to be able to achieve climate change targets through energy transition management. Presidential Regulation No. 112 of 2022 in October 2022 and the availability of new obligations from the G-20 conference, which will take place at its peak in November 2022, serve to underline this position even more. At the moment, financial institutions like banks manage a sizable amount of funding for initiatives and projects aimed at managing energy sources in the direction of the energy transition (Tempo Magazine, November 27, 2022 edition, pp. 73–76). The Financial Services Authority Regulation No. 60 of 2017 regulates one of them, the Green Bond (Fiscal Policy Agency, 2019). The local government and its local stakeholders play a significant role in reaching the transition energy targets as a result of the local nature of those renewable energy operations.

4 Indonesia NDC's Health for Climate Resilience

Due to COVID-19, which caused the world economy to contract in 2018, Indonesia increased the scope of its climate commitments. Indonesia's revised Nationally Determined Contribution (NDC), which was submitted to the UNFCCC on July 22, 2021, incorporates equitable emission reduction targets and improved alignment between the nation's development and climate goals. This is the result of lengthy multi-stakeholder assessments and consultation processes that have been conducted over the course of more than a year by Indonesia's Directorate General of Climate Change, a part of the country's Ministry of Environment and Forestry (DGCC MoEF).

The largest archipelagic nation in the world now pledges to increase its greenhouse gas emissions target by 2030 from the business as usual (BAU) scenarios of 834 $MtCO_2e$ and 1185 $MtCO_2e$ to 29% unconditionally and 41% conditionally (with international support). The updated NDC reflects progress beyond the existing NDC, particularly through:

- Enhancing ambition on adaptation
- Enhancing clarity on mitigation by adopting the Paris Agreement rule book (Katowice Package)
- Aligning the national context relating to the existing condition
- Enacting milestones along with national development for 2020–2024
- Providing indicative pathways toward Vision Indonesia 2045 and the Long-Term Strategy on Low Carbon and Climate Resilient Development 2050 (LTS-LCCR 2050), and
- Translating the Paris Agreement Rule Book (Katowice Package) into Indonesia's context.

Promoting climate resilience in food is one of Indonesia's NDC's strategic initiatives. The world's food security is already being impacted by climate change. Concerns about achieving food security have been highlighted by environmental changes such as variations in rainfall, droughts, warmer or colder temperatures (changing growing seasons), and changes in land cover (Alemu & Mengistu, 2019). A partial balancing approach model employed in recent empirical investigations reveals that crop yield losses brought on by climate change in agriculture are a substantial concern (Bandara & Cai, 2014). In 2017, the United Nations (UN) reported that the numbers of hungry people had increased for the first time in a decade, primarily because of conflict and climate change. In addition, the prevalence of micronutrient malnutrition has not been well measured, although it is assumed that this is also a significant problem (Raiten & Bremer, 2020). Child stunting reduction is the first of 6 goals in the Global Nutrition Targets for 2025 and a key indicator in the second Sustainable Development Goal of Zero Hunger. The prevalence of child stunting in Indonesia has remained high over the past decade, and at the national level is approximately 37%. Community and societal factors—particularly, poor access to health care and living in rural areas—have been repeatedly associated with child stunting. Published studies are lacking on how education; society and culture; agriculture and food systems; and water, sanitation, and the environment contribute to child stunting (Beal et al., 2018). Given the serious threats to food security, attention should shift to an action-oriented research agenda, where we see four key challenges: (a) changing the culture of research; (b) deriving stakeholder-driven portfolios of options for farmers, communities and countries; (c) ensuring that adaptation actions are relevant to those most vulnerable to climate change; (d) combining adaptation and mitigation (Wollenberg et al., 2016).

The tools must be developed with serious effort and dedication if numerous parties are to use them to enable multi-stakeholder participation in achieving Indonesia's NDC commitments. Indonesia released plans and strategies to address the effects of climate change on the health sector after taking into account the needs. Indonesia's most common climate change studies are focused on access to basic needs and climate-related disasters (Novita, 2020; Perdinan et al., 2019; UNDP, 2013). In the NDC Roadmap on Adaptation, the implications of climate change have been estimated with this focus in mind. To understand the contributing variables to climate change risks in the health sector, as required by MoH Regulation 35/2012, this document also supports the need to conduct a more thorough health risk assessment (WHO UNISDR, 2017). This knowledge will improve the ability to monitor illness incidences for risk factors, which will help to create an early warning system. Currently, the diseases that are affected by climate change are diarrhea, pneumonia, DHF, and malaria.

Studies linking health and climate change have also been developed in Indonesia. But unlike adaptation practices in other areas, these have not been specifically done for a Health Impact Assessment (Ebi et al., 2018). To ensure that the requirements of Indonesia's policymakers are met, involvement of the funders and researchers is also required in monitoring and responding to associated risk research (ASEAN Report, 2021). The main areas for developing assessment studies in Indonesia could be adaptation options for health-related morbidity and mortality, options for facilitating non-heat early warning response systems, multisectoral adaptation for nutrition, and vulnerability on mosquito-borne illnesses (Haryanto, 2016; IPCC, 2022a, 2022b). Furthermore, health research should focus on assessing acute, chronic, and ecotoxicological effects of global climate change (IPCC, 2007). The risks of illness, life expectancy, and reduction in nutrition are likely to influence vulnerability to specific hazards in different geographical and socio-political contexts (Masum, 2019). Additionally, the government must add the following to its action plans: multi-dimensional risk assessment, identification and evaluation of data points related to climate disasters that can be used for tracking early warning, inclusion of climate and disaster risk, and governance and risk management capabilities to all multi-sectoral movements (UNDRR, 2020). Studies or reports are required to address the issues of linked meteorological elements, adverse mental health outcomes, the link between heat and adverse occupational consequences, and adverse birth outcomes, as mentioned in the Compendium document (WHO, 2022).

Indonesia must also develop a communication plan to address its health and climate change problems. To engage the public and give information readiness and protection, effective indicators must be chosen. The other requirement is to drive and guide policy choices, prioritize ongoing investments in climate change-related health protection, and describe returns on health protection investment at various time horizons (IPCC, 2022a, 2022b). Indonesia can also adopt the climate and health vulnerability assessment steps to develop climate change risk assessment. This step can gain support from existing sources such as *SIDIK* (MoEF), Crisis Center (MoH), and *INARISK* (BNPB). The other subjects are related to the topics of (1) alteration of the geographical distribution of diseases, (2) new and emerging health issues, and (3) the effects of climate change on health and productivity in the workplace with implications for occupational health, safety, and social protection (UNFCCC, 2017).

Actions need to be taken for climate resilience in Indonesia should involve several important considerations as:

- The process and resources required to execute the NDC should be concentrated on achieving the goals that the nation has set. To fulfill the obligations, an implementation strategy should be devised.
- Collaboration is key to meeting the aim by 2030 for public–private partnerships to tackle climate change in terms of resources, money, and many other areas.
- To tackle the climate catastrophe, a community's disaster resilience is essential. To resolve the conflict between the need for livelihood and welfare and the preservation of the ecosystem, we must switch to renewable energy sources and fight to advance the growth of the nation.
- Three entry points to support the implementation of NDCs—disaster resilience, health and nutrition, and viable local economy.
- The approach to solving the climate issue and reaching the NDC is to incorporate local knowledge, tradition, and community strength on climate change adaptation as a part of research, development, and innovation.

- To address the local research and development needs and maximize possible synergies between national and local governments, partnerships in research and development with universities, local government, and young professionals should be prioritized.

Policies at the national level have been established, especially policies that are very closely related to efforts to achieve NDCs in Indonesia. The policies on SDGs, Climate Change, Energy Transition, and FOLU Net Sink issued in 2021 and 2022 are a unit of policies and laws, and regulations needed in implementing the National Mid-Term Development Plan or RPJMN (2020–2024 and beyond). By being integrated into the RPJMN, the budget allocation can be more guaranteed until the implementation in the regions, especially in implementing the NDC BAU (Business as Usual) scenario.

Awareness and readiness of the capacity of all stakeholders is a vital and strategic requirement in realizing the achievement of NDC targets. Among those that still need to be prepared is the SPM (Minimum Service Standards) by the Ministry of Home Affairs (MoHA) which explains the mechanism for achieving NDCs at the regional level. In addition, the instruments are needed for monitoring and evaluating the implementation and human resource capacity of its implementers, both formal and informal organizations. Moreover, because the achievement of the NDC is a comprehensive achievement for the Indonesian nation increased mass literacy is needed for the entire community.

References

Alemu, T., & Mengistu, A. (2019). Impacts of climate change on food security in Ethiopia: Adaptation and mitigation options: A review. In P. Castro, A. M. Azul, W. Leal Filho, U. M. Azeiteiro (Eds.), *Climate change-resilient agriculture and agroforestry: Ecosystem services and sustainability* (pp. 397–412). Springer International Publishing. Available from: https://doi.org/10.1007/978-3-319-75004-0_23

ASEAN. (2021). *ASEAN State of climate change report*. Jakarta Indonesia.

Bandara, J. S., & Cai, Y. (2014). The impact of climate change on food crop productivity, food prices and food security in South Asia. *Economic Analysis and Policy, 44*(4), 451–465.

Beal, T., Tumilowicz, A., Sutrisna, A., Izwardy, D., & Neufeld, L. M. (2018). A review of child stunting determinants in Indonesia. *Maternal & Child Nutrition, 14*.

BNPB. (2022a). *Indonesia disaster risk index 2021*. BNPB.

BNPB. (2022b). *Potential all disaster risks in Indonesia*. http://inarisk.bnpb.go.id/infografis/new downloaded on December 2, 2022b

DG of Climate Change Control. (2022). *Indonesia's challenges in dealing with climate change*. DG of Climate Change Control—MOEFA.

Ebi, K. L., Boyer, C., Bowen, K. J., Frumkin, H., & Hess, J. (2018). Monitoring and evaluation indicators for climate change-related health impacts, risks, adaptation, and resilience. *International Journal of Environmental Research and Public Health, 15*, 1943. https://doi.org/10.3390/ijerph15091943

European Commission (EC). (2010). *Climate risk assessment: An introduction*.

Fiscal Policy Agency. (2019). *Public funding for Indonesia's climate change control*. BKF—Ministry of Finance.

Haryanto, B. (2016). Health adaptation scenario and Dengue fever vulnerability assessment in Indonesia. In *Climate change and human health scenario in South and Southeast Asia*. Springer. https://doi.org/10.1007/978-3-319-23684-1_13

Indonesia Profile. (2022). *Indonesia.go.id downloaded* December 2, 2022.

IPCC. (2007). *FAR climate change: Impacts assessment of climate change Chapter 5: Human settlement; the energy, transport and industrial sectors; human health; air quality; and changes in ultraviolet-B radiation.*

IPCC. (2017). *Chair's vision paper*. IPCC—WMO—UNEP

IPCC. (2022a). *AR6 climate change 2022a: Impacts, adaptation, and vulnerability Chapter 7.*

IPCC. (2022b). *Climate change 2022b; Impacts, adaptation and vulnerability, working group II contribution to the sixth assessment report of the intergovernmental panel on climate change* (3056 pp) [In H. -O. Pörtner, D. C. Roberts, M. Tignor, E. S. Poloczanska, K. Mintenbeck, A. Alegría, M. Craig, S. Langsdorf, S. Löschke, V. Möller, A. Okem, B. Rama (Eds.)]. Cambridge University Press. https://doi.org/10.1017/9781009325844

KLHK. (2022). *Enhanced NDC: Komitmen Indonesia Untuk Makin Berkontribusi Dalam Menjaga Suhu Global*. https://www.menlhk.go.id/site/single_post/4983/enhanced-ndc-komitmen-indonesia-untuk-makin-berkontribusi-dalam-menjaga-suhu-global

Masripatin et al. (2017). *Strategi Implementasi NDC (Nationally Determined Contribution)*. Direktroat Jendral PPI, KLHK.

Masum, J. H. (2019). *Climate change risk assessment: Exploring the current state of knowledge to assess climate change risks from local to national level*. Coastal Development Partnership (CDP). ISBN: 978-984-34-7046-1

Ministry of EFA. (2019). NDC achievement in 2017 reached 24.4%. https://www.menlhk.go.id/site/single_post/1762. Downloaded December 2, 2022

Ministry of EFA. (2020). *Roadmap of National Determined Contribution (NDC) on climate change adaptation*. MOEFA and GIZ.

Novita, R. (2020). Literature review: The impact of climate change on the incidence of mosquito-borne diseases, especially lymphatic filariasis. *JHECDS 5*(1), 30–39. Available from: https://ejournal2.litbang.kemkes.go.id/index.php/jhecds/article/view/1583

Nurlambang, T., et al. (2021). *Policy, legality and utilization of drought data in Indonesia*. UI Press.

Nurlambang, T., & Tambunan, R. P. (2021). Strategic environmental studies. In J. Supriatna. Yayasan Obor Indonesia.

Perdinan, P., Atmaja, T., Adi, R. F., Estiningtyas, W. (2019). Climate change adaptation and food security: A review of initiatives and policies. *Indonesian Journal of Environmental Law, 5*(60). https://doi.org/10.38011/jhli.v5i1.75

Raiten, D. J., Bremer, A. A. (2020). Exploring the nutritional ecology of stunting: New approaches to an old problem. *Nutrients, 12*(2).

Simamora, P. (2019). *Energy transition: Developments and prospects in Indonesia, delivered at the national seminar on energy corner goes to Campus*. Institute for Essential Service Reform (IESR).

Tempo Magazine. (2022a). Lula's call from the red seaside, November 27, 2022a. *Tempo Magazine*, 80–82.

Tempo Magazine, (2022b). Flood of green economy fund, November 27, 2022b. *Tempo Magazine*, 73–76.

UNDP. (2013). *Climate risk assessment guide-Central Asia.*

UNDRR. (2020). Integrating disaster risk reduction and climate change adaptation in the UN sustainable development cooperation framework.

UNFCCC. (2017). *Human health and adaptation: Understanding climate impacts on health and opportunities for action.*

WHO. (2020). *WHO global strategy in health, environment and climate change.*

WHO. (2021). *The health argument for climate action.* Available at: https://drive.google.com/file/d/1X-axRWM-CZSlq8eSlfizVswA3DXqR36r/view?usp=sharing

WHO. (2022). *Compendium of WHO and other UN guidance on health and environment.*

WHO UNISDR. (2017). *Health emergency and disaster risk management: Climate risk management*.

Wollenberg, E., Vermeulen, S. J., Girvetz, E., Loboguerrero, A. M., Ramirez-Villegas, J. (2016). Reducing risks to food security from climate change. In *Global food security* (Vol. 11, pp. 34–43). Elsevier B.V.

Air Pollution in Urban Bangladesh from Climate Change and Public Health Perspectives

Palash Basak, Soma Dey, and K. Maudood Elahi

Abstract The air quality of Bangladesh, especially in the urban areas, has deteriorated in recent years. Dhaka, the country's capital, is often ranked as one of the worst cities in the world for its degraded air quality. Based on the analysis of ground and satellite-based air quality data and literature review, this chapter underscores the connection between air pollution and climate change and its combined effect from a public health perspective. Findings suggest that in 2022, Dhaka's annual average concentration of $PM_{2.5}$ was nearly 18 times higher than the WHO guideline threshold. A persistent NO_2 hotspot prevails in Dhaka and its surrounding districts. It is also found that variation of NO_2 is associated with climate variables such as temperature and rainfall. Densely populated urban areas will experience more intense air pollution if long spells of hot and dry days prevail due to climate change. The findings inform recommendations for framing coordinated actions to curb local sources of air pollution and global carbon emissions.

Keywords Air quality · Air pollution · $PM_{2.5}$ · NO_2 · Climate change · Public health · Urban · Bangladesh · Hotspots

---

P. Basak (✉)
Department of Geography and Environment, University of Dhaka, Dhaka, Bangladesh
e-mail: Palash.Basak@northsouth.edu

S. Dey
Department of Women and Gender Studies, University of Dhaka, Dhaka, Bangladesh
e-mail: Soma.Dey@du.ac.bd

K. M. Elahi
Department of Geography and Environment, Jahangirnagar University, Dhaka, Bangladesh
e-mail: iamelahi.km@gmail.com

National University, Dhaka, Bangladesh

Stamford University, Dhaka, Bangladesh

R. Akhtar (ed.), *Climate Change and Human Health Scenarios*, Global Perspectives on Health Geography, https://doi.org/10.1007/978-3-031-38878-1_9

1 Introduction

Located at the lower end of the Ganges–Brahmaputra-Meghna (GBM) river system and having proximity to the great Himalayan Mountain Range and the Bay of Bengal, Bangladesh is assumed to be one of the most severely impacted countries in the world due to climate change (Alamgir et al., 2020). Most of the 170 million people in Bangladesh will become potential victims of sea-level rise (Huq, 2001), which may cause large-scale inundation and increased salinity (Dasgupta et al., 2015). Climate change may also trigger a rise in the severity and frequency of extreme cyclones and floods in this tropical South Asian country (Bandyopadhyay et al., 2021). Variations in seasonal temperature and rainfall are likely to be heightened due to climate change (Shahid et al., 2016), which will hinder the country's agricultural production, alter local weather conditions, and deteriorate environmental quality, including air quality status.

Regarding air quality, Bangladesh is one of the worst-performing countries in the world (Raza et al., 2022). Over the recent years, Dhaka—the country's capital and a densely populated megacity—has been making national and international news headlines for its poor air quality (UNB Dhaka, 2023). In 2021, it was ranked as the third most polluted city in the world (Vanzo, 2022), with approximately 20 million people exposed to its detrimental health impacts.

Given this backdrop, it was hypothesized in this study that climate change-induced variations in monthly temperature and rainfall would further deteriorate the country's air quality, which in turn will severely affect public health. The primary aim of this chapter is to examine the potential impact of climate change with a focus on urban air quality and public health concerns in the urbanized areas of Bangladesh. Specific objectives of this study are to (a) analyze satellite and in situ air quality data; (b) explore the spatio-temporal distribution of air quality measures; (c) identify air pollutant hotspots in Bangladesh; (d) assess the potential connection of air pollutants with climate variables; and (e) underscore the potential impact of climate change-induced air pollution in urban Bangladesh from a public health perspective. Some possible mitigation measures have been proposed to prevent more severe air pollution and associated public health issues.

2 Data and Methodology

This study is based on the analysis of ground and satellite-based air quality data and a review of relevant scholarly works. It utilizes three datasets to examine the temporal and spatial variations of air pollution levels. The first dataset consisted of the U.S. Air Quality Index (AQI) and $PM_{2.5}$ concentration measurement from AirNow.gov (2023). A sensor at the US Embassy in Dhaka has been logging AQI since March 2016 and is still active. Hourly observations of $PM_{2.5}$ and AQI data of that sensor have been summarized to estimate daily, monthly, and yearly averages.

The second dataset came from the daily AQI archive of the Department of Environment, Bangladesh (DoE, 2023). DoE's daily AQI data were available for 13 urban locations in Bangladesh, including Dhaka, and were captured between February 2014 and February 2023. In this analysis, data for Savar, Narayanganj, and Gazipur were excluded because these locations fall in greater Dhaka.

The third dataset, Nitrogen Oxide (NO_2), was obtained from European Space Agency (ESA)'s Copernicus Sentinel-5P (S5P) satellite images (GEE, 2023). A JavaScript code was developed in Google Earth Engine (Gorelick et al., 2017) to process and extract S5P NO_2 data captured between July 2018 and January 2023 (Annex A). Two GEE apps were developed to visualize the monthly and weekly distribution of NO_2 over Bangladesh (Annex B).

Data from GEE were also used for additional analysis in R-Studio (Posit, 2023) and ArcGIS Pro (Esri, 2023). R-Studio was used for data manipulation, summarization, statistical analysis, and graph construction with R coding. ArcGIS Pro was used for map preparation and interpretation of the spatio-temporal aspects of the NO_2 distribution.

The satellite measurements of NO_2 were utilized to assess the spatial distribution of air quality in the entire country because $PM_{2.5}$ data are only available for a limited number of stations. The association between in situ (ground) measurement of AQI and satellite observation NO_2 was examined. Based on the association between NO_2 concentration and AQI, NO_2-based AQI categories for the whole of Bangladesh were estimated following U.S. AQI nomenclature. The association between NO_2 concentration and several climate variables (i.e., temperature, rainfall, and relative humidity) was also investigated. Climate data for that analysis came from the Bangladesh Department of Meteorology (BMD, 2023a, 2023b; Kaggle, 2023). Finally, relevant scholarly articles and reports were reviewed to understand the potential impact of air quality degradation on public health.

3 Findings

3.1 $PM_{2.5}$ and AQI in Urban Bangladesh

$PM_{2.5}$, a fine-grained, airborne particulate matter with 2.5 μ or less in width, is one of the key air quality parameters. The AQI is calculated based on $PM_{2.5}$ and several other air quality measures (i.e., PM_{10}, NO_2, CO_2, etc.), and the index provides an overall status of pollution level. The plot of Dhaka's daily average $PM_{2.5}$ concentration (as recorded at the US Embassy in Dhaka) varies from 50 to 300 $\mu g/m^3$ (Fig. 1). It was estimated that between April 2016 and January 2023 (almost seven years), the average daily mean concentration of $PM_{2.5}$ in Dhaka was 89.2 ± 0.5 $\mu g/m^3$.

The yearly average $PM_{2.5}$ concentration in Dhaka ranges from 79 to 98 $\mu g/m^3$, while the grand average for the six years (i.e., 2017 through 2022) is 88.6 $\mu g/m^3$ (Fig. 2). The annual average concentration of $PM_{2.5}$ in Dhaka has been rising. The

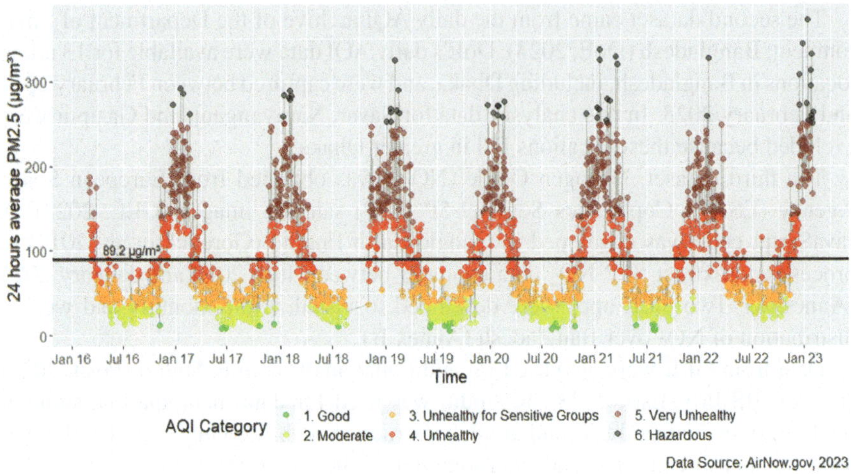

Fig. 1 Daily average $PM_{2.5}$ and AQI categories of Dhaka between April 2016 and January 2023

highest annual $PM_{2.5}$ was recorded in 2021 (98.3 ± 1 $\mu g/m^3$). Between 2017 and 2022, the annual average concentration of $PM_{2.5}$ increased by approximately 15 $\mu g/m^3$.

Figure 1 depicts a cyclic pattern for the intensity of $PM_{2.5}$ concentration in Dhaka. A higher level of $PM_{2.5}$ concentration can be observed between November and April. These months are associated with low rainfall. Typically, Dhaka's dry months are more polluted than the wet months (May–October).

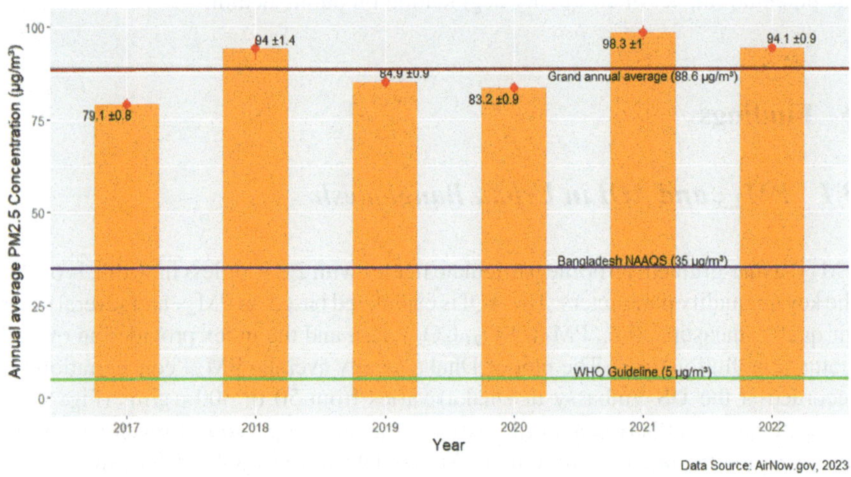

Fig. 2 Annual average $PM_{2.5}$ concentration in Dhaka between 2017 and 2022

In general, an AQI value of over 300 is considered hazardous, and people are advised to stay indoors (AirNow.gov, 2023). According to this guideline, January is one of the worst months for Dhaka in terms of air quality. During this month, air quality in Dhaka frequently crosses the hazardous level (Fig. 3). The daily average AQI of January 13, 2023 was 404, the highest number ever recorded by the sensor. As shown in Fig. 3, except for a few days in January 2020, the air quality in Dhaka falls under the unhealthy or worse category.

$PM_{2.5}$ and AQI measurements of Dhaka are highly associated, $r = 0.99$, $p < 0.001$. The rest of the analysis, therefore, had been conducted based on the AQI by considering the AQI as a proxy for $PM_{2.5}$ concentration and other related pollutants. Based on almost seven years of measurements (between March 2016 and January 2023), it was found that only 2 out of 100 days in Dhaka had 'Good' air quality (Table 1 and Fig. 1). Over 47% of days of a year reached 'Unhealthy' (25.4%), 'Very Unhealthy' (20.2), or 'Hazardous' (2.2%) air quality levels.

The DoE's daily AQI data (DoE, 2023) indicate that the long-term annual AQI averages (based on data recorded between February 2014 and February 2023) of Dhaka came under the unhealthy category (mean AQI = 171 ± 2, $n = 1944$). It is also evident that not only Dhaka but also other major cities of the country experience bad air quality. Among the selected ten major cities in Bangladesh, Dhaka had the highest level of air pollution in 2022 (Fig. 4). The annual averages of four out of these ten cities fell under the unhealthy category (AQI value between 151 and 200) in 2022. However, the air of Barisal, a southern divisional city of the country, was deemed less harmful.

Further analysis of the DoE's data reveals that there is a statistically significant strong positive association ($r \geq 0.82$, $p < 0.001$) between the daily AQI of Dhaka and the previously mentioned ten major cities in Bangladesh (Fig. 5). In other words, the periodicity of air quality in those cities follows a similar pattern as in Dhaka.

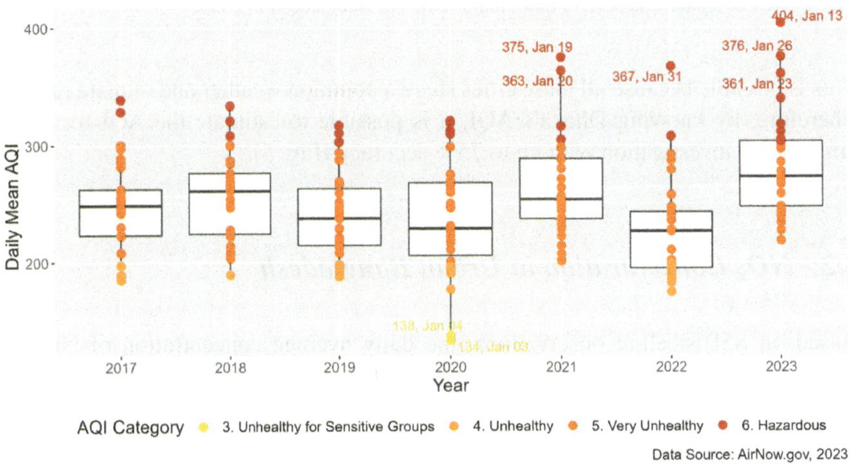

Fig. 3 AQI of Dhaka for January, since 2017

Table 1 Proportion of days based on AQI categories in Dhaka City between March 2016 and January 2023

AQI category	Number of days	Percentage (%)
Good	29	1.2
Moderate	535	22.2
Unhealthy for sensitive groups	695	28.8
Unhealthy	611	25.4
Very unhealthy	488	20.2
Hazardous	52	2.2
Total	2410	100.0

Data source AirNow.gov (2023)

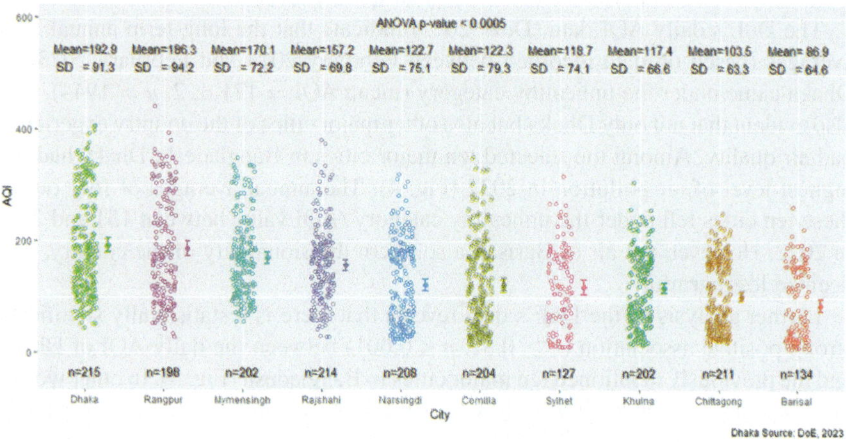

Fig. 4 AQI of Dhaka and several other major cities of Bangladesh in 2022

This is possible because all these cities share a common weather and climate region. Therefore, by knowing Dhaka's AQI, it is possible to estimate the AQI for other cities under investigation with up to 75% accuracy (Fig. 5).

3.2 NO_2 Concentration in Urban Bangladesh

Based on S5P satellite observations, the daily average concentration of NO_2 in Dhaka (at the US Embassy and its surrounding areas in Gulshan) ranges from 100 to 400 μmol/m^2. Between July 2018 and January 2023, the highest average NO_2 concentration was observed during January 2022 (400 μmol/m^2). Dhaka's average NO_2 concentration in 2022 was 223 $\pm$ 19 μmol/m^2. In general, a clear annual cycle is noticeable (Fig. 6), like the variation of $PM_{2.5}$ measures. NO_2 concentration becomes higher than the annual average during five months (October–March) in a year.

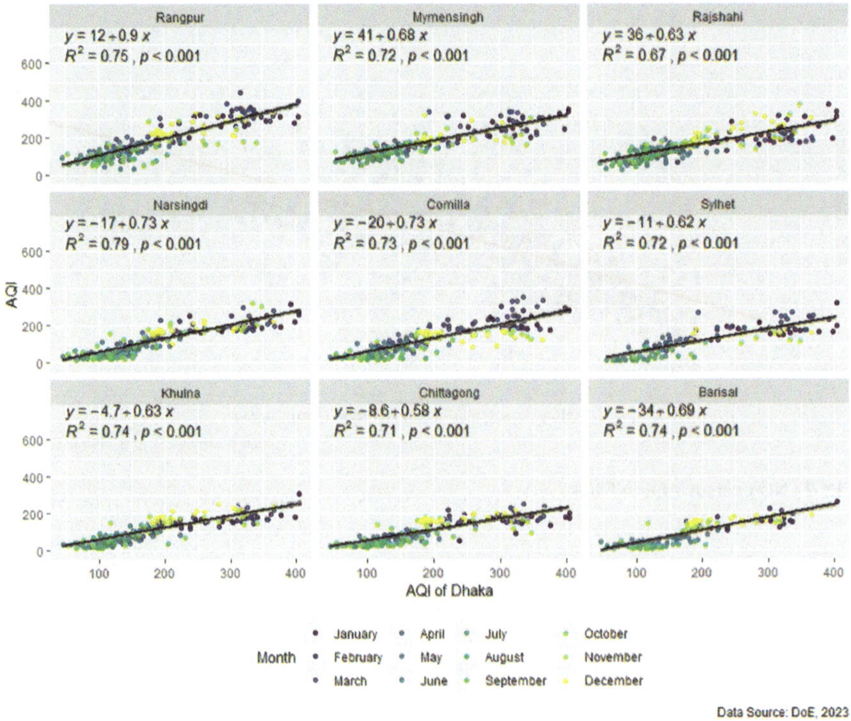

Fig. 5 Association of AQI values of Dhaka with other major cities of Bangladesh in 2022

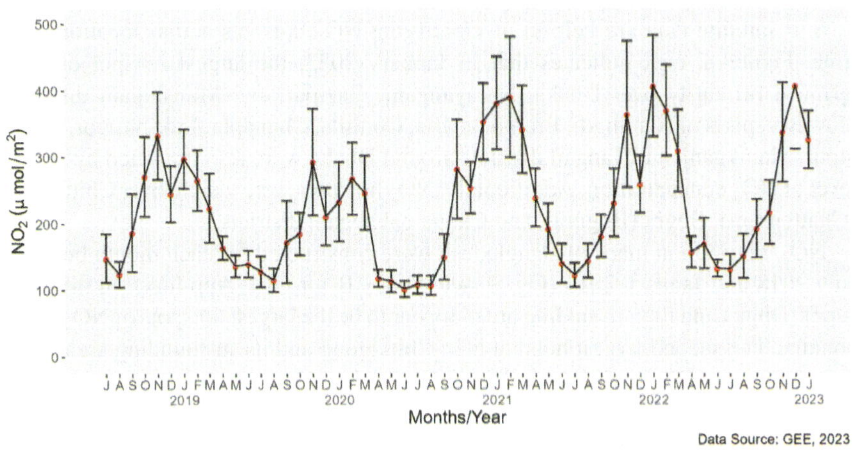

Fig. 6 Monthly mean concentration of topographic NO_2 in Dhaka, from July 2018 to January 2023, as measured within a one-kilometer radius of the US Embassy

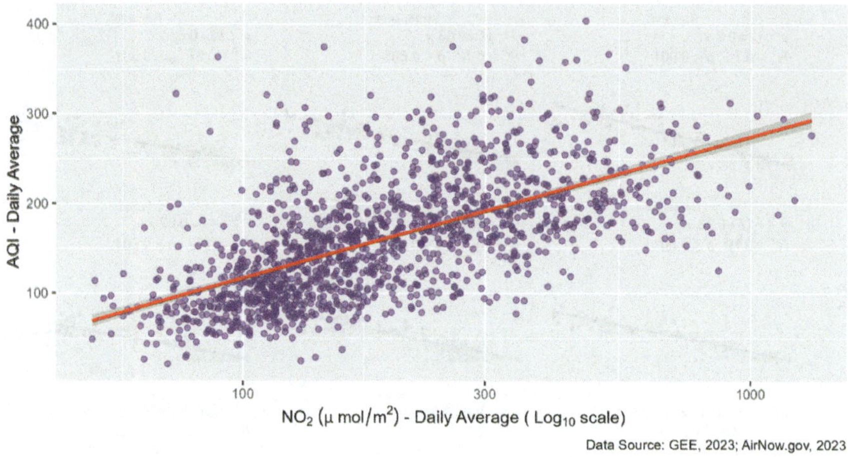

Fig. 7 NO_2 versus AQI in Dhaka

A significant positive correlation between daily average NO_2 concentration (log value) and daily average AQI value in Dhaka, $r = 0.62$, $p < 0.001$ (Fig. 7), was identified. That is, when NO_2 in air increases, AQI value also increases in the city. This association does make sense as NO_2 is one of the precursors of $PM_{2.5}$ in the air (Hodan & Barnard, 2004). As both $PM_{2.5}$ and NO_2 are associated with the AQI, the level of NO_2 was considered a proxy measure for air quality. Therefore, the spatial and temporal distribution of NO_2-based AQI for Bangladesh was estimated using satellite observations.

S5P satellite data are helpful in identifying NO_2 hotspots across locations and time. Figure 8, e.g., indicates that in January 2023, the largest hotspot of NO_2 spanned primarily over Dhaka, Narayanganj, Gazipur, and Munshiganj districts. However, parts of Narsingdi, Brahmanbaria, Comilla, Chandpur, Lakshmipur, Shariatpur, Manikganj, and Tangail districts around Dhaka were also impacted by higher level of NO_2 concentration. Additionally, NO_2 hotspots were visible over Chittagong (Chattogram), Bogra (Bogura), and Chapainababganj districts.

NO_2 distribution maps of Bangladesh were constructed for each month between July 2018 and January 2013 (Fig. 9, see Annex B for GEE App link). In this map series, Dhaka and its surrounding areas appear to be the largest hotspots of NO_2 in the country. The second largest hotspot is in Chittagong and its surrounding areas. The NO_2 hotspot in Dhaka shrinks in May–October but does not disappear completely. NO_2 hotspots are larger during dry months (November–April). All NO_2 hotspots are associated with urban agglomeration.

Observing the strong and significant association between NO_2 concentration level and AQI, NO_2 measurements-based AQI categories of the whole country were calculated. The mean NO_2 values for each AQI category were calculated and then a range was assigned for each category to classify the concentration of NO_2 as per AQI

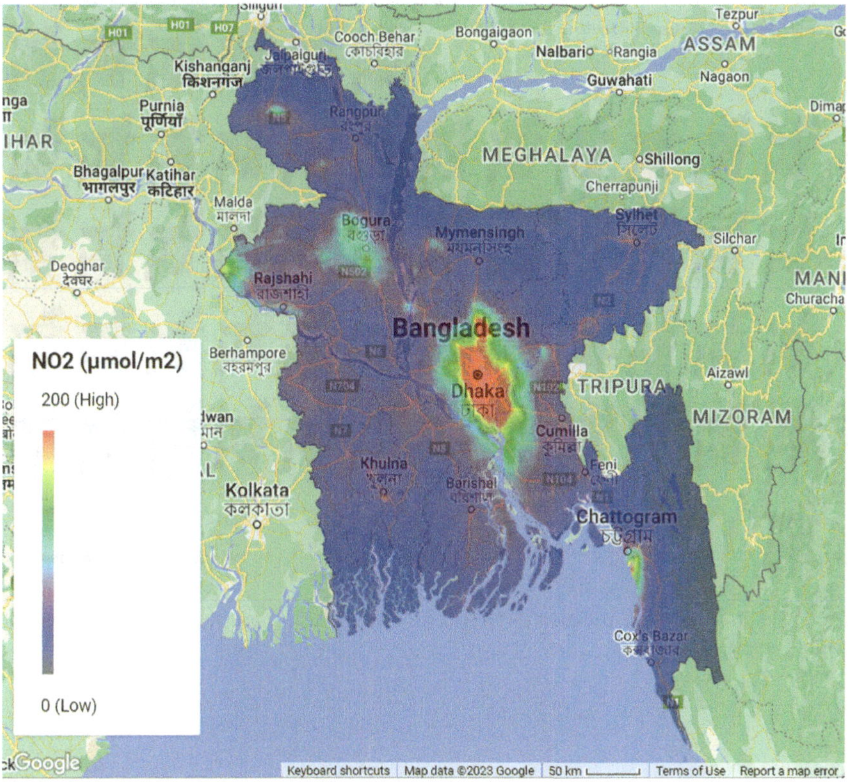

Fig. 8 NO_2 distribution in Bangladesh during January 2023, based on S5P satellite observation

nomenclature (Fig. 10). A similar classification scheme was developed for the AQI categories. A map of Bangladesh was constructed with classified AQI data (Fig. 11).

Figure 11 shows the spatial distribution of NO_2-based AQI categories in Bangladesh during January 2023. The legend of the figure shows the ranges of NO_2 values for each AQI category. According to the figure, a significant portion of Dhaka, Narayanganj, Gazipur, Munshiganj, Narsingdi, and Chandpur districts has been experiencing heightened levels of air pollution. It is estimated that over 20 million people in these districts are exposed to severe air pollution. There are other smaller hotspots in different parts of the country. However, due to the coarse resolution of S5P satellite images, those hotspots are not visible in Fig. 11.

Evidently, a significant part of the country remains less polluted. Except a small part of Chapainawabganj (Nawabganj) district, border areas of Bangladesh are free from NO_2 hotspot (Figs. 9 and 11). Rural areas of the country also exhibit lower levels of NO_2. Rural areas are generally free from air pollution unless pollutants from other urban and industrial places are transported by the wind. In line with these findings, it can be assumed that the sources of air pollution are mainly local and concentrated in areas with higher levels of urbanization and industrialization.

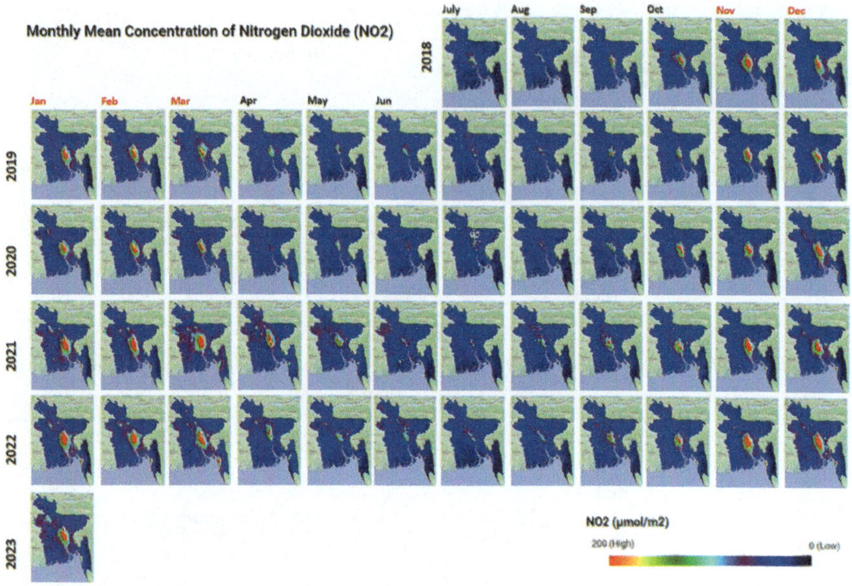

Fig. 9 Spatio-temporal distribution of NO_2 across Bangladesh, since July 2018, based on S5P satellite observation

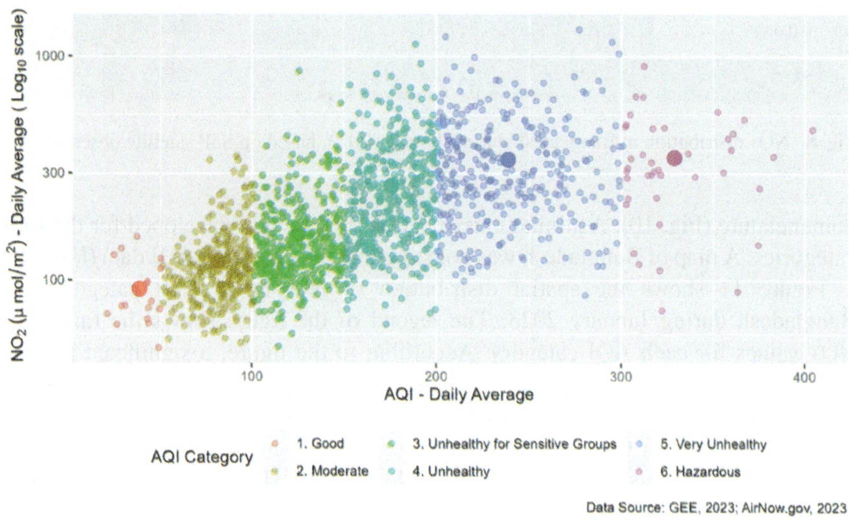

Fig. 10 AQI categories and NO_2 concentration in Dhaka

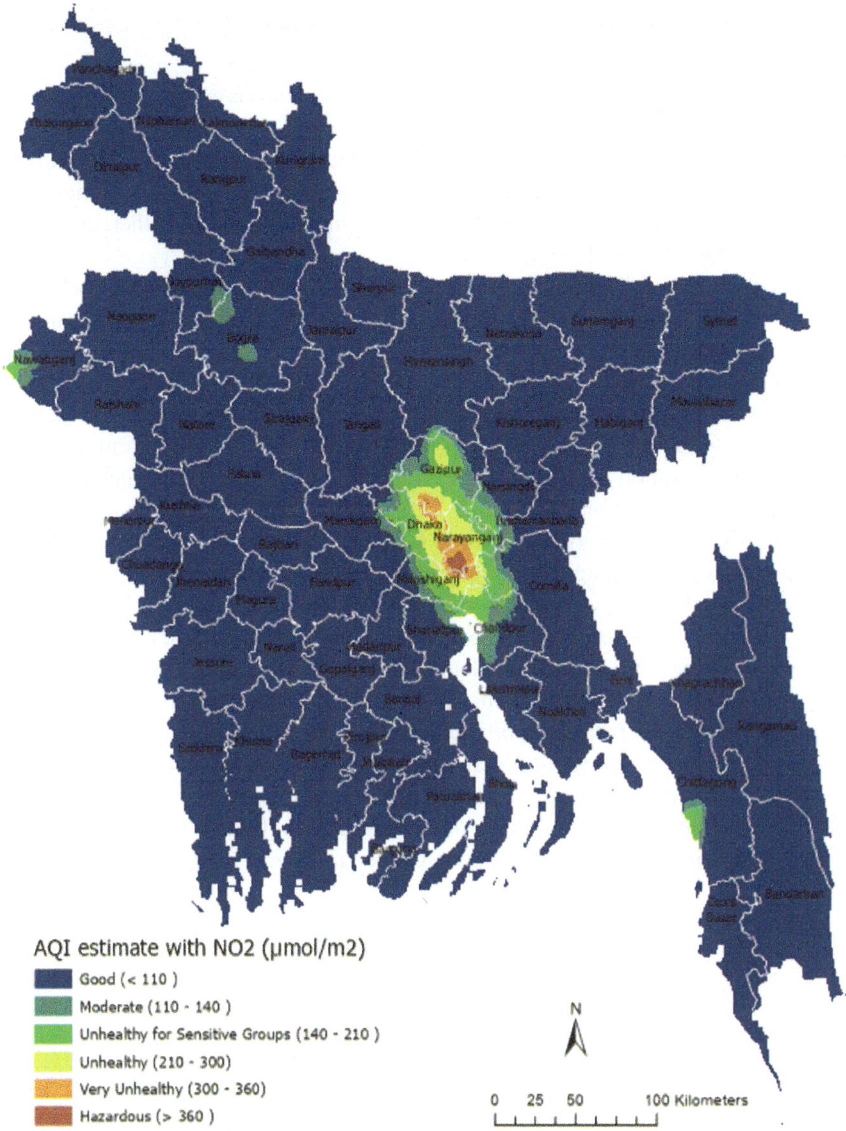

Fig. 11 Distribution of NO_2-based AQI categories in Bangladesh during January 2023, based on S5P satellite observation

Usually, the sources of air pollutants and greenhouse gases are often the same (World Bank, 2022c). Therefore, it can be assumed that air polluting sources of Bangladesh contribute to climate warming.

3.3 NO_2 Concentration Versus Climate Variables

A statistically significant negative association can be observed between long-term monthly rainfall patterns and NO_2 concentration, $r = -0.94$, $p < 0.001$. Figure 12 shows that a higher level of NO_2 pollution is associated with a lower level of rainfall. According to this model, rainfall can predict approximately 87% of the variation in NO_2 concentration for a given month (without controlling for any other variables), $F(1,10) = 77.8$, $p < 0.001$, $R^2_{\text{adj}} = 0.87$.

As with the rainfall pattern, the monthly average temperature of Dhaka is also associated with NO_2 distribution. Months with lower temperature experience higher NO_2 pollution (Fig. 13).

Analysis shows that in addition to rainfall and minimum temperature, NO_2 concentration is also negatively correlated to maximum temperature and relative humidity ($r \geq \pm 0.66$, Fig. 14).The correlation between monthly normal rainfall and NO_2 concentration is -0.94, whereas the correlation between normal minimum temperature and NO_2 concentration is -0.92. These findings confirm the fact that air quality in Dhaka and other cities of Bangladesh deteriorated the most during dry winter season. Climate change may trigger modification to this typical pattern, adding more polluted days during hot dry season.

It is evident in literature that temperature and rainfall patterns are likely to change in Bangladesh because of climate change (Shahid et al., 2016). The variations in both temperature and rainfall patterns, marked by the above analysis, will impact the NO_2 concentration and other air pollutants in Bangladesh. Studies indicate that hotter days associated with a warming climate can increase ground-level ozone (Nolte et al., 2018) and particulate matters, such as windblown dust from droughts (EPA, 2023). Since temperature is expected to be increasing in Bangladesh, hot and dry sunny days will potentially contribute to higher concentrations of ground-level ozone and

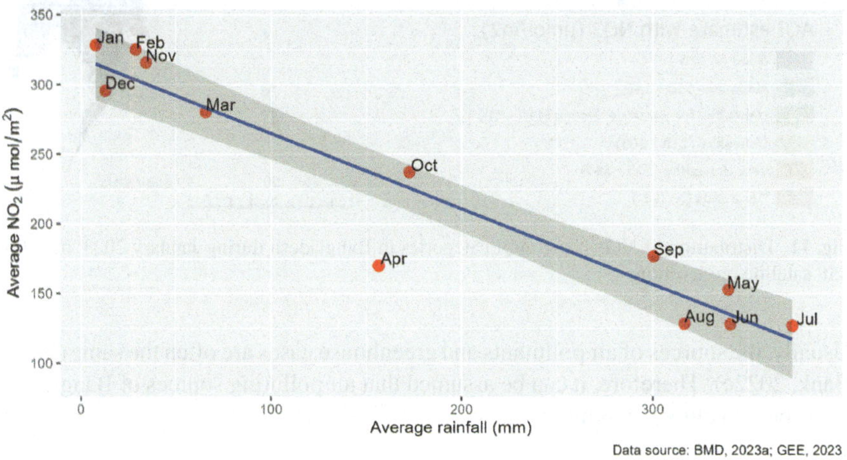

Fig. 12 Monthly rainfall versus NO_2 concentration in Dhaka

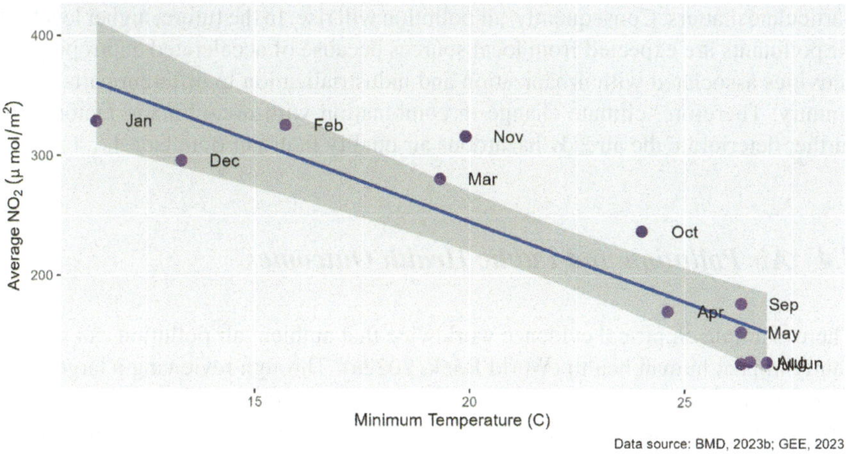

Fig. 13 Monthly minimum temperature versus NO_2 concentration in Dhaka

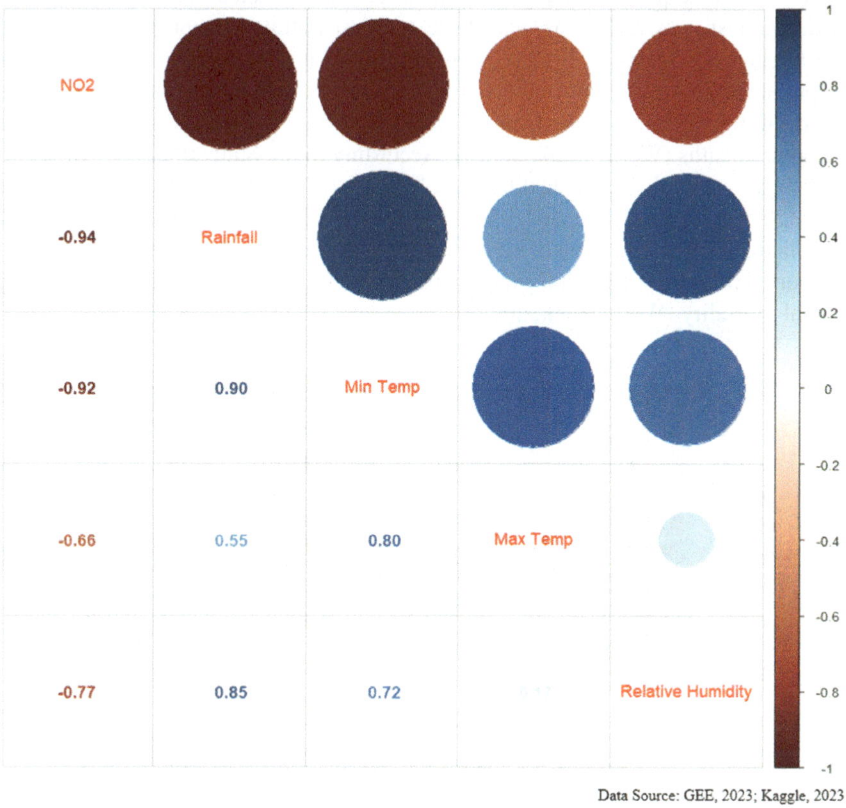

Fig. 14 Correlation matrix showing association between NO_2 and climate variables for Dhaka

particulate matters. Consequently, air pollution will rise. In the future, higher levels of air pollutants are expected from local sources because of accelerated anthropogenic activities associated with urbanization and industrialization in different parts of the country. Therefore, climate change in combination with non-climatic factors will further deteriorate the already hazardous air quality in urban Bangladesh.

3.4 Air Pollution and Public Health Outcomes

There is ample empirical evidence worldwide that ambient air pollution can significantly impact human health (World Bank, 2022a). Through reviewing a large body of research work, Raza et al. (2022) summarized air pollution's short-term and long-term effects on human health. The effects can range from short-term discomfort (i.e., irritation of eyes, nose skin, and throat; headache; dizziness; breathing difficulties; etc.) to more severe consequences (i.e., chronic obstructive pulmonary disease [COPD]; asthma; pneumonia; lung and heart problems; cancer; etc.).

Bangladesh was ranked as one of the topmost polluted countries in the world in 2021 and 2022 (Health Effects Institute, 2020; Fig. 15). Recent studies marked air pollution as the second highest risk factor (after malnutrition), causing death and disability in Bangladesh (Murray et al., 2020), accounting for about 20% of premature deaths (World Bank, 2022a). According to the State of Global Air report published by the US-based Health Effect Institute (2020), $PM_{2.5}$-attributable deaths in Bangladesh were 74 thousand (95% UI: 48,000–102,000) in 2019. The same study identified an increasing trend of air pollutant-related health impacts in the country. It is reported that every 10 ug/m^3 increase in PM_{10} will result in around 0.5% increase in total mortality (World Bank, 2022a).

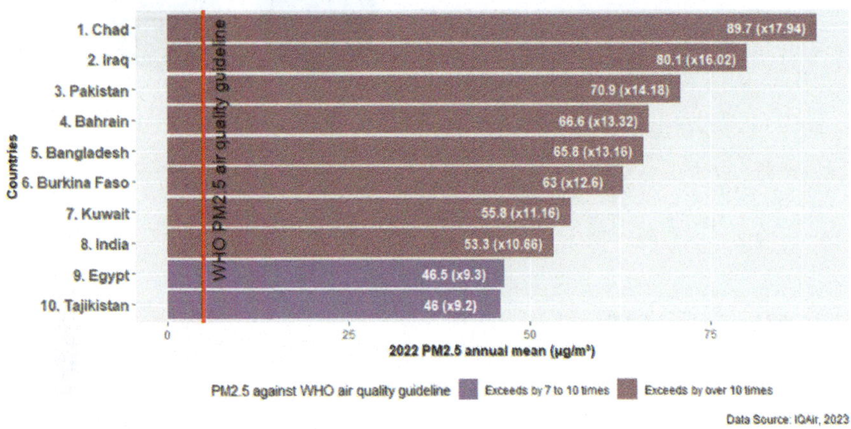

Fig. 15 Top 10 countries of the world with the most air pollution (IQAir, 2023)

Applying a gender lens, several scholars have attempted to examine the health effects of air pollution in Bangladesh. However, their focus primarily revolves around indoor air pollution. A study of this sort found a large difference between women's and men's exposure to indoor air pollution, locating less educated women from poor households and their children as the most exposed groups to indoor air pollution (Dasgupta et al., 2006). In Bangladesh, everyday women spend a significant time in the kitchen for cooking, and many use solid waste as fuel. According to Pitt et al. (2005), Bangladeshi women's typical everyday household responsibilities trigger higher exposure to air pollution, affecting their respiratory health and the young children they supervise.

It is also revealed that children exposed to solid fuel use (SFU) were 1.47 times more likely to have developmental delays (Rana et al., 2022). Infants and young children suffer the worst mortality and morbidity from indoor air pollution (Dasgupta et al., 2006). Collecting evidence from Bangladesh, India, and Pakistan, another study claims household air pollution brings severe health impacts for women, such as sterilization, termination, and duration of current pregnancy (Ahmed et al., 2022). Kurata et al. (2020) reported a positive association between the use of solid fuels and respiratory illness among Bangladeshi girls.

Women tend to face double exposure to air pollution—one from indoor air and another from ambient air. A recent study by Al Nahian et al. (2023) reports an alarming increase in unhealthy or extremely unhealthy air days in Dhaka. According to the study, women impacted by air pollution in Dhaka had babies with lower birth weights (LBW), and they gave preterm births (PTB) at a higher rate. The results also suggest that the second trimester could be the most vulnerable pregnancy period due to ambient air pollution. Overall, in the literature, home-bound women and their children appeared as the worst victims of air pollution-induced health hazards. So far, very little has been known about the exposure of men or other vulnerable groups (e.g., elderlies) to air pollution in their occupational settings or surrounding environment. Counting the worsening climate change scenario and its observed health effects worldwide, it may not be very hasty to predict that all segments of the rapidly expanding urban population of Bangladesh will suffer more in the future from indoor and outdoor air pollution.

4 Discussion and Recommendations

The findings of this study reiterate that air pollution and climate change are interrelated phenomena. These are two sides of the same coin (World Bank, 2022c). Air pollution affects climate change, and at the same time, climate change also affects air pollution (Nolte et al., 2018). Compared to the global average, per capita greenhouse gas (GHG) emission is relatively low in Bangladesh (World Bank, 2023). Hence, Bangladesh is not really impacting the global climate pattern with its GHG emission; rather, the country is assumed to be one of the prime victims of climate change (Huq, 2001). Yet, the local GHG emission is causing significant air pollution, which will intensify in the context of global climate change. In particular, black carbon—one of the key components of $PM_{2.5}$ in Dhaka's air (Salam et al., 2021) generated from incomplete combustion of fossil fuels, wood, and other fuels (Climate & Clean Air Coalition, 2023)—may amplify climate warming.

It is evident from data analysis that Dhaka and other major cities of the country rarely meet global or national air quality standards. The World Health Organization (WHO) threshold for a healthy annual $PM_{2.5}$ is 5 $\mu g/m^3$ (World Health Organization, 2021). However, as shown above, in 2022, the annual average concentration of $PM_{2.5}$ in Dhaka was nearly 18 times higher than the WHO guideline threshold. In certain months, $PM_{2.5}$ concentration in the city reaches up to 50 times or higher than the WHO guideline (IQAir, 2022).

In contrast to WHO standards, the regulatory requirements in Bangladesh are much more relaxed. For example, the Bangladesh National Ambient Air Quality Standard (NAAQS) threshold for an acceptable annual average of $PM_{2.5}$ is 35 $\mu g/m^3$ (GoB, 2022). It certainly allows higher exposure compared to the WHO guideline threshold. Still, the $PM_{2.5}$ of Dhaka in 2022 was almost two and a half (2.5) times higher than the national guideline threshold. The surrounding cities of Dhaka and several major cities across the country also have high concentrations of $PM_{2.5}$.

The spatial and temporal analysis of NO_2 distribution uncovers persistent air pollution hotspots on top of Dhaka and Chittagong city. The hotspots turn larger during the months with low rainfall every year, exposing millions of people to air pollutants. The distribution pattern of NO_2 hotspots indicates that local non-climatic sources are primarily responsible for air pollution, and rural areas of Bangladesh have a low NO_2 concentration.

The findings also reveal a yearly cycle of air pollution and its connection with the local climate variables. The air quality of Bangladesh, especially in the high-density urban areas, deteriorates between November and March (see Begum et al., 2013; Kayes et al., 2019; Rahman et al., 2019). These months typically have less rainfall and humidity (DoE, 2018). Under the climate change scenario, an increased temperature will cause higher ground-level ozone concentrations. Also, drought potential will increase (Alamgir et al., 2020). These changes will intensify air pollution in urban Bangladesh. The situation will be aggravated by non-climatic factors associated with rapid urbanization and industrialization.

As discussed above, exposure to high levels of indoor and ambient air pollution can trigger adverse health impacts. Globally, $PM_{2.5}$-related air pollution kills about 6.4 million people annually. It results in \$8.1 trillion of economic loss, equivalent to 6.1% of the global gross domestic product (GDP). Bangladesh replicates the global scenario. In 2019, the health cost of mortality and morbidity caused by exposure to $PM_{2.5}$ air pollution in this country was \$26.5 billion, equivalent to 8.8% of the GDP (World Bank, 2022b). If no effective action is taken, the geographical extent of air pollution hotspots will increase, and more people in Bangladesh will experience associated short and long-term health impacts.

Even though the air quality situation is very grave in Bangladesh, with sincere and tactful initiatives, it is possible to curb air pollutants. To minimize $PM_{2.5}$, NO_2, and other pollutants in the air, especially in the dry months, the Department of Environment (DoE) under the Ministry of Environment, Forest and Climate Change should initiate public awareness-building programs and strictly enforce the *Air Pollution (Control) Rules, 2022*[1] (GoB, 2022). Although, reportedly, the DoE failed to undertake adequate steps to curb air pollution despite repeated directives from the court ('You are killing us …', 2023), still, we hope DoE will be able to play a proactive role very soon.

In addition to the DoE, different government and local government agencies should take coordinated actions to reduce the intensity and extent of air pollution. Especially, city corporations and municipalities can adopt strategies for controlling and managing industrial establishments, transportation, and construction activities. Industrial and urban expansions must be regulated strictly based on air pollution considerations. Tree plantation and preservation of greeneries in and around urban centers will be effective measures for reducing particulate matter in the air. To minimize fossil fuel burning, Bangladesh should promote renewable energy and electric vehicle. Urban environmental organizations and civil societies should closely monitor government initiatives and raise their voices to protect vulnerable people.

The dire air quality situation in urban Bangladesh is not unique. Several megacities in nearby India and Pakistan, i.e., Delhi (Garg & Gupta, 2020), Kolkata (Dutta et al., 2021), and Lahore (Tabinda et al., 2020) have been facing serious air pollution problems. These cities also show a periodic variation in the intensity of air pollutants. In the context of global climate change, the problem is becoming more complex and severe. Global initiatives involving national governments and various stakeholders are a must to limit global carbon emissions. Without well-coordinated multi-scale initiatives, it is impossible to address the air pollution problem and save the communities at risk.

While satellite observations can provide important insight into the overall air quality situation, this sort of data is insufficient to pinpoint smaller air pollution hotspots and their sources. Therefore, additional in situ data capture and monitoring must be increased nationwide to detect and manage air pollution hotspots. A portion of the climate change adaptation fund can be allocated to reduce the sources of air pollution in Bangladesh. Further studies can be conducted to explore the gender differences in air pollution impact.

5 Conclusion

This study examined the air quality of urban Bangladesh in terms of three indicators (i.e., $PM_{2.5}$, AQI, and NO_2) and quantified their association with climate variables (i.e., temperature, rainfall, and humidity).The findings of this study indicate that the largest hotspot of air pollution in the country exists in Dhaka and its surrounding

[1] The Air Pollution (Control) Rules 2022, formulated under Section 20 of Bangladesh Environmental Conservation Act, 1995, aims to prevent, control, and reduce air pollution by punishing polluters and rewarding protesters of pollution and non-polluters.

urban areas. As a result, a large number of populations are exposed to serious health consequences from air pollutants.

It is revealed that the air quality of Bangladesh is associated with several climatic and non-climatic factors. Usually, months with low rainfall and less humidity have the highest level of pollutants in the air. While sources of air pollutants are primarily local, the extreme changes in the weather pattern due to global climate change can further deteriorate the country's air quality. Densely populated urban areas will be severely impacted if the spells of hot dry days extend around the year due to climate change.

Considering the dire consequences of air pollution on public health, Bangladesh should take immediate and drastic actions to curb the sources of indoor and ambient air pollution. Mitigation and adaptation strategies for climate change in Bangladesh should be developed with special attention to the rising air pollution crisis. At the same time, the global community should take necessary actions to prevent further escalation of the climate change phenomenon.

Supplementary Materials

Annex A: Google Earth Engine Code (JavaScript)

This script extracts near real-time nitrogen dioxide (NO_2) data for one-kilometer radius area centered at Dhaka's US Embassy. It exports the NO_2 measurement along with time of the measurement as a CSV file in Google Drive.

https://code.earthengine.google.com/5fcb16fe470b41c791a371c5c5b1e2b7

Annex B: Google Earth Engine Apps

Annex B.1: Monthly Distribution of NO_2 Across Bangladesh

This app shows the average monthly distribution of NO_2 concentrations over Bangladesh. Select a specific year and month to update the display. Wait for the layer to be re-drawn after changing year or month. Use the inspector tool to get value for a specific location. Note that Sentiel-5P NO_2 data are available only since July 2018.

https://palashbasak.users.earthengine.app/view/no2-bd

Annex B.2: Weekly Distribution of NO_2 Across Bangladesh

This app shows the average weekly distribution of NO_2 concentrations over Bangladesh. Select a date range to update the display. Wait for the layer to be re-drawn after changing the date range. Use the inspector tool to get value for a specific location. This app displays data for last four years.

https://palashbasak.users.earthengine.app/view/no2-bd-weekly

References

Ahmed, M., Shuai, C., Abbas, K., Rehman, F. U., & Khoso, W. M. (2022). Investigating health impacts of household air pollution on woman's pregnancy and sterilization: Empirical evidence from Pakistan, India, and Bangladesh. *Energy, 247*, 123562.

AirNow.gov. (2023). *The historical archive of U.S. Air Quality Index (AQI) for Dhaka US Consulate, Bangladesh*. Retrieved February 05, 2023, from https://www.airnow.gov/international/us-embassies-and-consulates/#Bangladesh$Dhaka

Al Nahian, M., Ahmad, T., Jahan, I., Chakraborty, N., Nahar, Q., & Streatfield, P. K. (2023). Air pollution and pregnancy outcomes in Dhaka, Bangladesh. *The Journal of Climate Change and Health, 9*, 100187. https://doi.org/10.1016/j.joclim.2022.100187

Alamgir, M., Khan, N., Shahid, S., Yaseen, Z. M., Dewan, A., Hassan, Q., & Rasheed, B. (2020). Evaluating severity–area–frequency (SAF) of seasonal droughts in Bangladesh under climate change scenarios. *Stochastic Environmental Research and Risk Assessment, 34*, 447–464.

Bandyopadhyay, S., Dasgupta, S., Khan, Z. H., & Wheeler, D. (2021). Spatiotemporal analysis of tropical cyclone landfalls in Northern Bay of Bengal, India and Bangladesh. *Asia-Pacific Journal of Atmospheric Sciences*, 1–17.

Begum, B. A., Hopke, P. K., & Markwitz, A. (2013). Air pollution by fine particulate matter in Bangladesh. *Atmospheric Pollution Research, 4*(1), 75–86.

BMD. (2023a). *Normal monthly rainfall*. Bangladesh Meteorological Department. Accessed on January 15, 2023a from http://live4.bmd.gov.bd/p/Normal-Monthly-Rainfall/

BMD. (2023b). *Temperature data*. Bangladesh Meteorological Department. Accessed on January 15, 2023b from https://live4.bmd.gov.bd/p/Temperature-Data

Climate & Clean Air Coalition. (2023). *Black carbon*. Accessed on April 18, 2023 from https://www.ccacoalition.org/en/slcps/black-carbon

Dasgupta, S., Hossain, M. M., Huq, M., & Wheeler, D. (2015). Climate change and soil salinity: The case of coastal Bangladesh. *Ambio, 44*, 815–826.

Dasgupta, S., Huq, M., Khaliquzzaman, M., Pandey, K., & Wheeler, D. (2006). Who suffers from indoor air pollution? Evidence from Bangladesh. *Health Policy and Planning, 21*(6), 444–458.

DoE. (2018). *Ambient air quality in Bangladesh*. Clean Air and Sustainable Environment Project, Department of Environment, Ministry of Environment, Forest, and Climate Change of the Government of Bangladesh. http://doe.portal.gov.bd/sites/default/files/files/doe.portal.gov.bd/page/cdbe516f_1756_426f_af6b_3ae9f35a78a4/2020-06-10-11-02-5a7ea9f58497800ec9f0cea00ce7387f.pdf

DoE. (2023). *AQI archives*. Department of Environment. Accessed on February 25, 2023 from http://case.doe.gov.bd/index.php?option=com_content&view=category&id=8&Itemid=32

Dutta, S., Ghosh, S., & Dinda, S. (2021). Urban air-quality assessment and inferring the association between different factors: A comparative study among Delhi, Kolkata and Chennai megacity of India. *Aerosol Science and Engineering, 5*, 93–111.

EPA. (2023). *Climate change impacts on air quality*. United States Environmental Protection Agency (EPA). Accessed on April 11, 2023 from https://www.epa.gov/climateimpacts/climate-change-impacts-air-quality

Esri. (2023). *ArcGIS Pro (version 3.1.0)*. Accessed on February 2023 from https://www.esri.com/en-us/arcgis/products/arcgis-pro/overview

Garg, A., & Gupta, N. C. (2020). The great smog month and spatial and monthly variation in air quality in ambient air in Delhi, India. *Journal of Health and Pollution, 10*(27).

GEE. (2023). *Sentinel-5P NRTI NO_2: Near real-time nitrogen dioxide* (Google Earth Engine dataset provided by European Union/ESA/Copernicus). Accessed on January 15, 2023 from https://developers.google.com/earth-engine/datasets/catalog/COPERNICUS_S5P_NRTI_L3_NO2#description

GoB. (2022). *Air pollution (control) rules, 2022*. Ministry of Environment, Forest and Climate Change, Government of Bangladesh (GoB). Accessed on February 05, 2023, from https://moef.portal.gov.bd/sites/default/files/files/moef.portal.gov.bd/page/6ee9d54b_b349_4e85_b0da_6df1225285cb/_compressed.pdf

Gorelick, N., Hancher, M., Dixon, M., Ilyushchenko, S., Thau, D., & Moore, R. (2017). Google earth engine: Planetary-scale geospatial analysis for everyone. *Remote Sensing of Environment, 202*, 18–27.

Health Effects Institute. (2020). *State of global air 2020*. Special Report. Health Effects Institute.

Hodan, W. M., & Barnard, W. R. (2004). Evaluating the contribution of $PM_{2.5}$ precursor gases and re-entrained road emissions to mobile source $PM_{2.5}$ particulate matter emissions. *MACTEC Federal Programs, Research Triangle Park, NC*.

Huq, S. (2001). Climate change and Bangladesh. *Science, 294*(5547), 1617–1617.

IQAir. (2022). *Air quality in Dhaka*. Accessed on December 15, 2022 from https://www.iqair.com/bangladesh/dhaka

IQAir. (2023). *World's most polluted countries and regions (historical data 2018–2022)*. Accessed on April 05, 2023 from https://www.iqair.com/us/world-most-polluted-countries

Kaggle. (2023). *65 years of weather data Bangladesh preprocessed* (Original data from Bangladesh Meteorological Department). Accessed on March 10, 2023 from https://www.kaggle.com/datasets/emonreza/65-years-of-weather-data-bangladesh-preprocessed

Kayes, I., Shahriar, S. A., Hasan, K., Akhter, M., Kabir, M. M., & Salam, M. A. (2019). The relationships between meteorological parameters and air pollutants in an urban environment. *Global Journal of Environmental Science and Management, 5*(3), 265–278. https://doi.org/10.22034/GJESM.2019.03.01

Kurata, M., Takahashi, K., & Hibiki, A. (2020). Gender differences in associations of household and ambient air pollution with child health: Evidence from household and satellite-based data in Bangladesh. *World Development, 128*, 104779. https://doi.org/10.1016/j.worlddev.2019.104779

Murray, C. J., Aravkin, A. Y., Zheng, P., Abbafati, C., Abbas, K. M., Abbasi-Kangevari, M., Abd-Allah, F., Abdelalim, A., Abdollahi, M., Abdollahpour, I., Hussein Abegaz, K., Abolhassani, H., Aboyans, V., Guimarães Abreu, L., Abrigo, M. R. M., Abualhasan, A., Jamal Abu-Raddad, L., Abushouk, A. I., Adabi, M., ... Borzouei, S. (2020). Global burden of 87 risk factors in 204 countries and territories, 1990–2019: A systematic analysis for the Global Burden of Disease Study 2019. *The Lancet, 396*(10258), 1223–1249.

Nolte, C. G., et al. (2018). Ch. 13: Air quality. In: *Impacts, risks, and adaptation in the United States: Fourth national climate assessment, volume II* (pp. 512–538). U.S. Global Change Research Program. https://doi.org/10.7930/NCA4.2018.CH13

Pitt, M. M., Rosenzweig, M. R., & Hassan, M. (2005). *Sharing the burden of disease: Gender, the household division of labor and the health effects of indoor air pollution*. CID Working Paper Series.

Posit. (2023). *RStudio desktop* (IDE for open source data science, version 2022.12.0). Accessed on February 02, 2023 from https://posit.co/download/rstudio-desktop/

Rahman, M. M., Mahamud, S., & Thurston, G. D. (2019). Recent spatial gradients and time trends in Dhaka, Bangladesh, air pollution and their human health implications. *Journal of the Air & Waste Management Association, 69*(4), 478–501. https://doi.org/10.1080/10962247.2018.1548388

Rana, J., Luna-Gutiérrez, P., Haque, S. E., Nazif-Muñoz, J. I., Mitra, D. K., & Oulhote, Y. (2022). Associations between household air pollution and early child development among children aged 36–59 months in Bangladesh. *Journal of Epidemiology and Community Health, 76*(7), 667–676.

Raza, W. A., Mahmud, I., Rabie, T. S. (2022). *Breathing heavy: New evidence on air pollution and health in Bangladesh.* International Development in Focus. World Bank. Retrieved on March 03, 2023, from https://openknowledge.worldbank.org/handle/10986/38289

Salam, A., Andersson, A., Jeba, F., Haque, M. I., Hossain Khan, M. D., & Gustafsson, O. (2021). Wintertime air quality in megacity Dhaka, Bangladesh strongly affected by influx of black carbon aerosols from regional biomass burning. *Environmental Science & Technology, 55*(18), 12243–12249.

Shahid, S., Wang, X. J., Harun, S. B., Shamsudin, S. B., Ismail, T., & Minhans, A. (2016). Climate variability and changes in the major cities of Bangladesh: Observations, possible impacts and adaptation. *Regional Environmental Change, 16*, 459–471.

Tabinda, A. B., Ali, H., Yasar, A., Rasheed, R., Mahmood, A., & Iqbal, A. (2020). Comparative assessment of ambient air quality of major cities of Pakistan. *Mapan, 35*, 25–32.

UNB, Dhaka. (2023, February 17). Dhaka's air hazardous, worst in the world this morning. *The Daily Star*. Retrieved March 2, 2023, from https://www.thedailystar.net/environment/pollution/air-pollution/news/dhakas-air-hazardous-worst-the-world-morning-3250346

Vanzo, T. (2022). *25 most polluted cities in the world (2023 rankings).* Accessed on February 07, 2023, from https://smartairfilters.com/en/blog/25-most-polluted-cities-world-2023-rankings/

World Bank. (2022a). *Striving for clean air: Air pollution and public health in South Asia.* Accessed on March 03, 2023 from https://www.worldbank.org/en/region/sar/publication/striving-for-clean-air

World Bank. (2022b). *The global health cost of $PM_{2.5}$ air pollution: A case for action beyond 2021.* The World Bank. Accessed on 12 April 2023 from https://openknowledge.worldbank.org/handle/10986/36501

World Bank. (2022c). *What you need to know about climate change and air pollution.* Accessed on April 11, 2023 from https://www.worldbank.org/en/news/feature/2022/09/01/what-you-need-to-know-about-climate-change-and-air-pollution

World Bank. (2023). *CO_2 emissions (metric tons per capita)—Bangladesh, Australia, United States, World, China, India.* Accessed on April 15, 2023 from https://data.worldbank.org/indicator/EN.ATM.CO2E.PC?locations=BD-AU-US-1W-CN-IN

World Health Organization. (2021). *WHO global air quality guidelines: Particulate matter ($PM_{2.5}$ and PM_{10}), ozone, nitrogen dioxide, sulfur dioxide and carbon monoxide.* World Health Organization. https://apps.who.int/iris/handle/10665/345329

'You are killing us by not curbing air pollution': HC tells DoE. (2023, Feb 5). *The Daily Star*. Retrieved March 6, 2023, from https://www.thedailystar.net/environment/pollution/air-pollution/news/you-are-killing-us-not-curbing-air-pollution-3239836

Heatwave Mortality and Adaptation Strategies in India

Rais Akhtar

Abstract Research studies conducted in both developed and developing countries, have asserted that global climate change is likely to be accompanied by an increase in the frequency and intensity of heat waves The WHO estimates that from 1998–2017, more than 166,000 people died due to heatwaves, including more than 70,000 who died during the last two weeks of August 2003 heatwave in Europe. Indian Meteorological department (IMD) predicts that most parts of India including those in peninsular and coasts of India will experience in increase in duration of heatwave by 12–18 days by 2060. The Lancet, medical journal asserts that . India saw a 55% rise in deaths due to extreme heat between 2000–2004 and 2017–2021, a recent study published has found (BBC, 2022). The study shows that mortality due to heatwaves started declining in the country from 2016.This is because various heat protection plans initiated by the governments both at the Centre and state levels and other organizations helped in the reduction of heatwave related deaths. The introduction of Ahmedabad Heat Action Plans (HAP) in some of the cities in Gujarat, Maharashtra, Odisha and Telangana have demonstrated that such action plans can improve resilience of the citizens and reduce the severe health impacts of Heat Wave.

Keywords IMD · Heatwave mortality · Ahmedabad Heat Action Plan · Vulnerability · Andhra Pradesh · Odisha · 2003 European heatwave

1 Introduction

Across the globe, covering both developed and developing countries, hot days are getting hotter and more frequent, while we're experiencing fewer cold days. Research studies conducted in both developed and developing countries have asserted that global climate change is likely to be accompanied by an increase in the frequency and intensity of heatwaves (Orindi, 2005). Orindy's prediction was based on 2003

R. Akhtar (✉)
Formerly Professor of Geography, University of Kashmir, Srinagar, Jammu & Kashmir, India
e-mail: raisakhtar@gmail.com

R. Akhtar (ed.), *Climate Change and Human Health Scenarios*, Global Perspectives on Health Geography, https://doi.org/10.1007/978-3-031-38878-1_10

heatwave that brought devastation in central and western Europe, and intense heatwave conditions in India in 2003. The WHO estimates that from 1998 to 2017, more than 166,000 people died due to heatwaves, including more than 70,000 who died during the last two weeks of August 2003 heatwave in Europe. The year 2022 was the fifth or sixth warmest year on record with the mean global temperature was 1.15 °C above the pre-industrial average, despite the rare third year of La Nina—a natural temporary cooling of parts of the Pacific Ocean that changes weather conditions worldwide (India Today Environment desk, 2023). A new study suggests that the intense heatwaves and subsequent wildfires in July-August 2023 in Southern Europe, Canada, California and Algeria is the result of climate change. Another report propounds that 1.5 °C goal likely to be breached by 2030 (Hindustan Times, 2022).

1.1 Intensity of Heatwave Impacts Became Evident in India

Heatwaves typically occur in India between March and June. May is the peak month of the heatwave over India. Every season, on average, two to three heatwave events are anticipated. Recent report from the Indian Meteorological department (IMD) predicts that most parts of India including those in peninsular and coasts of India will experience in increase in duration of heatwave by 12–18 days by 2060 (Hindustan Times, 2023).

Based on analysis of data from 1979 till 2022, it is interesting to note that mortalities due to heatwaves were highest in the year 2015 (Fig. 1).

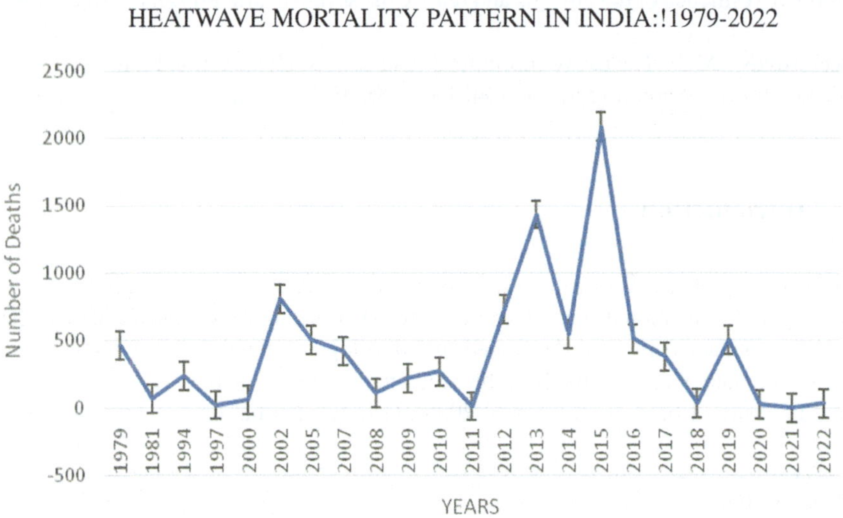

Fig. 1 Based on data for various years, from IMD, NCC, Pune

It is evident from Fig. 1 that mortality due to heatwaves started declining in the country. This is because various heat protection plans initiated by the governments both at the central and state levels helped in the reduction of heatwave-related deaths.

However, global climate change with extreme heatwave conditions has been seen not only in India, but also in other countries including France (Mora et al., 2017; WMO, 2018).

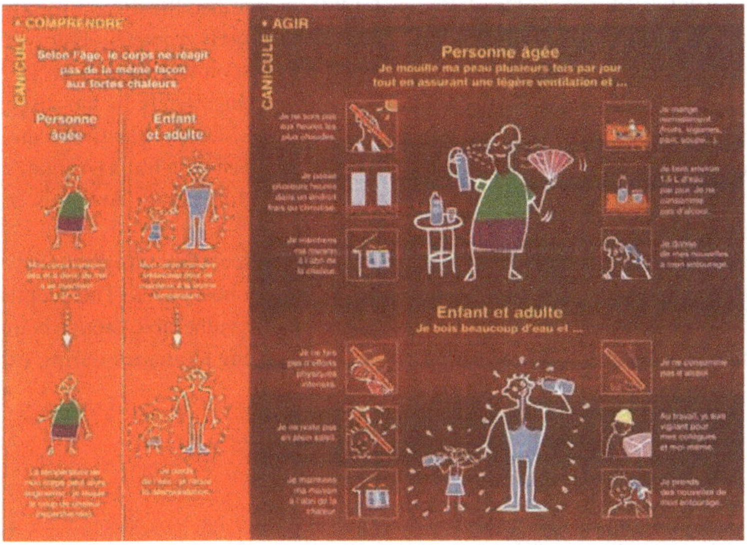

French Government circulars emphasizing the importance of spraying water several times on the body, and specifically instructing aged people to make them aware about heatstroke issued after 2003 heatwave.

(These pamphlets were obtained by the author during his visit to Paris Municipality in 2004)

In the national context, a number of studies in India show that the country has been experiencing extreme weather events for the past few decades, particularly after the 1990s. Srivastava, Dandekar et al. (2007) examined the discomfort trend in several Indian cities based on an index called "Thermo-Hygrometric Index (THI)", and showed that there was an increasing trend of discomfort seen from 20 April till June end. Many cities showed significant increasing trends in the discomfort indices, especially in May and June (Srivastava et al., 2007). Heatwave and cold wave conditions and consequent deaths have been increasing in India (Akhtar, 2010). Chandler opined in 2017 that deadly heat wave could hit South Asia this century (Chandler, 2017)

The objective of the chapter is to ascertain heatwave scenarios and pinpoint various adaptation strategies adopted by people in the event of heatwaves. The chapter also discusses ways through which communities, especially vulnerable groups belonging to different socio-economic strata, adapt to local-coping strategies during extreme events (DTE, 2018).

1.2 Heatwave Pattern in India Since 1998

Analysis of data collected on heatwave records suggest that the number of heatwave events and consequent mortality is on the rise. During 1979–88, 96 heatwave events occurred with 2098 deaths. These figures rose to 147 and 2441, respectively, for the period 1989–98. Data on the number of heatwave days and deaths for the period 1993–2008 further depict that the number of deaths has increased.

In India in the year 1998, major parts of north India and the northern parts of peninsular India experienced severe heatwave. During the second half of May, the heatwave was one of the severest ones seen in the last 50 years and led to deaths of more than 2600 people. This severe heatwave condition initially prevailed over north-west India and then extended south and south-east towards Odisha and costal Andhra Pradesh (De & Mukhopadhyay, 1998). In Odisha, which is one of the poorest states, the temperature rose to a scorching 49.5 °C and nearly 1300 people died. Other states, including Bihar, West Bengal, Andhra Pradesh, Uttar Pradesh, Maharashtra and Punjab, had temperatures soaring to 45–49 °C (Kumar, 1998). Not only North India, but also South India, in interiors of Karnataka and parts of Tamil Nadu, was under the grip of severe heatwave conditions during this period. Chennai recorded the second highest maximum temperature (44 °C) of the century on 24 May 1998, which was 8 °C above normal. The highest ever recorded maximum temperature (45 °C) in Chennai was earlier seen on 21 May 1910.

The people residing at a place for a sufficiently long time get adapted to the weather conditions of that place. Hence, the maximum temperature of north-west India, especially Rajasthan, though neared 50 °C, the death toll in Rajasthan and north-west parts of the country was less compared to that in Odisha. The large number of deaths in Odisha was perhaps due to the lack of adaptability to such extreme conditions. Adaptation measures commonly used by the population affected by heatwaves include increased consumption of drinking water and soft drinks and use of coolers and air conditioners. Further, changes are adopted in shopping behaviour schedule and office and school timings.

The heatwave also affected the environment. Mountain glaciers which shrank by 10% over the summer in Europe, and compared to the average course of the past decade, the 25 days of heatwaves in the Himalayan region in 2022 caused a glacier mass loss that corresponds to 56% of the overall mass loss experienced on average during summers 2010–2020, demonstrating the relevance of heatwaves for seasonal melt. Forest fires raged across western Europe California state and in eastern and southern part of Australia. The heat affected harvests as well: fodder and grain production declined.

1.3 Heatwave Mortality Pattern

The results of a scientific study revealed a sum total of 12,273 fatalities, which have been caused by 660 heatwave events resulting in 332 fatalities annually during the 37-year period (1978–2014). Heatwave events the study has shown a rising trend, whereas no significant rising or declining trend was observed in the heatwave fatalities. Majority of heatwave events and fatalities have occurred in the months of April, May and June. Maharashtra state with a human development index with 12 rank has experienced maximum count of events, while Andhra Pradesh with a human development index rank 24 has the highest count of fatalities. More than 80% of heatwave fatalities have occurred only in five states (Andhra Pradesh, Rajasthan, Odisha, Uttar Pradesh and Bihar). In relation to fatality rate per annum, topmost five states are Andhra Pradesh, Rajasthan, Odisha, Chandigarh and Punjab. In fatality density, topmost five states are Chandigarh, Delhi, Andhra Pradesh, Bihar and Odisha (Malik et al, 2021).

However, this study by Malik et al. is based on data from 1978 to 2014, and the country experienced most serious heatwave impacts during 2015 as evident in Fig. 1. Figures, shared in response to a Parliament Question, show that **2081** people had died due to heatwave in 2015 (Rajya Sabha, 2023). The number of casualties then started declining to 700, 375, 33, 305, 25, 0 and 30 in 2016, 2017, 2018, 2019, 2020, 2021 and in 2022, respectively. This was a tremendous achievement towards reducing heatwave mortality in India.

1.4 Measures to Combat Heat-Related Mortality

Mostly followed measures in different parts of the country include rescheduling of working hours for outdoor workers to avoid their exposure to extreme hot weather, changes school timings for Classes 1–8, due to heatwave, creation of drinking water kiosks, supply of water through tankers, erection of special shelter homes for poor and vulnerable people in every cities, increase in health facilities with special focus of heat related illnesses, stocking of ORS packets at health centres and the nearest *anganwadi* centres. A new report from the IMD released recently recommended a comprehensive response plan for heatwaves which includes cultural, institutional, technological and ecosystem-based adaptation strategies (Hindustan Times, 2023), besides early warning is key to climate adaptation.

Inter-agency coordination among the IMD, Ministry of Earth Science, Integrated Disease Surveillance Programme (IDSP) of the National Centre for Disease Control (NCDC), Ministry of Health and Family Welfare and other concerned ministries/departments has been mobilized. Despite heatwaves being a major challenge, the combined action taken by the central and state governments, district administrations, the forecast department, the health department and civil society in a planned way to monitor the situation has resulted in significant reduction in casualties particularly since 2016 (NDMA).

The India Meteorological Department (IMD) has been issuing impact-based warnings so that states can take the necessary mitigation measures and population embrace necessary adaptation and conduct Information Education Communication (IEC) activities for awareness of Do's and Don'ts among the public. These measures also focus on livestock/animal issues.

The efforts made/being made by the Government to build resilience infrastructure, develop early warning infrastructure and create public awareness about the impact of heatwaves in the country have been successful.

NDMA: National Disaster Management Authority (NDMA) has issued a manual on House Owners' Guide to alternate roof cooling solutions to build heat resilience infrastructure NDMA, year not mentioned. "The introduction of Ahmedabad Heat Action Plans (HAP) in some of the cities in Gujarat, Maharashtra, Odisha and Telangana has demonstrated that such action plans can improve resilience of the citizens and reduce the severe health impacts of heatwave" (Tyagi, 2017). A scientific study on "A Heat Vulnerability Index: Spatial Patterns of Exposure, Sensitivity and Adaptive Capacity for Urbanites of Four Cities of India" (Rathi et al., 2022) by a group of scientists focusing on computing vulnerability index taking into account exposure, sensitivity and adaptive capacity in four urban cities of India. There is need to conduct such studies in other geographically diverse vulnerable urban/rural areas mainly in south-eastern and north-western areas of the country to assess the level of vulnerability to extreme high temperature.

Recent report from India Meteorological department entitled "Heat and Cold waves in India:Processes and Predictability" include: improving India's buildings through ventilation and insulation; raising awareness about heat stress and resultant health problems. Thus, the above measures were adopted which helped reduced the fatalities since 2016 due to heatwaves in India.

References

Akhtar, R. (2010). ElNino related health hazards in India. *Current Science, 98*(2), 144–147.

BBC. (2022). India heatwave: High temperatures killing more Indians now. *Lancet Study Finds*. https://www.bbc.com/news/world-asia-india-63384167. https://doi.org/10.1016/S0140-6736(05)78823-1

Chandler, D. L. (2017). Deadly heat waves could hit South Asia this century: Action, climate change could devastate a region home to one-fifth of humanity, study finds. *MIT News Office*.

De, U. S., & Mukhopadhyay, R. K. (1998). Severe heat wave over the Indian sub-continents in 1998, in perspective of global climate. *Current Science, 75*(12), 1308–1311.

DTE. (2018). Unequal India: 101 billionaires thrive; 364 million poor struggle to survive. *Down to Earth*.

Hindustan Times. (2022). 1.50 C goal likely to be breached by 2030: Experts. *Hindustan Times*, New Delhi.

Hindustan Times. (2023). *Heatwave duration to rise in most areas by 2060*. IMD.

India Today Environment Desk. (2023). *2022 one of the warmest years on record, pace of rise in global sea level has doubled* (UN Climate Report, New Delhi, April 23)

Kumar, S. (1998) India's heatwave and rains result in more death toll. *The Lancet*. https://doi.org/10.1016/S0140-6736(05)78823-1

Malik, P., Bhardwaj, P., & Singh, O. (2021). Heat wave fatalities over India: 1978–2014. *Current Science, 120*(10).

Mora, C., et al. (2017). Global risk of deadly heat. *Nature Climate Change*, *7*, 501–506.

NDMA. (Year not mentioned). Beating the heat how India successfully reduced mortality due to heat waves. National Disaster Management Authority Ministry of Home Affairs, Govt. of India.

Rathi, S. K., Chakraborty, S., Mishra, S. K., Dutta, A., & Nanda, L. (2022). A heat vulnerability index: Spatial patterns of exposure, sensitivity and adaptive capacity for urbanites of four cities of India. *International Journal of Environmental Research and Public Health, 19*(1), 283. https://doi.org/10.3390/ijerph19010283

Rajya Sabha. (2023). Unstarred question No. 691 to be answered on the 08th February, 2023/Magha 19, 1944 (Saka) declaring heatwaves and coldwaves as natural disaster.

Srivastava, A. K., Dandekar, M. M., Kshirsagar, S. R., & Dikshit, S. K. (2007). Is summer becoming more uncomfortable at Indian cities? *Mausam, 58*(3), 335–344.

Tyagi, A. (2017). Personal communication.

WMO. (2018). *Summer sees heat and extreme weather*. WMO.

Climate Change and Human Health: Vulnerability, Impact and Adaptation in Hindu Kush Himalayan Region

Meghnath Dhimal, Dinesh Bhandari, and Mandira Lamichhane Dhimal

Abstract The Hindu Kush Himalayan (HKH) region occupies areas of eight countries, namely Afghanistan, Bangladesh, Bhutan, China, India, Myanmar, Nepal, and Pakistan. The HKH region is warming at a rate higher than the global average, and precipitation has also increased significantly over the last six decades, along with increased frequency and intensity of some extreme events. Changes in temperature and precipitation have affected and would like to affect the health and well-being of mountain people. This book chapter aims to document how climate change has impacted and will impact the health and well-being of the people in the HKH region and offers adaptation and mitigation measures to reduce the vulnerability and impacts of climate change on the health and well-being of the people.

Keywords Climate change · Human health · Adaptation · Hindu Kush Himalayan region · Disease burden · Policy response

1 Introduction

Mountains are significant for the ecological system, aesthetic values, and socioeconomics, not just for those who live in mountainous places but also for those who live in the nearby lowlands (Sharma, 2012). The health of almost 240 million people living in the Hindu Kush Himalayan (HKH) region (the cryosphere that feeds water to 10 major river basins directly supporting 2 million inhabitants of the mountain ranges of the Pamirs, the Tien Shan, and the Tibetan Plateau) is extremely

M. Dhimal (✉)
Nepal Health Research Council, Ramshah Path, Kathmandu, Nepal
e-mail: meghdhimal@gmail.com

D. Bhandari
School of Nursing and Midwifery, Monash University, Melbourne, Australia
e-mail: Dinesh.Bhandari@monash.edu

M. Lamichhane Dhimal
Policy Research Institute, Kathmandu, Nepal

R. Akhtar (ed.), *Climate Change and Human Health Scenarios*, Global Perspectives on Health Geography, https://doi.org/10.1007/978-3-031-38878-1_11

vulnerable to climate change (Krishnan et al., 2019). The livelihood of residents in HKH region is largely sustained by crop-livestock agriculture practice, and to some extent by tourism, which is strongly dependent on the availability of natural resource in this region. The observed environmental changes, including increasing temperature, changing seasonal patterns, reduction in snow cover extent, rapidly receding glaciers, and increased permafrost thaw, are exacerbating the decline in natural resource endowment in the region (Adler et al., 2022). As such, climate change is affecting the livelihood and well-being of people in the HKH region by degrading the quality of social and environmental determinants of health, including air quality, safe drinking water, adequate food, and safe haven (Dhimal et al., 2021a). Climate change is leading to the death and illness of people in HKH region from increasingly frequent extreme weather events, such as floods and landslides due to glacier lake outburst, an increase in zoonoses and food-, water-, and vector-borne diseases, non-communicable diseases, and mental health issues (Ebi et al., 2007).

An expanse of glacier in the HKH region serving as water towers to more than 22% of the global population in the past two decades has turned into a repository of pollutants due to an increased influx of tourists, mountaineers, and trekkers (Dhimal et al., 2021a). As a result of accelerated melting due to unprecedented warming and increasing anthropogenic activities in this region, pollutants are being released into drinking water, posing a threat to the well-being of people depending on the HKH cryosphere (Adler et al., 2022; Mayewski et al., 2020). The rapid warming of the HKH region (0.3 °C in the northwest Himalaya and 0.7 °C in the Karakoram region) compared to the warming rate of the rest of the world is shifting the spatial distribution of biota to higher elevation, which is increasing the risk of vector-borne diseases like zika virus infection, malaria, dengue, and leishmania in the non-endemic areas of high mountains in the HKH region (Dhimal et al., 2018, 2021b). While the health impacts of climate change on people living in tropical, lowland, and coastal regions are widely discussed in literature, the threat posed by climate change to the well-being of people living in mountains remains scarce. In this chapter, we discuss the plausible impacts on the health and well-being of people living in HKH region based on the review of published literature (Dhimal et al., 2021a).

1.1 Climate Change and Health Vulnerability in HKH Region

The Hindu Kush Himalayan (HKH) region is particularly vulnerable to climate change. The HKH is the youngest and one of the most diverse ecosystems among the global mountain biome, with extreme variations in flora and fauna, climate, and ecosystem resulting from altitudinal, latitudinal, and soil gradients (Xu et al., 2019). Over the past five to six decades, the HKH region has experienced a rising trend of extreme warmth, and in the future, even if global warming is kept to 1.5 °C, warming in the HKH region will likely be at least 0.3 °C higher and in the northern

Himalaya and Karakorum at least 0.7 °C higher than the global average. Such large warming could trigger a multitude of biophysical and socioeconomic impacts, such as biodiversity loss, increased glacial melting causing disasters like glacial lake outburst floods, less predictable water availability, and diminishing water quality, all of which will adversely impact the livelihood, health, and well-being of people. Furthermore, elevation dependent warming is apparently observed in the past as well as in future projections for the HKH region. The potential impact of climate change on the various resources of the HKH region is of great concern as a huge population (around 240 million people in the region) relies on monsoon rainfall (Krishnan et al., 2019), and a further 1.7 billion people in the downstream areas are benefitted from the goods and services of ecosystem services of the HKH region (Xu et al., 2019). The HKH region experiences strong climate change, which increases extreme events such as flooding, landslides, droughts, glacier melting, and river flow, resulting in water scarcity when it is needed and too much water at other times. As a result of climatic variability and change, there will be high vulnerability to climatic factors and less adaptive capacity (Roy et al., 2019). Beside climate change, other drivers such as land use and land cover change, pollution, solid waste, the introduction of invasive species, habitat degradation, and exploitation of resources are impacting biodiversity loss, disrupting ecosystem services, and ultimately affecting the livelihood, health, and social well-being of people (Xu et al., 2019). Hence, building social, ecological, and climate-resilient systems is of utmost importance for achieving the sustainable development goals (SDGs) in KHH region.

Vulnerability encompasses a variety of concepts and elements, including sensitivity or susceptibility to harm and a lack of capacity to cope and adapt (Pandey, 2019). The Intergovernmental Panel on Climate Change (IPCC) defines vulnerability as a function of exposure, sensitivity, and adaptive capacity. In order to assess the potential health impact of climate variability and change, comprehensive understanding of both the vulnerability of populations and their capacity to respond to new conditions is required. Hence, the vulnerability of human health to climate change is a function of (i) sensitivity to changes in weather and climate; (ii) exposure to weather- or climate-related hazards, including the character, magnitude, and rate of climate variation; and (iii) adaptation measures and actions in place to reduce the burden of a specific adverse health outcome. People in the HKH region face an increasing challenge in adapting to the health impacts of climate change (WHO, 2006). It is reported that the prevalence of poverty, exclusions, and gender disparities is high in the HKH region, and people experience differential vulnerabilities shaped by an intersection of class, caste, gender, age, marital status, health, profession, and others in the HKH region (Goodrich et al., 2019).

1.2 Climate Change and Health Impacts in HKH Region

Climate Change and Infectious Diseases in HKH Region

It is now well established that climate change is affecting the onset and distribution of almost 58% of infectious diseases confronted by human beings. Extant research reveals more than thousand different unique pathways through which climate change, via different transmission modes, can exacerbate the onset of infectious diseases (Mora et al., 2022). In this section, we will explore food-, water-, and vector-borne diseases that have been affecting the people residing in the HKH region.

1.3 Food- and Water-Borne Diseases in the HKH Region

Food- and water-borne diarrheal diseases are mainly transmitted through the consumption of food and water contaminated with human and animal fecal materials containing enteropathogenic bacteria, viruses, and protozoa (Bhandari et al., 2020a, 2020b, 2020c). Besides direct consumption, food- and water-borne diseases also spread due to poor hygienic conditions, inappropriate food storage and handling practices, and transmission facilitated by insect vectors like flies, which are all affected by changes in temperature, humidity, and rainfall patterns (Dhimal et al., 2021a). In the context of HKH regions, the quantity, variability, and timing of runoff from snowmelt and glaciers directly or indirectly influence the transmission of food- and water-borne diseases (WHO, 2006). Historically, the natural environment in the mountainous regions itself served as a freezer, and there was no risk of food contamination. However, the rapidly increasing temperature in the HKH region, particularly in the summer season, is creating a conducive environment for the growth of food spoilage organisms, which is likely to increase the chances of food poisoning among the residents and the travelers. Open defecation or improper disposal of fecal materials by the tourists, seasonal travelers, and mountaineers in the HKH region can contaminate the water sources and result in the outbreak of water-borne diseases. Although the HKH region is equally vulnerable to diarrheal diseases as the rest of the world (Dhimal et al., 2021a), limited evidence exit on the epidemiology and burden of diarrheal diseases among the people living KKH region. A recent national and subnational level study on the effect of climate change on diarrhea among children in Nepal reported a 3.42% increase in the incidence of diarrhea in the high mountains per 1 °C increase in mean temperature, which was surprisingly higher than the burden reported in the southern plain or the Terai region (1.46% per 1 °C increase in mean temperature) (Dhimal et al., 2022). Similar reports of increased burden of diarrhea incidence as a result of climate change induced increase in temperature and precipitation have been reported from the high mountains of Bhutan (Wangdi & Clements, 2017), Pakistan (Malik et al., 2012) and Afghanistan (Anwar et al., 2019).

2 Vector-Borne Disease in the HKH Region

The mechanism by which climate change influences the increased transmission of vector-borne disease (VBD) is widely established because VBDs are the most studied health problem in relation to climate change (Campbell-Lendrum et al., 2015). Briefly, increases in temperature and precipitation surge the growth rate of a disease vector, increase the biting rate, decrease the incubation period of the agent/pathogen inside the vector, and improve the rate of development of pathogen within the vector (Rocklöv & Dubrow, 2020). Recently, VBDs have been reported from the high mountains of the HKH regions, which were not endemic for these diseases. The major VBDs affecting people in the HKH include malaria, dengue, chikungunya, Japanese encephalitis, lymphatic filariasis, and visceral leishmaniasis. These VBDs are transmitted either by the bite of disease-carrying mosquitoes or sandflies (Dhimal et al., 2021b). The sixth assessment report of the International Panel on Climate Change states with high confidence that VBDs are now shifting to higher altitudes and will continue to do so under the future climate change scenarios (Cissé et al., 2022). In fact, a recent study on the genetic traits of *Aedes aegypti*, the disease vector of dengue fever, collected from high altitudes in Nepal reported that these vectors have genetically evolved to survive in a colder climate as a result of climate change adaptation (Kramer et al., 2023). Increased cases of VBDs like malaria and dengue fever have been reported from the highlands of Nepal (Acharya et al., 2018), Bhutan (Wangdi et al., 2020), India (Dhiman et al., 2011), Pakistan (Bouma et al., 1996), and the Tibetan plateau (Yi et al., 2019).

2.1 Impacts of Climate Change on Non-communicable Diseases and Mental Health in the HKH Region

There is increasing evidence about the impacts of climate change on non-communicable diseases (NCDs) and mental health. Climate change will exacerbate the incidence of NCDs such as cardiovascular disease, cancer, respiratory health injuries (Campbell-Lendrum & Prüss-Ustün, 2019), and mental health problems including suicide (Lawrence et al., 2022). A wide range of risk factors for NCDs are strongly linked to environmental exposures and to climate change; hence, the combination of climate change, air pollution, and NCDs is among the most serious threats to global health (Campbell-Lendrum & Prüss-Ustün, 2019). For example, climate change may increase air pollution and thus augment the risk of cardiovascular disease through three main exposure pathways: directly via air pollution and extreme temperatures and indirectly via changes to dietary options (Friel et al., 2011). Besides NCDs, mountain communities are also at risk of climate-induced mental health challenges carried by increase in natural disasters such as drought (Cianconi et al., 2020). Though, there are limited studies on the impacts of climate change on NCDs and mental health in the HKH region; existing evidence highlighting the immense impacts of climate change on the health and well-being of HKH populations points to a great need to pay attention to the issue.

Although the negative effects of climate change on physical health have been recognized for some time, the effects on mental health have been less well documented, but mental health and well-being are interconnected with climate change via multiple and diverse pathways (Lawrance et al., 2022). The recent IPCC sixth assessment report concludes that climate change has already harmed human physical and mental health in all geographical regions, which undermines efforts for inclusive development (Pörtner et al., 2022). Climate change disproportionately affects the mental health of the marginalized and vulnerable population, as climate change can worsen pre-existing mental health inequalities, especially where health care is inadequate (Romanello et al., 2022). Mental health challenges increase with warming temperatures, trauma associated with extreme weather, and loss of livelihoods and culture. Increasing temperatures and heatwaves have increased mortality and morbidity, with impacts that vary by age, gender, urbanization, and socioeconomic factors. Mental health problems are caused by climate-related ecological grief associated with environmental change (e.g., loss of green space), extreme weather and climate events, vicarious experience or anticipation of climate events, and climate-related loss of livelihoods and food insecurity. Vulnerability to the mental health effects of climate change varies by region and population, with evidence that indigenous peoples, agricultural communities, first responders, women, and members of minority groups experience greater impacts. As the KHH region is home to indigenous people, agricultural communities, and minority people, it is predicted that the impact of climate change is higher in the HKH region (Pörtner et al., 2022).

2.2 *Implication of Climate Change on Livelihood*

The lives and livelihoods of 240 million people are dependent directly on the HKH region (Sharma et al., 2019). Climate change is undermining many of the social determinants of good health, such as livelihoods, equality, access to health care, and social support structures (Climate Change and Health, n.d.). Climate change disproportionately impacts people according to their sociocultural, economic, and geographical contexts. The most vulnerable and disadvantaged are women, children, ethnic minorities, poor communities, migrants, older populations, and those with underlying health conditions and who have low coping capacity. The majority of the mountain people are dependent on agriculture and livestock rearing. It is evident that the increasing temperature in the HKH region has caused losses of glacier mass and area, changes in rainfall patterns, and erratic hailstorms, which have posed threats to crop productivity and are highly vulnerable to food security (Merrey et al., 2018). The changing climatic conditions are expected to decrease crop production as a result of floods, droughts, hail, frost, and diseases. A report published by ICIMOD on agriculture and food security in the Hindu Kush Himalayan region showed that more than 30% of the population suffers from food insecurity, 50% of the population faces malnutrition, with women and children suffering more from it. The impacts of climate change are adding challenges to water availability in river basins, reduced groundwater recharge, and drying of springs, which causes several socio-environmental changes. The water stress puts pressure on the production of the agricultural system and has serious implications for food and nutrition security (Rasul et al., 2019).

Mountain is an important source of freshwater. Changes to the cryosphere due to climate change have influenced human mobility and migration by altering freshwater availability and increasing disasters (Hock et al., 2019). In the HKH region, changes in climate variability, such as droughts and altered water cycles, have profound consequences for mountain agriculture, agrobiodiversity, and the resilience of crop diseases (Krishnan et al., 2019). It is resulting in problems for local communities, such as increased avalanches and landslides. Additionally, other factors like increased tourism and dependence on imported food are also impacting the way of life in these communities. The changes in snowfall patterns have disrupted local livelihoods and created cryosphere hazards such as avalanches and landslides (Tuladhar et al., 2021).

Similarly, Bangladesh revealed that the most severe issues affecting livelihood are food shortage, unemployment, income loss, and housing and sanitation problems due to the changing climate associated with disasters (Hossain et al., 2022). The literature on the Multidimensional Livelihood Vulnerability Index in three basins (Nepal, India, and Pakistan) demonstrates that the vulnerable population was caused by water insufficiency, the slope of agricultural land, and the lack of improved cooking fuel (Gerlitz et al., 2017). Climate change is challenging the water, food, and energy nexus (WEFE). The relationships among food, energy, and water are dynamic, and the interdependences among them are numerous and multidimensional. The HKH Region has major implications for water, food, and energy availability (Rasul & Sharma, 2015).

2.3 Adaptation to Climate Change impacts in HKH Region

The HKH region bears a significant burden of infectious disease, and climate change is further increasing the risk of the onset and transmission of these diseases among its residents (Dhimal et al, 2021a). However, the increased vulnerability posed by climate change in HKH region can be countered to some extent by improving the adaptation capacity of the region. The adaptation measure should focus on socioeconomic determinants such as population growth, urbanization, and employment in addition to climatic determinants to strengthen the capacity of the HKH region in withstanding the impacts of climate change (Dhimal et al, 2021a). Efforts should be focused on the development of climate-resilient health systems in this region to ensure continuity of essential service delivery even during the climate crisis or emergencies. The establishment of regional disease intelligence systems, such as an integrated disease surveillance unit via trans-border collaboration of interdisciplinary experts in the HKH region, can improve the adaptation capacity by issuing climate-based early warning messages to alert people about potential disease outbreaks based on the observed changes in climatic conditions favorable for disease propagation (Bhandari et al., 2020a, 2020b, 2020c). Similarly, climate change should be integrated as a national health priority in the national adaptation plan of each member countries of the HKH regions.

The adaptation and mitigation strategies of climate change on NCDs and mental health could be broadly classified into three sectors, namely energy, municipal planning, and food and agriculture. In the energy sector, clean energy such as electricity could be generated, which could reduce the use of biomass fuel. Additionally, improving home energy performance through efficient heating and cooling mechanisms will also reduce greenhouse gas emissions (GGE), leading to a decrease in indoor and outdoor pollution and reducing the risk of NCDs. Additionally, municipal planning could be improved by developing road lanes reserved for walking and cycling, thereby lowering greenhouse gas emissions, and building parks, promoting physical activity, which reduces obesity and respiratory illnesses and diseases. Lastly, food cultivation could be improved by decreasing consumption of animal products and supporting new food harvesting approaches and rural employment, food system heterogeneity, and investment in urban farming (Dhimal et al, 2021a).

3 Conclusion

The KHK region is witnessing impact of climate change on infectious diseases, food insecurity, non-communicable diseases, and mental health. Hence, intergraded adaptation and mitigation strategies are recommended to overcome likely impacts of climate change on human health in the KHK region.

References

Acharya, B. K., Cao, C., Xu, M., Khanal, L., Naeem, S., & Pandit, S. (2018). Present and future of dengue fever in Nepal: Mapping climatic suitability by ecological niche model. *International Journal of Environmental Research and Public Health, 15*(2), 187.

Adler, C., Wester, P., Bhatt, I. D., Huggel, C., Insarov, G., Morecroft, M., Muccione, V., Prakash, A., Pörtner, H., & Roberts, D. (2022). Cross-chapter paper 5: Mountains, climate change 2022: Impacts, adaptation and vulnerability. In *Contribution of working group II to the sixth assessment report of the intergovernmental panel on climate change*. Cambridge University Press, Cambridge, UK and New York, USA.

Anwar, M. Y., Warren, J. L., & Pitzer, V. E. (2019). Diarrhea patterns and climate: A spatiotemporal Bayesian hierarchical analysis of diarrheal disease in Afghanistan. *American Journal of Tropical Medicine and Hygiene, 101*(3), 525–533.

Bhandari, D., Bi, P., Dhimal, M., Sherchand, J. B., & Hanson-Easey, S. (2020c). Non-linear effect of temperature variation on childhood rotavirus infection: A time series study from Kathmandu, Nepal. *Science of the Total Environment, 748*, 141376.

Bhandari, D., Bi, P., Sherchand, J. B., Dhimal, M., & Hanson-Easey, S. (2020a). Assessing the effect of climate factors on childhood diarrhea burden in Kathmandu, Nepal. *International Journal of Hygiene and Environmental Health, 223*(1), 199–206.

Bhandari, D., Bi, P., Sherchand, J. B., Dhimal, M., & Hanson-Easey, S. (2020b). Climate change and infectious disease research in Nepal: Are the available prerequisites supportive enough to researchers? *Acta Tropica, 204*, 105337.

Bouma, M., Dye, C., & Van der Kaay, H. (1996). Falciparum malaria and climate change in the northwest frontier province of Pakistan. *The American Journal of Tropical Medicine and Hygiene, 55*(2), 131–137.

Campbell-Lendrum, D., Manga, L., Bagayoko, M., & Sommerfeld, J. (2015). Climate change and vector-borne diseases: What are the implications for public health research and policy? *Philosophical Transactions of the Royal Society London B: Biological Sciences 370*(1665).

Campbell-Lendrum, D., & Pruss-Ustun, A. (2019). Climate change, air pollution and noncommunicable diseases. *Bulletin of the World Health Organization, 97*(2), 160.

Cianconi, P., Betrò, S., & Janiri, L. (2020). The impact of climate change on mental health: A systematic descriptive review. *Frontiers in Psychiatry, 11*, 74. https://doi.org/10.3389/FPSYT.2020.00074/BIBTEX

Cissé, G., McLeman, R., Adams, H., Aldunce, P., Bowen, K., Campbell-Lendrum, D., Clayton, S., Ebi, K. L., Hess, J., Huang, C., Liu, Q., McGregor, G., Semenza, J., & Tirado, M. C. (2022). Health, wellbeing, and the changing structure of communities. In: *Climate change 2022: Impacts, adaptation, and vulnerability. Contribution of working group II to the sixth assessment report of the intergovernmental panel on climate change, Cambridge, UK and New York, NY, USA*.

Climate Change And Health. (n.d.). Retrieved April 1, 2023, from https://www.who.int/news-room/fact-sheets/detail/climate-change-and-health

Dhimal, M., Bhandari, D., Dhimal, M. L., Kafle, N., Pyakurel, P., Mahotra, N., Akhtar, S., Ismail, T., Dhiman, R. C., Groneberg, D. A., Shrestha, U. B., & Müller, R. (2021a). Impact of climate change on health and well-being of people in Hindu Kush Himalayan region: A narrative review. *Frontiers in Physiology, 12*.

Dhimal, M., Bhandari, D., Karki, K. B., Shrestha, S. L., Khanal, M., Shrestha, R. R. P., Dahal, S., Bista, B., Ebi, K. L., Cissé, G., Sapkota, A., & Groneberg, D. A. (2022). Effects of climatic factors on diarrheal diseases among children below 5 years of age at national and subnational levels in Nepal: An ecological study. *International Journal of Environmental Research and Public Health, 19*(10), 6138.

Dhimal, M., Dahal, S., Dhimal, M. L., Mishra, S. R., Karki, K. B., Aryal, K. K., Haque, U., Kabir, M. I., Guin, P., Butt, A. M., Harapan, H., Liu, Q. Y., Chu, C., Montag, D., Groneberg, D. A., Pandey, B. D., Kuch, U., & Müller, R. (2018). Threats of Zika virus transmission for Asia and its Hindu-Kush Himalayan region. *Infectious Diseases of Poverty, 7*(1), 40.

Dhimal, M., Kramer, I. M., Phuyal, P., Budhathoki, S. S., Hartke, J., Ahrens, B., Kuch, U., Groneberg, D. A., Nepal, S., & Liu, Q.-Y. (2021b). Climate change and its association with the expansion of vectors and vector-borne diseases in the Hindu Kush Himalayan region: A systematic synthesis of the literature. *Advances in Climate Change Research, 12*(3), 421–429.

Dhiman, R. C., Chavan, L., Pant, M., Pahwa, S. (2011). National and regional impacts of climate change on malaria by 2030. *Current Science* 372–383.

Ebi, K. L., Woodruff, R., von Hildebrand, A., & Corvalan, C. (2007). Climate change-related health impacts in the Hindu Kush-Himalayas. *EcoHealth, 4*(3), 264–270.

Friel, S., Bowen, K., Campbell-Lendrum, D., Frumkin, H., McMichael, A. J., & Rasanathan, K. (2011). Climate change, noncommunicable diseases, and development: The relationships and common policy opportunities. *32*, 133–147. https://doi.org/10.1146/Annurev-Publhealth-071910-140612

Gerlitz, J. Y., Macchi, M., Brooks, N., Pandey, R., Banerjee, S., & Jha, S. K. (2017). The multidimensional livelihood vulnerability index—An instrument to measure livelihood vulnerability to change in the Hindu Kush Himalayas. *Climate and Development, 9*(2), 124–140. https://doi.org/10.1080/17565529.2016.1145099

Goodrich, C. G., Prakash, A., & Udas, P. B. (2019). Gendered vulnerability and adaptation in Hindu-Kush Himalayas: Research insights. *Environmental Development, 31*, 1–8.

Hock, R., Rasul, G., Adler, C., Cáceres, B., Gruber, S., Hirabayashi, Y., Jackson, M., Kääb, A., Kang, S., Kutuzov, S., Milner, Al., Molau, U., Morin, S., Orlove, B., & Steltzer, H. (2019). High mountain areas. In H.-O. Pörtner, D. C. Roberts, V. Masson-Delmotte, P. Zhai, M. Tignor, E. Poloczanska, K. Mintenbeck, A. Alegría, M. Nicolai, A. Okem, J. Petzold, B. Rama & N. M. Weyer (Eds.), *IPCC special report on the ocean and cryosphere in a changing climate* (pp. 131–202). Cambridge University Press.

Hossain, B., Shi, G., Ajiang, C., Sarker, M. N. I., Sohel, M. S., Sun, Z., & Yang, Q. (2022). Climate change induced human displacement in Bangladesh: Implications on the livelihood of displaced riverine island dwellers and their adaptation strategies. *Frontiers in Psychology, 13*, 5975. https://doi.org/10.3389/FPSYG.2022.964648/BIBTEX

Kramer, I. M., Pfenninger, M., Feldmeyer, B., Dhimal, M., Gautam, I., Shreshta, P., Baral, S., Phuyal, P., Hartke, J., Magdeburg, A., Groneberg, D. A., Ahrens, B., Müller, R., & Waldvogel, A.-M. (2023). Genomic profiling of climate adaptation in *Aedes aegypti* along an altitudinal gradient in Nepal indicates nongradual expansion of the disease vector. *Molecular Ecology, 32*(2), 350–368.

Krishnan, R., Shrestha, A. B., Ren, G., Rajbhandari, R., Saeed, S., Sanjay, J., Syed, M. A., Vellore, R., Xu, Y., You, Q., & Ren, Y. (2019). Unravelling climate change in the Hindu Kush Himalaya: Rapid warming in the mountains and increasing extremes. In P. Wester, A. Mishra, A. Mukherji, A. B. Shrestha (Eds.), *The Hindu Kush Himalaya assessment: Mountains, climate change, sustainability and people* (pp. 57–97). Springer International Publishing.

Lawrance, E. L., Thompson, R., Newberry Le Vay, J., Page, L., & Jennings, N. (2022). The impact of climate change on mental health and emotional wellbeing: A narrative review of current evidence, and its implications. *International Review of Psychiatry, 34*(5), 443–498.

Malik, S. M., Awan, H., & Khan, N. (2012). Mapping vulnerability to climate change and its repercussions on human health in Pakistan. *Global Health, 8*, 31.

Mayewski, P. A., Perry, L. B., Matthews, T., & Birkel, S. D. (2020). Climate change in the Hindu Kush Himalayas: Basis and gaps. *One Earth, 3*(5), 551–555.

Merrey, D. J., Hussain, A., Tamang, D. D., Thapa, B., & Prakash, A. (2018). Evolving high altitude livelihoods and climate change: A study from Rasuwa District, Nepal. *Food Security, 10*(4), 1055–1071. https://doi.org/10.1007/S12571-018-0827-Y/TABLES/9

Mora, C., McKenzie, T., Gaw, I. M., Dean, J. M., von Hammerstein, H., Knudson, T. A., Setter, R. O., Smith, C. Z., Webster, K. M., Patz, J. A., & Franklin, E. C. (2022). Over half of known human pathogenic diseases can be aggravated by climate change. *Nature Climate Change, 12*(9), 869–875.

Pandey, A. (2019). Glossary. In P. Wester, A. Mishra, A. Mukherji, & A. B. Shrestha (Eds.), *The Hindu Kush Himalaya assessment: Mountains, climate change, sustainability and people* (pp. 579–604). Springer International Publishing.

Pörtner, H. O., Roberts, D. C., Adams, H., Adler, C., Aldunce, P., Ali, E., … Ibrahim, Z. Z. (2022). Technical summary. In *Climate change 2022: Impacts, adaptation and vulnerability* (p. 3056). *Contribution of working group II to the sixth assessment report of the intergovernmental panel on climate change* (pp. 37–120). Cambridge University Press.

Rasul, G., Saboor, A., Tiwari, P. C., Hussain, A., Ghosh, N., & Chettri, G. B. (2019). Food and nutrition security in the Hindu Kish Himalaya: Unique challenges and niche opportunities. In P. Wester, A. Mishra, A. Mukherji, & A. B. Shrestha (Eds.), *The Hindu Kush Himalaya assessment: Mountains, climate change, sustainability and people* (pp. 301–338). Springer International Publishing.

Rasul, G., & Sharma, B. (2015). The nexus approach to water–energy–food security: an option for adaptation to climate change. *16*(6), 682–702. https://doi.org/10.1080/14693062.2015.1029865

Rocklöv, J., & Dubrow, R. (2020). Climate change: An enduring challenge for vector-borne disease prevention and control. *Nature Immunology, 21*(5), 479–483.

Romanello, M., Di Napoli, C., Drummond, P., Green, C., Kennard, H., Lampard, P., Scamman, D., Arnell, N., Ayeb-Karlsson, S., Ford, L. B., Belesova, K., Bowen, K., Cai, W., Callaghan, M., Campbell-Lendrum, D., Chambers, J., van Daalen, K. R., Dalin, C., Dasandi, N., … Costello, A. (2022). The 2022 report of the lancet countdown on health and climate change: Health at the mercy of fossil fuels. *The Lancet, 400*(10363), 1619–1654.

Roy, J., Moors, E., Murthy, M. S. R., Prabhakar, S. V. R. K., Khattak, B. N., Shi, P., Huggel, C., & Chitale, V. (2019). Exploring futures of the Hindi Kush Himalaya: Scenarios and pathways. In P. Wester, A. Mishra, A. Mukherji & A. B. Shrestha (Eds.), *The Hindu Kush Himalaya assessment: Mountains, climate change, sustainability and people* (pp. 99–126). Springer International Publishing.

Sharma, E., Molden, D., Rahman, A., Khatiwada, Y. R., Zhang, L., Singh, S. P., Yao, T., & Waster, P. (2019). Introduction to the Hindu Kush Himalaya assessment. In P. Wester, A. Mishra, A. Mukherji & A. B. Shrestha (Eds.), *The Hindu Kush Himalaya assessment: Mountains, climate change, sustainability and people* (pp. 1–16). Springer International Publishing.

Sharma, R. (2012). Impacts on human health of climate and land use change in the Hindu Kush-Himalayan region. *32*(4), 480–486. https://doi.org/10.1659/MRD-JOURNAL-D-12-00068.1

Tuladhar, S., Pasakhala, B., Maharjan, A., & Mishra, A. (2021). Unravelling the linkages of cryosphere and mountain livelihood systems: A case study of Langtang, Nepal. *Advances in Climate Change Research, 12*(1), 119–131. https://doi.org/10.1016/J.ACCRE.2020.12.004

Wangdi, K., & Clements, A. C. (2017). Spatial and temporal patterns of diarrhoea in Bhutan 2003–2013. *BMC Infectious Diseases, 17*(1), 1–9.

Wangdi, K., Xu, Z., Suwannatrai, A. T., Kurscheid, J., Lal, A., Namgay, R., Glass, K., Gray, D. J., & Clements, A. C. (2020). A spatio-temporal analysis to identify the drivers of malaria transmission in Bhutan. *Scientific Reports, 10*(1), 1–10.

World Health Organization. (2006). *Human health impacts from climate variability and climate change in the Hindu Kush-Himalaya region: Report of an inter-regional workshop.*

Xu, J., Badola, R., Chettri, N., Chaudhary, R. P., Zomer, R., Pokhrel, B., Hussain, S. A., Pradhan, S., & Pradhan, R. (2019). Sustaining biodiversity and ecosystem services in the Hindu Kush Himalaya. In P. Wester, A. Mishra, A. Mukherji, & A. B. Shrestha (Eds.), *The Hindu Kush Himalaya assessment: Mountains, climate change, sustainability and people* (pp. 127–166). Springer International Publishing.

Yi, L., Xu, X., Ge, W., Xue, H., Li, J., Li, D., Wang, C., Wu, H., Liu, X., & Zheng, D. (2019). The impact of climate variability on infectious disease transmission in China: Current knowledge and further directions. *Environmental Research, 173*, 255–261.

Health Impacts of Global Climate Change in the Middle East; Vulnerabilities

Hasan Bayram, Nur Konyalilar, and Muge Akpinar-Elci

Abstract Middle East is among the mostly affected areas of the world by climate change, and by the end of this century, this region is expected to have an increased mean temperature about 3–5 °C. Heat waves cause significant mortality and morbidity, whereas hotter and drier climates lead to wildland fire seasons, impairing air quality. Land degradation and desertification are promoting dust pollution and impairing food production. An association between increased risk for death, hospital admissions and emergency room visits due to respiratory diseases such as asthma and COPD, and desert dust storms and increased daily temperatures in the Middle East has been reported. Climate change may also affect the emergence and spread of viral infections including Severe Acute Respiratory Syndrome-related Coronavirus 2 (SARS-CoV-2) through deterioriation in the environment and ecosystem.

Keywords Middle East · Global climate change · Dust storms · Air pollution · Human health outcomes · Respiratory disease

1 Introduction

In today's world, where population growth and industrialization are increasing rapidly, environmental pollution and related problems are among the most critical global concerns to focus on. Increased greenhouse gas (GHG) emissions leading to climate change, disruption of the ozone layer, and air pollution cause irreversible damage to the Earth, which has also a detrimental effect on human health (Gavurova et al., 2021). Intergovernmental Panel on Climate Change (IPCC) describes climate

H. Bayram (✉)
Koc University School of Medicine, Istanbul, Turkey
e-mail: habayram@ku.edu.tr

H. Bayram · N. Konyalilar
Koc University Research Center for Translational Medicine (KUTTAM), Istanbul, Turkey

M. Akpinar-Elci
University of Nevada, Reno, VA, USA

R. Akhtar (ed.), *Climate Change and Human Health Scenarios*, Global Perspectives on Health Geography, https://doi.org/10.1007/978-3-031-38878-1_12

change as temperature and sea level changes worldwide together with precipitation pattern altering and enhanced incidence of extreme weather events (IPCC, 2021). The Middle East countries including Turkey are among the most affected regions due to topological and anthropological factors (Bayram & Öztürk, 2021). Global action must be taken as soon as possible to keep this region inhabitable (Zittis et al., 2022).

This chapter reviews the published data including the governmental and nongovernmental reports on global climate change-related parameters including changes in temperature, GHG emissions, desertification and their consequences on sandstorms, water use, and loss of biodiversity, and their adverse effects on the human health in the Middle East countries including Turkey.

2 Green House Gases and the Emission in Middle East Countries

GHG emission is the main driver of climate change. Unfortunately, there is a continuous increase in the level of CO_2, which is the main GHG, in the atmosphere. Indeed, the highest monthly record was measured in June 2022 at Mauna Loa at 420.99 ppm (CO_2.Earth, 2022).

According to United Nations Environment Program (UNEP) Emission Gap Report 2022, total global GHG emissions reached 54 gigatons of CO_2 equivalent ($GtCO_2e$) in 2020. Total emissions included CO_2 from fossil fuel and industry (fossil CO_2), CO_2 emissions from land use, land-use change, and forestry (LULUCF), methane (CH_4), nitrous oxide (N_2O), and fluorinated gases (F-gases). Although 2021 data is not complete yet, it is expected to exceed current values. Emissions are suggested to be reduced by 45% to prevent a global catastrophe (UNEP, 2022). It is known that more than 70% of global emissions originates from energy consumption that has doubled in the last 30 years (UNEP, 2019). This is followed by agriculture (Climate Watch, 2019).

Middle East is the major source of gas and oil production and consumption both locally and globally, contributing to CO_2 emissions significantly (Shokoohi et al., 2022). According to the Global Carbon Atlas 2020 values, there are two Middle Eastern countries in the top ten of fossil fuel emissions. Iran ranks 6th in the world with 745 metric tons of carbon dioxide ($MtCO_2$), while Saudi Arabia ranks 8th with 626 $MtCO_2$. Turkey ranks 14th with 393 $MtCO_2$ (Global Carbon Atlas, 2020). In the last decade, these countries were responsible for 73% of total CO_2 emissions in that region. Moreover, Iran, Iraq, and Saudi Arabia were responsible for 53% of CH_4 emissions in the same period (Zittis et al., 2022).

Under the Paris Agreement, which entered into force in 2016, countries collectively accepted to reduce their emission levels. The goal of the Paris agreement is to limit global warming to well below 2 °C, preferably to 1.5 °C (Skjærseth et al., 2021). However, if the current policies continue to be implemented as it is now, an increase of 2.8 °C is expected by the end of this century (UNCC, 2022). In order to reach the desired limits, a rapid and effective plan must be implemented on the global scale.

3 Climate Change and Related Events, and Impacts on Human Health

Climate change affects human health directly by changes in temperature and meteorological parameters such as humidity, wind speeds, and atmospheric pressure, or indirectly by increasing risk factors such as extreme weather conditions (hurricanes, floods, storms, etc.), wildfires, air pollution, distribution of aero-allergens, desertification, transmission of respiratory pathogens, and socioeconomic vulnerability of the population, which are summarized in Table 1. Respiratory health effects of climate change include exacerbations of chronic lung diseases, development of certain acute respiratory diseases including infectious diseases, increased allergic response, and premature mortality. Climate change has a significant impact on cardiorespiratory health by directly exacerbating the respiratory diseases or expanding the risk factors (Bayram et al., 2017; D'Amato et al., 2014). Millions of people of all age groups in the MENA region suffer from chronic lung diseases such as COPD and asthma (Ben Abdallah et al., 2011).

Table 1 Summary of potential health impacts associated with climate change and related events

Climate change parameters	Health impacts
Direct weather effects *Heat stress* *Drought* *Desertification*	Increased mortality and morbidity due to acute and chronic cardiopulmonary diseases, spontaneous abortions, and toxemia of pregnancy; malnutrition; increased risk for infectious diseases such as SARS-CoV-2
Indirect weather effects *Loss of Biodiversity*	Malnutrition, negative effect on pharmacology, elevated infectious disease risks
Air pollution and allergens	Increased mortality and morbidity due to asthma, rhinitis, COPD, and LRTIs, lung cancer and cardiac pathologies, elevated premature mortality, excess risk for respiratory viral infections such as SARS-CoV-2
Extreme weather conditions (hurricanes, floods, storms, wildfires)	Deteriorated living conditions, injuries, severe casualties, increased risk for contagious infections, and allergic airway diseases

COPD, chronic obstructive pulmonary disease; LRTIs, lower respiratory tract infections; SARS-CoV-2, severe acute respiratory syndrome-related Coronavirus 2

3.1 Increased Temperature

Based on the National Centers for Environmental Information, National Oceanic and Atmospheric Administration (NOAA) Annual 2021 Global Climate Report, land and ocean temperature has risen at an average rate of 0.08 °C per decade since 1880; however, the average rate of increase has doubled since 1981. The year of 2021 became the sixth warmest year of the globe with a temperature increase that was 0.84 °C above the twentieth-century average (NOAA, 2021). Since 1990, the temperature has risen 0.7 °C globally, while it was twice of that rise in the Middle East (Fig. 1, Stepanyan et al., 2022). This region is one of the top climate change hotspots. It has been warmed drastically in recent years, especially in the last 50 years, when compared with other populated locations, (Zittis et al., 2022).

Pal and Eltahir (2015) showed that this region will cease to be habitable for humans by 2100, using a high-resolution regional climate model simulation. Global circulation models are the most effective tools to create climate scenarios and they are used to simulate the planet's response to GHG concentrations. Temperature is predicted to increase on an average of 0.8 and 1.6 °C between 2020 and 2049 in the Middle East North Africa (MENA) countries including Algeria, Bahrain, Egypt, Iran, Iraq, Israel, Jordan, Kuwait, Lebanon, Libya, Morocco, Oman, Palestine, Qatar, Saudi Arabia, Syria, Tunisia, Turkey, the United Arab Emirates, and Yemen, and it is expected to increase further by an excess temperature of 1.2 and 3.3 °C between 2050 and 2079 (Majdi et al., 2022; Stepanyan et al., 2022).

The dramatic increase in temperature started to cause critical record-breaking weather events. Last year, Nawasib, a city in Kuwait reached 53.2 °C, which was the highest temperature in the inhabited world (World Meteorological Organization,

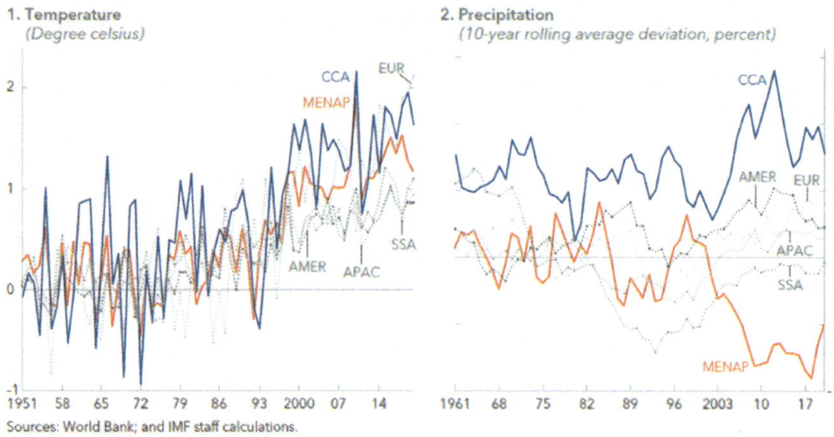

Fig. 1 Climate trends, 1951–2019. (Mean deviation of countries' annual weather averages from their respective 1901–50 averages). (AMER: America, APAC: Asia and Pacific, CCA: Caucasus and Central Asia, EUR: Europe, MENAP: Middle East, North Africa, Afghanistan, and Pakistan, SSA: Sub-Saharan Africa; Stepanyan et al., 2022)

2021). A study conducted at the University of East Anglia using the Climate Research Unit data set (CRU-TS4.04) showed that the Eastern Mediterranean and Middle East (EMME) region warmed 0.45 °C/decade for 1981–2019, while the trend was 0.27 °C/decade for the rest of the globe (Harris et al., 2020). Another research performed in Iraq which is one of the most vulnerable countries in the Arab region demonstrated that the temperature of that region increased two to seven times faster than in the rest of the globe (Salman et al., 2017). In Türkiye, the mean annual temperature for 2021 was 14.9 °C, which was 1.4 °C above the 1981–2010 mean annual temperature of 13.5 °C (Fig. 2, Turkish State Meteorological Service, 2021).

Heat stress and extreme weather patterns have adverse effects on human health (Ebi et al., 2021). Epidemiological studies reported a significant association between increased temperatures and mortality and morbidity due to acute and chronic diseases including pulmonary diseases such as asthma, COPD, lower respiratory tract infections (LRTIs), and pulmonary emboli (Anderson et al., 2013; Bogan et al., 2022; Cevik et al., 2015; Lin et al., 2009; Zhao et al., 2019). Humidity has also been shown to be associated with asthma and COPD exacerbations (Ayres et al., 2009; Venkatesan, 2022). There are several studies reporting an association between high temperatures and mortality and morbidity in Middle East countries such as Kuwait and Lebanon (Alahmad et al., 2020; El-Zein et al., 2004). More recently, we demonstrated a significant association between increased daily temperatures and deaths and hospitalization due to LRTIs in the Southeast Türkiye (Bogan et al., 2022).

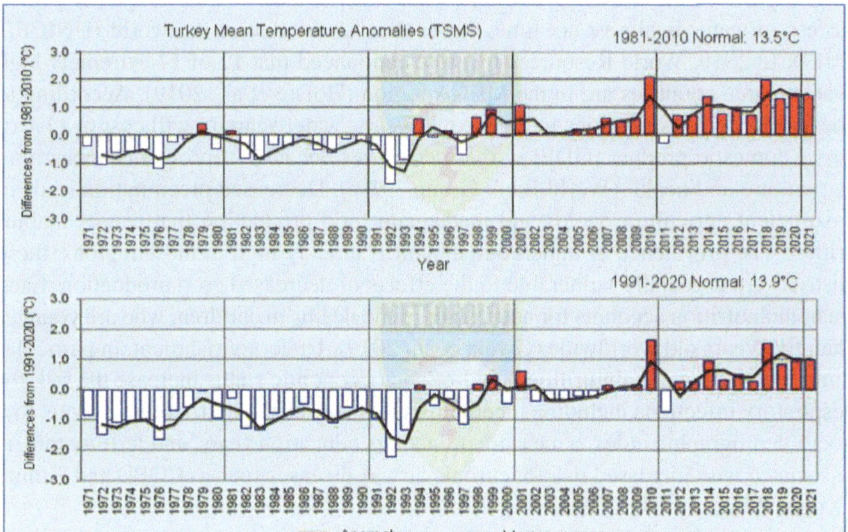

Fig. 2 Annual mean temperature anomalies according to two normal periods of years of 1981–2010 and 1991–2020 in Türkiye (Turkish State Meteorological Service, 2021). The mean annual temperature for 2021 was 14.9 °C, which was 1.4 °C above the 1981–2010 mean annual temperature of 13.5 °C

Furthermore, increased temperatures were associated with emergency room visits, hospitalization, and deaths due to acute coronary syndrome (Al et al., 2018). Interestingly, increased temperatures were associated with morbidity due to spontaneous abortions and toxemia of pregnancy (Bogan et al., 2021).

3.2 *Decreased Precipitation*

Annual precipitation has become less and less predictable, and precipitation is falling faster in the Middle East than in the rest of the world (Fig. 1) (Stepanyan et al., 2022; Zittis et al., 2022). For instance, at the end of twenty-first century rainfall is expected to drop between 10 and 37% in Jordan (Abdulla, 2020). Egypt is another country with significant precipitation decline, which has been decreased by 22% over the past 30 years (World Bank Climate Change Portal, 2021). Türkiye areal mean precipitation in 2021 was 524.8 mm, which was about 9% below from 1981 to 2010's normal (574 mm) (Turkish State Meteorological Service, 2021). A climate model suggests that the precipitation in MENA countries will decrease by 10–20% in the forthcoming decades compared with the 1961–1990 period (Lange, 2019).

Furthermore, there is a massive growth in population together with industrialization in MENA that leads to increased water scarcity, which is one of the major problems in this region. About 6% of the global population lives in this area, which has only 2% of the world's renewable fresh water sources. The average per-person access to water is almost six times lower than in the rest of the world (UNICEF, 2017). In 2019, World Resources Institute announced that 12 of 17 extremely high water-scarce countries are in the MENA region (Hofste et al., 2019). According to the World Bank estimations in the next 30 years, water scarcity will cost 6–14% of gross domestic product (GDP) in these countries because of problems concerning agriculture and health (World Bank Group, 2016). Decreased precipitation leads to insufficient agriculture yields and inadequate food production that causes malnutrition. The prevalence of undernourishment is already high in these regions, these districts are especially vulnerable to the effects of decreased crop production. Each year, malnutrition accounts for nearly half of the deaths in children, who are younger than five years old worldwide (Gwela et al., 2019). Undernourishment, in particular child and maternal malnutrition, and low socioeconomic status increase the risk for respiratory infections including tuberculosis (Murray et al., 2020). Furthermore, low socio-demographic index, a measure of income, education years, and fertility rate, is associated with increased risk for chronic airway diseases such as COPD and asthma (Kuiper-Makris et al., 2021).

3.3 *Desertification*

There is an intensified water cycle worldwide, increased evaporation due to rising temperatures can result in more frequent and extreme storms while causing drier weather conditions in some lands, so that storm-affected areas get heavy precipitation, which may cause flooding, while dry locations exposed to the decreased precipitation levels have the risk of drought (NASA, 2022). The Middle East is the region with one of the highest risks of desertification, triggering a critical water crisis soon (Fig. 3) (Mirzabaev et al., 2019). Drought periods are expected to increase by about 90% in most parts of this region due to extraordinary heat waves (Tabari & Willems, 2018; Zittis et al., 2016). Recent studies suggest that there is an increase in the frequency of dust outbreaks in the Middle East, which has a significant impact on human health and poor air quality for the ecosystem (Akpinar-Elci et al., 2021; Bayram & Öztürk, 2021; Goudie, 2009).

Desert dust can cause headache, fatigue, and flu like symptoms in healthy individuals. However, it may cause much severe health effects in susceptible population such as elderly, children, and individuals with chronic health conditions (Bayram & Öztürk, 2021; Bayram et al., 2017; D'Amato et al., 2015). Studies reported an association between total mortality and increased concentrations of desert particles (Tobías et al., 2011). Middle East desert dust particles led to an excess of hospital admissions for COPD and an excess of the respiratory mortality in Iran (Khaniabadi et al., 2017). A study from Kuwait reported a significant association between dust storms and same-day respiratory and asthma admissions, which were particularly evident in children (Thalib & Al-Taiar, 2012). More recently, we demonstrated that Middle East desert dust storms were associated with increased risk of asthma deaths in Southeast Turkey. Furthermore, there were increases in emergency room visits and hospitalization due to asthma, COPD, and LRTIs (Bogan et al., 2022). Similarly, dust storms were associated with hospitalization and mortality due to acute coronary syndrome in adults (Al et al., 2018). Asian dust storms were reported to be associated with increased risk for acute myocardial infarction. Interestingly, Middle East

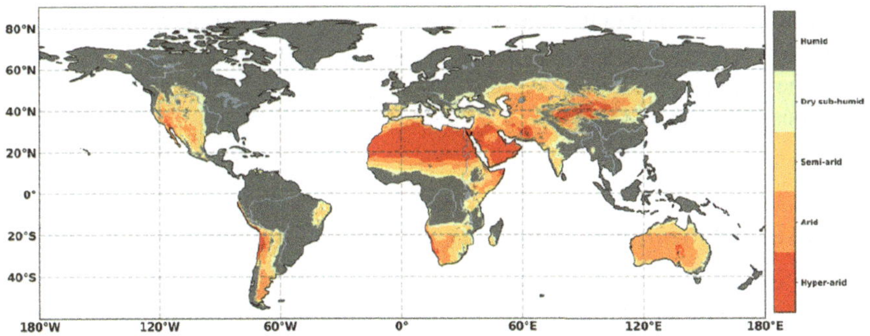

Fig. 3 Geographical distribution of drylands, delimited based on the aridity index (AI). The classification of AI is: humid AI > 0.65, dry sub-humid 0.50 < AI ≤ 0.65, semi-arid 0.20 < AI ≤ 0.50, arid 0.05 < AI ≤ 0.20, hyper-arid AI < 0.05. Data: TerraClimate precipitation and potential evapotranspiration (1980–2015) (Mirzabaev et al., 2019)

dust storms even increased risk for hospital admissions and hospitalizations due to toxemia of pregnancy (Bogan et al., 2021). Mechanistic studies suggest that desert dust causes inflammatory changes, cellular toxicity and induces cellular permeability both in vivo (Wilfong et al., 2011) and in vitro models (Ghio et al., 2014).

3.4 Loss of Biodiversity

Climate change influences decreased genetic diversity due to migration and directional selection. It modifies the web of interaction between species (Bellard et al., 2012). Although deserts occupy a very large area in the terrestrial area, the habitats in the Middle East are very rich, especially the marine ecosystem is among the richest on Earth (Krupp et al, 2009). Many endemic species live especially in the Red Sea and the Gulf (Fernandez et al., 2022). For instance, Socotra, a Yemeni Island in the Indian Ocean, is one of the islands with the highest number of endemic plant species per square kilometer (UNESCO, 2020). In 2008, UNESCO recognized 75% of Socotra's land as a World Heritage Site (UNEP, 2022). Unfortunately, rapid economic development, growing population, and expanded demand for renewable sources in the Middle East have resulted in a great stress on the ecosystem and biodiversity. Habitat destruction, such as deforestation, hunting, overgrazing, and degradation of rangelands together with limited freshwater resources and intense air pollution leads to critical environmental problems. However, the conservation in the region receives relatively little attention (Krupp et al., 2009).

Human health is directly or indirectly dependent on ecosystem resources and what they provide, such as clean water, nutrients and fuel resources. The loss of biodiversity affects these services. For instance, sustainable soil productivity for crops and livestock depends on the extent of biodiversity, which is required for adequate nutrient intake (WHO, 2023). It promotes resistance and resilience to environmental stresses. Moreover, the biological diversity of microorganisms is essential component of the medicine and pharmacological sciences (Myers et al., 2013). Also, biodiversity loss can elevate the risk or emergence of infectious diseases in humans, animals, and plants. Disturbance of the ecosystem due to human activities leads to a reduction in the abundance of some species, while it causes growth in others. This imbalance alters the interaction between organisms and this may make some specious more infectious (Hales et al., 2005).

3.5 Air Pollution and Allergens

There is a close interaction between climate change and air pollution. It has been shown that increased temperatures are associated with increased O_3 and PM levels (Joshi et al., 2020; Schnell & Prather, 2017). Climatological factors can also affect the dispersion, transportation, and suspension of these pollutants in the air together with the vectors of infectious diseases and allergens (Joshi et al., 2020). A recent study conducted in the Middle East with outdoor and indoor bioaerosols demonstrated that particle types and bioaerosol concentrations of the particle components vary by climate changes and meteorological factors (Soleimani et al., 2020). Increased periods of drought contribute to the annual dust emissions. Agriculture and deforestation activities are anthropogenic contributor of wind erosions (Soleimani et al., 2020). A study conducted in Ahvaz, Iran showed that there was an initial peak of ambient particulate matter $\leq 10\ \mu m$ in diameter (PM_{10}) concentration prior to severe dust storms (Maleki et al., 2022). On the other hand, wildfires contribute to ambient PM concentrations (Joshi et al., 2020), and they have been remarkably increased since 1970s (Wittenberg & Kutiel, 2016). Israel, Lebanon, and Turkey are among the countries, which are most affected by wild fires due to mixed impact of droughts, high temperatures, and strong winds especially in summer seasons (Koutsias et al., 2013; Turco et al., 2017).

There is an increased body of evidence reporting a close association between indoor and outdoor air pollution and mortality and morbidity, due to cardiopulmonary and cerebrovascular diseases (Landrigan et al., 2018). It has been estimated that indoor and outdoor air pollution cause more than 7 million premature deaths worldwide (WHO, 2014). Silva and colleagues (2017) demonstrated increased premature mortality due to air pollution attributable to climate change by using multiple global chemistry-climate models. Studies reported that increased levels of air pollutants were associated with increases in hospital visits, hospitalization due to pulmonary diseases such as asthma, COPD, and LRTIs, and cardiac pathologies (Bayram et al., 2017; Boğan et al., 2022; Schraufnagel et al., 2019). Middle East is estimated to have one of the greatest increases in mortality rates with 50,400 deaths per year in 2100 (Silva et al., 2017). Also, air pollutants such as PM and DEP have been shown to rise allergenicity by increasing permeability, easier infiltration of allergens into the mucus membranes, and interaction with immune cells that can lead to enhanced risk for hospital visits due to allergic airway diseases such as asthma and rhinitis (Bayram et al., 2002; Pacheco et al., 2021). Mechanistic studies demonstrated that DEP could decrease ciliary beat frequency of, increase inflammatory response, and alter cell cycle and apoptosis of airway epithelial cells, mechanisms by which air pollutants including DEP, ozone, and nitrogen dioxide play a role in the pathogenesis of chronic respiratory diseases such as asthma, COPD, and lung cancer (Bayram et al., 1998a, 2006, 2013). Furthermore, cells from patients with asthma were more susceptible to the deleterious effects of air pollutants (Bayram et al., 1998b, 2001, 2002).

Increased ambient carbon dioxide together with increased temperature change the flowering season that could elevate the pollen count, time, and frequency of pollen release and extend the allergy season so that exacerbating the allergic respiratory diseases (D'Amato et al., 2014). As a long-term effect of global warming, changes in plant habitat and species may also alter the risk for allergic diseases (Pacheco et al., 2021). Furthermore, some studies suggest that pollens may prone human beings to viral infections by impairing the antiviral immune response (Bergougnan et al., 2020; Gilles et al., 2019).

3.6 Extreme Weather Conditions

Intensified water cycles and raised evaporation together with annual precipitation anomalies can cause extreme storms, hurricanes; these torrential rain events can trigger the floods (NASA, 2022). In addition to the devastating effects on the environment, economy, and infrastructure, these extreme events can deteriorate living conditions and cause injuries and severe casualties (Zittis et al., 2022). The damage to the infrastructure and living conditions let people to gather and live in more crowded conditions, which increase the risk for contagious diseases including respiratory infections and makes access to health services much difficult (Pacheco et al., 2021). For instance, the strongest tropical cyclone over the Arabian Sea was observed in 2007, and it was the greatest natural disaster in Oman leading to enormous economic losses (Fritz et al., 2010). Cyclone Chapala, the second-strongest cyclone on record in the Arabian Sea hit Yemen in 2015 (Access Science, 2015). These natural disasters caused many deaths, as well as serious socioeconomic consequences in these countries (Sarker, 2018). Finally, such floods and rain thunderstorms may lead to increased pollen release, fungal spore, and aeroallergen production that could cause enhanced hospital visit for allergic airway diseases such as asthma and rhinitis (Mitchell et al., 2012; Pacheco et al., 2021; Venkatesan, 2022).

3.7 Respiratory Infections Including SARS-CoV-2

Furthermore, global warming influences prevalence, total period, and severity of common respiratory infectious epidemics such as influenza and respiratory syncytial virus (Towers et al., 2013; Weinberger et al., 2015). It has been shown that PM of various sizes can carry viruses in the air, therefore they may be effective in the spread of measles, RSV, influenza, and even SARS viruses (Chen et al., 2010, 2017; Kayalar et al., 2021; Ye et al., 2016).

In recent years, air pollution has been a hot topic of the COVID-19 pandemic too. Many publications around the world correlate COVID-19 cases, death rates, and air pollution (Ogen, 2020; Pansini & Fornacca, 2021; Travaglio et al., 2021). Most of the studies are conducted in Europe, US, or Asia showing a correlation with especially PM and nitrogen dioxide (NO_2) concentrations. Several studies from the Middle East investigated the associations between air pollution and COVID-19 pandemic. Aykaç and Etiler (2021) found a moderate correlation between COVID-19 mortality rate per 100,000 population in Istanbul, Tıurkey and levels of PM10, sulfur dioxide (SO_2), and NO_2. There was also an association between COVID-19 mortality and lower socioeconomic status. Khorsandi et al. (2021) showed an association between air pollutants including $PM_{2.5}$, PM_{10}, and O_3, and COVID-19 hospitalization/mortality in Tehran, Iran during warm seasons. During summer, 2020, the prevalence of COVID-19 was relatively high in Riyadh, Jeddah, and Makkah, which were the most affected cities in Saudi Arabia. PM_{10}, NO_2, and O_3 levels were positively correlated to daily cases of COVID-19. Fluctuations in temperature difference and wind speed had an impact on the number of COVID-19 cases, too. This may be one of the reasons why the number of cases is increased in Middle Eastern countries, where the temperature difference between day and night is high. Also, extreme temperature encourages people to stay inside, where the ventilation is poor and heavily air-conditioned (Ben Maatoug et al., 2021). Moreover, a case study found a correlation between population density and city/town total population and COVID-19 morbidity in Israel (Levi & Barnett-Itzhaki, 2021). Studies from Italy (Setti et al., 2020) and Turkey (Kayalar et al., 2021) showed that SARS-CoV-2 can be found and transported on PM in different sizes. However, whether these were live viruses or not has not been investigated. Furthermore, studies suggest that air pollution may increase the risk of viral infections by impairing the human natural defense against airborne viruses (Ciencewicki & Jaspers, 2007).

4 Adaptation and Mitigation Strategies

Adaptation to climate change, natural disasters, and crisis management have come to the forefront, as it has become clear that recent studies to minimize climate change and mitigation strategies have not been sufficient globally. After the Paris Climate Agreement in 2015, the Conference of the Parties (COP) 26 agreed on carbon markets, and they decided to start reducing coal-fired energy and removing subsidies for other fossil fuels rather than taking any radical action. However, urgent global and national policies are required to reduce CO_2 emissions and protect society's most vulnerable members from the health consequences of climate change.

Despite efforts to mitigate the worst potential impacts of climate change, societies must become more resilient to the climate change impacts, which are currently unavoidable (e.g., heat waves, dust storms, floods, wildfires). Unfortunately, these effects usually disproportionately affect low-income communities and vulnerable populations; therefore, equitable resilience planning and resource allocation are very important. Greening inner-city heat islands, for example, and the establishment of cooling/clean air centers may aid in adapting to extreme heat. Additionally, measures can be taken against chronic events such as poor air quality, deforestation, desertification, and population migration.

Physicians can educate patients about effects of climate change (i.e., heat, smoke, allergens, or other exposure) on their health. This can help patients to protect themselves from the adverse health effects of climate change and improve understanding of climate change as a health hazard. Nongovernmental organizations (NGOs) may play a role in educating physicians and other healthcare providers about the effects of climate change on health, allowing them to better care for their patients. Physicians also play an important role as policy advocates, explaining the health benefits of policies, which can reduce climate-related events.

5 Conclusion

Global climate change is a serious human and environmental health concern worldwide, and some regions such as the Middle East are more vulnerable. As one of the major global GHG emitter regions, Middle East is warming almost twice more than rest of the globe (Zittis et al., 2022). The region is under the increased risk for heat waves, water shortage, desertification, dust storms, loss of biodiversity, and their health impacts. In addition to their support to the global action against climate change, resident countries must take local and regional adaptation and mitigation measures. Furthermore, more research is required to comprehend the scope of the problem, regional vulnerabilities, and its implications for human health.

References

Abdulla, F. (2020). 21st century climate change projections of precipitation and temperature in Jordan. *Procedia Manufacturing, 44*, 197–204. https://doi.org/10.1016/j.promfg.2020.02.222

Akpinar-Elci, M., Berumen-Flucker, B., Bayram, H., & Al-Taiar, A. (2021). Climate change, dust storms, vulnerable populations, and health in the Middle East: A review. *Journal of Environmental Health, 84*, 8–15.

Al, B., Bogan, M., Zengin, S., Sabak, M., Kul, S., Oktay, M. M., Bayram, H., & Vuruskan, E. (2018), Effects of dust storms and climatological factors on mortality and morbidity of cardiovascular diseases admitted to ED. *Emergency Medicine International, 2018*, 1–7. https://doi.org/10.1155/2018/3758506

Alahmad, B., Tomasso, L. P., Al-Hemoud, A., James, P., & Koutrakis, P. (2020). Spatial distribution of land surface temperatures in Kuwait: Urban heat and cool islands. *International Journal of Environmental Research and Public Health, 17*(9), 2993. https://doi.org/10.3390/ijerph17092993

Anderson, G. B., Dominici, F., Wang, Y., McCormack, M. C., Bell, M. L., & Peng, R. D. (2013). Heat-related emergency hospitalizations for respiratory diseases in the Medicare population. *American Journal of Respiratory and Critical Care Medicine, 187*(10), 1098–1103. https://doi.org/10.1164/rccm.201211-1969oc

Aykaç, N., & Etiler, N. (2021). Covid-19 mortality in Istanbul in association with air pollution and socioeconomic status: An ecological study. *Environmental Science and Pollution Research, 29*(9), 13700–13708. https://doi.org/10.1007/s11356-021-16624-1

Ayres, J. G., Forsberg, B., Annesi-Maesano, I., Dey, R., Ebi, K. L., Helms, P. J., Medina-Ramon, M., Windt, M., & Forastiere, F. (2009). Climate change and respiratory disease: European Respiratory Society Position statement. *European Respiratory Journal, 34*(2), 295–302. https://doi.org/10.1183/09031936.00003409

Bayram, H., Bauer, A. K., Abdalati, W., Carlsten, C., Pinkerton, K. E., Thurston, G. D., Balmes, J. R., & Takaro, T. K. (2017). Environment, global climate change, and Cardiopulmonary Health. *American Journal of Respiratory and Critical Care Medicine, 195*(6), 718–724. https://doi.org/10.1164/rccm.201604-0687pp

Bayram, H., Devalia, J. L., Khair, O. A., Abdelaziz, M. M., Sapsford, R. J., Sagai, M., & Davies, R. J. (1998b). Comparison of ciliary activity and inflammatory mediator release from bronchial epithelial cells of nonatopic nonasthmatic subjects and atopic asthmatic patients and the effect of diesel exhaust particles in Vitro. *Journal of Allergy and Clinical Immunology, 102*(5), 771–782. https://doi.org/10.1016/s0091-6749(98)70017-x

Bayram, H., Devalia, J. L., Sapsford, R. J., Ohtoshi, T., Miyabara, Y., Sagai, M., & Davies, R. J. (1998a). The effect of diesel exhaust particles on cell function and release of inflammatory mediators from human bronchial epithelial cells in vitro. *American Journal of Respiratory Cell and Molecular Biology, 18*(3), 441–448. https://doi.org/10.1165/ajrcmb.18.3.2882

Bayram, H., Fakili, F., Gögebakan, B., Bayraktar, R., Öztuzcu, S., Dikensoy, Ö., & Chung, K. F. (2013). Effect of serum on diesel exhaust particles (DEP)-induced apoptosis of airway epithelial cells in vitro. *Toxicology Letters, 218*(3), 215–223. https://doi.org/10.1016/j.toxlet.2013.02.006

Bayram, H., Ito, K., Issa, R., Ito, M., Sukkar, M., & Chung, F. (2006). Regulation of human lung epithelial cell numbers by diesel exhaust particles. *European Respiratory Journal, 27*(4), 705–713. https://doi.org/10.1183/09031936.06.00012805

Bayram, H., & Öztürk, A. B. (2021). Global climate change, desertification, and its consequences in Turkey and the Middle East. *Climate Change and Global Public Health*, 445–458.https://doi.org/10.1007/978-3-030-54746-2_21

Bayram, H., Rusznak, C., Khair, O. A., Sapsford, R. J., & Abdelaziz, M. M. (2002). Effect of ozone and nitrogen dioxide on the permeability of bronchial epithelial cell cultures of non-asthmatic and asthmatic subjects. *Clinical Experimental Allergy, 32*(9), 1285–1292. https://doi.org/10.1046/j.1365-2745.2002.01435.x

Bayram, H., Sapsford, R. J., Abdelaziz, M. M., & Khair, O. A. (2001). Effect of ozone and nitrogen dioxide on the release of proinflammatory mediators from bronchial epithelial cells of nonatopic nonasthmatic subjects and atopic asthmatic patients in vitro. *Journal of Allergy and Clinical Immunology, 107*(2), 287–294. https://doi.org/10.1067/mai.2001.111141

Bellard, C., Bertelsmeier, C., Leadley, P., Thuiller, W., & Courchamp, F. (2012). Impacts of climate change on the future of biodiversity. *Ecology Letters, 15*(4), 365–377. https://doi.org/10.1111/j.1461-0248.2011.01736.x

Ben Abdallah, F. C., Taktak, S., Chtourou, A., Mahouachi, R., & Kheder, A. B. (2011). Burden of chronic respiratory diseases (CRD) in Middle East and North Africa (MENA). *World Allergy Organization Journal, 4*(Supplement). https://doi.org/10.1097/wox.0b013e3182093cbb

Ben Maatoug, A., Triki, M. B., & Fazel, H. (2021). How do air pollution and meteorological parameters contribute to the spread of covid-19 in Saudi Arabia? *Environmental Science and Pollution Research, 28*(32), 44132–44139. https://doi.org/10.1007/s11356-021-13582-6

Bergougnan, C., Dittlein, D. C., Hümmer, E., Riepl, R., Eisenbart, S., Böck, D., Griesbaum, L., Weigl, A., Damialis, A., Hartwig, A., Neumann, A. U., Zenk, J., Traidl-Hoffmann, C., & Gilles, S. (2020). Physical and immunological barrier of human primary nasal epithelial cells from non-allergic and allergic donors. *World Allergy Organization Journal, 13*(3), 100109. https://doi.org/10.1016/j.waojou.2020.100109

Bogan, M., Al, B., Kul, S., Zengin, S., Oktay, M., Sabak, M., Gümüşboğa, H., & Bayram, H. (2021). The effects of desert dust storms, air pollution, and temperature on morbidity due to spontaneous abortions and toxemia of pregnancy: 5-year analysis. *International Journal of Biometeorology, 65*(10), 1733–1739. https://doi.org/10.1007/s00484-021-02127-8

Boğan, M., Kul, S., Al, B., Oktay, M. M., Akpinar Elçi, M., Pinkerton, K. E., & Bayram, H. (2022). Effect of desert dust storms and meteorological factors on respiratory diseases. *Allergy, 77*(7), 2243–2246. https://doi.org/10.1111/all.15298

Cevik, Y., Doğan, N. Ö., Daş, M., Ahmedali, A., Kul, S., & Bayram, H. (2015). The association between weather conditions and stroke admissions in Turkey. *International Journal of Biometeorology, 59*(7), 899–905. https://doi.org/10.1007/s00484-014-0890-9

Chen, G., Zhang, W., Li, S., Williams, G., Liu, C., Morgan, G., Jaakkola, J., & Guo, Y. (2017). Is short-term exposure to ambient fine particles associated with measles incidence in China? A multi-city study. *Environmental Research, 156*, 306–311. https://doi.org/10.1016/j.envres.2017.03.046

Chen, P., Tsai, F., Lin, C., Yang, C., Chan, C., Young, C., & Lee, C. (2010). Ambient influenza and avian influenza virus during dust storm days and background days. *Environmental Health Perspectives, 118*, 1211–1216. https://doi.org/10.1289/ehp.0901782

Ciencewicki, J., & Jaspers, I. (2007). Air pollution and respiratory viral infection. *Inhalation Toxicology, 19*(14), 1135–1146. https://doi.org/10.1080/08958370701665434

CO_2.Earth. (2022). *Annual CO_2 data.* [cited 2022 Nov 2022]. Available from: https://www.co2.earth/annual-co2

CO_2 Emissions|Global Carbon Atlas. (2020). *Global carbon atlas.* Retrieved October 31, 2022, from http://www.globalcarbonatlas.org/en/CO2-emissions

D'Amato, G., Cecchi, L., D'Amato, M., & Annesi-Maesano, I. (2014). Climate change and respiratory diseases. *European Respiratory Review, 23*(132), 161–169. https://doi.org/10.1183/09059180.00001714

D'Amato, G., Holgate, S. T., Pawankar, R., Ledford, D. K., Cecchi, L., Al-Ahmad, M., Al-Enezi, F., Al-Muhsen, S., Ansotegui, I., Baena-Cagnani, C. E., Baker, D. J., Bayram, H., Bergmann, K. C., Boulet, L.-P., Buters, J. T. M., D'Amato, M., Dorsano, S., Douwes, J., Finlay, S. E., … Annesi-Maesano, I. (2015). Meteorological conditions, climate change, new emerging factors, and asthma and related allergic disorders. A statement of the World Allergy Organization. *World Allergy Organization Journal, 8*, 25. https://doi.org/10.1186/s40413-015-0073-0

Ebi, K. L., Vanos, J., Baldwin, J. W., Bell, J. E., Hondula, D. M., Errett, N. A., Hayes, K., Reid, C. E., Saha, S., Spector, J., & Berry, P. (2021). Extreme weather and climate change: Population health and health system implications. *Annual Review of Public Health, 42*(1), 293–315. https://doi.org/10.1146/annurev-publhealth-012420-105026

El-Zein, A., Tewtel-Salem, M., & Nehme, G. (2004). A time-series analysis of mortality and air temperature in Greater Beirut. *Science of the Total Environment, 330*(1–3), 71–80. https://doi.org/10.1016/j.scitotenv.2004.02.027

Fernandez, S., Ardura, A., Georges, O., Martinez, J. L., Suarez-Menendez, M., Planes, S., & Garcia-Vazquez, E. (2022). Non-native species in the North Gulf of aqaba (Red Sea) revealed from environmental DNA. *Frontiers in Marine Science, 9*.https://doi.org/10.3389/fmars.2022.940889

Fritz, H. M., Blount, C. D., Albusaidi, F. B., & Al-Harthy, A. H. M. (2010). Cyclone Gonu storm surge in Oman. *Estuarine, Coastal and Shelf Science, 86*(1), 102–106. https://doi.org/10.1016/j.ecss.2009.10.019

Gavurova, B., Rigelsky, M., & Ivankova, V. (2021). Greenhouse gas emissions and health in the countries of the European Union. *Frontiers in Public Health, 9*.https://doi.org/10.3389/fpubh.2021.756652

Ghio, A. J., Kummarapurugu, S. T., Tong, H., Soukup, J. M., Dailey, L. A., Boykin, E., Ian Gilmour, M., Ingram, P., Roggli, V. L., Goldstein, H. L., & Reynolds, R. L. (2014). Biological effects of desert dust in respiratory epithelial cells and a murine model. *Inhalation Toxicology, 26*(5), 299–309. https://doi.org/10.3109/08958378.2014.888109

Gilles, S., Blume, C., Wimmer, M., Damialis, A., Meulenbroek, L., Gökkaya, M., Bergougnan, C., Eisenbart, S., Sundell, N., Lindh, M., Andersson, L. M., Dahl, Å., Chaker, A., Kolek, F., Wagner, S., Neumann, A. U., Akdis, C. A., Garssen, J., Westin, J., … Traidl-Hoffmann, C. (2019). Pollen exposure weakens innate defense against respiratory viruses. *Allergy, 75*(3), 576–587.https://doi.org/10.1111/all.14047

Goudie, A. S. (2009). Dust storms: Recent developments. *Journal of Environmental Management, 90*(1), 89–94. https://doi.org/10.1016/j.jenvman.2008.07.007

Greenhouse gas (GHG) emissions|climate watch. (n.d.). Retrieved December 5, 2022, from https://www.climatewatchdata.org/ghg-emissions?end_year=2019;start_year=1990

Gwela, A., Mupere, E., Berkley, J. A., & Lancioni, C. (2019). Undernutrition, host immunity and vulnerability to infection among young children. *The Pediatric Infectious Disease Journal, 38*(8), e175–e177. https://doi.org/10.1097/INF.0000000000002363

Hales, S., McMichael, A. J., & Butler, C. (2005). *Ecosystems and human well-being—Health synthesis a report of the Millennium ecosystem assessment.* World Health Organization.

Harris, I., Osborn, T. J., Jones, P., & Lister, D. (2020). Version 4 of the CRU TS monthly high-resolution gridded multivariate climate dataset. *Scientific Data, 7*(1), 1–18. https://doi.org/10.1038/s41597-020-0453-3

Hofste, R. W., Reig, P., & Schleifer, L. (2019). *17 countries, home to one-quarter of the world's population, face extremely high water stress*. World Resources Institute. Retrieved November 1, 2022, from https://www.wri.org/insights/17-countries-home-one-quarter-worlds-population-face-extremely-high-water-stress

IPCC, Intergovernmental Panel on Climate Change. (n.d.). *Climate change 2021: The physical science basis*. Retrieved November 1, 2022, from https://www.ipcc.ch/report/ar6/wg1/

Joshi, M., Goraya, H., Joshi, A., & Bartter, T. (2020). Climate change and respiratory diseases. *Current Opinion in Pulmonary Medicine, 26*(2), 119–127. https://doi.org/10.1097/mcp.0000000000000656

Kayalar, Ö., Arı, A., Babuççu, G., Konyalılar, N., Doğan, Ö., Can, F., Şahin, Ü. A., Gaga, E. O., Kuzu, S. L., Arı, P. E., Odabası, M., Taşdemir, Y., Cindoruk, S. S., Esen, F., Sakın, E., Çalışkan, B., Tecer, L. H., Fıçıcı, M., Altın A., … Bayram, H. (2021). Existence of SARS-CoV-2 RNA on ambient particulate matter samples: A nationwide study in Turkey. *Science of Total Environment, 789*, 147976. https://doi.org/10.1016/j.scitotenv.2021.147976

Khaniabadi, Y. O., Daryanoosh, S. M., Amrane, A., Polosa, R., Hopke, P. K., Goudarzi, G., Mohammadi, M. J., Sicard, P., & Armin, H. (2017). Impact of Middle Eastern Dust storms on human health. *Atmospheric Pollution Research*, 606–613.https://doi.org/10.1016/j.apr.2016.11.005

Khorsandi, B., Farzad, K., Tahriri, H., & Maknoon, R. (2021). Association between short-term exposure to air pollution and COVID-19 hospital admission/mortality during warm seasons. *Environmental Monitoring and Assessment, 193*(7). https://doi.org/10.1007/s10661-021-09210-y

Koutsias, N., Xanthopoulos, G., Founda, D., Xystrakis, F., Nioti, F., Pleniou, M., Mallinis, G., & Arianoutsou, M. (2013). On the relationships between forest fires and weather conditions in Greece from long-term national observations (1894–2010). *International Journal of Wildland Fire, 22*(4), 493. https://doi.org/10.1071/wf12003

Krupp, F., Khalaf, M., Malek, M., Streit, B., & Al-Jumaily, M. (2009). The Middle Eastern biodiversity network: Generating and sharing knowledge for ecosystem management and conservation. *ZooKeys, 31*, 3–15. https://doi.org/10.3897/zookeys.31.371

Kuiper-Makris, C., Selle, J., Nüsken, E., Dötsch, J., & Alejandre Alcazar, M. A. (2021), Perinatal nutritional and metabolic pathways: Early origins of chronic lung diseases. *Frontiers in Medicine, 8*.https://doi.org/10.3389/fmed.2021.667315

Landrigan, P. J., Fuller, R., Acosta, N. J., Adeyi, O., Arnold, R., Basu, N. N., Baldé, A. B., Bertollini, R., Bose-O'Reilly, S., Boufford, J. I., Breysse, P. N., Chiles, T., Mahidol, C., Coll-Seck, A. M., Cropper, M. L., Fobil, J., Fuster, V., Greenstone, M., Haines, A., … Zhong, M. (2018), The lancet commission on pollution and health. *The Lancet, 391*(10119), 462–512.https://doi.org/10.1016/s0140-6736(17)32345-0

Lange, M. A. (2019). Impacts of climate change on the Eastern Mediterranean and the Middle East and North Africa region and the Water-Energy Nexus. *Atmosphere, 10*(8), 455. https://doi.org/10.3390/atmos10080455

Levi, A., & Barnett-Itzhaki, Z. (2021). Effects of chronic exposure to ambient air pollutants, demographic, and socioeconomic factors on COVID-19 morbidity: The Israeli case study. *Environmental Research, 202*, 111673. https://doi.org/10.1016/j.envres.2021.111673

Lin, S., Luo, M., Walker, R. J., Liu, X., Hwang, S.-A., & Chinery, R. (2009). Extreme high temperatures and hospital admissions for respiratory and cardiovascular diseases. *Epidemiology, 20*(5), 738–746. https://doi.org/10.1097/ede.0b013e3181ad5522

Majdi, F., Hosseini, S. A., Karbalaee, A., Kaseri, M., & Marjanian, S. (2022). Future projection of precipitation and temperature changes in the Middle East and North Africa (MENA) region based on CMIP6. *Theoretical and Applied Climatology, 147*(3–4), 1249–1262. https://doi.org/10.1007/s00704-021-03916-2

Maleki, H., Sorooshian, A., Alam, K., Fathi, A., Weckwerth, T., Moazed, H., Jamshidi, A., Babaei, A. A., Hamid, V., Soltani, F., & Goudarzi, G. (2022). The impact of meteorological parameters on PM10 and visibility during the Middle Eastern dust storms. *Journal of Environmental Health Science and Engineering, 20*(1), 495–507. https://doi.org/10.1007/s40201-022-00795-1

Mirzabaev, A., Wu, J., Evans, J., García-Oliva, F., Hussein, I. A. G., Iqbal, M. H., Kimutai, J., Knowles, T., Meza, F., Nedjraoui, D., Tena, F., Türkeş, M., Vázquez, R. J., Weltz, M. (2019). Desertification. In P. R. Shukla, J. Skea, E. Calvo Buendia, V. Masson-Delmotte, H. -O. Pörtner, D. C. Roberts, P. Zhai, R. Slade, S. Connors, R. van Diemen, M. Ferrat, E. Haughey, S. Luz, S. Neogi, M. Pathak, J. Petzold, J. Portugal Pereira, P. Vyas, E. Huntley … J. Malley (Eds.), *Climate change and land: An IPCC special report on climate change, desertification, land degradation, sustainable land management, food security, and greenhouse gas fluxes in terrestrial ecosystems*. In press.

Mitchell, H., Cohn, R. D., Wildfire, J., Thornton, E., Kennedy, S., El-Dahr, J. M., Chulada, P. C., Mvula, M. M., Grimsley, L. F., Lichtveld, M. Y., White, L. A. E., Sterling, Y. M., Stephens, K. U., & Martin, W. J. (2012). Implementation of evidence-based asthma interventions in Post-Katrina New Orleans: The head-off environmental asthma in Louisiana (HEAL) study. *Environmental Health Perspectives, 120*(11), 1607–1612. https://doi.org/10.1289/ehp.1104242

Murray, C. J., Aravkin, A. Y., Zheng, P., Abbafati, C., Abbas, K. M., Abbasi-Kangevari, M., Abd-Allah, F., Abdelalim, A., Abdollahi, M., Abdollahpour, I., Abegaz, K. H., Abolhassani, H., Aboyans, V., Abreu, L. G., Abrigo, M. R., Abualhasan, A., Abu-Raddad, L. J., Abushouk, A. I., Adabi, M., … Lim, S. S. (2020). Global burden of 87 risk factors in 204 countries and territories, 1990–2019: A systematic analysis for the global burden of disease study 2019. *The Lancet, 396*(10258), 1223–1249.https://doi.org/10.1016/s0140-6736(20)30752-2

Myers, S. S., Gaffikin, L., Golden, C. D., Ostfeld, R. S., Redford, K. H., Ricketts, T. H., Turner, W. R., & Osofsky, S. A. (2013). Human health impacts of ecosystem alteration. *Proceedings of the National Academy of Sciences, 110*(47), 18753–18760.https://doi.org/10.1073/pnas.1218656110

NASA. (n.d.). *How does climate change affect precipitation?* NASA. Retrieved November 1, 2022, from https://gpm.nasa.gov/resources/faq/how-does-climate-change-affect-precipitation

NOAA National Centers for Environmental Information, State of the Climate: Global Climate Report for 2021, online January 2021, retrieved on June 28, 2022, from https://www.ncdc.noaa.gov/sotc/global/202113

Ogen, Y. (2020). Assessing nitrogen dioxide (NO_2) levels as a contributing factor to coronavirus (COVID-19) fatality. *Science of the Total Environment, 726*, 138605. https://doi.org/10.1016/j.scitotenv.2020.138605

Pacheco, S. E., Guidos-Fogelbach, G., Annesi-Maesano, I., Pawankar, R., D'Amato, G., Latour-Staffeld, P., Urrutia-Pereira, M., Kesic, M. J., & Hernandez, M. L. (2021). Climate change and global issues in allergy and immunology. *Journal of Allergy and Clinical Immunology, 148*(6), 1366–1377.https://doi.org/10.1016/j.jaci.2021.10.011

Pal, J. S., & Eltahir, E. A. (2015). Future temperature in Southwest Asia projected to exceed a threshold for human adaptability. *Nature Climate Change, 6*(2), 197–200. https://doi.org/10.1038/nclimate2833

Pansini, R., & Fornacca, D. (2021). Early spread of covid-19 in the air-polluted regions of eight severely affected countries. *Atmosphere, 12*(6), 795. https://doi.org/10.3390/atmos12060795

Salman, S. A., Shahid, S., Ismail, T., Chung, E.-S., & Al-Abadi, A. M. (2017). Long-term trends in daily temperature extremes in Iraq. *Atmospheric Research, 198*, 97–107. https://doi.org/10.1016/j.atmosres.2017.08.011

Sarker, M. A. (2018). Numerical modelling of waves and surge from Cyclone Chapala (2015) in the Arabian Sea. *Ocean Engineering, 158*, 299–310. https://doi.org/10.1016/j.oceaneng.2018.04.014

Schnell, J. L., & Prather, M. J. (2017). Co-occurrence of extremes in surface ozone, particulate matter, and temperature over eastern North America. *Proceedings of the National Academy of Sciences, 114*(11), 2854–2859. https://doi.org/10.1073/pnas.1614453114

Schraufnagel, D. E., Balmes, J. R., Cowl, C. T., De Matteis, S., Jung, S.-H., Mortimer, K., Perez-Padilla, R., Rice, M. B., Riojas-Rodriguez, H., Sood, A., Thurston, G. D., To, T., Vanker, A., & Wuebbles, D. J. (2019). Air pollution and noncommunicable diseases. *Chest, 155*(2), 409–416. https://doi.org/10.1016/j.chest.2018.10.042

Setti, L., Passarini, F., De Gennaro, G., Barbieri, P., Licen, S., Perrone, M. G., Piazzalunga, A., Borelli, M., Palmisani, J., Di Gilio, A., Rizzo, E., Colao, A., Piscitelli, P., & Miani, A. (2020). Potential role of particulate matter in the spreading of covid-19 in Northern Italy: First observational study based on initial epidemic diffusion. *BMJ Open, 10*(9), e039338. https://doi.org/10.1136/bmjopen-2020-039338

Shokoohi, Z., Dehbidi, N. K., & Tarazkar, M. H. (2022). Energy intensity, economic growth and environmental quality in populous Middle East countries. *Energy, 239*, 122164. https://doi.org/10.1016/j.energy.2021.122164

Silva, R. A., West, J. J., Lamarque, J.-F., Shindell, D. T., Collins, W. J., Faluvegi, G., Folberth, G. A., Horowitz, L. W., Nagashima, T., Naik, V., Rumbold, S. T., Sudo, K., Takemura, T., Bergmann, D., Cameron-Smith, P., Doherty, R. M., Josse, B., MacKenzie, I. A., Stevenson, D. S., & Zeng, G. (2017). Future global mortality from changes in air pollution attributable to climate change. *Nature Climate Change, 7*(9), 647–651. https://doi.org/10.1038/nclimate3354

Skjærseth, J. B., Andresen, S., Bang, G., & Heggelund, G. M. (2021). The Paris Agreement and key actors' domestic climate policy mixes: Comparative patterns. *International Environmental Agreements: Politics, Law and Economics, 21*(1), 59–73. https://doi.org/10.1007/s10784-021-09531-w

Soleimani, Z., Teymouri, P., Darvishi Boloorani, A., Mesdaghinia, A., Middleton, N., & Griffin, D. W. (2020). An overview of bioaerosol load and health impacts associated with dust storms: A focus on the Middle East. *Atmospheric Environment, 223*, 117187. https://doi.org/10.1016/j.atmosenv.2019.117187

Stepanyan, V., Abdih, M. Y., Al-Hassan, A., Gerling, M. K., Saksonovs, M. S., Baum, M. A., Anderson, G., Agoumi, L., Andaloussi, M. B., Chen, C., Sakha, S., Saliba, F., Sanchez, J., & Duenwald, M. C. (2022). Feeling the heat: Adapting to climate change in the Middle East and Central Asia. *International Monetary Fund*.

Tabari, H., & Willems, P. (2018). More prolonged droughts by the end of the century in the Middle East. *Environmental Research Letters, 13*(10), 104005. https://doi.org/10.1088/1748-9326/aae09c

Thalib, L., & Al-Taiar, A. (2012). Dust storms and the risk of asthma admissions to hospitals in Kuwait. *Science of the Total Environment, 433*, 347–351. https://doi.org/10.1016/j.scitotenv.2012.06.082

Tobías, A., Pérez, L., Díaz, J., Linares, C., Pey, J., Alastruey, A., & Querol, X. (2011) Short-term effects of particulate matter on total mortality during Saharan dust outbreaks: A case-crossover analysis in Madrid (Spain). *Science of The Total Environment, 412–413*, 386–389. https://doi.org/10.1016/j.scitotenv.2011.10.027

Towers, S., Chowell, G., Hameed, R., Jastrebski, M., Khan, M., Meeks, J., Mubayi, A., & Harris, G. (2013). Climate change and influenza: The likelihood of early and severe influenza seasons following warmer than average winters. *PLoS Currents*. https://doi.org/10.1371/currents.flu.3679b56a3a5313dc7c043fb944c6f138

Travaglio, M., Yu, Y., Popovic, R., Selley, L., Leal, N. S., & Martins, L. M. (2021). Links between air pollution and covid-19 in England. *Environmental Pollution, 268*, 115859. https://doi.org/10.1016/j.envpol.2020.115859

Tropical cyclones Chapala and Megh (Arabian Sea, 2015). (n.d.). AccessScience. https://doi.org/10.1036/1097-8542.br1110151

Turco, M., Levin, N., Tessler, N., & Saaroni, H. (2017). Recent changes and relations among drought, vegetation and wildfires in the Eastern Mediterranean: The case of Israel. *Global and Planetary Change, 151*, 28–35. https://doi.org/10.1016/j.gloplacha.2016.09.002

Turkish State Meteorological Service, Ministry of Environment, Urbanization and Climate Change, Republic of Türkiye. State of Türkiye's Climate in 2021. https://mgm.gov.tr/eng/Yearly-Climate/State_of_the_Climate_in_Turkey_in_2021.pdf

UN Environment Programme. (2022). *Emissions gap report 2022*. Retrieved October 31, 2022, from https://www.unep.org/resources/emissions-gap-report-2022

UNEP. (2019). *Emissions gap report 2019*. Retrieved August 14, 2021, from https://wedocs.unep.org/bitstream/handle/20.500.11822/30797/EGR2019.pdf

UNEP. (n.d.). Yemen: Protecting Socotra's dwindling biodiversity. Retrieved December 6, 2022, from https://www.unep.org/news-and-stories/story/yemen-protecting-socotras-dwindling-biodiversity

UNESCO. (2020, February 27). *Connect 2 socotra*. Retrieved February 2, 2023, from https://en.unesco.org/connect2socotra

UNICEF Middle East and North Africa. (2017). *WASH: Water, sanitation and hygiene providing safe water to children around the region*. Retrieved November 1, 2022, from https://www.unicef.org/mena/wash

United Nations Climate Change (UNCC). (2022). *The Paris agreement, what is the Paris agreement?*. Retrieved October 31, 2022, from https://unfccc.int/process-and-meetings/the-paris-agreement/the-paris-agreement

Venkatesan, P. (2022). Epidemic thunderstorm asthma. *The Lancet Respiratory Medicine, 10*(4), 325–326. https://doi.org/10.1016/s2213-2600(22)00083-2

Weinberger, D. M., Klugman, K. P., Steiner, C. A., Simonsen, L., & Viboud, C. (2015). Association between respiratory syncytial virus activity and pneumococcal disease in infants: A time series analysis of US hospitalization data. *PLoS Medicine, 12*(1). https://doi.org/10.1371/journal.pmed.1001776

Wilfong, E. R., Lyles, M., Rietcheck, R. L., Arfsten, D. P., Boeckman, H. J., Johnson, E. W., Doyle, T. L., & Chapman, G. D. (2011). The acute and long-term effects of Middle East sand particles on the rat airway following a single intratracheal instillation. *Journal of Toxicology and Environmental Health, Part A, 74*(20), 1351–1365. https://doi.org/10.1080/15287394.2010.516239

Wittenberg, L., & Kutiel, H. (2016). Dryness in a Mediterranean-type climate—Implications for wildfire burnt area: A case study from Mount Carmel, Israel. *International Journal of Wildland Fire, 25*(5), 579–591. https://doi.org/10.1071/WF15135

WMO, World Meteorological Organization. (2021). *State of the global climate 2021: WMO provisional report*. Retrieved from https://library.wmo.int/index.php?lvl=notice_display%26id=21982

World Bank Climate Change Knowledge Portal. (2021). *Climatology|climate change knowledge portal*. Retrieved December 10, 2022, from https://climateknowledgeportal.worldbank.org/country/egypt/climate-data-historical

World Bank Group. (2016). High and dry: Climate change, water, and the economy. *World Bank*. Retrieved November 1, 2022, from https://www.worldbank.org/en/topic/water/publication/high-and-dry-climate-change-water-and-the-economy

World Health Organization. (2014). *7 million premature deaths annually linked to air pollution*. World Health Organization. Retrieved February 2, 2023, from https://www.who.int/news/item/25-03-2014-7-million-premature-deaths-annually-linked-to-air-pollution

World Health Organization. (2023). *Biodiversity and health*. Retrieved February 2, 2023, from https://www.who.int/news-room/fact-sheets/detail/biodiversity-and-health

Ye, Q., Fu, J., Mao, J., & Shang, S. (2016). Haze is a risk factor contributing to the rapid spread of respiratory syncytial virus in children. *Environmental Science and Pollution Research*, 20178–20185.https://doi.org/10.1007/s11356-016-7228-6

Zhao, Y., Huang, Z., Wang, S., Hu, J., Xiao, J., Li, X., Liu, T., Zeng, W., Guo, L., Du, Q., & Ma, W. (2019). Morbidity burden of respiratory diseases attributable to ambient temperature: A case study in a subtropical city in China. *Environmental Health, 18*(1). https://doi.org/10.1186/s12940-019-0529-8

Zittis, G., Almazroui, M., Alpert, P., Ciais, P., Cramer, W., Dahdal, Y., Fnais, M., Francis, D., Hadjinicolaou, P., Howari, F., Jrrar, A., Kaskaoutis, D. G., Kulmala, M., Lazoglou, G., Mihalopoulos, N., Lin, X., Rudich, Y., Sciare, J., Stenchikov, G., … Lelieveld, J. (2022). Climate change and weather extremes in the Eastern Mediterranean and Middle East. *Reviews of Geophysics, 60*(3). https://doi.org/10.1029/2021rg000762

Zittis, G., Hadjinicolaou, P., Fnais, M., & Lelieveld, J. (2016). Projected changes in heat wave characteristics in the Eastern Mediterranean and the Middle East. *Regional Environmental Change, 16*(7), 1863–1876. https://doi.org/10.1007/s10113-014-0753-2

Europe

Possible Implications of Annual Temperature and Precipitation Changes in Tick-Borne Encephalitis and West Nile Virus Incidence in Italy, Between 2010 and 2020

Alessandra di Masi, Cristiano Pesaresi, Stefano Di Bella, and Cosimo Palagiano

Abstract According to the Istat data, between 2010 and 2020 in Italy some relevant changes in Average Annual Temperature (°C) and Total Annual Precipitation (mm) have been recorded, with consequent effects on hydrological drought, environmental aspects, economic activities, and human health. To provide an overall dynamic picture, an analysis has been conducted about the mean calculated on the twenty Italian regional capitals and single regional capitals. After providing a synthetic framework, also evidencing specific cases, the main aim was to examine and integrate the up-to-date knowledge on the impacts of specific changes on tick-borne encephalitis virus (TBEV) and West Nile virus (WNV) transmission. Indeed, during the last two decades we experienced a changing epidemiology for some arboviral infections in Italy, especially for TBEV and WNV. Here, we discussed the most relevant peer-reviewed studies that deal with linkages between meteorological and climatic elements and TBEV or WNV diffusion, as well as vector ecology (i.e., ticks and mosquitoes, respectively).

Even if the paper was devised together by the Authors, Cosimo Palagiano wrote the Introduction and coordinated the research, Cristiano Pesaresi wrote paragraphs 2 and 3, Alessandra di Masi and Stefano Di Bella wrote all the other paragraphs.

A. di Masi
Roma Tre University, Rome, Italy

Centro Linceo Interdisciplinare Beniamino Segre, Rome, Italy

C. Pesaresi
Sapienza University of Rome, Rome, Italy

S. Di Bella
Trieste University, Trieste, Italy

C. Palagiano (✉)
Sapienza University of Rome, Rome, Italy
e-mail: cosimo.palagiano@uniroma1.it

R. Akhtar (ed.), *Climate Change and Human Health Scenarios*, Global Perspectives on Health Geography, https://doi.org/10.1007/978-3-031-38878-1_13

Keywords Italy · Infectious diseases · Arbovirosis · Tick-borne encephalitis virus · West Nile virus · Temperature · Precipitation

1 Introduction

The goal of this research is to investigate the impact of Average Annual Temperature (°C) and Total Annual Precipitation (mm) changes and fluctuations on selected infectious diseases in Italy. Particularly, we wish to provide some prime considerations about the possible relationship between these meteorological elements and changes and some infectious diseases in Italy between 2010 and 2020.

In fact, changes in Average Annual Temperature (°C) and Total Annual Precipitation (mm) can lead to the onset of vector-borne diseases such as tick-borne encephalitis (TBE), malaria, dengue, and West Nile by altering their rates, ranges, distribution and seasonality (Ebi et al., 2013; Patz et al., 2005; Semenza & Menne, 2009; Tabachnick, 2010; Walther et al., 2002).

Vector-borne diseases are dynamic systems with complex ecology, which tend to adjust continually to environmental changes in multifaceted ways. Weather and climatic conditions—temperature and precipitations in particular—can affect the survival and reproduction rates of the vectors, their habitat suitability, distribution and abundance (Rogers & Randolph, 2006) Additionally, elements like temperature and precipitation can impact the intensity and temporal activity of the vector throughout the year and affect the rates of development, reproduction, and survival of pathogens within the vectors (Rogers & Randolph, 2006; Semenza & Menne, 2009). The Intergovernmental Panel on Climate Change (IPCC) (IPCC, 2013) lists vector-borne diseases among the consequences most likely to change due to global warming. Moreover, as these diseases are particularly sensitive to weather fluctuations, they could serve as an alert to focus attention on climate change threats.

In this contribution, we provided a framework regarding Average Annual Temperature (°C) and Total Annual Precipitation (mm) data from 2010 to 2020 in Italy, also focusing on the twenty Italian regional capitals. Then, we examine and integrate the up-to-date knowledge on the impacts of weather and climatic changes on tick-borne encephalitis virus (TBEV) and West Nile virus (WNV) transmission. Indeed, in Italy TBEV and WNV are under specific surveillance plans by the National Surveillance System of Arboviral Diseases of the Italian National Institute of Health (https://www.epicentro.iss.it/en/arboviral-diseases/surveillance; https://www.epicentro.iss.it/en/west-nile-fever/surveillance). Therefore, for these arboviral diseases, particular attention should be given to change of parameters, like temperature and precipitation, affecting their diffusion. Here, we have reported and discussed the most relevant peer-reviewed studies that deal with linkages between these (and other) elements and TBEV or WNV diffusion, as well as vector ecology (i.e., ticks and mosquitoes, respectively). Special attention has been paid to recent publications that highlight regional meteorological and climatic change impacts on vector population dynamics and on disease transmission.

2 The Average Annual Temperature and Total Annual Precipitation in Italy Between 2010 and 2020

In this work, we have considered some Istat (The Italian National Institute of Statistics) data collected for the 2022 Report "Climate changes: statistical measures" (*I cambiamenti climatici: misure statistiche*) referred to the 2010–2020 period (https://www.istat.it/it/archivio/268615). To provide a synthetic and dynamic framework, based on the data available, we have selected the Average Annual Temperature (°C) and Total Annual Precipitation (mm) and we have calculated the means on the twenty Italian regional capitals, with the highlighting of the trends. Next, to underline the main variations recorded in terms of specific territorial contexts, we have also considered the Average Annual Temperature and Total Annual Precipitation for the single Italian regional capitals. The analysis of the data regarding the mean calculated on the twenty Italian regional capitals makes it possible to put in evidence specific anomalies at national level, while the focus on the single Italian regional capitals shows the cases with the highest increases (as far as concerns temperature) and decreases (precipitation).

In particular, the mean calculated on the twenty Italian regional capitals for the Average Annual Temperature (°C) in temporal perspective (Fig. 1) underlines an increase from 14.8 to 15.9 °C. The year 2010 recorded the lowest value of the series, that went sharply to 15.8 °C in 2011 and 15.9 in 2012; successively, the dynamic shows a fluctuation and the temperature decreased to 15.4 °C in 2013 and then increased once again. In fact, the 2014–2015 biennium recorded high values, equal to 16.1 °C, and then the temperature remained steady at around 15.9 and 15.8 °C, reaching 16.2 °C in 2018 which was the year with the highest value. Finally, the values are 16.0 and 15.9 °C, respectively, in 2019 and 2020, with a general trend that shows an increase of 1.1 °C from 2010 to 2020.

On the other hand, the mean calculated on the twenty Italian regional capitals for the Total Annual Precipitation (mm) in diachronic perspective (Fig. 2) highlights a significant decrease (with an undulating dynamic) of 385 mm, from 1.047.9 to 662.9 mm. The year 2010 recorded the highest value of the series, falling sharply to 632.7 mm in 2011, the penultimate value of the period examined. The values then increased to 738.9 in 2012, 928.9 in 2013 and 983.6 mm in 2014, the second highest value of the series. Successively, the precipitation showed a new drop to 703.5 mm in 2015, a slight increase to 774.5 in 2016, and the lowest value of the series in 2017, equal to 586.3 mm. After a biennium of considerable increased values (in 2018 and 2019, with 856.3 and 867.4 mm, respectively), a new very low value was recorded in 2020, equal to 662.9 mm, with a general trend which underlines a relevant decrease.

The analysis conducted on the single Italian regional capitals makes it possible to recognize specific details in terms of diachronic variations and territorial contexts mainly involved in processes of significant changes in temperature and precipitation. As far as the Average Annual Temperature (°C), the most significant variations were recorded in Milan and Bologna (Table 1).

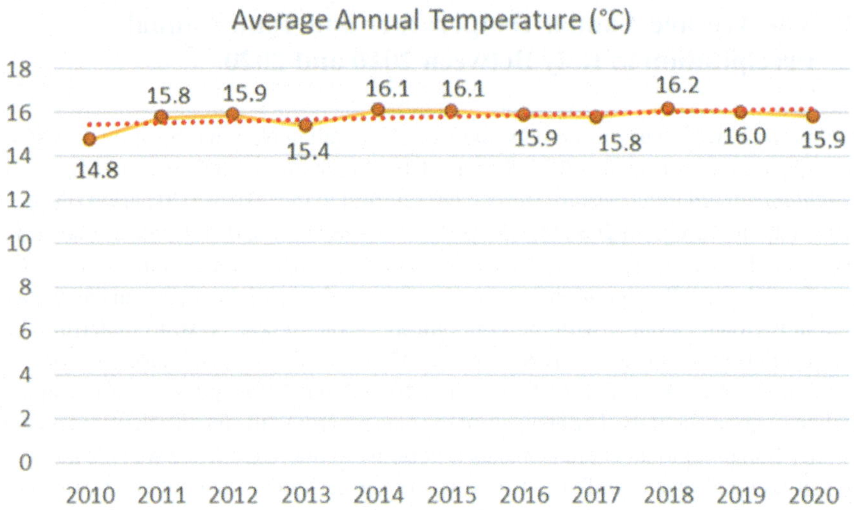

Fig. 1 Mean calculated on the twenty Italian regional capitals for the average annual temperature (°C) in temporal perspective, between 2010 and 2020. Authors' elaboration on ISTAT (2022)

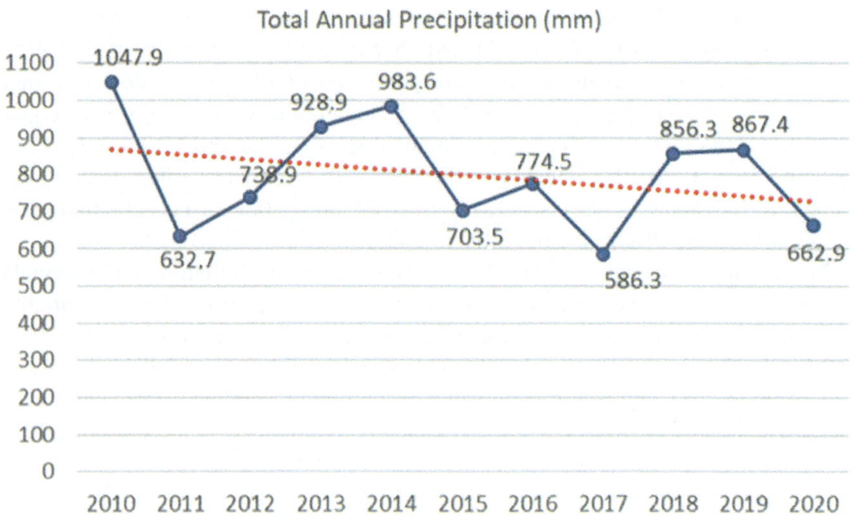

Fig. 2 Mean calculated on the twenty Italian regional capitals for the total annual precipitation (mm) in diachronic perspective, between 2010 and 2020. Authors' elaboration on ISTAT (2022)

Particularly, starting from 13.4 °C in 2010, Milan went to 15.9 °C in 2020, with an increase of 2.5 °C. It is worthy of note that in the period examined, the years which recorded the maximum values were 2014 and 2015, with 16.7 °C, followed by 2011, with 16.6 °C. Therefore, between 2010 and 2011 a very sharp increase of 3.2 °C was recorded and followed by an undulating dynamic.

Table 1 Average annual temperature (°C) in the twenty Italian regional capitals between 2010 and 2020 and its variation

Average Annual Temperature (°C)												
Regional capitals	2010	2011	2012	2013	2014	2015	2016	2017	2018	2019	2020	*Variation 2010-2020*
Milan	13.4	16.6	16.4	14.4	16.7	16.7	16.1	16.1	16.0	16.1	15.9	*2.5*
Bologna	14.2	15.5	15.6	14.9	15.9	14.8	15.3	16.4	16.3	16.5	16.4	*2.2*
Aosta	10.6	12.3	12.2	11.5	12.2	12.7	12.1	12.1	12.8	12.6	12.3	*1.7*
Venice	13.7	15.1	15.0	14.7	16.0	15.6	15.2	15.1	16.2	15.5	15.2	*1.5*
Perugia	14.2	15.5	15.5	14.9	15.4	15.8	15.3	15.6	15.8	15.9	15.6	*1.4*
Rome	16.5	17.2	17.3	17.0	17.5	18.3	18.2	18.5	18.0	17.9	17.8	*1.3*
Trento	12.4	13.7	13.7	13.2	13.8	13.9	13.4	13.5	14.0	13.9	13.7	*1.3*
Turin	13.9	15.4	15.0	14.4	15.3	15.5	15.1	15.4	15.5	15.4	15.2	*1.3*
Cagliari	17.6	18.3	18.7	18.2	19.0	18.9	19.0	18.6	18.7	18.6	18.8	*1.2*
Genoa	15.7	16.9	16.5	15.9	16.9	17.3	16.9	16.8	17.5	17.1	16.9	*1.2*
Campobasso	12.6	13.6	14.0	13.2	13.7	13.6	13.4	13.6	13.7	13.6	13.6	*1.0*
Florence	15.6	16.5	16.3	15.9	16.8	16.7	16.5	16.4	16.7	16.7	16.5	*0.9*
Catanzaro	16.7	16.9	17.3	17.2	17.3	17.4	17.2	17.1	17.5	17.5	17.5	*0.8*
Palermo	18.3	18.8	18.8	19.0	19.8	19.0	19.4	18.4	18.8	18.8	19.1	*0.8*
Potenza	12.8	13.5	13.9	13.6	14.0	14.9	13.7	13.8	13.9	13.6	13.6	*0.8*
Naples	16.5	17.2	17.4	17.4	17.7	17.5	17.7	17.1	17.3	17.3	17.2	*0.7*
Trieste	15.5	16.3	16.1	15.6	16.4	16.2	16.1	15.7	16.7	16.4	16.1	*0.6*
Bari	16.8	17.2	17.6	17.4	17.6	17.5	17.4	17.2	17.9	17.7	17.3	*0.5*
L'Aquila	12.4	12.9	13.2	12.9	13.3	13.4	13.1	12.6	12.8	12.8	12.8	*0.4*
Ancona	16.2	16.6	17.3	17.1	16.9	16.5	16.6	16.6	17.2	16.8	16.2	*0.0*

Authors' elaboration on ISTAT (2022)

Starting from 14.2 °C in 2010, Bologna went to 16.4 °C in 2020, with an increase of 2.2 °C. In this case, the hottest period was recorded between 2017 and 2020, with values always higher than 16 °C: 16.4 in 2017, 16.3 in 2018, 16.5 (the maximum value) in 2019 and again 16.4 in 2020.

The other regional capitals have been conventionally grouped into classes with temperature increases: between 1.1 and 2 °C and this class comprises eight regional capitals above all in central-northern Italy (encompassing Rome and with the exception of Cagliari, Sardinia); between 0.1 and 1 °C, with nine regional capitals above all in central-southern Italy (with the exception of Trieste, Friuli-Venezia Giulia, and also including Palermo, Sicily). Lastly, Ancona is the only regional capital without variation (even if all the years between 2011 and 2019 have shown higher values).

As far as the Total Annual Precipitation (mm), the most notable variations were recorded in Naples and Catanzaro (Table 2).

Starting from 1533.6 mm in 2010, Naples went to 536.5 mm in 2020, with a decrease of 997.1 mm. Moreover, it must be considered that in the period examined, there have been years which recorded even lower values, since, in 2018, 426.4 mm were recorded and in 2015 and 2017, respectively, 451 and 495.5 mm. These aspects can be useful in terms of possible situations of a critical drop which can have tangible repercussions for water supply, agriculture and field irrigation and daily use for primary needs. Furthermore, it is worthy of note that from 2010 to 2011 a steep decrease was seen (− 849.7 mm). Therefore, there was a very sharp decrease between 2010 and 2011, from 1533.6 to 683.9 mm, starting a dynamic which has shown

Table 2 Total annual precipitation (mm) in the twenty Italian regional capitals between 2010 and 2020 and its variation

Total Annual Precipitation (mm)												
Regional capitals	2010	2011	2012	2013	2014	2015	2016	2017	2018	2019	2020	*Variation 2010-2020*
Naples	1,533.6	683.9	809.6	887.2	791.8	451.0	595.3	495.5	426.4	1,073.6	536.5	*-997.1*
Catanzaro	1,417.2	1,169.0	1,063.4	1,296.2	1,027.0	1,247.8	1,030.6	828.8	1,575.8	840.2	698.0	*-719.2*
Florence	1,087.4	538.7	848.3	962.9	1,003.1	568.0	820.0	666.2	936.3	830.8	593.4	*-494.0*
Turin	1,317.6	1,162.4	871.4	1,035.4	1,150.2	894.2	975.4	485.8	1,080.8	1,203.4	842.8	*-474.8*
Trieste	1,292.7	659.1	750.8	1,104.1	1,209.5	722.6	916.6	1,223.6	723.8	1,167.3	845.0	*-447.7*
Potenza	1,139.2	595.4	678.4	932.4	677.6	567.8	778.8	442.6	858.8	722.2	696.4	*-442.8*
L'Aquila	932.6	456.6	654.6	929.6	760.6	547.4	628.6	431.2	826.0	525.2	492.4	*-440.2*
Bologna	979.3	502.0	609.4	911.1	916.1	896.8	735.6	612.4	786.3	829.3	543.2	*-436.1*
Milan	1,183.2	427.6	768.8	1,018.0	1,302.0	659.0	920.0	553.2	726.6	791.8	759.8	*-423.4*
Rome	947.7	600.2	691.5	945.4	1,098.6	724.4	667.9	526.7	984.2	860.1	600.8	*-346.9*
Ancona	957.0	631.4	931.6	902.2	890.8	715.6	774.4	717.0	598.6	678.2	610.8	*-346.2*
Trento	1,320.5	719.4	1,229.9	1,206.6	1,612.9	683.0	894.6	852.4	993.6	1,329.0	984.2	*-336.3*
Perugia	1,266.0	543.2	968.2	1,125.8	1,069.0	537.8	982.8	557.8	826.4	1,101.0	929.8	*-336.2*
Genoa	1,288.4	822.1	928.8	1,302.3	2,102.6	784.7	830.5	594.8	1,269.2	1,458.9	986.9	*-301.5*
Venice	917.2	476.0	487.0	896.0	1,069.1	525.8	704.1	612.8	621.8	785.2	627.9	*-289.3*
Campobasso	908.4	744.6	592.5	834.2	777.0	1,114.1	874.6	549.2	841.5	945.2	675.1	*-233.3*
Cagliari	604.8	344.8	313.2	468.4	344.8	466.2	367.0	221.2	775.6	471.6	376.0	*-228.8*
Palermo	700.6	539.2	619.6	673.8	676.4	951.0	451.6	452.4	829.0	569.2	530.4	*-170.2*
Bari	631.8	452.8	487.8	493.6	644.2	551.0	937.2	450.0	694.2	593.8	508.0	*-123.8*
Aosta	532.7	585.2	472.8	653.3	548.3	461.3	604.2	452.1	750.5	572.9	421.1	*-111.6*

Authors' elaboration on ISTAT (2022)

dramatic shifts, like also the recent drop between 2019 (the second year with the highest value) and 2020, from 1073.6 to 536.5 mm.

Starting from 1417.2 mm in 2010, Catanzaro went to 698 mm in 2020, with a decrease of 719.2 mm. In this case, until 2016 the dynamic had shown less exacerbated reductions and a situation with some upswings; 2017 and 2019 were instead years with low values (828.8 and 840.2 mm, respectively) interrupted by the year with the maximum value: 2018 recorded 1575.8 mm. Finally, 2020 was the year with the minimum value, equal to 698 mm.

The other regional capitals have been conventionally grouped into classes with precipitation decreases: (i) between – 1 and – 300 mm, comprising six regional capitals; and (ii) between – 301 and – 600 mm, comprising twelve regional capitals. No regional capitals recorded an increase in the period examined.

3 An Overall Contextualization

The combined analyses of the data concerning the Average Annual Temperature (°C) and Total Annual Precipitation (mm) as mean calculated on the twenty Italian regional capitals and the same data for single regional capitals show a composite framework with an overall increase of the temperature and a notable decrease of the precipitation, with some cases having particular prominence. This situation may cause repercussions on different spheres of daily life, relevant both for economic-productive and socio-health questions, and for natural, vegetation and environmental elements.

In terms of physical and environmental aspects, the general recent configuration with increased temperature and decreased precipitation must be considered bearing in mind a series of consequences also in terms of meteorological and hydrological drought which can have a more or less marked environmental impact on the basis of the persisting of the altered state (Mariani et al., 2018, p. 1). One effect is the decrease in river flow which sometimes also affects the main Italian rivers. The situation of the Po River—the longest in Italy, which rises from the North-Eastern flank of the Monviso (in the Western Alps of Piedmont) at Pian del Re—has been one the subjects of public attention during some periods of the last years. A number of animations based on the European State Agency satellite imagery (ESA) have shown part of its course, highlighting how it has shrunk considerably, i.e., from June 2020 and June 2022. In fact, the water level in the Po Valley has reached record-low levels, partly due to the decrease of rainfall, as well as the increase of temperatures with the consequent scarcity of snow in the mountains from which the river is fed (https://www.esa.int/ESA_Multimedia/Images/2022/06/Po_River_dries_up). Various consequences of the drought events regarding the Po River have been discussed in recent years (Musolino et al., 2017).

But probably, the most widely underlined effect of the general configuration with increased temperature and decreased precipitation is the glacier retreat and collapse in the Alpine area (Diolaiuti et al., 2012; Fugazza et al., 2018), quantified and analyzed also through specific data of snow cover in temporal series, elaboration of detailed maps and integration of information and images coming from various survey techniques, unmanned aerial vehicles (UAV) and terrestrial photogrammetry. This phenomenon is often used as a thermometer able to evidence and measure the variations underway, estimating ice volume, glacial tongue length and mass changes (Di Rita et al., 2020).

Moreover, regarding urban contexts and possible contributing and inclining factors, the general situation must be placed in a framework in which (ISTAT, 2022, pp. 5–7):

(i) the high level of urbanization and urban sprawl exert heavy pressure on the natural environment, increasing the consumption of the soil, natural resources, energy and emissions of polluting gases that in turn increase the urban heat island phenomenon and the vulnerability of cities, which should adopt specific virtuous mitigation actions, toward a sustainability perspective;
(ii) in Milan the frequency of exceedances regarding the concentration of different pollutants (above all PM_{10} and $PM_{2.5}$) has often been considerably higher than the average of the regional capitals, even if in the last years the dynamic tends to follow the one of the other contexts;
(iii) Milan is penalized by the poor presence of green areas (13.8% of the municipality surface), which instead increase the resilience of an urban context, while

Rome is characterized by a much higher overall coverage (35.8% of the municipality surface) and under this perspective the situation is better also in Naples (31.5% of the municipality surface);

(iv) an important problem is represented by the often high motorization rate and obsolete vehicle fleet. Particularly, Rome records a very high number of cars (621) per 1000 inhabitants (Naples: 605; Milan: 497) but an extremely problematic situation appears in Naples, both due to the obsolescence of the vehicle fleet and because, unlike Milan and Rome, the motorization rate shows no sign of stabilizing.

At the same time the analysis of the data must be carried out considering the possible effects on a wide spectrum of activities and aspects such as the following (Mercogliano et al., 2022, pp. 83–84; 104–105; 110–111).

(i) Agriculture sector, since the variations recorded in the thermo-pluviometric regime can cause great economic damage in terms of reduction of the value of aggregate production and related difficulties for water supply; similar variations can also determine a shifting of the production areas toward new suitable latitudes and altitudes.
(ii) Tourist sector where the economic damage can be associated with a possible contraction of the supply both in the summer season, due to the expected thermal discomfort, and winter season, due to the reduction of snow cover, with relevant losses in terms of arrivals and presences of international and national tourists.
(iii) Health system, since important repercussions can be recorded—due to heatwaves, summers of extreme heat, tropical nights (days with a minimum temperature of > 20 °C), etc.—on mental health and vector-related diseases, on people and above all the elderly who suffer from cardiovascular or neurological diseases, children and other vulnerable people who for example live in conditions of socioeconomic hardship.

In the perspective of the repercussions on people's health, in medical geography it is known that meteorological elements can cause relevant effects which can concur to even serious morbid conditions, e.g., in the case of turbulence and the sudden passage of warm or cold air fronts, for myocardial infarction and cardiac collapse and also asthma (Palagiano & Pesaresi, 2011, pp. 132, 167, 171). Moreover, from the match between temperature and humidity derive thresholds beyond which conditions of poor psycho-physical well-being give way in the human organism (Agostini et al., 2005, p. 142). Various analyses and considerations have been made concerning the possible effects of heatwaves by examining the data of the first years of 2000, with a special focus on the very hot August of 2003 in Italy (and France), which seems to be associated with a considerable over-mortality also in the successive months as a repercussion of the stressful summer events (Pinna & Macchia, 2009).

The data available and analyzed herein would require further integrative data for example concerning:

(i) data more widespread on the territory, involving, i.e., provincial capitals but also non-urban areas and inner areas in order to have a broader picture of the overall situation;
(ii) seasonal data because increases in temperature and decreases in precipitation would require specific in-depth examinations, since it is useful to know if the annual increase in temperature is strictly related to specific months or seasons or is overall, and it is useful to know if the annual decrease in precipitation is due to a general loss and whether there are brief periods with sudden downpours;
(iii) data about a longer retrospective period based on the same survey methods in order to provide a study able to quantify variations and recognize actual and potential changes worthy of note according to a continuous series.

However, this analysis aims to provide a first overview based on the data collected between 2010 and 2020, able to support some possible considerations about relational connections with the incidence/prevalence of infectious diseases due to specific pathogens, with reference to tick-borne encephalitis virus (TBEV) and West Nile virus (WNV).

4 Vectors-Borne Infectious Diseases

Transmission of infections occurs when there is an overlap of activities between reservoir, vector, and humans. The trasmission process differs according to the pathogen and the location. When studying the distribution of infectious diseases, it is important to consider variables such as the environment, the climate, the weather, the animal reservoir, the vector, and the intermediate host when implied. Having a clear picture of the variables and their trend over time could allow to predict epidemiological changes in terms of human infections and therefore introducing/implementing preventive measures or simply increasing awareness in regions previously unaffected by the problem. Arboviruses represent a public health concern in Italy, as demonstrated by the epidemic outbreaks reported in the last few decades, with the potential of introducing new species previously confined to tropical and sub-tropical regions. Based on the evolving epidemiological situation and due to the increasing number of cases, the Italian Ministry of Health has activated a national surveillance system for arboviral diseases. This integrated surveillance system, coordinated by the Italian National Institute of Health (Istituto Superiore di Sanità, ISS) and the Zooprophylactic Institute of Abruzzo and Molise (IZS-AM), annually publishes bulletins to trace and tackle arboviral infections and to ensure early detection of autochthonous and imported cases and contain any possible spread (Istituto Superiore di Sanità, 2022). During the last two decades, Italy experienced a changing epidemiology for some arboviral infections, in particular TBEV and WNV. Here follows the description of the TBE and WNV biology and ecology, with particular attention to their distribution and diffusion in Italy.

4.1 *Tick-Borne Encephalitis (TBE)*

TBE is an emblematic human infection in the field of complex interrelationships between different factors. TBE is an emerging zoonotic neurological disease caused by the TBEV, first isolated in Eastern Russia in 1937 (Lindquist & Vapalahti, 2008), which belongs to the genus flavivirus family *Flaviviridae* (Riccardi et al., 2019). Among the known TBEV subtypes (European, Siberian, Far-Eastern, Baikalian and Himalayan), the European one is the less virulent subtype, is the most diffused in Italy, and is transmitted by *Ixodes ricinus* ticks (Alfano et al., 2020). All the other TBEV subtypes are transmitted by *Ixodes persulcatus* (Lindquist & Vapalahti, 2008). *I. ricinus* is present in most of Europe, while *I. persulcatus* is mostly diffused in Asia; both subtypes co-circulate in the Western part of the former Soviet Union. TBEV is transmitted from the saliva of an infected tick within minutes from the tick bite but can occasionally be transmitted with ingestion of dairy products from infected animals (Kohl et al., 1996). In the natural environment, TBEV circulates between ticks and several wild vertebrates, such as rodents, hedgehogs, and moles that act as reservoirs and amplifier hosts (Riccardi et al., 2019) (Fig. 3). Horizontal transmission between ticks and vertebrates is necessary for sustained endemicity of TBE (Reiss, 2016). TBE in humans follows an incubation period of eight days (range 4–28) after the tick bite, and the clinical course is typically biphasic (Kaiser, 1999). Among symptomatic patients, which are approximately the 33% of those infected with TBEV, the most common clinical presentation is aseptic meningitis (50%), followed by meningoencephalitis (40%) and meningoencephalomyelitis (10%). Sequelae are common (Riccardi et al., 2019).

In Europe, tick activity begins in spring when the temperature is consistently above 6 °C until November when the temperature drops. From an epidemiologic point of view, milder winters and warmer springs may affect and prolong the time of human's exposure to ticks, as well as ticks activity in TBEV endemic areas. A recent paper by Nahmood et al. showed that the incidence of TBE has increased in Norway, Sweden and Finland from 1979 to 2020, although the bias of increasing awareness among physicians cannot be excluded (Mahmood et al., 2022).

4.2 *TBE in Italy*

The first cases of TBE in Italy were identified in Tuscany in the 1970s (Amaducci et al., 1976). To date, most cases of TBE occur in North-Eastern Italy. A survey conducted in North-Eastern Italy from 2000 to 2013 reported 367 total TBE cases. Researchers found a strong seasonality effect, with the highest number of cases occurring in July and in locations between 400 and 600 m of altitude (Rezza et al., 2015). Moreover, researchers found a significant temporal increase during the study period, with annual incidence rates increasing from 0.18 per 100,000 in the year 2000 to 0.59 per 100,000 inhabitants in 2013 ($p < 0.01$) (Rezza et al., 2015). In the

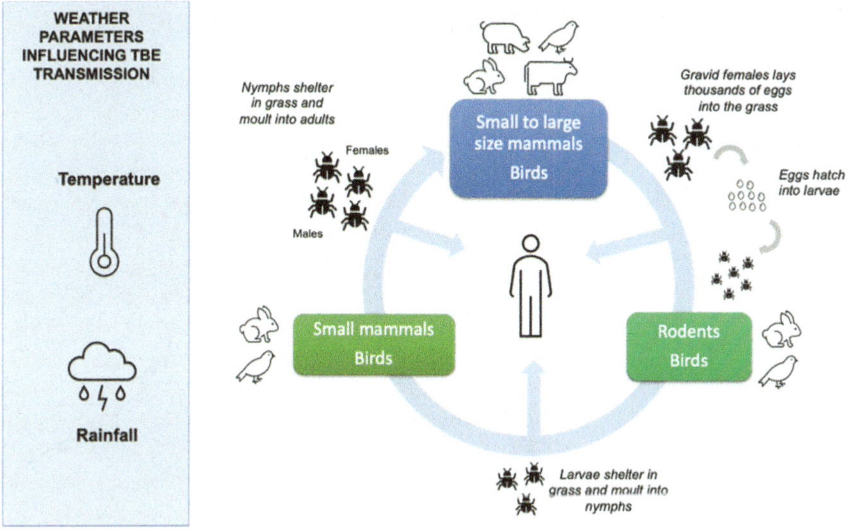

Fig. 3 Tick-borne encephalitis cycle and weather parameters influencing the virus transmission. Authors' elaboration

following decade, another increasing trend has been noticed in Italy with incidence reaching 1.42 cases per 100,000 inhabitants in 2020 (Global Health Press, 2021). The province of Belluno (Veneto), with its notification rate of TBE of 5.95 per 100,000 inhabitants, has been classified by the World Health Organization as a highly endemic area (Panatto et al., 2022) (Fig. 4). Overall, three endemic foci where risk of TBEV infection is likely high have been identified in Veneto, Friuli-Venezia Giulia and Trentino Alto-Adige (Rezza et al., 2015; Rossi et al., 2023). A recent paper by Rossi et al. reported a yearly range of TBE cases in Italy from 5 to 55 (mean 33 cases per year) in the period 2010–2022 (Rossi et al., 2023). When analyzing the causes of TBE changing epidemiology, the variable "vaccination" should also be considered. In Italy, official data on TBE vaccination rates are lacking; however, some reports suggest that vaccination coverage is low also among high-risk categories (Panatto et al., 2022).

4.3 *Effects of Weather Changes on Ticks Ecology*

I. ricinus and *I. persulcatus* are particularly susceptible to environmental conditions since in their nonparasitic phases they require a microclimatic relative humidity of at least 80% to avoid fatal desiccation. They are restricted to areas of moderate-to-high rainfall with a good cover of vegetation, so that the soil surface remains humid during the driest periods of the year. In areas where reduced summer precipitation coincides with raised summer temperatures, the survival, the activity, and the

Fig. 4 TBEV circulation (exemplary) in geographic areas close to Italy, from 1999 to 2016 (modified from Mahmood et al., 2022). Authors' elaboration

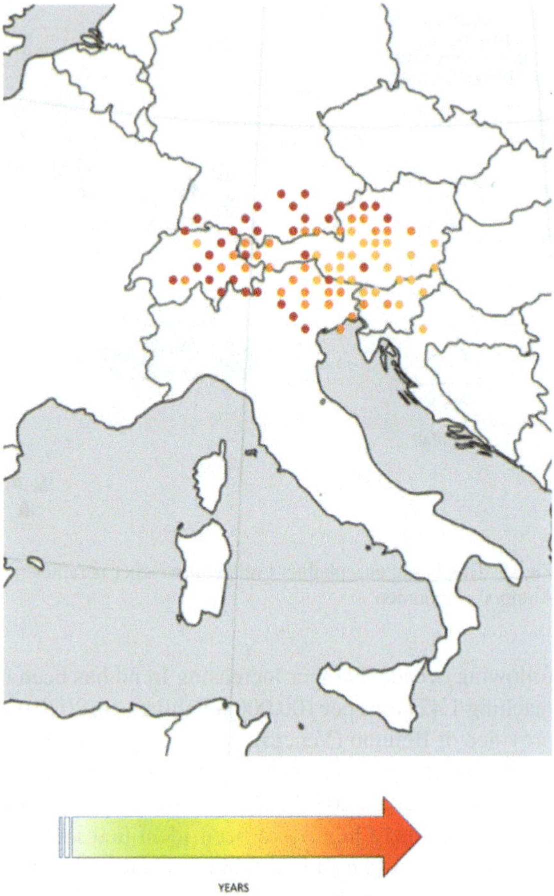

distribution of *I. ricinus* and *I. persulcatus* are likely to be reduced because of their vulnerability to desiccation. Ticks are not active during warm sunny days, nor during rainfall. However, in spring and autumn the extended tick activity periods commonly coincide with extended periods of outdoor human activity. Overall, the milder climate in Europe has contributed not only to the increase in TBE incidence, but also to an increased incidence of other diseases transmitted by *I. ricinus*, such as Lyme borreliosis and human ehrlichiosis (Ebi et al., 2013).

These tick species acquire their hosts by ambushing them from the vegetation and a significant number of large animals must be present in the habitat to feed the adult females, thus maintaining the tick populations. During the immature tick stages (larvae and nymph), reptiles, small and medium-sized mammals, and birds represent the favorite hosts which contribute significantly to the circulation of diverse pathogens between the tick and host populations. Climate changes may, therefore, exert a major influence on both tick abundance and disease prevalence by affecting faunal diversity. Although changes in the weather and in the length of the different seasons directly affect tick survival, activity, and development, there is no clear evidence that rising temperatures result in a greater abundance of ticks by simply increasing rates of development. Indirect effects of weather changes may impact the number of infected ticks by affecting vegetation, survival, and abundance of tick maintenance hosts (i.e., medium and large mammals) and pathogen-reservoir hosts (i.e., small mammals and birds).

The magnitude of weather change effects in endemic areas depends on local conditions and vulnerability and is determined not only by ecological conditions but may be influenced also by socioeconomic factors. Indeed, weather changes may influence disease risk by affecting human migration and settlement, human cultural, social, and behavioral patterns, immunity in the population, ecosystems, biodiversity, and migrating patterns of birds.

Overall, a complex chain of processes exists, making the precise factors responsible for changes in disease incidence often difficult to determine (Ebi et al., 2013; Gray et al., 2009). Moreover, since *I. ricinus*, the predominant tick species in Europe, is extremely flexible and adaptable and can exhibit different seasonal activity even in adjacent parts of its geographical range, the prediction of future scenarios is really hard (Gray et al., 2009).

4.4 West Nile Virus

Among human infectious diseases that are experiencing a significantly changing epidemiology in Italy, WNV is one of the most important. WNV is a flavivirus (like TBEV) and can cause neurological disorders in humans. WNV human transmission occurs through mosquitoes, primarily of the genus *Culex*, family *Culicidae*. Birds are the main reservoir and become highly viremic when infected (amplifying host), allowing WNV transmission to *Culex* mosquitoes. Infected mosquitoes can spread the virus to non-avian vertebrate hosts such as horses and humans (dead-end host), who develop a low-level viremia, usually insufficient to transmit the virus to other mosquitoes (Delbue et al., 2014) (Fig. 5).

The incubation period for WNV infection is 2–14 days; most infected individuals are asymptomatic, while 20–40% develop symptoms. Symptomatic WNV human infections consist of fever, headache, malaise, myalgia, fatigue, skin rash, lymphadenopathy, vomiting, and diarrhea (Kramer et al., 2007). Less than 1% of infected individuals develop severe neuro-invasive diseases that can be classified into three main clinical syndromes: West Nile meningitis, West Nile encephalitis, and acute flaccid paralysis (Kramer et al., 2007). Candidate vaccines for West Nile exist, but none are currently in late stage of clinical trials (Ulbert, 2019).

Migratory birds on their flyway from Africa introduced WNV into Europe where the virus overwinters in mosquitoes (Rudolf et al., 2017). Hence, WNV hitted many European countries with the peak notification rate of 2083 cases in 2018 (0.3 cases per 100,000 inhabitants) (Historical data by year—West Nile virus seasonal surveillance,

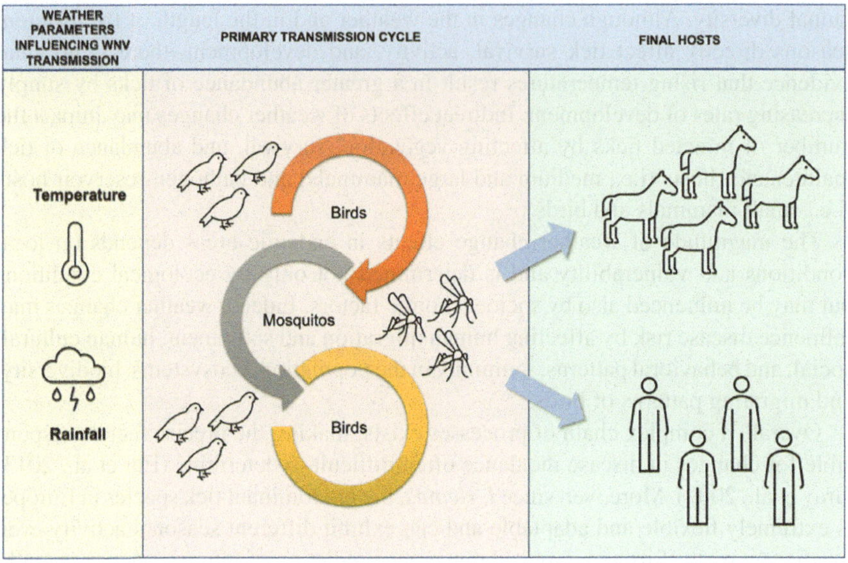

Fig. 5 West Nile virus cycle and weather parameters influencing its diffusion. Authors' elaboration

2019). High precipitation in late winter/early spring was found as one of the key predictors of WNV outbreaks in Europe (Marcantonio et al., 2015). Heavy rainstorms and flash floods are projected to be more frequent in Europe in the next future (Semenza & Paz, 2021).

4.5 WNV in Italy

Human cases of WNV-associated fever and neurological disorders have been reported in Italy since 2008 (Delbue et al., 2014). Between 2012 and 2020, 1145 WNV infections have been reported in Italy with the annual peak ranging from August to September (Riccò et al., 2021). The year 2018 registered the incidence peak of WNV infections and West Nile neuro-invasive disease also in Italy, with 1.009 and 0.377 reported cases per 100,000 inhabitants, respectively (Riccò et al., 2021). In 2021 it was reported the co-circulation in Italy of two distinct WNV lineages (1 [newly introduced] and 2) (Barzon et al., 2022). Rossi et al. reported in 2022 more WNV infection cases than in 2018, especially in summer, but data were reported as absolute numbers, without incidence mention (Rossi et al., 2023). This phenomenon is sustained by: (i) biotic factors, such as migratory birds that brought the virus to new regions and high density of *Culex* mosquitoes, and (ii) abiotic factors, such as higher temperatures in spring and summer (Rossi et al., 2023). Of note, the data here reported regarding the Average Annual Temperature increase in several regional capital requires attention with respect to possible arbovirosis diffusion. Riccò et al.

conducted an analysis taking into account, among others, air temperature, relative humidity, and precipitations, observing that West Nile neuro-invasive disease in Italy found a significant effector in the environmental temperatures (32% increase for a + 1 °C increase in the air temperature compared with the average estimates) (Riccò et al., 2021). Moirano et al. demonstrated an association between weekly total precipitations, recorded 1–4 weeks before WNV infection, and a subsequent higher incidence of WNV infections (Moirano et al., 2018).

4.6 *Effects of Weather Changes on Mosquitoes Ecology*

Weather changes increase the risk of human exposure to WNV. Studies show that warmer temperatures associated with climate change can accelerate mosquito development, biting rates, and incubation of the disease within a mosquito. The effect of climate change on the timing of bird migration and breeding patterns may also contribute to changes in long-range virus movement. Mild winters and drought have been associated with West Nile virus disease outbreaks (Hahn et al., 2015; Historical data by year—West Nile virus seasonal surveillance, 2019).

Weather conditions have direct and indirect influences on the vector ability to acquire, maintain and transmit the virus, as well as on the vector population dynamics and on virus replication rate within the mosquito, which are mostly weather-dependent. The importance of elements like temperature, precipitation, relative humidity, and winds as drivers in WNV epidemiology is increasing under conditions of climate change. Indeed, recent changes in climatic conditions, particularly increased ambient temperature and fluctuations in rainfall amounts, contributed to the endemization process of WNV in various parts of Southern Europe, Western Asia, Eastern Mediterranean, Canadian Prairies, parts of the USA, and Australia (Paz, 2015).

Temperature plays an important role in WNV replication rates and transmission by affecting the length of extrinsic incubation, the seasonal phenology of mosquito host populations, and the geographical variations in human case incidence (Paz, 2015). Higher temperatures determine a rapid increase in the growth rates of mosquitoes, a decrease in the interval between blood meals, an acceleration of the virus evolution rate and an increased viral transmission efficiency to birds (Kilpatrick et al., 2008; Paz & Albersheim, 2008; Paz et al., 2013; Ruiz et al., 2010). In the last years, several bird species have been migrating to their breeding grounds earlier as a result of an early increase of the mean temperatures during spring (Van Doren, 2022). In turn, this phenomenon could influence the diffusion of the WNV in locations near or along migration routes.

It is well-known that precipitations influence the abundance of mosquitoes and, consequently, increase the potential WNV disease diffusion in humans (Landesman et al., 2007). However, the influence of rainfall on WNV diffusion might differ in different geographical regions as it also depends on mosquitoes ecology. Indeed, frequent precipitations cause an increase of the standing water surface that is necessary for mosquito larval development (Landesman et al., 2007; Paz, 2015). Besides, heavy rainfalls dilute the nutrients, thus decreasing the development rate of larvae (Chevalier et al., 2013). On the other hand, below-average precipitations can facilitate population outbreaks of some species of mosquitoes because the drying of wetlands disrupts the aquatic food-web interactions that limit larval mosquito populations (Paz, 2015). During the periods of drought, avian hosts and mosquitoes are in close contact around the available water sources; in turn, this accelerates the virus (Asgarian et al., 2021; Shaman et al., 2005). Very little is known on the relationship between relative humidity and WNV diffusion. Overall, it seems that air temperature is a better predictor for increasing disease cases than air humidity (Paz, 2015; Walsh et al., 2008; Walther et al., 2002).

Wind might contribute to WNV spread by their impact on wind-blown mosquitoes (Ritchie & Rochester, 2001) and by affecting bird migration through changes in the patterns of storm tracks. In recent years, significant changes in the location and intensity of storms have been shown on a regional basis, as a part of climate change observations and scenarios (Bengtsson et al., 2006; Paz, 2015; Ritchie & Rochester, 2001; Van Den Broeke & Gunkel, 2021).

5 Crimean-Congo Hemorrhagic Fever: A Remote or Possible Eventuality in Italy?

Apart from arbovirosis that has become endemic in Italy (e.g., TBE and West Nile), there are others arbovirosis, that have the potential to become epidemic/endemic in Italy. One of these, that we selected also for its clinical severity, is Crimean-Congo hemorrhagic fever (CCHF). CCHF is a viral infection usually transmitted by ixodic ticks, usually *Hyalomma* spp., especially *Hyalomma marginatum* (Gargili et al., 2017; *Hyalomma marginatum*—current known distribution: January 2019). Apart from the transmission from ticks, a transmission by contact with acute CCHF patients or by contact with blood or tissues from viremic livestock is possible (Whitehouse, 2004). The viral reservoir usually consists of wild vertebrates, but birds are thought to have a role in the transportation of CCHF virus-infected ticks between different countries (Ergönül, 2006). CCHF is distributed in Asia, Africa, Eastern Europe, and the middle east (Shahhosseini et al., 2021). According to the geographic range, CCHF represents the most extensive medically important tick-borne virus (Ergönül, 2006). CCHF is endemic in the Balkans (Portillo et al., 2021) and birds migrating from the Balkans were suggested to be the cause of the 2002 outbreak in Turkey (Karti et al., 2004).

Approximately one of every five infected people develop CCHF (Goldfarb et al., 1980). The mortality reported in the Turkish outbreak 2002–2005 was 5% (Ergönül, 2006). The incubation following tick bite is usually 3–7 days and the clinical has four distinct phases: incubation, prehemorrhagic, hemorrhagic and convalescence (Ergönül, 2006).

In 1985 on a sample of 258 Portuguese humans screened, two had positive serology for CCHF (Filipe et al., 1985). In 2013, the first CCHF case in the UK was reported but was an Afghan migrant (Barr et al., 2013). In 2017 two autochthonous CCHF cases were reported in Spain: the first was a patient who reported a tick bite and the second was the nurse caring for the patient (Negredo et al., 2017). Of note, CCHF virus was already identified in *Hyalomma lusitanicum* ticks in Spain in 2010 (Estrada-Peña et al., 2012). Overall, from 2013 to 2021 10 human cases of CCHF have been diagnosed in Spain (Lorenzo Juanes et al., 2023) and retrospectively (stored serum) researchers were able to trace back that to the fact that the first case occurred in 2013 (Negredo et al., 2021).

Until today, no human CCHF case has been reported in Italy; however Italy is considered a country at medium risk of entry, at high risk of exposure, and at medium risk of occurrence (Fanelli & Buonavoglia, 2021). In addition, Italy is the only European CCHF-free country with a widely distributed *H. marginatum* resident population (*Hyalomma marginatum*—current known distribution: January 2019). In 2022, Fanelli et al. demonstrated that 21 out of 794 tested Italian bovines had a positive CCHF serology (Fanelli et al., 2022).

6 Conclusions

Vector-borne diseases are considered strongly related to climate/weather changes; however modeling is complex since many variables (some poorly known in numerical terms) are implied. To date, we are still not able to predict the possible impacts of weather change on the ecology of TBEV and WNV. The combined analyses of data here reported, concerning the Average Annual Temperature and the Total Annual Precipitation between 2010 and 2020 for the twenty Italian regional capitals and for some single regional capitals, show an overall increase of the temperature and a decrease of the precipitation, with the cities that show relevant changes of both these meteorological elements requiring particular attention. These data suggest the necessity of continuous surveillance especially in the Italian regions already defined as endemic foci. Data from long-term studies on disease incidence, tick and mosquitoes biology and ecology, hosts abundance and distribution, and ecosystems and biodiversity, specifically in relation to climate change, are required. Such data will allow the development of models to predict future TBE and West Nile disease scenarios, which should consider dynamic biological processes instead of simple processes such as the likelihood of occurrence of climate suitability for specific vector species. For sure, apart from climatic factors, other drivers contribute to the geographical spread of TBEV and WNV in Italy such as host migration patterns, vectors spread by international transportation, landscape features and land use. Nevertheless, as weather elements have significant direct and indirect influences on the TBEV and

WNV endemization, the impacts of the changing climate must be considered in any evaluation of TBEV and WNV transmission in the coming years.

In 2022, we observed the peak of notified cases of West Nile infections in Italy, with even more cases compared with the 2018 former peak. Looking at the human side, it looks like a plateau is still to be reached, even because temperature increase is known to decrease the interval between mosquitoes blood meals. Perhaps, it is early to say if this increase is still the adjustment of an endemization process or is rather secondary to climate/weather changes, or both. Regarding TBE, it is likely that the supposedly "weather hostile" trend for ticks (e.g., increased temperatures with decreased precipitation) is counterbalanced by the fact that milder temperatures for longer periods allow a high number of outdoor activities, therefore enhancing human exposure to ticks. Moreover, both diseases' trends may have been affected by COVID-19 restrictions in the last years. Therefore, it would be interesting to look at the trend with a "zoom-out" in the next few years to put together all the puzzle pieces.

References

Agostini, G., Pinna, M., Pinna, S., & Russo, F. (2005). *Bioclimatologia umana*. UTET.

Alfano, N., Tagliapietra, V., Rosso, F., Ziegler, U., Arnoldi, D., & Rizzoli, A. (2020). Tick-borne encephalitis foci in northeast Italy revealed by combined virus detection in ticks, serosurvey on goats and human cases. *Emerging Microbes & Infections, 9*(1), 474–484. https://doi.org/10.1080/22221751.2020.1730246

Amaducci, L., Arnetoli, G., Inzitari, D., Balducci, M., Verani, P., & Lopes, M. C. (1976). Tick-borne encephalitis (TBE) in Italy: report of the first clinical case. *Rivista Di Patologia Nervosa E Mentale, 97*(2), 77–80. https://www.ncbi.nlm.nih.gov/pubmed/1028140

Asgarian, T. S., Moosa-Kazemi, S. H., & Sedaghat, M. M. (2021). Impact of meteorological parameters on mosquito population abundance and distribution in a former malaria endemic area, central Iran. *Heliyon, 7*(12), e08477. https://doi.org/10.1016/j.heliyon.2021.e08477

Barr, D. A., Aitken, C., Bell, D. J., Brown, C. S., Cropley, I., Dawood, N., Hopkins, S., Jacobs, M., Jeffs, B., MacConnachie, A., Mulvaney, D. W., Nicol, E., & Fox, R. (2013). First confirmed case of Crimean-Congo haemorrhagic fever in the UK. *The Lancet, 382*(9902), 1458. https://doi.org/10.1016/S0140-6736(13)61718-3

Barzon, L., Montarsi, F., Quaranta, E., Monne, I., Pacenti, M., Michelutti, A., Toniolo, F., Danesi, P., Marchetti, G., Gobbo, F., Sinigaglia, A., Riccetti, S., Dal Molin, E., Favero, L., Russo, F., & Capelli, G. (2022). Early start of seasonal transmission and co-circulation of West Nile virus lineage 2 and a newly introduced lineage 1 strain, northern Italy, June 2022. *Euro Surveillance: Bulletin Europeen Sur Les Maladies Transmissibles = European Communicable Disease Bulletin, 27*(29). https://doi.org/10.2807/1560-7917.ES.2022.27.29.2200548

Bengtsson, L., Hodges, K. I., & Roeckner, E. (2006). Storm tracks and Climate Change. *Journal of Climate, 19*(15), 3518–3543. https://doi.org/10.1175/jcli3815.1

Chevalier, V., Tran, A., & Durand, B. (2013). Predictive modeling of West Nile virus transmission risk in the Mediterranean Basin: How far from landing? *International Journal of Environmental Research and Public Health, 11*(1), 67–90. https://doi.org/10.3390/ijerph110100067

Delbue, S., Ferrante, P., Mariotto, S., Zanusso, G., Pavone, A., Chinaglia, M., L'Erario, R., Monaco, S., & Ferrari, S. (2014). Review of West Nile virus epidemiology in Italy and report of a case of West Nile virus encephalitis. *Journal of Neurovirology, 20*(5), 437–441. https://doi.org/10.1007/s13365-014-0276-0

Di Rita, M., Fugazza, D., Belloni, V., Diolaiuti, G., Scaioni, M., & Crespi, M. (2020). Glacier volume change monitoring from UAV Observations: Issues and potentials of state-of-the-art techniques. *The International Archives of the Photogrammetry, Remote Sensing and Spatial Information Sciences, XLIII– B2–B2020*, 1041–1048. https://doi.org/10.5194/isprs-archives-xliii-b2-2020-1041-2020

Diolaiuti, G., Bocchiola, D., D'agata, C., & Smiraglia, C. (2012). Evidence of climate change impact upon glaciers' recession within the Italian Alps. *Theoretical and Applied Climatology, 109*(3–4), 429–445. https://doi.org/10.1007/s00704-012-0589-y

Ebi, K. L., Lindgren, E., Suk, J. E., & Semenza, J. C. (2013). Adaptation to the infectious disease impacts of climate change. *Climatic Change, 118*(2), 355–365. https://doi.org/10.1007/s10584-012-0648-5

Ergönül, O. (2006). Crimean-Congo haemorrhagic fever. *The Lancet Infectious Diseases, 6*(4), 203–214. https://doi.org/10.1016/S1473-3099(06)70435-2

Estrada-Peña, A., Palomar, A. M., Santibáñez, P., Sánchez, N., Habela, M. A., Portillo, A., Romero, L., & Oteo, J. A. (2012). Crimean-Congo hemorrhagic fever virus in ticks, Southwestern Europe, 2010. *Emerging Infectious Diseases, 18*(1), 179–180. https://doi.org/10.3201/eid1801.111040

Fanelli, A., & Buonavoglia, D. (2021). Risk of Crimean Congo haemorrhagic fever virus (CCHFV) introduction and spread in CCHF-free countries in southern and Western Europe: A semi-quantitative risk assessment. *One Health (Amsterdam, Netherlands), 13*, 100290. https://doi.org/10.1016/j.onehlt.2021.100290

Fanelli, A., Buonavoglia, D., Lanave, G., Monaco, F., Quaranta, V., Catanzariti, R., Ruiz-Fons, F., & Buonavoglia, C. (2022). First serological evidence of Crimean–Congo haemorrhagic fever virus in transhumant bovines in Italy. *Transboundary and Emerging Diseases, 69*(6), 4022–4027. https://doi.org/10.1111/tbed.14710

Filipe, A. R., Calisher, C. H., & Lazuick, J. (1985). Antibodies to Congo-Crimean haemorrhagic fever, Dhori, Thogoto and Bhanja viruses in southern Portugal. *Acta Virologica, 29*(4), 324–328. https://www.ncbi.nlm.nih.gov/pubmed/2864836

Fugazza, D., Scaioni, M., Corti, M., D'Agata, C., Azzoni, R. S., Cernuschi, M., Smiraglia, C., & Diolaiuti, G. A. (2018). Combination of UAV and terrestrial photogrammetry to assess rapid glacier evolution and map glacier hazards. *Natural Hazards and Earth System Sciences, 18*(4), 1055–1071. https://doi.org/10.5194/nhess-18-1055-2018

Gargili, A., Estrada-Peña, A., Spengler, J. R., Lukashev, A., Nuttall, P. A., & Bente, D. A. (2017). The role of ticks in the maintenance and transmission of Crimean-Congo hemorrhagic fever virus: A review of published field and laboratory studies. *Antiviral Research, 144*, 93–119. https://doi.org/10.1016/j.antiviral.2017.05.010

Global Health Press. (2021, May 25). *The TBE book* (4th ed.) available now! Issuewire. https://www.issuewire.com/the-tbe-book-4th-edition-available-now-1700727362541107

Goldfarb, L. G., Chumakov, M. P., Myskin, A. A., Kondratenko, V. F., & Reznikova, O. Y. (1980). An epidemiological model of Crimean hemorrhagic fever. *The American Journal of Tropical Medicine and Hygiene, 29*(2), 260–264. https://doi.org/10.4269/ajtmh.1980.29.260

Gray, J. S., Dautel, H., Estrada-Peña, A., Kahl, O., & Lindgren, E. (2009). Effects of climate change on ticks and tick-borne diseases in europe. *Interdisciplinary Perspectives on Infectious Diseases, 2009*, 593232. https://doi.org/10.1155/2009/593232

Hahn, M. B., Monaghan, A. J., Hayden, M. H., Eisen, R. J., Delorey, M. J., Lindsey, N. P., Nasci, R. S., & Fischer, M. (2015). Meteorological conditions associated with increased incidence of West Nile virus disease in the United States, 2004–2012. *The American Journal of Tropical Medicine and Hygiene, 92*(5), 1013–1022. https://doi.org/10.4269/ajtmh.14-0737

Historical data by year—West Nile virus seasonal surveillance (2019). Accessed on February 27, 2023, from https://www.ecdc.europa.eu/en/west-nile-fever/surveillance-and-disease-data/historical

Hyalomma marginatum—current known distribution: January 2019. (2019). Accessed on March 2, 2023 from https://www.ecdc.europa.eu/en/publications-data/hyalomma-marginatum-current-known-distribution-january-2019

IPCC. (2013). Climate change 2013: The physical science basis. In: T. F. Stocker, D. Qin, G.-K. Plattner, M. Tignor, S. K. Allen, J. Boschung, A. Nauels, Y. Xia, V. Bex, & P. M. Midgley (Eds.), Contribution of working group I to the fifth assessment report of the intergovernmental panel on climate change (1535 p.). Cambridge, United Kingdom and New York, NY, USA: Cambridge University Press.

ISTAT. (2022). *I cambiamenti climatici: Misure statistiche* (Anno 2020. Report).

Istituto Superiore di Sanità. (2022). *EpiCentro—Epidemiology for public health.* Istituto Superiore Di Sanità. Accessed on April 18, 2023 from https://www.epicentro.iss.it/en/

Kaiser, R. (1999). The clinical and epidemiological profile of tick-borne encephalitis in southern Germany 1994–98. *Brain, 122*(11), 2067–2078. https://doi.org/10.1093/brain/122.11.2067

Karti, S. S., Odabasi, Z., Korten, V., Yilmaz, M., Sonmez, M., Caylan, R., Akdogan, E., Eren, N., Koksal, I., Ovali, E., Erickson, B. R., Vincent, M. J., Nichol, S. T., Comer, J. A., Rollin, P. E., & Ksiazek, T. G. (2004). Crimean-Congo hemorrhagic fever in Turkey. *Emerging Infectious Diseases, 10*(8), 1379–1384. https://doi.org/10.3201/eid1008.030928

Kilpatrick, A. M., Meola, M. A., Moudy, R. M., & Kramer, L. D. (2008). Temperature, viral genetics, and the transmission of West Nile virus by Culex pipiens mosquitoes. *PLoS Pathogens, 4*(6), e1000092. https://doi.org/10.1371/journal.ppat.1000092

Kohl, I., Kozuch, O., Elecková, E., Labuda, M., & Zaludko, J. (1996). Family outbreak of alimentary tick-borne encephalitis in Slovakia associated with a natural focus of infection. *European Journal of Epidemiology, 12*(4), 373–375. https://doi.org/10.1007/BF00145300

Kramer, L. D., Li, J., & Shi, P.-Y. (2007). West Nile virus. *Lancet Neurology, 6*(2), 171–181. https://doi.org/10.1016/S1474-4422(07)70030-3

Landesman, W. J., Allan, B. F., Langerhans, R. B., Knight, T. M., & Chase, J. M. (2007). Interannual associations between precipitation and human incidence of West Nile virus in the United States. *Vector Borne and Zoonotic Diseases, 7*(3), 337–343. https://doi.org/10.1089/vbz.2006.0590

Lindquist, L., & Vapalahti, O. (2008). Tick-borne encephalitis. *The Lancet, 371*(9627), 1861–1871. https://doi.org/10.1016/S0140-6736(08)60800-4

Lorenzo Juanes, H. M., Carbonell, C., Sendra, B. F., López-Bernus, A., Bahamonde, A., Orfao, A., Lista, C. V., Ledesma, M. S., Negredo, A. I., Rodríguez-Alonso, B., Bua, B. R., Sánchez-Seco, M. P., Muñoz Bellido, J. L., Muro, A., & Belhassen-García, M. (2023). Crimean-Congo hemorrhagic fever, Spain, 2013–2021. *Emerging Infectious Diseases, 29*(2), 252–259. https://doi.org/10.3201/eid2902.220677

Mahmood, S., Erber, W., Yi, Z., Dobler, G., & Schmitt, H. -J. (2022). The changing epidemiology of tick-borne encephalitis (TBE). *VacciReview*. https://doi.org/10.33442/vr220601

Marcantonio, M., Rizzoli, A., Metz, M., Rosà, R., Marini, G., Chadwick, E., & Neteler, M. (2015). Identifying the environmental conditions favouring West Nile Virus outbreaks in Europe. *PLoS ONE, 10*(3), e0121158. https://doi.org/10.1371/journal.pone.0121158

Mariani, S., Braca, G., Romano, E., Lastoria, B., & Bussettini, M. (2018). *Linee guida sugli indicatori di siccità e scarsità idrica da utilizzare nelle attività degli osservatori permanenti per gli utilizzi idrici.* CReIAMO PA.

Mercogliano, P., Barbato, G., Padulano, R., Ellena, M., Bosello, F., Lanzi, E., & Forzieri, G. (2022). Gli impatti dei cambiamenti climatici in Italia. In Mims (Ministero delle Infrastrutture e della Mobilità Sostenibili). (Eds.), *Cambiamenti climatici, infrastrutture e mobilità. Soluzioni e strategie per gli investimenti infrastrutturali in un contesto di adattamento ai cambiamenti climatici e di mitigazione delle emissioni di gas-serra* (pp. 82–118). Rapporto della "Commissione cambiamenti climatici, infrastrutture e mobilità sostenibili".

Moirano, G., Gasparrini, A., Acquaotta, F., Fratianni, S., Merletti, F., Maule, M., & Richiardi, L. (2018). West Nile Virus infection in Northern Italy: Case-crossover study on the short-term effect of climatic parameters. *Environmental Research, 167*, 544–549. https://doi.org/10.1016/j.envres.2018.08.016

Musolino, D., de Carli, A., & Massarutto, A. (2017). Evaluation of socio-economic impact of drought events: The case of Po river basin. *European Countryside, 9*(1), 163–176. https://doi.org/10.1515/euco-2017-0010

Negredo, A., de la Calle-Prieto, F., Palencia-Herrejón, E., Mora-Rillo, M., Astray-Mochales, J., Sánchez-Seco, M. P., Bermejo Lopez, E., Menárguez, J., Fernández-Cruz, A., Sánchez-Artola, B., Keough-Delgado, E., Ramírez de Arellano, E., Lasala, F., Milla, J., Fraile, J. L., Ordobás Gavín, M., Martinez de la Gándara, A., López Perez, L., Diaz-Diaz, D., ... Crimean Congo Hemorrhagic Fever@Madrid Working Group. (2017). Autochthonous Crimean-Congo hemorrhagic fever in Spain. *The New England Journal of Medicine, 377*(2), 154–161. https://doi.org/10.1056/NEJMoa1615162

Negredo, A., Sánchez-Ledesma, M., Llorente, F., Pérez-Olmeda, M., Belhassen-García, M., González-Calle, D., Sánchez-Seco, M. P., & Jiménez-Clavero, M. Á. (2021). Retrospective identification of early autochthonous case of Crimean-Congo hemorrhagic fever, Spain, 2013. *Emerging Infectious Diseases, 27*(6), 1754–1756. https://doi.org/10.3201/eid2706.204643

Palagiano, C., & Pesaresi, C. (2011). *La salute nel mondo: geografia medica e qualità della vita.* https://books.google.com/books/about/La_salute_nel_mondo.html?hl=&id=skshkgAACAAJ

Panatto, D., Domnich, A., Amicizia, D., Reggio, P., & Iantomasi, R. (2022). Vaccination against Tick-Borne Encephalitis (TBE) in Italy: Still a long way to go. *Microorganisms, 10*(2). https://doi.org/10.3390/microorganisms10020464

Patz, J. A., Campbell-Lendrum, D., Holloway, T., & Foley, J. A. (2005). Impact of regional climate change on human health. *Nature, 438*(7066), 310–317. https://doi.org/10.1038/nature04188

Paz, S. (2015). Climate change impacts on West Nile virus transmission in a global context. *Philosophical Transactions of the Royal Society of London. Series B, Biological Sciences, 370*(1665). https://doi.org/10.1098/rstb.2013.0561

Paz, S., & Albersheim, I. (2008). Influence of warming tendency on Culex pipiens population abundance and on the probability of West Nile fever outbreaks (Israeli Case Study: 2001–2005). *EcoHealth, 5*(1), 40–48. https://doi.org/10.1007/s10393-007-0150-0

Paz, S., Malkinson, D., Green, M. S., Tsioni, G., Papa, A., Danis, K., Sirbu, A., Ceianu, C., Katalin, K., Ferenczi, E., Zeller, H., & Semenza, J. C. (2013). Permissive summer temperatures of the 2010 European West Nile fever upsurge. *PLoS ONE, 8*(2), e56398. https://doi.org/10.1371/journal.pone.0056398

Pinna, S., & Macchia, P. (2009). *Clima e mortalità in Italia. Uno sguardo geografico delle relazioni fra temperature medie mensili e decessi nel periodo 1969–2005.* https://books.google.com/books/about/Clima_e_mortalit%C3%A0_in_Italia_Uno_sguardo.html?hl=&id=ntyFPgAACAAJ

Portillo, A., Palomar, A. M., Santibáñez, P., & Oteo, J. A. (2021). Epidemiological aspects of Crimean-Congo hemorrhagic fever in Western Europe: What about the future? *Microorganisms, 9*(3). https://doi.org/10.3390/microorganisms9030649

Reiss, C. S. (2016). *Neurotropic viral infections: Volume 1: Neurotropic RNA viruses.* Springer. https://play.google.com/store/books/details?id=nx8DDQAAQBAJ

Rezza, G., Farchi, F., Pezzotti, P., Ruscio, M., Lo Presti, A., Ciccozzi, M., Mondardini, V., Paternoster, C., Bassetti, M., Merelli, M., Scotton, P. G., Luzzati, R., Simeoni, J., Mian, P., Mel, R., Carraro, V., Zanin, A., Ferretto, R., Francavilla, E., & TBE Virology Group. (2015). Tick-borne encephalitis in north-east Italy: a 14-year retrospective study, January 2000 to December 2013. *Euro Surveillance: Bulletin Europeen Sur Les Maladies Transmissibles = European Communicable Disease Bulletin, 20*(40). https://doi.org/10.2807/1560-7917.ES.2015.20.40.30034

Riccardi, N., Antonello, R. M., Luzzati, R., Zajkowska, J., Di Bella, S., & Giacobbe, D. R. (2019). Tick-borne encephalitis in Europe: A brief update on epidemiology, diagnosis, prevention, and treatment. *European Journal of Internal Medicine, 62*, 1–6. https://doi.org/10.1016/j.ejim.2019.01.004

Riccò, M., Peruzzi, S., & Balzarini, F. (2021). Epidemiology of West Nile Virus Infections in Humans, Italy, 2012–2020: A summary of available evidences. *Tropical Medicine and Infectious Disease, 6*(2). https://doi.org/10.3390/tropicalmed6020061

Ritchie, S. A., & Rochester, W. (2001). Wind-blown mosquitoes and introduction of Japanese encephalitis into Australia. *Emerging Infectious Diseases, 7*(5), 900–903. https://doi.org/10.3201/eid0705.017524

Rogers, D. J., & Randolph, S. E. (2006). Climate change and vector-borne diseases. *Advances in Parasitology, 62*, 345–381. https://doi.org/10.1016/S0065-308X(05)62010-6

Rossi, B., Barreca, F., Benvenuto, D., Braccialarghe, N., Campogiani, L., Lodi, A., Aguglia, C., Cavasio, R. A., Giacalone, M. L., Kontogiannis, D., Moccione, M., Malagnino, V., Andreoni, M., Sarmati, L., & Iannetta, M. (2023). Human Arboviral infections in Italy: Past, current, and future challenges. *Viruses, 15*(2). https://doi.org/10.3390/v15020368

Rudolf, I., Betášová, L., Blažejová, H., Venclíková, K., Straková, P., Šebesta, O., Mendel, J., Bakonyi, T., Schaffner, F., Nowotny, N., & Hubálek, Z. (2017). West Nile virus in overwintering mosquitoes, central Europe. *Parasites & Vectors, 10*(1), 452. https://doi.org/10.1186/s13071-017-2399-7

Ruiz, M. O., Chaves, L. F., Hamer, G. L., Sun, T., Brown, W. M., Walker, E. D., Haramis, L., Goldberg, T. L., & Kitron, U. D. (2010). Local impact of temperature and precipitation on West Nile virus infection in Culex species mosquitoes in northeast Illinois, USA. *Parasites & Vectors, 3*(1), 19. https://doi.org/10.1186/1756-3305-3-19

Semenza, J. C., & Menne, B. (2009). Climate change and infectious diseases in Europe. *The Lancet Infectious Diseases, 9*(6), 365–375. https://doi.org/10.1016/S1473-3099(09)70104-5

Semenza, J. C., & Paz, S. (2021). Climate change and infectious disease in Europe: Impact, projection and adaptation. *The Lancet Regional Health. Europe, 9*, 100230. https://doi.org/10.1016/j.lanepe.2021.100230

Shahhosseini, N., Wong, G., Babuadze, G., Camp, J. V., Ergonul, O., Kobinger, G. P., Chinikar, S., & Nowotny, N. (2021). Crimean-Congo hemorrhagic fever virus in Asia, Africa and Europe. *Microorganisms, 9*(9). https://doi.org/10.3390/microorganisms9091907

Shaman, J., Day, J. F., & Stieglitz, M. (2005). Drought-induced amplification and epidemic transmission of West Nile virus in southern Florida. *Journal of Medical Entomology, 42*(2), 134–141. https://doi.org/10.1093/jmedent/42.2.134

Tabachnick, W. J. (2010). Challenges in predicting climate and environmental effects on vector-borne disease episystems in a changing world. *The Journal of Experimental Biology, 213*(6), 946–954. https://doi.org/10.1242/jeb.037564

Ulbert, S. (2019). West Nile virus vaccines—Current situation and future directions. *Human Vaccines & Immunotherapeutics, 15*(10), 2337–2342. https://doi.org/10.1080/21645515.2019.1621149

Van Den Broeke, M. S., & Gunkel, T. J. (2021). The influence of isolated thunderstorms and the low-level wind field on nocturnally migrating birds in central North America. *Remote Sensing in Ecology and Conservation, 7*(2), 187–197. https://doi.org/10.1002/rse2.179

Van Doren, B. M. (2022). How migratory birds might have tracked past climate change. *Proceedings of the National Academy of Sciences of the United States of America, 119*(3). https://doi.org/10.1073/pnas.2121738119

Walsh, A. S., Glass, G. E., Lesser, C. R., & Curriero, F. C. (2008). Predicting seasonal abundance of mosquitoes based on off-season meteorological conditions. *Environmental and Ecological Statistics, 15*(3), 279–291. https://doi.org/10.1007/s10651-007-0056-6

Walther, G.-R., Post, E., Convey, P., Menzel, A., Parmesan, C., Beebee, T. J. C., Fromentin, J.-M., Hoegh-Guldberg, O., & Bairlein, F. (2002). Ecological responses to recent climate change. *Nature, 416*(6879), 389–395. https://doi.org/10.1038/416389a

Whitehouse, C. A. (2004). Crimean-Congo hemorrhagic fever. *Antiviral Research, 64*(3), 145–160. https://doi.org/10.1016/j.antiviral.2004.08.001

Climate Change, Air Pollution and Respiratory Health

Gennaro D'Amato and Maria D'Amato

Abstract The impact of climate change on the environment, biodiversity and the biosphere has become more evident in the past years. Climate change affects the quantity, intensity and frequency of precipitation type as well as extreme climate change events such as heat waves, droughts, thunderstorms, floods and hurricanes. Respiratory health can be particularly affected by climate change, which can contribute to the development of asthma and allergic respiratory diseases. Pollen allergens have been shown to trigger the release of immunomodulatory and proinflammatory mediators that accelerate the onset of allergy. Allergy to pollen and pollen season at its beginning, in duration and intensity are altered by change climate. Studies show that plants exhibit enhanced photosynthesis and reproductive effects and produce more pollen as a response to high atmospheric levels of carbon dioxide (CO_2). Pollen allergy is generally used to evaluate the interrelation between air pollution and allergic respiratory diseases, such as rhinitis and asthma. Lightning storms during pollen seasons can cause exacerbation of respiratory allergy and asthma in patients with pollinosis.

Keywords Respiratory allergy · Allergenic pollen · Climate change and allergy · Biodiversity and allergy · Pollen allergy · Severe allergic asthma · Thunderstorm asthma · Seasonal allergy

---

G. D'Amato (✉)
Division of Respiratory and Allergic Diseases, Department of Chest Diseases, High Speciality A. Cardarelli Hospital, Napoli, Italy
e-mail: gdamatomail@gmail.com

Medical School of Specialization in Respiratory Diseases, University of Naples Federico II, Napoli, Italy

Via Rione Sirignano, 10, 80121 Napoli, Italy

M. D'Amato
First Division of Pneumology, High Speciality Hospital 'V. Monaldi' and University 'Federico II' Medical School Naples, Napoli, Italy

R. Akhtar (ed.), *Climate Change and Human Health Scenarios*, Global Perspectives on Health Geography, https://doi.org/10.1007/978-3-031-38878-1_14

1 Introduction

Climate change is a physic meteorological fact that, among other aspects, will also affect human health. Allergies appear to be at the front line of the sequelae of climate change in addition to global health (D'Amato & Akdis, 2020). Evidence is accumulating that climate change affects besides climate, food supplies, water, soil and air quality. Present-day knowledge on the relationship between allergic respiratory diseases, asthma and environmental factors such as meteorological variables, airborne allergens and air pollution is taken from experimental and epidemiological studies (Bielory et al., 2012; D'Amato et al., 2016a, 2016b; Hegerl et al., 2007; Singer et al., 2005).

Urbanization with its high levels of vehicle emissions, and a westernized lifestyle, is linked to the rising frequency of respiratory allergic diseases and bronchial asthma observed over recent decades in most industrialized countries (D'Amato et al., 2016a). Climatic factors (thunderstorms, temperature, humidity and wind speed) can affect both interaction components (chemical and biological) (D'Amato et al., 2016a).

2 Climate Change, Why and How?

Today, millions of tons of CO_2 are produced by burning thousands of hectares of forests around the world each year, which is an important factor in the greenhouse effect (Gent et al., 2003; http://www.fire.ca.gov/index.php; McDonnell et al., 1999).

By an increased on carbon dioxide [CO_2] concentration, plant growth is affected in various ways leading to prolonged pollination periods.

The climate change pattern varies regionally depending on latitude, altitude, rainfall and storms, land-use patterns, urbanization, transportation and energy production. The greatness of climate change and related increases in allergic disease will be affected by how ambitiously greenhouse gas mitigation strategies are followed up on (Bielory et al., 2012). CO_2, primarily emitted from burning fossil fuels, is the predominant greenhouse gas. Other greenhouse gasses include methane (CH_4), nitrous oxide (NO_2) and fluorinated gases (United States Environmental Protection Agency (EPA), 2019).

Studies on plant responses to high atmospheric levels of CO_2 demonstrate that plants exhibit enhanced photosynthesis, reproductive effects and produce more pollen (Rogers et al., 1994).

Human activities have increased the natural concentration of carbon dioxide in our atmosphere, increasing Earth's natural greenhouse effect. In 1870, prior to the industrial revolution, carbon dioxide was 280 ppm and grew by 2.87 parts per million (ppm) at the mountain top observatory during 2018, shooting from an average of 407.05 ppm on January 1, 2018, to 409.92 on January 1, 2019, according to a new analysis of air samples collected by NOAA's Global Monitoring Division (GMD) (Oceanic et al., 2018).

As stipulated in the intergovernmental panel on climate change (IPCC) (Hegerl et al., 2007), it is important to reduce anthropogenic CO_2 emissions. Several measures to reduce greenhouse gas emissions can have positive health benefits. Unfortunately, after CO_2 emissions are reduced and atmospheric concentrations stabilize, the surface air temperature will continue to rise for a period of a century or more. This rise in temperature leads to an increase in the concentrations of ozone and particles at ground level (due to droughts, forest fires, desertification and a greater use of coal energy to produce energy to cool).

3 The Peculiarity of Allergenic Pollen and Pollen Allergy

The increasing prevalence of allergic respiratory symptoms due to the inhalation of pollen grains and the associated increasing costs make pollen allergy a public health problem. From 8 to 35% of European young adults show serum IgE antibodies to grass pollen allergens. Changes in pollen patterns can result in an allergy that causes difficulty working, disabilities, medical consultations and medication requirements with a significant impact on the cost of health care (Lindsey, 2018).

In natural pollination, mature pollen grains dehydrate when they are released by the anthers at the time of dispersion, and after coming into contact with a wet surface, the pollen grains absorb water and undergo a rapid metabolic change. Subsequently, when pollen grains penetrate the conjunctival, nasal or oral mucosa, the pollen allergens are released quickly, which induces symptoms of pollinosis in the ocular and respiratory mucous membranes of sensitized patients (Oceanic et al., 2018; Wayne et al., 2002).

Under an osmotic shock, pollen grains can explode, releasing cytoplasmic allergens into the atmosphere. In particular, fresh birch pollen can break under high humidity conditions and release an aerosol characterized by fragments of pollen cytoplasm in microdroplets. In this sense, Taylor et al. (2004) describe that approximately 65% of pollen grains grew in a pollen tube up to 300 μm long before rupture and that the release of their cytoplasmic content was under intense humidity.

The released particles, such as the cytoplasm of the fragmented pollen, form an ultrafine aerosol. Grass anthers are a pollen-breaking site and a source of fine particle aerosols containing pollen allergens (Taylor et al., 2002). There are multiple associations between allergen exposure, inflammation of the upper and lower respiratory tract and clinical symptoms. In allergic patients, the severity of symptoms depends on the amount of allergenic pollen; however, other factors besides allergens seem to be involved. Although pollen grains penetrate the upper respiratory tract, they rarely reach the bronchi, since their size is always greater than 10 μm in diameter. Nevertheless, bronchial asthma and its equivalents, such as irritative cough, are not uncommon in patients who are allergic to pollen.

Although it is believed that rain removes pollen from the air, some studies have shown that allergens can be released from the pollen's surface in only a few seconds, when they come in contact with water. The hypothesis is that during thunderstorms and precipitation, pollen releases allergens that carry particles much smaller than pollen grains (paucimicronic particles). The paucimicronic particles there are small granules of 1–5 μm developed from anther tissues, loaded with allergens that may play a role in allergic asthma (D'Amato et al., 2016c).

4 Effect of Climate Change on Pollen Allergens and Allergy Trigger

Changes in pollen allergens are correlated to climate change, due to an increasing concentration of CO_2 in the atmosphere that is capable of inducing a faster and larger growth of plants, a rise in the potency of the pollen allergen and an increase in the intensity and time of flowering (Bielory et al., 2012).

In addition, climate change produces greater exposure and sensitivity to subtropical grasses. An advance in the initiation of the pollen season and its peak are most pronounced in the species that bloom in early spring and species that respond more to a warmer temperature. Likewise, species in urban areas flourish earlier than in rural areas, with an earlier pollination by approximately 2–4 days (D'Amato et al., 2007; Singer et al., 2005; Wayne et al., 2002). Ziska et al. showed that along the gradient, average daytime CO_2 concentration went up by 21% and maximum (daytime) and minimum (nighttime) daily temperatures went up by 1.6 and 3.3 °C, respectively, in an urban relative to a rural location. The urban-induced environmental changes that were seen were consistent with the majority of short-term (~50 year) global change scenarios regarding CO_2 concentration and air temperature. Productivity, defined as final above-ground biomass, and maximum plant height were positively affected by daytime and soil temperatures as well as enhanced [CO_2], increasing 60 and 115% for the suburban and urban sites, respectively, relative to the rural site (Ziska et al., 2004). Singer et al. demonstrated that recent and projected increases in [CO_2] could directly increase the allergenicity of ragweed pollen and consequently the prevalence and/or severity of seasonal allergic disease (Singer et al., 2005). Wayne et al. observed that the doubling of the atmospheric concentration of CO_2 potentiated the production of pollen from ragweed by 61% more per plant. In addition, the ambrosia pollen collected along high traffic roads showed greater allergenicity than pollen in vegetative areas (Wayne et al., 2002).

As a result of climate change and the habitat patterns of the plants and the distribution of their species are likely to change, with a gradual migration northward in the northern hemisphere and further south in the southern hemisphere. Changes in land use also play an important role for some disseminated allergenic species, such as grasses (Beggs, 2004; Cecchi et al., 2006).

Plants can be classified into two groups, and these are C3 and C4 according to the metabolic differences related to carbon fixation. The members of the subfamily Pooideae are C3 grasses (temperate), while the subfamilies Chloridoideae and Panicoideae are C4 (subtropical) grasses (D'Amato et al., 2015). In seasonal climates when both grass types are present, C3 grasses increase in winter and bloom in spring, while C4 grasses grow and bloom in summer and early fall (Beggs, 2004; D'Amato et al., 2015). Peaks for pollen of grasses in the air at the end of summer were observed at the same time as the flowering of subtropical grasses (Davies, 2014; Osborne et al., 2017).

The three subfamilies are well represented in the southern hemisphere in countries such as Brazil, Argentina, Australia and Uruguay; however, Pooideae are the most present (Biganzoli & Zuloaga, 2015; Soreng et al., 2015). Research has shown an allergic sensitivity to subtropical pollen with climatic and geographic differences, especially in the southern hemisphere (Cherrez-Ojeda et al., 2018; Davies et al., 2016; Ramon et al., 2018; Soreng et al., 2017).

Studies on the response of plants to high levels of CO_2 in the atmosphere show that plants exhibit improved photosynthesis and reproductive effects and produce more pollen (Albertine et al., 2014; Rogers et al., 2006).

Currently, not only climate change is favoring the size of the population of subtropical grass and its expansion to areas where previously infrequent, but agriculture is also exerting a positive pressure (Augustine et al., 2017; Blumenthal et al., 2014; Sorokin et al., 2017). Australia and Argentina were the main countries in emerging organic agricultural lands (Willer & Lernoud, 2016).

A greater presentation of respiratory allergy caused by pollen in patients living in urban areas compared to those living in rural areas is linked to high levels of vehicle emissions, urbanization and having a western lifestyle (D'Amato & Cecchi, 2008).

Because of environmental pollutant effects, which do not only act as irritants to skin and mucous membranes, allergen carriers such as pollen can be altered in the atmosphere and release allergens resulting in allergen-containing aerosols in the ambient air. Pollen has been demonstrated not only to be an allergen carrier, but also to release highly active lipid mediators (pollen-associated lipid mediators), which have proinflammatory and immunomodulating effects in allergy diseases (D'Amato & Akdis, 2020).

5 Effect of Climate Change on Chemical Air Pollution

Climate change together with exposure to chemical air pollutants has been shown to have alarming consequences for human health and be responsible for some episodes of asthma exacerbations.

The effects of ozone on the respiratory system are well known. Ozone inhalation has been associated with an acute decrease in lung function, an increase in airway responsiveness mediated by airway injury, inflammation and systemic oxidative stress (Gent et al., 2003; Islam et al., 2007; McConnell et al., 2002; McDonnell et al., 1999).

Gent et al. analyzed the concomitant effects of fine particulate matter (PM 2.5) and ozone at levels below the standards of the EPA (the US Environmental Protection Agency) on the respiratory symptoms presented daily and the use of oral rescue or crisis medications in asthmatic children. In particular, the ozone was significantly associated with the presentation of respiratory symptoms and the need to use rescue medications in asthmatic children who use maintenance medications. A 1-h increase of 50 parts per billion (ppb) of ozone was associated with the presence of wheezing (35%) and chest tightness (47%). Higher ozone levels (averages of 1–8 h) were associated with increased dyspnea and the need for rescue or emergency medication (Gent et al., 2003).

Allergens derived from pollen or other parts of plants that reach the peripheral airways through inhaled air can induce asthma in patients with hypersensitivity. Air pollution by ozone, particulate matter (PM) and diesel exhaust particles (DEP), nitrogen dioxide and sulfur dioxide increase the permeability of the respiratory tract, facilitate the penetration of allergens into mucous membranes and cause interaction with the cells of the immune system. As a result, air pollution plays an inflammatory role in the airways of predisposed patients (McConnell et al., 2002).

Air pollutants adhere to the surface of pollen grains and paucimicronic-sized plant particles and change not only the morphology of these antigen-bearing agents, but also their allergenic potential. In addition, by producing airway inflammation increases the permeability of this pathway and pollutants overcome the mucosal barrier and could be responsible for enhancing the responses of pollinosis induced by allergens in atopic patients (D'Amato et al., 2005).

Published data suggest an increasing effect of aeroallergics in allergic patients, which generates a higher probability of developing an allergic respiratory disease in sensitized patients and an exacerbation in patients who are already symptomatic (Burney et al., 1997; D'Amato et al., 2016c, 2021; Mayaux et al., 2005; Schappi et al., 1999; Taylor et al., 2002, 2004; Traidl-Hoffmann et al., 2003).

6 Association Between Respiratory Allergies, Urban Environmental Factors and Climate Change

The effect of climate change on changes in wind patterns, time and intensity of rainfall and temperature increase can affect the frequency and severity of air pollution episodes and also anthropogenic emissions such as increase in energy demand for space cooling or heating. The urban heat island effect can increase some pollutants such as ozone and indirectly natural sources of air pollutant emissions (forest fires, vegetation breakdown, vehicle emission and soil erosion). The tropospheric ozone (O_3) originates in the presence of bright sunlight due to the reaction between volatile organic compounds (VOC) and nitrogen oxides (NOx), emitted from natural and anthropogenic sources. A link between tropospheric ozone concentrations and temperature has been demonstrated from measurements in outdoor smog chambers and measurements in ambient air (Singh et al., 2017). Birch pollen exposed to higher levels of ozone induces larger wheals and erythema in the skin prick tests compared to pollen exposed to lower amounts of ozone, suggesting an allergenic effect of ozone (Beggs, 2004; Cecchi et al., 2006; D'Amato et al., 2007; Taylor et al., 2004; Ziska et al., 2004).

Changes in precipitation and temperature can also increase the severity of forest fires which can lead to respiratory diseases. Higher temperatures can not only prolong the periods of time when ozone levels rise, but can also further aggravate maximum ozone concentrations.

Modifications in wind patterns can increase long-distance transport of pollutants and pollen grains, making this transport mechanism as important as the local one (Kellogg et al., 2013).

7 Severe Asthma Induced by the Extreme Weather Phenomenon of Lightning Storms

Asthma related to thunderstorms is one of the phenomena that represent a threat to human health. Thunderstorms that occur during the pollen season have been observed to induce severe asthma attacks and also deaths in pollen allergic patients (D'Amato et al., 2012). Asthma by thunderstorm is a term used to describe an increase in cases of acute bronchospasm after the appearance of thunderstorms in an area. Exacerbations of asthma due to a thunderstorm are characterized, at the beginning of the storms, by a rapid increase in visits to the general practitioner or in the emergency services of hospitals due to asthma. Patients without asthma symptoms, but who suffer from seasonal rhinitis, may have an asthma attack. There is a strong association with the elevation of atmospheric concentrations of pollen grains, such as grasses or other species of allergenic plants. A possible explanation for asthma related to thunderstorms involves the role of rainwater promoting the release of inhalable particles. Thunderstorms occur at the end of spring and in the summer when there are high levels of pollen grains in the air. In the first 20–30 min of a thunderstorm, patients with pollen allergy can inhale a high concentration of allergens that disperse in the atmosphere (Andrew et al., 2017). Exacerbations and asthma epidemics related to thunderstorms have been described in several cities, mainly in Europe (Birmingham and London in the UK and Naples in Italy) and Australia (Melbourne and Wagga Wagga) (Andrew et al., 2017; Final Report, 2017; Lindstrom et al., 2017) (Fig. 1).

Fig. 1 Greenhouse effect on forests

8 Climate Change and Its Impact on Infectious Respiratory Disease (SARS-CoV-2)

An extensive body of literature shows climate change's impact on incidence and severity of infectious respiratory diseases through the modifications of host immune response, exposure to fungal and mycobacterial species, vectors vitality and the spread of novel viruses (Fig. 2) (Mirsaeidi et al., 2016). Recently, studies on climate change have considered its influence on the outbreak of pandemics of novel pathogenic species such as COVID-19 caused by the emergence of the new coronavirus SARS-CoV-2.

Dramatic temperature shifts can lead to an increased exposure to environments where vector-borne pathogens thrive. Rise in temperatures is able to increase vectors' vitality and therefore the risk of disease spread (Rossati, 2017). This has been shown for example in rodents that are reservoirs for Hantaviruses, a virus known for regional outbreaks manifesting as pneumonia and diffuse systemic disease (Bayard et al., 2004).

Furthermore, desertification, expansion of drylands and dust storms have contributed to the release and diffusion of fungal dust-borne spores commonly found on soil that can cause respiratory infections, as observed in Southwestern USA with

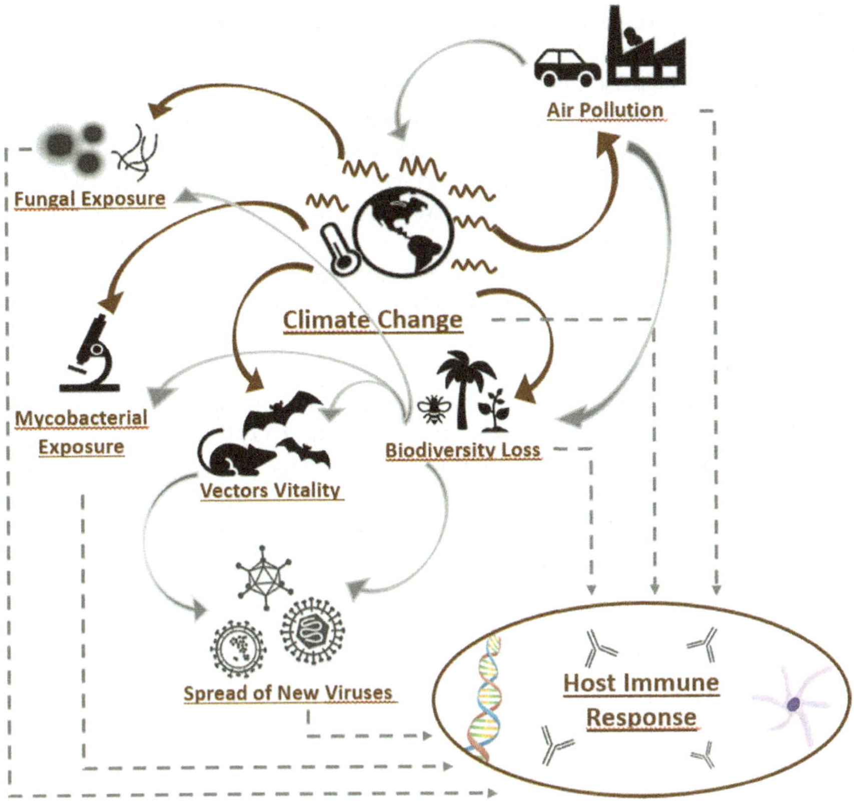

Fig. 2 Climate change's impact on infectious respiratory diseases through direct (red arrow) and indirect (gray arrows) effects on microbial exposure and spread and its effect on the host immune response

Coccidiomycosis (Park et al., 2005; Rhijn & Bromley, 2021; Schneider et al., 1997; Williams et al., 1979).

Another example is the geographic spread of *Cryptococcus gattii*, causal agent of Cryptococcosis, a disease that most commonly affects immunocompromised human hosts. This respiratory disease, originally only present in subtropical areas, is expanding in Mediterranean regions of Europe and in Pacific Northwest regions of the USA and it has been hypothesized that trees and livestock trading, flocks of migratory birds, anomalous atmospheric events (e.g., tsunamis) and human interactions have substantially contributed to the diffusion of this pathogen (Mak et al., 2010; Mitchell et al., 1995).

A similar case is Histoplasma capsulatum, an endemic fungus transmitted through inhalation in areas with bird or bat droppings in northern parts of the USA and known to cause severe pneumonia in immunocompromised hosts. Changes in animal behavior and geographic distribution due to global warming likely have had an impact on the diffusion of this disease (Rodrigues et al., 2020).

Not only fungal respiratory infections but also the spread of mycobacteriosis is intertwined with climate change. It has been demonstrated that hurricanes, which number is lately increased due climate change, contribute to an increase in non-tuberculous mycobacteria (NTM) disease (Waddell et al., 2021; Woodward & Samet, 2018). Differently from TB, NTM lung diseases are typically conveyed through environmental sources, such as municipal water and soil (Griffith & Aksamit, 2016), and environment cross-contamination by NTM is greatly favored by hurricanes (Honda et al., 2015; Kambali et al., 2021).

It is evident to all how public health and safety are threatened and damaged by emerging viral diseases such as Ebola, severe acute respiratory syndrome (SARS), the avian flu and novel viruses in the coronavirus family.

Climate change must be considered as a co-factor in their outbreak and spreading. Notably, both biodiversity decrease and air pollution increase, caused by climate change, might favor the onset and diffusion of the COVID-19 pandemic. The rising in air pollution not only modifies respiratory tracts' permeability through the oxidative stress and over-expression of Angiotensin-converting enzyme 2 (ACE-2), but also triggers a chronic inflammatory status and promote respiratory co-morbidities that greatly increase the risk of severe course and mortality of COVID-19 (Annesi-Maesano et al., 2021).

Finally, it is known that exposure to high temperatures and pollution, as direct effects of climate change, can affect host immune system (McMichael et al., 2006). Therefore, the fight against fossil fuel emissions and air pollutants release can prevent from the outbreak of new viral diseases and therefore new epidemics, but also limit the damage on societies and health systems caused by these diseases.

9 Conclusions

Climate change has important effects on the origin of hypersensitivity and pollen allergy. Climate change causes an increase in the production of pollen and a change in the characteristics that increase their allergenic properties. Through the effects of climate change in the future, plant growth can be altered so that the new pollen produced is modified and affects human health. Similarly, although data are sparse, climate change impacts on molds' proliferation through precipitations; increase and floods. As a consequence, an increase in the incidence of allergic diseases secondary to pollen and molds is expected in the medium and long terms.

The education of the population and the emergence of governmental measures to prevent environmental pollution and climate change are urgent measures to be dealt with throughout the world. Table 1 describes adaptation and mitigation measures that can be undertaken to limit climate change impacts on chemical air pollution, pollen and molds. Mitigation addresses the causes of climate change (accumulation of greenhouse gases in the atmosphere), whereas adaptation addresses the impacts of climate change. On the other hand, adaptation will not be able to eliminate all negative impacts and mitigation is crucial to limit changes in the climate system.

Extreme weather phenomena such as thunderstorms trigger exacerbations of asthma and severe asthma, with an important socio-economic impact, which have also to be prevented by meteorological broadcasting. Finally, the general population and in particular patients with asthma and pollen allergies should be educated about the health risk related to climate change.

Conflict of Interest None

References

Albertine, J. M., Manning, W. J., DaCosta, M., Stinson, K. A., Muilenberg, M. L., & Rogers, C. A. (2014). Projected carbon dioxide to increase grass pollen and allergen exposure despite higher ozone levels. *PLoS ONE, 9*, e111712. https://doi.org/10.1371/journal.pone.0111712. [PMID: 25372614].

Andrew, E., Nehme, Z., Bernard, S., Abramson, M. J., Newbigin, E., Piper, B., Dunlop, J., & Smith, K. (2017). Stormy Weather: A retrospective analysis of demand for emergency medical services during epidemic thunderstorm asthma. *BMJ, 359*, j5636. https://doi.org/10.1136/bmj.j5636. [PMID: 29237604].

Annesi-Maesano, I., Maesano, C. N., D'Amato, M., & D'Amato, G. (2021). Pros and cons for the role of air pollution on COVID-19 development. *Allergy, 76*, 2647–2649. https://doi.org/10.1111/all.14818. [PMID: 33733491].

Augustine, D. J., Derner, J. D., Milchunas, D., Blumenthal, D., & Porensky, L. M. (2017). Grazing moderates increases in C3 grass abundance over seven decades across a soil texture gradient in shortgrass steppe. *Journal of Vegetation Science, 28*, 562–572. https://doi.org/10.1111/jvs.12508

Bayard, V., Kitsutani, P. T., Barria, E. O., Ruedas, L. A., Tinnin, D. S., Muñoz, C., de Mosca, I. B., Guerrero, G., Kant, R., Garcia, A., Caceres, L., Gracio, F. G., Quiroz, E., de Castillo, Z., Armien, B., Libel, M., Mills, J. N., Khan, A. S., Nichol, S. T., … Peters, C. J. (2004). Outbreak of hantavirus pulmonary syndrome, Los Santos, Panama, 1999–2000. *Emerging Infectious Diseases, 10*, 1635–1642. https://doi.org/10.3201/eid1009.040143. [PMID: 15498167].

Beggs, P. J. (2004). Impacts of climate change on aeroallergens: Past and future. *Clinical and Experimental Allergy, 34*, 1507–1513. https://doi.org/10.1111/j.1365-2222.2004.02061.x. [PMID: 15479264].

Bielory, L., Lyons, K., & Goldberg, R. (2012). Climate change and allergic disease. *Current Allergy and Asthma Reports, 12*, 485–494. https://doi.org/10.1007/s11882-012-0314-z. [PMID: 23065327].

Biganzoli, F., & Zuloaga, F. (2015). Analysis of Poaceae biodiversity in austral South America. *Rodriguésia, 66*, 337–351. https://doi.org/10.1590/2175-7860201566205

Blumenthal, D. M., Kray, J. A., Ziska, L., & Dukes, J. (2014). Climate change, plant traits and invasion in natural and agricultural ecosystems. *Invasive Species Global Climate Change, 4*, 62.

Burney, P., Malmberg, E., Chinn, S., Jarvis, D., Luczynska, C., & Lai, E. (1997). The distribution of total and specific serum IgE in the European community respiratory health survey. *The Journal of Allergy and Clinical Immunology, 99*, 314–322. https://doi.org/10.1016/s0091-6749(97)70048-4. [PMID: 9058686].

Californian Department of Forestry and Fire Protection. http://www.fire.ca.gov/index.php

Cecchi, L., Morabito, M., Domeneghetti, M. P., Crisci, A., Onorari, M., & Orlandini, S. (2006). Long distance transport of ragweed pollen as a potential cause of allergy in central Italy. *Annals of Allergy, Asthma & Immunology, 96*, 86–91. https://doi.org/10.1016/s1081-1206(10)61045-9. [PMID: 16440538].

Cherrez-Ojeda, I., Ramon, G. D., Barrionuevo, L. B., Arango, N., Long, M. A., & Viego, V. (2018). Prevalence of skin sensitivity to temperate and subtropical grasses in patients with seasonal allergic rhinitis in Bahía Blanca, Argentina. *Journal of Allergy and Clinical Immunology, 141*, AB128. https://doi.org/10.1016/j.jaci.2017.12.408

D'Amato, G., & Akdis, C. (2020). Global warming, climate change, air pollution and allergies. *Allergy, 75*, 2158–2160. https://doi.org/10.1111/all.14527. [PMID: 32738058].

D'Amato, G., Annesi-Maesano, I., Urrutia-Pereira, M., Del Giacco, S., Rosario Filho, N. A., Chong-Neto, H. J., Solé, D., Ansotegui, I., Cecchi, L., Sanduzzi Zamparelli, A., Tedeschini, E., Biagioni, B., Murrieta-Aguttes, M., & D'Amato, M. (2021). Thunderstorm allergy and asthma: State of the art. *Multidisciplinary Respiratory Medicine, 16*, 806. https://doi.org/10.4081/mrm.2021.806. [PMID: 35003735].

D'Amato, G., & Cecchi, L. (2008). Effects of climate change on environmental factors in respiratory allergic diseases. *Clinical and Experimental Allergy, 38*, 1264–1274. https://doi.org/10.1111/j.1365-2222.2008.03033.x. [PMID: 18537982].

D'Amato, G., Cecchi, L., & Annesi-Maesano, I. (2012). A trans-disciplinary overview of case reports of thunderstorm-related asthma outbreaks and relapse. *European Respiratory Reviews, 21*, 82–87. https://doi.org/10.1183/09059180.00001712. [PMID: 22654079].

D'Amato, G., Cecchi, L., Bonini, S., Nunes, C., Annesi-Maesano, I., Behrendt, H., Liccardi, G., Popov, T., & van Cauwenberge, P. (2007). Allergenic pollen and pollen allergy in Europe. *Allergy, 62*, 976–990. https://doi.org/10.1111/j.1398-9995.2007.01393.x. [PMID: 17521313].

D'Amato, G., Holgate, S. T., Pawankar, R., Ledford, D. K., Cecchi, L., Al-Ahmad, M., Al-Enezi, F., Al-Muhsen, S., Ansotegui, I., Baena-Cagnani, C. E., Baker, D. J., Bayram, H., Bergmann, K. C., Boulet, L. P., Buters, J. T., D'Amato, M., Dorsano, S., Douwes, J., Finlay, S. E., …, Annesi-Maesano, I. (2015). Meteorological conditions, climate change, new emerging factors, and asthma and related allergic disorders. A statement of the World Allergy Organization. *World Allergy Organization Journal, 8*, 25. https://doi.org/10.1186/s40413-015-0073-0. [PMID: 26207160].

D'Amato, G., Liccardi, G., D'Amato, M., & Holgate, S. (2005). Environmental risk factors and allergic bronchial asthma. *Clinical and Experimental Allergy, 35*, 1113–1124. https://doi.org/10.1111/j.1365-2222.2005.02328.x. [PMID: 16164436].

D'Amato, G., Pawankar, R., Vitale, C., Lanza, M., Molino, A., Stanziola, A., Sanduzzi, A., Vatrella, A., & D'Amato, M. (2016a). Climate change and air pollution: effects on respiratory allergy. *Allergy, Asthma & Immunology Research, 8*, 391–395. https://doi.org/10.4168/aair.2016.8.5.391. [PMID: 27334776].

D'Amato, G., Vitale, C., D'Amato, M., Cecchi, L., Liccardi, G., Molino, A., Vatrella, A., Sanduzzi, A., Maesano, C., & Annesi-Maesano, I. (2016c). Thunderstorm-related asthma: What happens and why. *Clinical and Experimental Allergy, 46*, 390–396. https://doi.org/10.1111/cea.12709. [PMID: 26765082].

D'Amato, G., Vitale, C., Lanza, M., Molino, A., & D'Amato, M. (2016b). Climate change, air pollution, and allergic respiratory diseases: An update. *Current Opinion in Allergy and Clinical Immunology, 16*, 434–440. https://doi.org/10.1097/ACI.0000000000000301. [PMID: 27518837].

Davies, J. M. (2014). Grass pollen allergens globally: The contribution of subtropical grasses to burden of allergic respiratory diseases. *Clinical and Experimental Allergy, 44*, 790–801. https://doi.org/10.1111/cea.12317. [PMID: 24684550].

Davies, J., Timbrell, V., Reibelt, L., Simmonds, C., Solley, G., Smith, W. B., Mclean-Tooke, A., Nunen, S., Smith, P., Upham, J., & Langguth, D. (2016). Regional variation in allergic sensitivity to subtropical and temperate grass pollen allergens; outcomes of the multicenter cross-sectional Grass Pollen Allergy Survey (GPAS). *European Journal of Immunology, 46*, 841. [Corpus ID: 89724373].

Final Report: Literature review on thunderstorm asthma and its implications for public health advice. Queensland University of Technology, Brisbane, Australia. Contracted by: Department of Health and Human Services, Victorian State Government, May 19, 2017.

Gent, J. F., Triche, E. W., Holford, T. R., Belanger, K., Bracken, M. B., Beckett, W. S., & Leaderer, B. P. (2003). Association of low-level ozone and fine particles with respiratory symptoms in children with asthma. *JAMA, 290*, 1859–1867. https://doi.org/10.1001/jama.290.14.1859. [PMID: 14532314].

Griffith, D. E., & Aksamit, T.R. (2016). Understanding nontuberculous mycobacterial lung disease: It's been a long time coming. *F1000Research, 5*, 2797. https://doi.org/10.12688/f1000research.9272.1. [PMID: 27990278].

Hegerl, G. C., Zwiers, F. W., Braconnot, P., Gillett, N. P., Luo, Y., Marengo, J. A., Orsini, J. A. M., Nicholls, N., Penner, J. E., Stott, P. A., Allen, M., & Ammann, C. (2007). Understanding and attributing climate change. In: S. Solomon, D. Qin, M. Manning, Z. Chen, M. Marquis, K. B. Averyt, M. Tignor, & H. L. Miller (Eds.), *Climate change 2007: The physical science basis. Contribution of the Working Group I to the Fourth Assessment Report of the Intergovernmental Panel on Climate Change* (pp. 663–746). Cambridge University Press.

Honda, J. R., Bernhard, J. N., & Chan, E. D. (2015). Natural disasters and nontuberculous mycobacteria: A recipe for increased disease? *Chest, 147*, 304–308. https://doi.org/10.1378/chest.14-0974. [PMID: 25644904].

Islam, T., Gauderman, W. J., Berhane, K., McConnell, R., Avol, E., Peters, J. M., & Gilliland, F. D. (2007). Relationship between air pollution, lung function and asthma in adolescents. *Thorax, 62*, 957–963. https://doi.org/10.1136/thx.2007.078964. [PMID: 17517830].

Kambali, S., Quinonez, E., Sharifi, A., Shahraki, A. H., Kumar, N., Dushyantha, J., & Mirsaeidi, M. (2021). Pulmonary nontuberculous mycobacterial disease in Florida and association with large-scale natural disasters. *BMC Public Health, 21*, 2058. https://doi.org/10.1186/s12889-021-12115-7. [PMID: 34758787].

Kellogg, E. A., Camara, P. E., Rudall, P. J., Ladd, P., Malcomber, S. T., Whipple, C., & Doust, A. N. (2013). Early inflorescence development in the grasses (Poaceae). *Frontiers in Plant Science, 4*, 250. https://doi.org/10.3389/fpls.2013.00250. [PMID: 23898335].

Lindsey, R. (2018). *Climate change: Atmospheric carbon dioxide*. Accessed August 17, 2019.

Lindstrom, S. J., Silver, J. D., & Sutherland, M. F. (2017). Thunderstorm asthma outbreak of November 2016: A natural disaster requiring planning. *Medical Journal of Australia, 207*, 235–237. https://doi.org/10.5694/mja17.00285. [PMID: 28899321].

Mak, S., Klinkenberg, B., Bartlett, K., & Fyfe, M. (2010). Ecological niche modeling of *Cryptococcus gattii* in British Columbia, Canada. *Environmental Health Perspectives, 118*, 653–658. https://doi.org/10.1289/ehp.0901448. [PMID: 20439176].

Mayaux, P., Holmgren, P., Achard, F., Eva, H., Stibig, H., & Branthomme, A. (2005). Tropical forest cover change in the 1990s and options for future monitoring. *Philosophical Transactions of the Royal Society B: Biological Sciences, 360*, 373–384. https://doi.org/10.1098/rstb.2004.1590. [PMID: 15814351].

McConnell, R., Berhane, K., Gilliland, F., London, S. J., Islam, T., Gauderman, W. J., Avol, E., Margolis, H. G., & Peters, J. M. (2002). Asthma in exercising children exposed to ozone: A cohort study. *Lancet, 359*, 386–391. https://doi.org/10.1016/S0140-6736(02)07597-9. [PMID: 11844508].

McDonnell, W. F., Abbey, D. E., Nishino, N., & Lebowitz, M. D. (1999). Long-term ambient ozone concentration and the incidence of asthma in nonsmoking adults: The AHSMOG Study. *Environmental Research, 80*, 110–121. https://doi.org/10.1006/enrs.1998.3894. [PMID: 10092402].

McMichael, A. J., Woodruff, R. E., & Hales, S. (2006). Climate change and human health: Present and future risks. *Lancet, 367*, 859–869. https://doi.org/10.1016/S0140-6736(06)68079-3. [PMID: 16530580].

Mirsaeidi, M., Motahari, H., Taghizadeh Khamesi, M., Sharifi, A., Campos, M., & Schraufnagel, D. E. (2016). Climate change and respiratory infections. *Annals of the American Thoracic Society, 13*, 1223–1230. https://doi.org/10.1513/AnnalsATS.201511-729PS. [PMID: 27300144].

Mitchell, D. H., Sorrell, T. C., Allworth, A. M., Health, C. H., McGregor, A. R., Papanaoum, K., Richards, M. J., & Gottlieb, T. (1995). Cryptococcal disease of the CNS in immunocompetent hosts: Influence of cryptococcal variety on clinical manifestations and outcome. *Clinical Infectious Diseases, 20*, 611–616. https://doi.org/10.1093/clinids/20.3.611. [PMID: 7756484].

National Oceanic and Atmospheric Administration. Global carbon dioxide growth in 2018 reached 4th highest in record. Accessed August 17, 2019.

Osborne, N. J., Alcock, I., Wheeler, B. W., Hajat, S., Sarran, C., Clewlow, Y., McInnes, R. N., Hemming, D., White, M., Vardoulakis, S., & Fleming, L. E. (2017). Pollen exposure and hospitalization due to asthma exacerbations: Daily time series in a European city. *International Journal of Biometeorology, 61*, 1837–1848. https://doi.org/10.1007/s00484-017-1369-2. [PMID: 28500390].

Park, B. J., Sigel, K., Vaz, V., Komatsu, K., McRill, C., Phelan, M., Colman, T., Comrie, A. C., Warnock, D. W., Galgiani, J. N., & Hajjeh, R. A. (2005). An epidemic of coccidioidomycosis in Arizona associated with climatic changes, 1998–2001. *Journal of Infectious Diseases, 191*, 1981–1987. [PMID: 15871133].

Ramon, G. D., Viego, V., Arango, N., Long, M. A., Kahn, A., & Barrionuevo, L. B. (2018). Allergy to *Cynodon dactylon* (Cyn d) pollen in seasonal rhinitis, comparison between skin tests and component-resolved diagnosis (CRD) in Bahía Blanca (Argentina). *Journal of Allergy and Clinical Immunol*ogy, *141*, AB129. https://doi.org/10.1016/j.jaci.2017.12.412

Rodrigues, A.M., Beale, M. A., Hagen, F., Fisher, M.C., Terra, P. P. D., de Hoog, S., Brilhante, R. S. N., de Aguiar Cordeiro, R., de Souz Collares Maia Castelo-Branc, D., Rocha, M. F. G., Sidrim, J. J. C., & de Camargo, Z. P. (2020). The global epidemiology of emerging *Histoplasma* species in recent years. *Studies in Mycol*ogy, *97*, 100095–100095. https://doi.org/10.1016/j.simyco.2020.02.001. [PMID: 33335607].

Rogers, C. A., Wayne, P. M., Macklin, E. A., Muilenberg, M. L., Wagner, C. J., Epstein, P. R., & Bazzaz, F. A. (2006). Interaction of the onset of spring and elevated atmospheric CO_2 on Ragweed (*Ambrosia artemisiifolia* L.) Pollen production. *Environmental Health Perspectives, 114*, 865–869. https://doi.org/10.1289/ehp.8549. [PMID: 16759986].

Rogers, H. H., Runion, G. B., & Krupa, S. V. (1994). Plant responses to atmospheric CO_2 enrichment with emphasis on roots and the rhizosphere. *Environmental Pollution, 83*, 155–189. https://doi.org/10.1016/0269-7491(94)90034-5. [PMID: 15091762].

Rossati, A. (2017). Global warming and its health impact. *International Journal of Occupational and Environmental Med*icine, *8*, 7–20. https://doi.org/10.15171/ijoem.2017.963. [PMID: 28051192].

Schappi, G., Taylor, P. E., Staff, I. A., Rolland, J. M., & Suphioglu, C. (1999). Immunologic significance of respirable atmospheric granules containing major birch allergen bet v 1. *Allergy, 54*, 478–483. https://doi.org/10.1034/j.1398-9995.1999.00838.x. [PMID: 10380779].

Schneider, E., Hajjeh, R. A., Spiegel, R. A., Jibson, R. W., Harp, E. L., Marshall, G. A., Gunn, R. A., McNeil, M. M., Pinner, R. W., Baron, R. C., Burger, R. C., Hutwagner, L. C., Crump, C., Kaufman, L., Reef, S. E., Feldman, G. M., Pappagianis, D., & Werner, S. B. (1997). A coccidioidomycosis outbreak following the Northridge, Calif, earthquake. *JAMA, 277*, 904–908. [PMID: 9062329].

Singer, B. D., Ziska, L. H., Frenz, D. A., Gebhard, D. E., & Straka, J. G. (2005). Increasing Amb a 1 content in common ragweed (*Ambrosia artemisiifolia*) pollen as a function of rising atmospheric CO_2 concentration. *Functional Plant Biology, 32*, 667–670. https://doi.org/10.1071/FP05039. [PMID: 32689165].

Singh, R. P., Singh, A., & Srivastava, V. (2017). *Environmental issues surrounding human overpopulation* (pp. 233–236). IGI Global, Cap.14. https://doi.org/10.4018/978-1-5225-1683-5

Soreng, R. J., Peterson, P. M., Romaschenko, K., Davidse, G., Zuloaga, F. O., Judziewicz, E. J., Filguerias, T. S., Davis, J. I., & Morrone, O. (2015). A worldwide phylogenetic classification of the Poaceae (Gramineae). *Journal of Systematics and Evolution, 53*, 117–137. https://doi.org/10.1111/jse.12150

Soreng, R. J., Peterson, P. M., Romaschenko, K., Davidse, G., Teisher, J. K., Clark, L. G., Barberá, P., Gillespie, L. J., & Zuloaga, F. O. (2017). A worldwide phylogenetic classification of the Poaceae (Gramineae) II: An update and a comparison of two 2015 classifications. *Journal of Systematics and Evolution, 55*, 259–290. https://doi.org/10.1111/jse.12262

Sorokin, Y., Zelikova, T. J., Blumenthal, D., Williams, D. G., & Pendall, E. (2017). Seasonally contrasting responses of evapotranspiration to warming and elevated CO_2 in a semiarid grassland. *Ecohydrology, 10*, e1880. https://doi.org/10.1002/eco.1880

Taylor, P. E., Flagan, R., Valenta, R., & Glovsky, M. (2002). M Release of allergens in respirable aerosols: A link between grass pollen and asthma. *The Journal of Allergy and Clinical Immunology, 109*, 51–56. https://doi.org/10.1067/mai.2002.120759. [PMID: 11799365].

Taylor, P. E., Flagan, R. C., Miguel, A. G., Valenta, R., & Glovsky, M. M. (2004). Birch pollen rupture and the release of aerosols of respirable allergens. *Clinical and Experimental Allergy, 34*, 1591–1596. https://doi.org/10.1111/j.1365-2222.2004.02078.x. [PMID: 15479275].

Traidl-Hoffmann, C., Kasche, A., Menzel, A., Jakob, T., Thiel, M., Ring, J., & Behrendt, H. (2003). Impact of pollen on human health: More than allergen carriers? *International Archives of Allergy and Immunology, 131*, 1–13. https://doi.org/10.1159/000070428. [PMID: 12759483].

United States Environmental Protection Agency (EPA). Overview of Greenhouse Gases. Accessed on August 17, 2019.

van Rhijn, N., & Bromley, M. (2021). The consequences of our changing environment on life threatening and debilitating fungal diseases in humans. *Journal of Fungi, 7*, 367. https://doi.org/10.3390/jof7050367. [PMID: 34067211].

Waddell, S. L., Jayaweera, D. T., Mirsaeidi, M., Beier, J. C., & Kumar, N. (2021). Perspectives on the health effects of hurricanes: A review and challenges. *International Journal of Environmental Research and Public Health, 18*, 2756. https://doi.org/10.3390/ijerph18052756. [PMID: 33803162].

Wayne, P., Foster, S., Connolly, J., Bazzaz, F., Epstein, P. (2002). Production of allergenic pollen by ragweed (*Ambrosia artemisiifolia* L.) is increased in CO_2-enriched atmospheres. *Annals of Allergy, Asthma & Immunology, 88*, 279–282. https://doi.org/10.1016/S1081-1206(10)62009-1. [PMID: 11926621].

Willer, H., & Lernoud, J. (Eds.) (2016). *The world of organic agriculture. Statistics and emerging trends 2016* (17 ed.). Research Institute of Organic Agriculture FiBL and IFOAM Organics International, Frick and Bonn.

Williams, P. L., Sable, D. L., Mendez, P., & Smyth, L. T. (1979). Symptomatic coccidioidomycosis following a severe natural dust storm. An outbreak at the Naval Air Station, Lemoore, Calif. *Chest, 76*, 566–570. https://doi.org/10.1378/chest.76.5.566. [PMID: 498830].

Woodward, A. J., & Samet, J. M. (2018). Climate Change, Hurricanes, and Health. *American Journal of Public Health, 108*, 33–35. https://doi.org/10.2105/AJPH.2017.304197. [PMID: 29211542].

Ziska, L. H., Bunce, J. A., & Goins, E. W. (2004). Characterization of an urban-rural CO_2/temperature gradient and associated changes in initial plant productivity during secondary succession. *Oecologia, 139*, 454–458. https://doi.org/10.1007/s00442-004-1526-2. [PMID: 15021982].

Climate Catastrophe and the Consequences for Health in the UK

Tom Douglass

Abstract This chapter discusses climate change in the United Kingdom in relation to the health of its people. I begin with a brief discussion of the nature of the existing and predicted impacts of climate change in the UK alongside a history of the politics of climate change in the UK and an overview of government climate change mitigation policy (and particularly its inadequacies). The chapter then concerns itself with the present and future impacts of climate change on human health in the UK. I examine the impacts of exposure to extreme heat, flooding, and disruption to health services as key examples of the negative impacts of climate change on health. Relatedly, I discuss the uneven distribution of the health risks associated with climate change according to socio-economic status. Finally, I consider benefits for human health that emerge from climate change mitigation strategies and the associated decarbonisation of the economy.

Keywords Climate change mitigation · Extreme weather · Health policy · Health impacts · Socioeconomic inequalities · United Kingdom

1 Introduction

A vast body of evidence suggests that climate change is impacting many dimensions of the societies we inhabit (see Linden, 2022). The world faces a climate emergency including rising temperatures, frequent and severe storms, rising sea levels, floods and drought (Somerville, 2021). Driven by a build-up of greenhouse gases, a result of the burning of fossil fuels for energy by humans, global temperatures have risen. Climate scientists fear that should global temperatures continue to rise unabated and surpass preindustrial temperature levels by 1.5 °C (and worse, reach much higher levels), then there will be devastating consequences globally for health, access to food and water, security and economic growth (IPCC, 2018). Environmental conditions are closely tried to human health and climate change threatens our ability to prosper and survive

T. Douglass (✉)
University of Birmingham, Birmingham, UK
e-mail: t.douglass@bham.ac.uk

R. Akhtar (ed.), *Climate Change and Human Health Scenarios*, Global Perspectives on Health Geography, https://doi.org/10.1007/978-3-031-38878-1_15

(Office for Health Improvement and Disparities, 2022). In other words, detrimental health impacts from climate change are emerging as a result of "changing exposure to heat and cold, air pollution, pollen, food safety risks, disruptions to access to and functioning of health services and facilities, emerging infections, flooding and other reasons such as water-borne diseases and increased exposure to UV radiation, as well as indirectly via for example changing prices of and access to food and energy" (Paavola, 2017: 61).

In the UK, since preindustrial times, changes to the climate include rising temperatures of about 0.25 °C per decade and decrease in summer rainfall with increases in winter rainfall. A report by the Health Protection Agency (2012) emphasises that these trends are set to worsen. Indeed, climate change predictions indicate that there will be temperature rises in the UK between 2 and 5 °C by 2080 (with larger increases expected in the south). Relatedly, heatwaves are likely to become more frequent. Winter and summer rainfalls are also predicted to rise and fall further, respectively. Resulting from the initial impacts of climate change, there have been several notable extreme weather events in the UK in recent years—including flooding in the East Midlands, West Midlands and Yorkshire in 2019 that displaced some residents for over a year (BBC News, 2020). The UK also experienced record-breaking temperatures over 40 °C during heatwaves in the summer of 2022 (Stylianou et al., 2022).

Despite growing evidence of the extent and harm of climate change, and the damage contributed by UK emissions[1] both historically and presently (Oxfam, 2020; Somerville, 2021), western authorities like the UK government have displayed prevarication and insufficient action on climate change (Linden, 2022). The 2008 Climate Change Act committed by law the UK government to the reduction of greenhouse gas emissions initially by 80% (compared with 1990 levels) by 2050. At the time, this was a globally unique commitment. However, since this legislation, Somerville (2021) argues that the government's climate change mitigation policy has been failing by its own standards and is characterised by complacency—with many economic sectors failing to significantly decarbonise. The largest political parties in the UK (the Conservatives, Labour and the Liberal Democrats) have all (at least for a period of time) shown wide-ranging rhetorical support for climate change mitigation and environmentalism. At the same time, the policy response to climate change since the 1990s has been inconsistent and insufficient (Somerville, 2021). It has variously emphasised technological innovation, reductions in energy demand, carbon pricing and playing down UK responsibility. In this regard, policy has not radically challenged or altered the focus within capitalist societies like the UK on economic growth and the associated reliance on fossil fuels to drive this growth.

Climate change rapidly ascended the political agenda and enjoyed cross-party support during the later years of the New Labour (1997–2010) government which

[1] This chapter is dedicated to the policy response and the impacts of climate change on health specifically in the UK. The climate change catastrophe is a global crisis and the causes and solutions to climate change are necessarily interconnected. The rest of this book is dedicated to climate change health outcomes in an international context.

resulted in the 2008 Climate Change Act (Carter, 2014). However, despite growing public concern about climate change and increased climate and environmental activism—with most British adults now believing that climate change poses high or fairly high risks to health (see The Climate Coalition, 2021)—under the subsequent coalition and then Conservative governments of the 2010s climate change policy became an increasingly partisan issue.[2] Many on the political right revealed their hostility to what they call 'green taxes' and 'unwarranted' environmental interventionism by the government (Carter, 2014). With funding cuts, government support for fracking, devolution of responsibility and a negative approach to regulation and associated deference to market forces and reliance on behaviour change, Somerville (2021) argues that in the recent years, climate change policy has gone backwards under Conservative rule in the UK. Even the lessons of the COVID-19 pandemic (in a sense a parallel crisis) that have emphasised the need for wide-ranging government intervention and a timely policy response to mitigate the worst consequences of global disasters (Klenert et al., 2020; see also Calnan & Douglass, 2022) have not mobilised the government into an appropriate climate crisis response. Indeed, at the time of writing, the government have even given approval to open the UK's first new coal mine in 30 years despite domestic protest and international astonishment (Harvey, 2022).

However, several published reports show that UK governments have nevertheless been conscious of the health impacts of climate change and the associated consequences for productivity and growth. The Department of Health and Social Care published one of the first reports of its kind internationally in the early 2000s concerned with the health impacts of climate change; the Health Protection Agency then worked to update these reports based on the latest evidence and predictions in 2008 and 2012 (see Health Protection Agency, 2008, 2012). Under the coalition and Conservative governments of more recent years, there have also been three Climate Change Risk Assessments published as required by the 2008 Climate Change Act which reference health impacts amongst broader economic and security concerns (see HM Government, 2022).

2 The Health Impacts of Climate Change

Drawing on a range of published evidence, this section of the chapter reviews the present and future impacts of climate change on the health of the people residing in the UK. I examine the impacts of exposure to extreme heat, pollution, flooding and the associated disruption to health services as key examples of the negative impacts of climate change on health. I also discuss the uneven distribution of the health risks associated with climate change according to socioeconomic status.

A report by The Climate Coalition (2021) suggests that the health effects that climate change is posing to the health of people in the UK take three forms. First,

[2] Not to the same extent as in the USA, however.

direct impacts on health (such as death or injury), which can be caused by flooding or exposure to extreme heat. Second, indirect impacts via ecosystems, which relate to health impacts resulting from air pollution or infectious diseases. Finally, indirect via societal systems, which are health impacts reflecting declining productivity and pressures on healthcare systems meaning that people find it more difficult to access healthcare products and services.

Arguably, the biggest threat to health from climate change in the UK is increased flooding (ibid). Nearly two million people in the UK live in areas at considerable risk of flooding—and this number could increase by 40% in less than two decades if the effects of climate change continue unabated. The immediate risk from flooding is death or injury. There are other possible indirect health impacts too, including mental health problems. Nearly one in three people who have been the victims of house flooding has suffered with post-traumatic stress disorder and flooding victims are four times more likely to suffer other mental health conditions (such as depression and anxiety). Additionally, standing water from flooding can become a breeding ground for a range of diseases that will also prosper as a result of increasingly warm temperatures—particularly diseases spread by insects and ticks such as malaria, dengue fever, West Nile fever, Lyme disease and tick-borne encephalitis. Standing water resulting from flooding also heightens the risk of water-borne diseases such as typhoid fever, hepatitis A and cholera (The Climate Coalition, 2021; Health Protection Agency, 2012).

Temperatures above 35 °C are becoming common in the UK, as are 'tropical nights' where night-time temperatures stay above 20 °C. This level of heat in the past was very rare in the UK and can be harmful to health (see Health Protection Agency, 2012). The UK is a vulnerable country to the health consequences of rising temperatures because of the nature of its ageing population, prevalence of chronic disease and pace of urbanisation (HM Government, 2022; Romanello et al., 2022; see also Public Health England 2020). Extreme heat and warmer summers expose the population to conditions such as sunstroke, dehydration, adverse pregnancy outcomes, worsened sleeping patterns, deteriorating cardiovascular and respiratory disease—and thus higher rates of mortality (Romanello et al., 2022). The worst impacts on health are during heatwaves where heat stress makes it more difficult for humans to keep their core temperature within optimal levels (The Climate Coalition, 2021). Research suggests that by 2050, there could be more than a 250% increase in heat-related deaths (Hajat et al., 2014). There are also indirect risks to health associated with increasing heat. For example, warmer summers and milder winters caused by climate change may allow pathogens like salmonella to grow more easily in food and conditions will be more favourable for flies and other insects that can affect the safety of food consumed by humans (Health Protection Agency, 2012). In a different example of the impact of increasing heat on human mortality and mental health, the risk of suicide has been found to increase by 3.8% with each one-degree Celsius rise above 18 °C (The Climate Coalition, 2021). In response to these wide-ranging threats, the government developed a heatwave plan first published in 2003 which aims to prepare, alert and prevent harm to the public's health from severe heat. The plan details action that should be taken before and after a period of severe heat, including strategic planning and the operation of an alert system in the summer months, as well as outlining preparations that individuals and institutions can take to reduce health risks, including targeted measures to protect vulnerable groups (UK Health Security Agency, 2022).

2.1 The Socioeconomic and Health Inequalities Exacerbated by Climate Change

Paavola (2017) provides a comprehensive review of how socioeconomic inequalities shape the impacts of UK climate change. This evidence suggests that older people are more likely to experience adverse effects from exposure to heat during hotter summers and heat waves. This reflects more limited ability to thermoregulate, but it is also associated with increased likelihood of having other medical conditions. Paavola also discusses how older people's capacity to adapt to heat may be restricted by mobility, isolation or lack of access to information. Low levels of autonomy and preparedness of care staff in care/nursing homes may additionally limit adaption to heat as well as a lack of regulation of temperatures in hospitals and care homes (The Climate Coalition, 2021). Indeed, in the years between 2004 and 2018, heat-related mortality increased by 21% in people older than 65, and as noted above, particularly considering that the UK's population is aging, by 2050 there could be a 250% increase in heat-related mortality in the over 65 s without action to halt climate change (ibid, 2021).

Additionally, the impacts of increasing temperatures are geographically shaped (ibid). The south and east of the UK and people living in urban and particularly deprived urbans areas (due to these areas being densely built, with limited green space) are more exposed due to the Urban Heat Island effect. Climate change may also be associated with elevated air pollution (such as by nitrogen dioxide) (ibid). The impacts of pollution are greater on urban environments where deprived people and ethnic minorities are overrepresented. In turn this can lead to increasingly negative health impacts on people in urban areas (such as increased vulnerability to respiratory infection and heightened cardiovascular mortality and morbidity). Urban areas that have higher levels of vegetation and trees experience lower temperatures, and there is evidence that urban vegetation can modify the risk of heat exposure (Murage et al., 2020). Relatedly, to combat this sort of problem, the government in the UK, at the time of writing, has announced plans to ensure that all citizens lived within 15 min' walk of green space or water (Briggs, 2023).

Continuing, Paavola (2017) argues that the risk of negative health impacts from flooding is unequal in the UK population. Coastal areas at an increased risk of flooding with 1.63 million people are currently at threat from a one metre rise in sea level (The Climate Coalition, 2021). Coastal areas have greater numbers of socioeconomically disadvantaged people and people living in affordable housing. Older people also disproportionately live in coastal areas and thus are at greater risk of the health impacts of flooding (ibid). The capacities of people in these areas to recover after flooding, in this regard, are likely to be lower than in more affluent areas because of low incomes or lack of insurance. Other groups, such as people with disabilities or chronic illnesses, as well as people dependent on public transport are also at heightened risk of negative health consequences from increased flooding. Moreover, the outcomes of emerging infectious disease are also shaped by socioeconomic status. Indeed, heightened risk from emerging infectious disease reflects age disparities and/or income which in disease-specific ways relate to heightened sensitivity to infection and/or lower capacity to adapt/protect oneself.

Extreme weather caused by climate change is likely to impact the functioning of the healthcare system in the UK. Impacts may occur through negative effects on infrastructure (including physical, institutional and social infrastructures) and can increase demand through the impacts of extreme weather on health (Curtis et al., 2017). The heightened risk of disruption to health and social care systems from extreme weather events resulting from climate change will be unequally experienced by socioeconomic status. For example, older people are more greatly exposed to negative health impacts because of reduced access to health and social care and the associated greater likelihood of experiencing pre-existing health conditions with some of these conditions, such as cardiovascular and respiratory disease, directly worsened by climate change. Finally, people living in rural areas are more likely to be impacted by disturbances to health and social care caused by and the associated health impacts of cold weather, whilst people living in urban areas are more likely to experience disturbance and the associated negative consequences from heat waves (Paavola, 2017).

3 The Co-Benefits of De-Carbonisation

In the final part of this chapter, it is interesting to consider the co-benefits for human health that emerge from climate change mitigation and the associated decarbonisation of the economy (see Jennings et al., 2020; Romanello et al., 2022; The Academy of Medical Sciences and The Royal Academy, 2021; The Climate Coalition, 2021). Most obviously, reducing harmful air pollutants generated by burning fossil fuels will save many thousands of life years with 40,000 premature deaths caused each year by air pollution. Additionally, increasing exercise levels in the everyday lives of the population could have the co-benefits of lessening greenhouse gas emissions related to travel in polluting vehicles whilst also increasing the activity level of a population in a country where sedentary lifestyles and physical inactivity are the fourth leading cause of death (e.g. through cycling or walking to work rather than driving). Alongside reductions in emissions, other health benefits related to lessening the use of motor vehicles and increasing walking and cycling include lower levels of injury from road traffic accidents (Woodcock et al., 2009) and decreased costs to the NHS relating to treatment of diabetes, dementia, ischaemic heart disease, breast cancer, depression, colorectal cancer and cerebrovascular disease (Jarrett et al., 2012). In a similar manner, increasing the amount of green space in urban areas can have physical and mental health benefits for people whilst also reducing extreme temperatures and decreasing greenhouse gas emissions (Jennings et al., 2020). Research also suggests that walkers and cyclists are exposed to lower levels of pollution than car drivers (Williams et al., 2018). Moreover, measures taken to make homes more energy efficient can make it easier for low-income households to heat their homes and provide sufficient food and resources for their families. This in turn can then improve household nutrition and lower health risks associated with living in a cold home (such as asthma and bronchitis in children) (Jennings et al., 2020). Finally, changes in diet reflecting decarbonisation, particularly limiting use of animal products, could also lead to beneficial health outcomes associated with less saturated fat consumption and reducing levels of obesity (Health Protection Agency, 2012; The Academy of Medical Sciences and The Royal Academy, 2021; see also Romanello et al., 2022). Research estimates suggest that if UK diets were improved to achieve World Health Organisation nutritional recommendations, there would be a 17% decline in greenhouse gas emissions and 7 million life years would be saved (Milner et al., 2015).

As it stands, the healthcare system is itself contributing considerably to greenhouse gas emissions and thus climate change. The health sector in the UK contributes over 5% within the picture of total national emissions. However, the National Health Service (NHS) committed in 2020 to becoming the world's first net zero healthcare system by 2040 (along the way the system wants to reduce greenhouse gas emissions by 80% by the year 2032) (see NHS England, 2020). This will include not only decarbonising its own buildings, services and vehicles but also decarbonising its supply chains[3] and the travel of patients and staff. Compared with 1990, the NHS in England has achieved a 26% reduction in emissions. In 2019, 62% of the emissions came from the supply chain, 24% from delivery of care directly, 10% from patient and staff travel to healthcare facilities and 4% from private services commissioned by the NHS (Tennison et al., 2021). The lengthy timeline for decarbonising the NHS and its supply chain might be questioned in a crisis context. There are also uncertainties about what a fully decarbonised psychiatry, oncology or general practice might look like (The Climate Coalition, 2021). However, the process of decarbonising the NHS and the broader healthcare sector will lower emissions with positive consequences for reducing the impacts of climate change on health.

[3] The pharmaceutical industry and the processes for developing and testing medicines to treat a range of diseases has had a considerable environmental impact (Bartolo et al., 2021). Decarbonising the development and supply of pharmaceuticals is necessarily part of attempts to decarbonise the NHS' supply chains.

4 Conclusion

This chapter has explored the evidence about the present and future impacts of climate change on the health of people living in the UK—including discussion of the co-benefit for human health through climate change mitigation. The growing risk of flooding, exposure to extreme heat amongst other problems resulting from climate change are a growing and grave threat to human health (both directly and indirectly) in the UK. This chapter has additionally discussed how the health risks from climate change are not equally shouldered by different population groups. Older people, the chronically ill, low-income households and people living in deprived urban and coastal communities are already beginning to experience the threat of the health impacts from climate change. This chapter has also discussed how there are a range of co-benefits for human health that are associated with decarbonisation and climate change mitigation action—including reducing death and diseases from pollution, physical inactivity, energy-inefficient homes and high saturated fat diets. The NHS is also working to decarbonise itself over the next two decades. However, the inadequacies in government climate change mitigation policy generally (both at home in the UK but also globally) suggest that we are only at the beginning of associated climate change and health catastrophes. The evidence presented in this chapter clearly shows that (further) death and disease will result from worsening, unabated climate change. At the time of writing, it remains to be seen whether governments can quicken the pace of their response to mitigate the very worst consequences.

References

Bartolo, N. S., Azzopardi, L. M., & Serracino-Inglott, A. (2021). Pharmaceuticals and the environment. *Early Human Development, 155*, 105218.

BBC Newsm. (2020). *South Yorkshire floods: Some victims still not home a year on.* https://www.bbc.co.uk/news/uk-england-south-yorkshire-54802049

Briggs, H. (2023). Everyone to live 15 minutes from green space or water in England under plans. BBC News. https://www.bbc.co.uk/news/science-environment-64456455

Calnan, M., & Douglass, T. (2022). *Power, policy and the pandemic: A sociological analysis of COVID-19 policy in England*. Bingley.

Carter, N. (2014). The politics of climate change in the UK. *Wiley Interdisciplinary Reviews: Climate Change, 5*(3), 423–433.

Curtis, S., Fair, A., Wistow, J., Val, D. V., & Oven, K. (2017). Impact of extreme weather events and climate change for health and social care systems. *Environmental Health, 16*(1), 23–32.

Hajat, S., Vardoulakis, S., Heaviside, C., & Eggen, B. (2014). Climate change effects on human health: Projections of temperature-related mortality for the UK during the 2020s, 2050s and 2080s. *Journal of Epidemiology and Community Health, 68*(7), 641–648.

Harvey, F. (2022). UK's first new coalmine for 30 years gets go-ahead in Cumbria. *The Guardian*. https://www.theguardian.com/environment/2022/dec/07/uk-first-new-coalmine-for-30-years-gets-go-ahead-in-cumbria

Health Protection Agency. (2008). *Health effects of climate change in the UK*. https://webarchive.nationalarchives.gov.uk/ukgwa/20130104181103/http://www.dh.gov.uk/en/Publicationsandstatistics/Publications/PublicationsPolicyAndGuidance/DH_080702

Health Protection Agency. (2012). *Health effects of climate change in the UK*. https://assets.publishing.service.gov.uk/government/uploads/system/uploads/attachment_data/file/371103/Health_Effects_of_Climate_Change_in_the_UK_2012_V13_with_cover_accessible.pdf

HM Government. (2022). *UK climate change risk assessment 2022*. https://assets.publishing.service.gov.uk/government/uploads/system/uploads/attachment_data/file/1047003/climate-change-risk-assessment-2022.pdf

IPCC. (2018). *Global warming of 1.5 °C*. https://www.ipcc.ch/sr15/

Jarrett, J., Woodcock, J., Griffiths, U. K., Chalabi, Z., Edwards, P., Roberts, I., & Haines, A. (2012). Effect of increasing active travel in urban England and Wales on costs to the National Health Service. *The Lancet, 379*(9832), 2198–2205.

Jennings, N., Fecht, D., & De Matteis, S. (2020). Mapping the co-benefits of climate change action to issues of public concern in the UK: A narrative review. *The Lancet Planetary Health, 4*(9), e424–e433.

Klenert, D., Funke, F., Mattauch, L., & O'Callaghan, B. (2020). Five lessons from COVID-19 for advancing climate change mitigation. *Environmental and Resource Economics, 76*(4), 751–778.

Linden, E. (2022). *Fire and flood: A people's history of climate change from 1979 to the present*. Allen Lane.

Milner, J., Green, R., Dangour, A. D., Haines, A., Chalabi, Z., Spadaro, J., Markandya, A., & Wilkinson, P. (2015). Health effects of adopting low greenhouse gas emission diets in the UK. *British Medical Journal Open, 5*(4), e007364.

Murage, P., Kovats, S., Sarran, C., Taylor, J., McInnes, R., & Hajat, S. (2020). What individual and neighbourhood-level factors increase the risk of heat-related mortality? A case-crossover study of over 185,000 deaths in London using high-resolution climate datasets. *Environment International, 134*, 105292.

NHS England. (2020). *Delivering a 'net zero' National Health Service*. https://www.england.nhs.uk/greenernhs/publication/delivering-a-net-zero-national-health-service/

Office for Health Improvement and Disparities. (2022). *Climate and health: Applying All Our Health*. https://www.gov.uk/government/publications/climate-change-applying-all-our-health/climate-and-health-applying-all-our-health

Oxfam. (2020). *Confronting carbon inequality*. https://oxfamilibrary.openrepository.com/bitstream/handle/10546/621052/mb-confronting-carbon-inequality-210920-en.pdf

Paavola, J. (2017). Health impacts of climate change and health and social inequalities in the UK. *Environmental Health, 1*(1), 61–68.

Public Health England. (2020). *Heatwave mortality monitoring report*. https://www.gov.uk/government/publications/phe-heatwave-mortality-monitoring/heatwave-mortality-monitoring-report-2020

Romanello, M., Di Napoli, C., Drummond, P., Green, C., Kennard, H., Lampard, P., Scamman, D., Arnell, N., Ayeb-Karlsson, S., Ford, L.B., Belesova, K., & Costello, A. (2022). The 2022 report of the Lancet Countdown on health and climate change: Health at the mercy of fossil fuels. *The Lancet, 400*(10363), 1619–1654.

Somerville, P. (2021). The continuing failure of UK climate change mitigation policy. *Critical Social Policy, 41*(4), 628–650.

Stylianou, N., Dale, B., Rivault, E., & Tauschinski, J. (2022). *Climate change: Summer 2022 smashed dozens of UK records*. BBC News. https://www.bbc.co.uk/news/science-environment-63244353

Tennison, I., Roschnik, S., Ashby, B., Boyd, R., Hamilton, I., Oreszczyn, T., Owen, A., Romanello, M., Ruyssevelt, P., Sherman, J. D., Smith, A. Z., & Eckelman, M. J. (2021). Health care's response to climate change: a carbon footprint assessment of the NHS in England. *The Lancet Planetary Health, 5*(2), e84–e92.

The Academy of Medical Sciences and The Royal Society. (2021). *A healthy future—tackling climate change mitigation and human health together*. https://acmedsci.ac.uk/file-download/94272758

The Climate Coalition. (2021). This report comes with a health warning. https://www.theclimatecoalition.org/health-report

UK Health Security Agency. (2022). *Heatwave plan for England: Protecting health and reducing harm from severe heat and heatwaves*. https://assets.publishing.service.gov.uk/government/uploads/system/uploads/attachment_data/file/1096593/heatwave-plan-for-England-2022-5-August-2022.pdf

Williams, M. L., Lott, M. C., Kitwiroon, N., Dajnak, D., Walton, H., Holland, M., Pye, S., Fecht, D., Toledano, M. B., & Beevers, S. D. (2018). The Lancet Countdown on health benefits from the UK Climate Change Act: a modelling study for Great Britain. *The Lancet Planetary Health, 2*(5), e202-e213.

Woodcock, J., Edwards, P., Tonne, C., Armstrong, B. G., Ashiru, O., Banister, D., Beevers, S., Chalabi, Z., Chowdhury, Z., Cohen, A., & Franco, O. H. (2009). Health and Climate Change 2 Public health benefits of strategies to reduce greenhouse-gas emissions: urban land transport. *Lancet, 374*(9705), 1930–1943.

Living with Climate Change in France: A Health Opportunity

Isabelle Roussel

Abstract Initially, climate control focused on energy and CO_2 not being among the most toxic pollutants, the link between climate and health was little highlighted despite all the French research on bioclimatology. After the deadly nature of the 2003 heat wave and the occurrence of numerous brutal, progressive or systemic climatic disasters, the French became aware that the predictions made by the models had become reality, but the issues raised by mitigation and adaptation go beyond the political field to question values such as justice, solidarity and sobriety. However, the insurance system used to repair disasters is at the end of its tether and only prevention, combined with mitigation are necessary for avoiding to fall into the trap of bad-adaptation: injustice, rebound effect. Only a very integrated territorial policy, following the "one health" concept can enable the French to improve their health by fighting against climate change.

Keyword Hot waves · "one health" · Bioclimatology · Environmental inequalities · France

If, during the summer of 2022, France experienced a succession of climatic disasters: heat waves, droughts, forest fires, this country took time to recognize the climatic threat and implement the necessary measures. The variability of the usual weather in a temperate climate has made it possible to neglect the results of climatologists' work who have estimated the warming since 1900 at more than 1.7 degrees (IPCC 2013).

But since the 2000s, a number of events such as the 2003 heat wave have accelerated awareness of the need to adapt, especially as the consequences of global warming are gradually threatening many sectors (health, biodiversity, tourism, agriculture, etc.). Thus, public opinion has evolved both under the influence of the occurrence of unusual extreme phenomena but also of that of scientific knowledge shared by French people who, in increasing numbers, fear for their health and that of the planet. Fear

I. Roussel (✉)
University of Lille, Lille, France
e-mail: appa.irou@gmail.com

R. Akhtar (ed.), *Climate Change and Human Health Scenarios*, Global Perspectives on Health Geography, https://doi.org/10.1007/978-3-031-38878-1_16

and anxiety generate a perceptible malaise, especially among young people. Given the rapidity of the worsening of the characteristics of the climate, the time has come to carry out transformations in the modes of functioning of society to cope with the magnitude of the catastrophic effects announced. But the new orientations to be implemented are based on the notions of justice, sobriety, and well-being which go beyond the legislative competence of the state. The health dimension of climate policies is particularly important through more integrated territorial policies.

1 The Mitigation and Application in France of International and European Treaties

1.1 It Was, Initially, Through International Commitments that France Took Climate Change into Account

The creation of the UNEP (1972) that of the IPCC or the Tokyo (1997) and Paris (2015) agreements represents considerable progress in the direction of an awareness of the planetarization of the issues.

It was the French state that was in charge of the negotiation[1] and then the signing of the international and European protocols. Moreover, in France, a centralized state, energy policy was the responsibility of the state. Initially, territories and individuals invested little in the implementation of these state commitments. For the elected officials, as for all the inhabitants, the question of energy did not arise, it was limited to the respect of contracts with "historical operators"; it was enough to turn a switch or fill the tank of the boiler or the car for a small price.

1.2 International Commitments Have Required the Adoption of a Legislative Framework

By signing the Kyoto Protocol in 1997, France undertook to reduce its emissions after adopting a national program to combat climate change (PNLCC) in 2000. The main objective of this was, according to a planning strategy quite characteristic of French policies, to establish assessments of carbon emissions in the various sectors of the economy.

In 2007, the "Grenelle de l'Environnement" concluded that more ambitious objectives were needed, which could be achieved more quickly.

[1] Because of the Peace of Westphalia, signed in 1648, imposes the concept of nation-state as the basis of international law.

The energy transition law of August 2015 aims to reduce final energy consumption by 50% in 2050 compared to the 2012 reference and to increase the share of renewable energies which should reach 23% of the gross final consumption of energy.

Years	Total	Total emissions in Mt eq CO_2 without taking agricultural data into account (LULUCF, land use, land use change, and forestry)
1990	547.18	
2000	553	
2010	512.75	
2020	400	
2021	418.2	

In 2017, a climate plan contracted with local authorities announced the objectives that France must set itself on the threshold of a new presidential five-year term to "accelerate the implementation of the Paris Agreement".

The implementation of an increasingly restrictive legislative framework results in a reduction in emissions (see table) during the decade 2010–2020.

The change in the distribution of GHG emissions by sector between 1990 and 2020 (Fig. 1) shows the relative decline in industrial emissions, which corresponds to the adoption of more energy-efficient processes but also to the de-industrialization of France. The French exception vis-à-vis nuclear energy explains the low share held by the energy sector.

The share of agriculture and transport tends to increase while that of emissions related to the building sector decreases slightly. To achieve the objectives envisaged, the reduction in emissions would have to be much faster. In particular, the agricultural and transport sectors are still too large consumers of energy despite the carbon sinks represented by soils and forests.

In this perspective, France, in conjunction with Europe (the green deal adopted in 2019 and the "Fit-for-55" project), has implemented a strategy to achieve "net zero carbon".

The low carbon strategy (SNBC) adopted in 2015 and revised in 2018–2019 and 2020

The SNBC is based on a scenario of achieving carbon neutrality by 2050 from a reference model covering the years 2015–2019.

This strategy is combined with the investments made to revive the economy after the pandemic. The objective is to limit the use of fossil fuels, the excessive dependence of which on the outside has been highlighted by the war in Ukraine.

1.3 French Advantages Become a Handicap

In this international context, France has not shown a pro-active attitude because it thought that it would benefit from the "advantage" brought by nuclear energy and

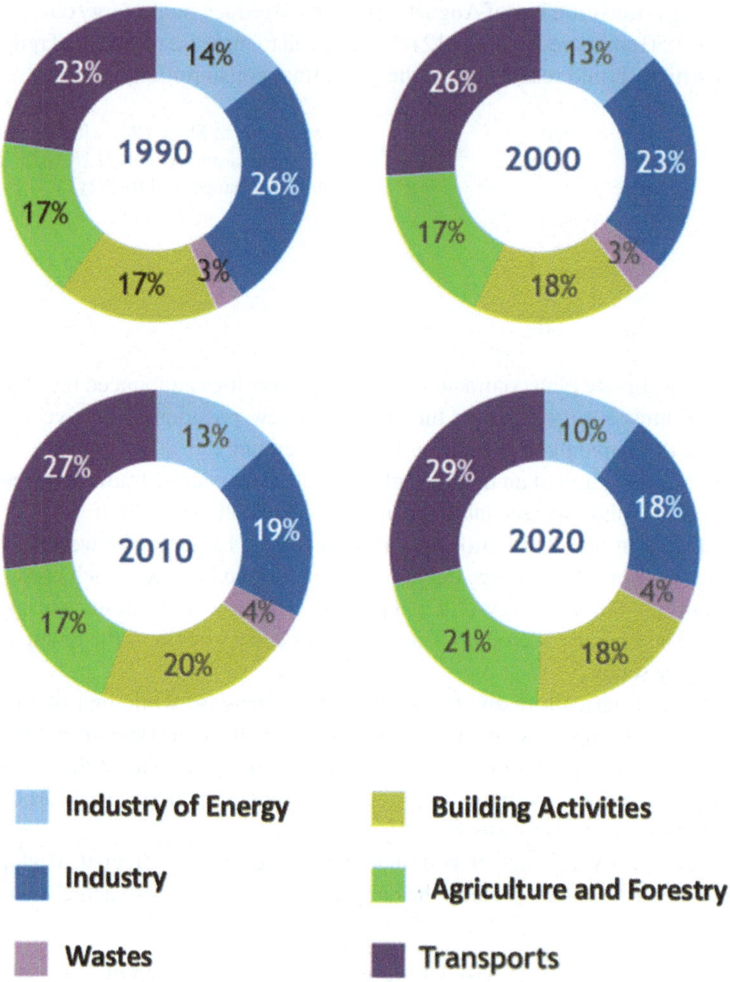

Fig. 1 Breakdown of GHG emissions in France by sector (Citepa)

that of a temperate climate. However, these assets which have above all had the consequence of delaying France's de-investment in fossil fuels. France is the European country that has invested the least in renewable energies. The renewed interest in aging nuclear power plants is recent, because the green parties are fundamentally opposed to nuclear power plants; however, the investments to be mobilized are very heavy to ensure the renewal of the fleet and its upgrading to standards according to a high level of safety. The power plants must undergo regular maintenance while some of them must solve corrosion problems.

But this delay is also linked to a top/down methodology which did not sufficiently involve the territories and the inhabitants.

The considerable investments to be implemented to achieve the objectives set must correspond to the co-benefits induced by a carbon-free economy, particularly in terms of health, since the combustion of gas and oil is largely responsible for polluting emissions and the 48,000 deaths per year attributed to air pollution in France. However, the health benefits and the quality of life observed in a decarbonized world are essential reasons for involving the population in a vast transformation of society.

2 The Excesses of the Climate Impose Themselves in Different Forms in France

Public awareness crystallizes under the effect of extreme weather events allowing the population to awaken to the climate cause through the logic of "I feel and I see" which has replaced that of "I know but it is far away in time and space".

2.1 Catastrophic Climatic Events: Raising Awareness

The awareness of the local impacts of global warming dates from the 2003 heat wave, the characteristics of which have already been described (Roussel, 2020). Other heat waves[2] have followed one another since 2003 (Fig. 2). Thanks to new techniques for calculating probabilities, Aurélien Ribes,[3] through studies carried out on the origin of these extreme events, can affirm that in the absence of greenhouse gas emissions linked to human activities, the heat wave that occurred in France in July 2019 would probably never have happened and that the thermometer recorded 2.1° more.

Heat waves do not benefit from the insurance status of natural disaster despite the negative effects they have on houses whose walls crack due to the desiccation of clays. On the other hand, they weigh heavily on mortality: between June 1 and August 22, 2022, an excess of 2800 deaths is attributed to the heat wave according to Public Health France.

Between 2001 and 2020, France and the overseas departments suffered 1,964 natural disasters, causing the death of 30,824 people and costing nearly 49 billion euros in damage.[4] About 49.7% of these events are caused by weather hazards that trigger storms, cyclones, bad weather, tornadoes, storms, hail, and snow. About 30.2% are climatic with forest fires, drought, heat waves, or cold spells. These climatic

[2] A heat wave is detected when a daily value of the thermal indicator at national level reaches or exceeds 25.3 °C and remains high for at least 3 days.

[3] https://www.climat-en-questions.fr/reponse/mecanismes-devolution/etudes-detection-attribution-par-aurelien-ribes.

[4] Assessment carried out by Ubyrisk Consultants from the "BD CATNAT" database.

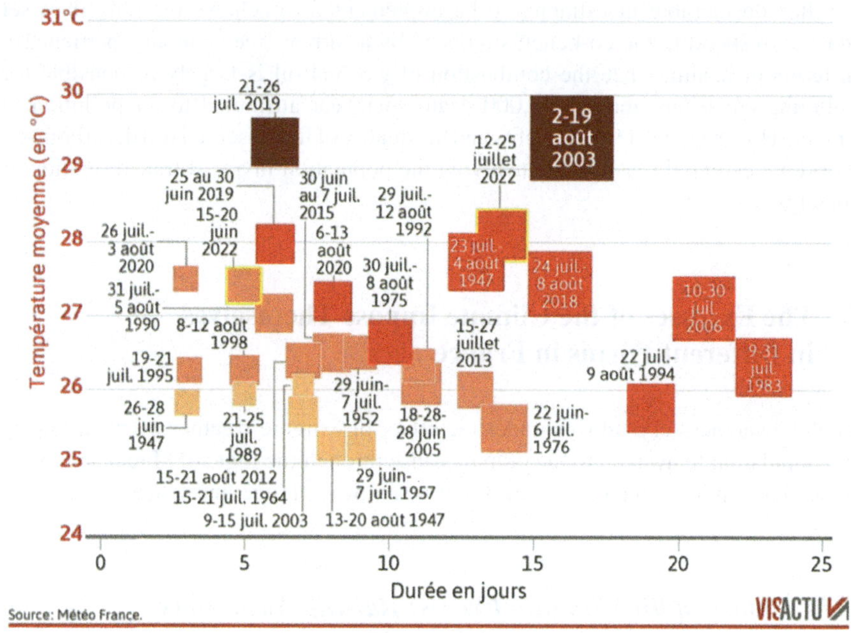

Fig. 2 Heat waves in France between 1947 and 2022. The size of the squares symbolizes the intensity of the phenomenon (Meteo France)

disasters cause a lot of damage; for example, the cost of the storm Xynthia which took place on February 28, 2010, is estimated at one billion euros.

These spectacular and catastrophic events should not mask the silent transformations caused by climate change.

2.2 *Silent Transformations*

Climate change does not only result in extreme events but also in a profound modification of the major biochemical cycles that ensure the maintenance of life on earth. The manifestations of these transformations carried out quietly for decades are visible today. The climatic profile of the French regions is changing with a rise in the Mediterranean climate towards the north of France (Météo France, 2022). In 2050, Paris could have the climate profile of Marseille or Seville with regular temperatures, in summer, of 40° in the shade, and Lyon, that of Madrid. The evolution of the date of the harvest which is carried out earlier and earlier or the melting of the glaciers is the result of an accumulated warming.

The fauna and flora follow these imperceptible modifications. Areas presenting risks for certain vector-borne diseases, such as malaria, Lyme disease, or dengue fever, tend to slide northward like the spread of the tiger mosquito (vector of dengue fever) which has been detected in 70 departments.

In addition to the decrease in the number of pollinators, the pollination calendar is modified. With global warming, flowering and pollination dates are getting earlier and earlier. The length of the pollen season is increasing as well as the quantity of pollen emitted by the most allergenic species. Pollen allergy or pollinosis concerns 20–30% of the French population.

The evolution of the coastline and the sea level is also part of the insidious effects of climate change which increases the vulnerability of coastal areas. Mediterranean coastal areas are experiencing sea-level rise which is currently accelerating (4.8 cm for 10 years).

2.3 *Everything is Linked: Systemic Risks*

Disasters are often the result of a combination of factors; for example, drought-related damage depends on temperature and rainfall, but also on cropping practices and soil water reserves. The effects of the climate cannot be interpreted outside of a systemic vision of the planet because "everything is linked". "The development of current research shows to what extent our planetary environment must be understood as a unique system, where the interactions between the atmosphere, the oceans, the glaciers and the sea ice, the oceanic or continental biosphere are permanent" Le Treut (1992). Thus, the megafires which developed thanks to the combination of heat and drought ravaged 72,000 hectares during the summer of 2022, six times more than the average of the last ten years. These fires release tons of CO_2 and dust into the atmosphere, which contribute to the deterioration of the climate.

Among these complex events that affect France, questions about the decrease in biodiversity incriminate climate change which, by introducing high variability in temperatures and precipitation, disturbs the balance of ecosystems by modifying the rhythm of the seasons and the growth of plants and trees; it therefore affects the reproductive cycle of many species. The degradation of biodiversity also affects human health, whose dependence on nature is increasingly well known. In addition, biodiversity is a source of medicine.

These findings illustrate the disruption of the earth/atmosphere system. The influences on the economy and society are evident as shown by the unease felt by the population.

2.4 *The Concern of the French*

The French are worried about their health and that of the planet because "everything is linked" and global warming has repercussions on areas far removed from the energy on which the first efforts were concentrated. As early as 2020, before the pandemic,

J. Fourquet, following an international survey, indicated that 65% of French people who are the biggest consumers of psychotropic drugs in the world have adhered to "collapsology".[5] This result means that the French consider the future of France to be bleak. This defeatism is explained by the feeling of downgrading of the country. On the other hand, paradoxically, they are rather satisfied with their particular situation. The green party, with its very dark and alarmist remarks, maintains this feeling, and therefore, the French do not consider it credible to act positively. Internal dissensions within the party mean that it is not considered a credible governing.

The Mental health and Climate Change Policy Brief states[6]: "However, climate change also exacerbates many social and environmental risk factors for mental health and psychosocial problems, and can lead to emotional distress, the development of new mental health conditions and a worsening situation for people already living with these conditions". Political inertia is of particular concern to younger generations who know they will be most affected by future disasters. The fear of the future on the part of young people who have followed higher education is reflected in the refusal to bring children into the world "no kids" (Depaulo,[7] 2920) and by their involvement in movements such as "extinction-rebellion" which engage in radical actions.

This existential concern also has economic foundations because sectors of activity as essential as agriculture and tourism are challenged by climate change. The extent of the transformations to be carried out in these two sectors of the economy is only the reflection of part of the upheavals induced by climate change.

According to an ADEME survey,[8] 58% of French people are aware that their way of life must change significantly and are ready to accept strong measures. They are more likely to accept this option than those who expect a solution from the international politics (17%) or the development of technical progress (13%). The response of resignation ("there is nothing to do") receives 11%.

3 Management to be Rethought

Preventing the effects of the climate presupposes profound individual and collective transformations at different scales. The interpenetration of the local and the global is particularly relevant for implementing a cross-cutting approach modeled on the "one health" concept.

[5] https://www.ifop.com/publication/la-france-patrie-de-la-collapsologie/.

[6] https://www.who.int/publications/i/item/9789240045125.

[7] https://www.psychologytoday.com/us/blog/living-single/202001/21-truths-about-people-who-dont-have-children.

[8] https://fr.calameo.com/read/00459949906e330410548.

3.1 *From Reparation to Prevention*

Given the increase in the number of climatic disasters and their scale,[9] the insurance coverage system, which is considered satisfactory in France, seems to be out of breath and raises the question of the use of royalties to invest in prevention.

One form of prevention involves "feedback" from extreme events to implement adaptation policies while knowing that the unimaginable is possible. The heat wave plan was developed based on feedback from the heat waves of 2003 in Paris and that of 1995 in Chicago, which was finely analyzed by Klinenberg (2017). The author shows how much the social fabric, the animation of the districts play a role as important as the characteristics of the housing to avoid the "heat strokes" of the isolated people. Indeed, beyond or outside a certain comfort zone,[10] excessive thermal differences in a dwelling have deleterious effects on the health of the inhabitants (Ezratty, 2009, 2015). Individuals, especially the most vulnerable, namely children and the elderly, spend more than 80% of their time in homes, especially on days with extreme temperatures, which are not favorable to outdoor activities. The insulation of houses, essential for energy savings, is also essential for health reasons provided that aeration and ventilation are respected to maintain the quality of indoor air.

3.2 *Adaptation Practices Are Necessary not Without a Certain Vigilance*

Adaptation aims to limit the risks in the face of climate change, while resilience is a broader concept that takes into account the ability of territories to face different risks.

Adaptation requires a lot of vigilance to avoid perverse effects called "bad-adaptation": the tools used to compensate for the inconveniences of the weather and increase immediate comfort can generate additional consumption costs (rebound effect) such as the use of air conditioning or induce pollution such as wood burning. Some "greenwashing" practices have the effect of encouraging consumption, such as the sale of air conditioners.

The adaptation of organisms: the physiological and ecological profile of human and non-human organisms is adapted to a climatic envelope inside a well-defined perimeter. Any displacement of this perimeter necessarily leads to a response from the organism or a geographical displacement such as the migrations that the vulnerability of the territories imposes. Sorre (1943) had insisted on the time necessary for adaptation by showing the difficulties encountered by the colonists for one or

[9] The average annual cost of climatic claims has risen from just over one billion euros in the 1980s to more than three billion over the past five years, according to the French Insurance Federation (FFA).

[10] According to the WHO guidelines of 1980, the temperature interval between 18.8° and 24.8° in dwellings is generally considered favorable to health.

two generations before adapting to the variations of the climate. According to him, peoples living in severe climatic conditions reveal, through their practices, even their physiology, a secular adaptation to climatic conditions, while tourists, fearing inadequacy, carry with them their way of living.

Does accelerating climate change leave enough time to adapt? Mitigation remains essential since adaptation can only be a slow process, sometimes generating adverse effects.

3.3 *Who Can Carry Out These Policies?*

The concept of adaptation was enshrined in "the Grenelle law".[11] Various national plans followed one another to lead to the revised PNACC2 in 2018, but adaptation depends heavily on local contexts for which the state is powerless. This is why it has been the subject of several complaints filed against its inaction and several challenges from young people or scientists. The citizens' convention, 150 citizens drawn by lot, had the mission of finding a consensus on which climate policies should be based. But the "climate and resilience" law, passed in 2021, which was fed by the results of this convention, is highly contested because it is considered too lax by some or too punitive by others. President Macron had the imprudence to declare that the recommendations given by the convention would be adopted "without filter"; however, in a democratic country, only the parliament is authorized to promulgate the laws.

The changes to be made are based on moral considerations. According to political scientist B. Badie[12]: "For the first time in history, the [climate] challenge is not a political challenge but it is a global and multi-sector challenge in which social issues are just as important. A government is not in a position to govern behavior".

3.4 *Injustice and Consumerism*

Climate change has the characteristic of generating injustice because it is the poorest populations, who emit less CO_2, who are likely to suffer the most not only from the consequences of disasters but also from inequitable adaptation policies.

Solidarities to be implemented are both territorial and social

The European mechanism for a just transition supposes the acceptance of differentiated financial efforts in favor of the most vulnerable territories or to help the necessary transformations of the agricultural world.

[11] http://www.developpement-durable.gouv.fr/-L-adaptation-au-changement-.html.

[12] France culture, l'esprit public le 14-11-2021.

The implementation of ZFE (low emission zones) is essential to transform the transport sector and reduce urban CO_2 emissions, but this system requires a certain territorial solidarity because it is the peri-urban inhabitants who suffer the greatest constraints.

The richest people, who represent 1% of the population, emit twice as much CO_2 and have more resources to adapt to climate change than the poorest half of the population. According to CREDOC,[13] the propensity to invest in an energy-efficient device increases significantly with income. The average and lowest incomes favor the solution of reducing their consumption and therefore their comfort and their health. The cost of the renovation work to be done is inaccessible for people with low incomes despite the aid in place. Thus, fuel poverty, which affected 3.5 million households in 2019, reinforces social inequalities in health.

This structural inequality serves to boost and justify growth through the myth of the link between abundance and happiness.

The injustice of consumerism

If emissions are falling (Fig. 1), consumption continues to increase and generates a flow of imported carbon that imperfectly reflects the carbon footprint.

The analysis of material flows in the Ile-de-France region for 2015 shows that the regional territory consumes 3.4 tons of fossil fuels and derived products (plastics, bituminous mixes) per inhabitant per year (Fig. 3), whereas that only 1.2 tons are of local origin. In other words, the "fossil footprint" of the region is almost tripled by integrating indirect consumption.

Reducing CO_2 emissions means reducing consumption.

End consumerism

We must not confuse the benefit maintained since the dawn of time by trade with consumerism which is an addiction destroying all the altruistic benefits of the exchange. Jean Baudrillard described, how the "society of objects" imposes, through advertising, the consumption of material goods to meet personal satisfactions (Baudrillard, 1986).

The ruse of consumerism consists in imposing the superfluous as necessary, thus encouraging waste.

The attraction of consumption goes so far as to monetize intangible goods, leisure but also nature and health. Biodiversity is valued by the "ecosystem services" it can provide. Commodification has also affected health, as the economist Eloi Laurent (2020) observes: "The so-called cost–benefit analysis techniques applied to human life, at the heart of this approach, induce a monetization of the living (human or non-human life) which by construction produces considerable ethical damage".

Giving up the addiction to "always more" is a personal choice that can be experienced as a real liberation and a return to more humanistic values which, most often, are accompanied by good health.

---

[13] Source CRÉDOC (2010), "French living conditions and aspirations" survey, January, in CGDD (2010), Barometer of opinion on energy and climate in 2010, October.

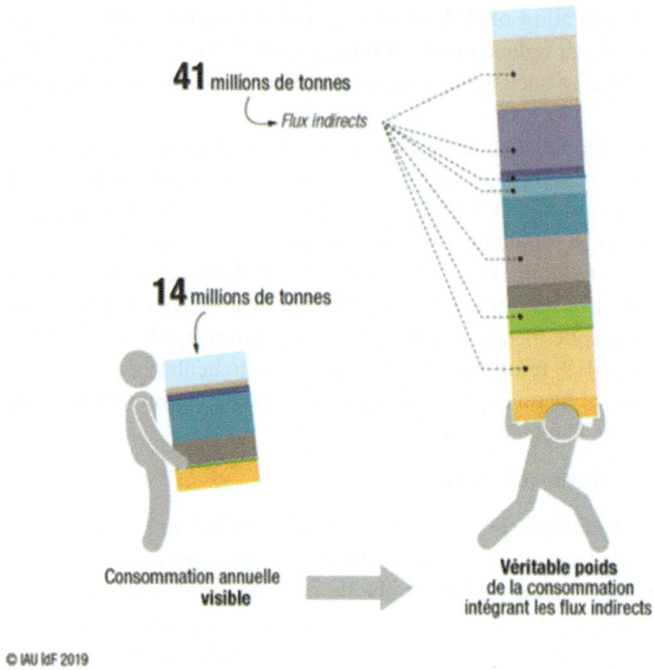

Fig. 3 Direct and indirect consumptions of fossil fuels in Ile-de-France in 2019. *Source* IAU IDF

3.5 The Importance of Health and Well-Being Considerations

If climate change had negative effects on health: emerging diseases, heat waves, it can also become an opportunity to build a more peaceful low-carbon society.

Controlling CO_2 can have health benefits. The link between the qualitative requirements of consumers and the evolution of agriculture toward practices using fewer inputs goes in the direction of improving the diet and health of the French. Reducing meat consumption makes it possible to limit pollution linked to livestock farming and promotes the prevention of many pathologies.

The link between atmospheric pollutants and the climate has already been mentioned (Roussel, 2020). The drastic reduction in automotive emissions during the first period of lockdown (BEH 13, 2021) prevented approximately 2300 Particle Matter (PM)-related deaths and nearly 1200 NO_2-related deaths. By casting opprobrium on fossil fuels, the fight against climate change reinforces the urgency of controlling air quality, the deleterious consequences of which we discover every day.

The development of cycling and active modes of transport by increasing physical effort is an interesting means of prevention.

It is at the territorial level that the integration of the various health, economic, and environmental policies is most fruitful, as shown by the implementation of the "one health" concept.

4 Integrated Territorial Policies

The climate policy was rolled out at the territorial level in 2004[14] and then in 2010 with the creation of PCAETs (territorial climate/air/energy plans). Nevertheless, they remain on the sidelines of the health policies set out at the regional level by the regional environmental health plan. It is the development of these documents which, in the various French regions, concentrate the reflections and recommendations on the climate which still too often take place in "silos". Only the most autonomous cities have the possibility of developing more integrated air–climate–energy–health policies while avoiding the perverse effects of injustice.

It is the increase in the resilience of the territories that explains why the number of victims linked to extreme phenomena has decreased significantly.

Max Sorre in 1933 had taken up the ideas of neo-Hippocratism to show the importance of adaptation to the climate in the development of societies. Foreshadowing medical ecology and the discovery of zoonoses, he had broadened the notion of adaptation to the whole environment by inventing the notion of "pathogenic complex", making it possible to better target the conditions of emergence of infectious elements within interactions between man and his environment. It thus foreshadowed the "one health" concept, whose systemic thinking is increasingly needed to shed light on territorial management. This concept was institutionalized in 2011.[15] It is based on a plurality of scientific disciplines to integrate all the dimensions of the problem raised by the risk of human infectious diseases of animal origin (zoonoses). The study of the behavior of animal reservoirs and vectors of transmission is essential because viruses and bacteria are part of the living world that we must learn to know better, as the current discoveries on the microbiota indicates.

Knowledge of ecosystems is also essential to optimize the development of the only habitable planet. However, by using the techniques available in town planning and agriculture to overcome the climate and its vagaries (Berdoulay and Soubeyran, 2020), the technocratic paradigm developed since the nineteenth century has cut humanity off from contact with nature and with the reality of the weather, a notion that is nevertheless essential for adapting. According to Anna Maria Lammel, the holistic thinking developed by certain peoples is better able to grasp the complexity of the earth/atmosphere system than a more analytical way of thinking, more conceptually deprived and therefore more worried (Kozakai and Lammel, 2005).

Beyond this "fight against or live with" alternative, there is emerging in today's society an ancestral need to maintain links with nature and its resources in a less utilitarian mode than that of "ecosystem services".

[14] Law 2010–788 of July 12, 2010 on the national commitment to the environment.

[15] It is supported by the World Health Organization (WHO), the World Organization for Animal Health (OIE), and the United Nations Food and Agriculture Agency (FAO).

5 Conclusion

The technocratic management of the climate crisis through the prism of energy is over. The predictions made by the models have become reality and the issues raised by mitigation and adaptation go far beyond the political field to question values such as justice, solidarity, and sobriety. The only possible way out is to mobilize the creativity of all of civil society and territories to transform housing, modes of travel, and production in such a way as to respect the recommendations of the IPCC (2021) and legislative commitments. The challenge of climate management depends on the ability to build bridges between the different actors to control any adverse effects and generate a peaceful society less dependent on fossil fuels. It is a question of integrating the French in a pro-active approach in which they find an immediate benefit in terms of health and quality of life.

References

Baudrillard, J. (1986). *La société de consommation* (320 p.). Galimard.

BEH 13, 2021, impact of air pollution on mortality in metropolitan France: reduction related to the spring 2020 lockdown and long-term impact for 2016 (pp. 232–242) (2019). http://beh.santepubliquefrance.fr/beh/2021/13/pdf/2021_13_2.pdf

Berdoulay, V., Soubeyran, O. (2020). *L'aménagement face au changement climatique: le défi de l'adaptation*, UGA ed. (242 p.).

Ezratty, V. (2009). Liens entre l'efficacité énergétique du logement et la santé des résidents: Résultats de l'étude européenne LARES. *Environnement, Risques & Santé, 8*, 497–506. https://doi.org/10.1684/ers.2009.0303

Ezratty, V. (2015). Thermal discomfort in housing—a threat to health (part 1). *Environ Risque Santé, 14*, 215–220. https://doi.org/10.1684/ers.2015.0784

GIEC—IPCC. (2013). *Changement climatique: les éléments scientifiques*, 5e rapport d'évaluation du GIEC. https://www.ipcc.ch/report/ar5/wg1/

GIEC—IPCC. (2021). *Climate change: The physical science basis. Contribution of working group I to the sixth assessment report of the intergovernmental panel on climate change* (2391 pages). Cambridge University Press.

Klinenberg, E. (2017). *Canicule, Chicago, été 1995 : autopsie sociale d'une catastrophe*, École urbaine de Lyon (415 p.).

Kozakai, T., & Lammel, A. (2005). Mécanisme de défense collectif et acculturation volontaire: l'exemple de l'occidentalisation du Japon. *Pratiques Psychologiques, 11*, 60–83.

Laurent, E. (2020). *Et si la santé guidait le monde* ? Les liens qui libèrent (192 p.).

Le Treut, H. (1992). Que nous apprennent les modèles du climat? La Recherche n°243, mai 1992.

Météo France. (2022). *Changement climatique: à quoi ressembleront les hivers parisiens?* https://meteofrance.com/actualites-et-dossiers/actualites/climat/changement-climatique-quoi-ressembleront-les-hivers-parisiens

Roussel, I. (2020). In: Akhtar, R. (Ed.) *Extrem weather events and human health* (pp. 59–78). Springer.

Sorre, M. (1933). Complexes pathogènes et géographie médicale. *Annales De Géographie, 42*, 1–18.

Sorre, M. (1943). *Les fondements biologiques de la géographie humaine* (440p.). Armand Colin.

Impact of Climate Change and Human Health in Spain. The First Approach to the State of the Art

José María Senciales-González, Lucía Echevarría-Lucas, and Jesús Rodrigo-Comino

Abstract Spain is reaching the forecasts set by the Intergovernmental Panel on Climate Change (IPCC) since 1990–1992. To get a consensus and reach a minimum governmental awareness of the problem, numerous global meetings were necessary in Spain, like in other countries. However, it was clear that there is a need to transfer this reality to society clearly, concisely and forcefully, influencing changes in social norms, political priorities and cultural values. The scientific literature agrees that the most important climate change events affecting human health are: high temperatures, heat waves and ultraviolet radiation, as well as air, soil and water pollution. In addition, torrential rains, droughts, forest fires, diminishing water resources, coastal phenomena and endangered habitats could be also included. The aim of this chapter is to present the state of the art on the effects of climate change on health in Spain. So, methodologically, diseases exacerbated by climate change detected in Spain were organized according to medical specialities and climatic elements, analysing morbidity and mortality. Spain increased its population from 2000 to 2020 by 16.6% and stabilised its mortality at 9.01‰ (omitting Covid-19). Other reasons aside, increases in morbidity or mortality above these demographic values may be due to the effects of climate change. Thus, the data consulted indicate that 26.7% of mortality is due to cancers, which increase in women (26% between 2000 and 2020) and stabilise in men. This is followed by heart disease (18.8%), which has fallen since 2000; digestive diseases (11.8%), which have increased by 20.3%; and

Electronic Supplementary Material The online version of this chapter (https://doi.org/10.1007/978-3-031-38878-1_17) contains supplementary material, which is available to authorized users.

J. M. Senciales-González
Department of Geography, University of Malaga, 29010 Málaga, Spain

L. Echevarría-Lucas
Ophthalmology Service of Axarquía Hospital, 29700 Vélez-Málaga, Spain

J. Rodrigo-Comino (✉)
Department of Regional Geographical Analysis and Physical Geography, Faculty of Philosophy and Letters, Campus of Cartuja, University of Granada, 18071 Granada, Spain
e-mail: geo.jrc@gmail.com

R. Akhtar (ed.), *Climate Change and Human Health Scenarios*, Global Perspectives on Health Geography, https://doi.org/10.1007/978-3-031-38878-1_17

respiratory and neurological diseases (13.1% and 12%, respectively), which have stabilised since the effect of Covid-19 has been cancelled out; this zoonotic disease.

Keywords Human health · Climate change · Spain · Control measures · Regional studies

1 Introduction

The outlook for climate change presented by the consecutive IPCC reports from 1992 (IPCC, 1992) to 2022 (Pörtner et al., 2022) shows evidences that in Spain, the forecasts of this panel are being reached. There are ample references to this from various evolutionary analyses focused both on climatological variables published by Spanish national agencies and collectives (e.g. AEMETblog, 2019) and also with respect to the measurement of greenhouse gas emissions (e.g. CC.OO., 2019). Without wishing to repeat what has already been pointed out by various institutions and authors (Amblar et al., 2017; Galacho et al., 2021; Mestre et al., 2019; Rodrigo & Senciales, 2021; Sanz & Galán, 2020), it is essential to make the Spanish society aware of the close relationship between the alteration of climate variables and diseases, both infectious and many other types. Thus, it is necessary to draw up key analysis, in which through regionalized medical and climatological tests, the severity of the problem is revealed.

Special emphasis should also be placed on the need for human health experts and climatologists to work together in order to find the risk attributable to each variable. In fact, it is common for the different specialists to work separately, with no mutual information other than that provided by published scientific papers. However, it is worthy to highlight that joint analysis of data from both sources (medical and climatological ones) could lead to significant improvements and depths of results (Echevarría et al., 2021).

Numerous world meetings at the highest level from the first Stockholm Summit in 1972 (United Nations, 1972) to the last one again in Stockholm (United Nations, 2022) were celebrated in order to reach a "minimum awareness" of the problem, at least at governmental level. In addition, together with these conferences, normally spaced in time, have been added the different annual Conferences of the Parties (COP), which have been organized every year since 1995 (COP1, Berlin, 1995), the first of the meetings in which climate change is addressed as a real problem (United Nations, 1995).

Thus, a "minimum awareness" needs to be transferred to the Spanish society as a whole through a clear, concise, but forceful, presentation of the reality of this issue (human health and climate change), which will ultimately influence changes in social norms, political priorities and cultural values (González Gaudiano, 2012). This transfer of knowledge is urgent, given the proliferation of social movements calling for change (e.g. Friday for Future, Greenpeace, WWF), but also political and social movements, as well as public figures who deny the problem in their proclamations and electoral programmes (Montañez, 2021).

However, our purpose is not to praise some and vilify others, but rather to examine what is being carried out in a wrong way by scientists in: (i) presenting climate change as a distant reality both in time and complex effects (Kindelán, 2013; Meira, 2009) and (ii) providing clear information on the consequences of this process in the immediate reality of the country of the authors of this chapter (Spain), especially with regard to the most important thing we have: human health.

The authors of this chapter have already emphasized in a previous research (Echevarría et al., 2021) the need to consider "disease sinks", defined as any climate change mitigation action that reduces the incidence or morbidity of a disease. Therefore, a first step in engaging the public is not so much to warn of the evils of climate change, but about the benefits we can obtain from combating it by creating disease sinks. In fact, the apocalypse narratives do not help to fight climate change; on the contrary, it makes it more difficult to combat the problem (Fagan, 2017).

In summary, the objectives of this book chapter are to: (i) show the trend of climate variables in Spain over the last 30 years, based on the data available through the scientific literature; (ii) show the trend of climate change-related diseases in Spain in recent times (2000–2020) and (iii) present summary panels of causes and effects, as well as protective measures, where possible (Tables 1, 2 and 3, as well as Supplementary Material).

2 The State of the Art in Spain

Before addressing the effect of climate change on human health in Spain, addressing the evolution of environmental variables in recent years, regionalising where necessary, given the different climates in our country is necessary (Table 1). Multiple factors through which climate change affects health can be found, both in terms of morbidity and mortality and, thus, in terms of life quality (for more information, see Supplementary Material). It affects both directly (e.g. heat waves, floods, droughts, etc.) and indirectly (e.g. pollution, allergens, food security, etc.) (Díaz et al., 2020).

The World Health Organization (WHO, 2018) estimates that climate change could cause an additional 250,000 deaths per year between 2030 and 2050 due to changes in pathology patterns, especially, for vector-borne diseases (i.e. dengue or malaria), which are very sensitive to changes in temperature and precipitation. Table 2 shows a breakdown of the diseases and disorders most likely to increase due to climate change, according to leading scientific journals in each medical specialty. This also shows the climate change-related factors (grouping some of them together), which were cross-referenced with the various medical specialties that address pathologies for which the scientific literature has found some evidence of a cause–effect relationships. Thus, consultation of leading scientific journals in each medical specialty for terms such as "climate change", "global warming", "environmental factors" or, more specifically, for each environmental factor known to be exacerbated or modified by climate change, has led to the production of Table 2. These environmental factors, in turn, have been consulted in global or national climate change reports. So,

Table 1 Trends of environmental factors in Southern Spain

Environmental factors	Trend	Additional details
Temperatures	A trend towards a sustained increase in mean, minimum and maximum annual values has been identified, at least since 1980 especially in the interior and east of the Iberian peninsula (MITECO, 2022). Although some stations studied show increases of 0.6 °C in 35 years (1986–2021) (datosmundial.com), others, such as Malaga airport (southern Spain), have shown variations of 0.38 °C/decade from 1972 (1.34 °C in 35 years) (modified from Echevarría et al., 2021)	Although there is an increase in hot nights in Andalusia (Echevarría et al., 2021), this is not generalizable to the rest of Spain
Heat waves	Heat waves are increasing, especially in Levante, Balearic and Canary Islands (Amblar et al., 2017). AEMET studies indicate that although the number of heatwave episodes remains stable, the maximum duration and number of days of each heatwave have increased considerably since 1975, having been reduced in the Canary Islands (AEMET, 2022)	Heat waves cause thousands of deaths per year. In example in Spain alone, there were 4700 deaths in 2022 due to heat waves (EL MUNDO, 2022; RTVE, 2022a)
Humidity	While reductions of 5% in relative humidity are expected in Spain (MITECO, 2022), reductions of 0.85% per decade have been detected in some places (Malaga Airport) in the period 1968–2021 (Echevarría et al., 2021)	There are no evolutionary data on humidity along the Spanish national territory
Floods	There is no clear trend in recent times, nor can it be generalized to the whole Spanish national territory. In many cases, they depend more on human exposure in inappropriate places than on an increasing trend. The upward trend is particularly noticeable in northern Spain, in common with the rest of Europe (IEEE, 2021)	The number of deaths due to the effects of floods in Spain since the beginning of the XXI century amounts to 300 people (Gregori, 2019)

(continued)

Table 1 (continued)

Environmental factors	Trend	Additional details
Droughts	A higher frequency of droughts has been documented from 1961 to 2000 and is expected to be generalized for the whole peninsula, although less clear or fainter in the Levante, Canary Islands and Cantabrian coast (CEDEX, 2017)	Droughts tend to be frequent throughout the Spanish Mediterranean, although they are particularly severe on the Atlantic, where they are less expected. On the other hand, irrigated crops are particularly extensive in the Mediterranean area, and meteorological drought frequently turns not only into hydrological drought but also into agricultural drought (MITECO, 2022)
Forest burnings	There is increased vulnerability due to the longer dry season (Amblar et al., 2017) and higher temperatures, evaporation and, therefore, higher flammability; this is evidenced by the increasing frequency of fires in Spain (Aymerich et al., 2021). Between 2001 and 2021, the communities with the largest burned areas were Galicia, Castilla-León and Andalusia, with no clear trend of increase at national level, but with a clear trend of increase in severity (RTVE, 2022b)	Plant growth, especially shrub growth, has increased due to higher CO_2 concentrations. In turn, the increased dryness increases the risk of fires. Forest fires are associated with higher PM_{10} concentrations, responsible for premature deaths (Johnston et al., 2012; Littell et al., 2016; Manojkumar & Srimuruganandam, 2019; Moya et al., 2017)
Electric Storms	In the period 1995–2015, 52 people were killed by lightning in Spain; however, since 2000, the number of lightning strikes in Spain has shown a clear downward trend (AEMET, 2019)	In the period 1995–2015, 52 people were killed by lightning in Spain
Dust storms	Although there is no clear trend in dust storms, in the south of the peninsula and the Levant, dust storms occur more than 20 times a year, especially in summer (Alonso y Vázquez, 2014)	Dust storms are related to droughts and high temperatures. They are common during low pressures episodes in North Africa

(continued)

Table 1 (continued)

Environmental factors	Trend	Additional details
Pouring rain and associated erosive processes	Except for the Canary Islands, Asturias and Cantabria, the rest of the country shows a trend towards an increase in heavy rainfall since 1970 (Amblar et al., 2017). The comparison between the number of torrential events from 1965 to 1991 and those from 1992 to 2020 clearly shows an increasing trend throughout Spain (MITECO, 2022)	It is common for high temperatures to be associated with increased torrential events and, with it, increased flooding and erosion (Sillero et al., 2020)
Pressure	There are no data on air pressure evolution at the national level. However, the higher frequency of droughts is usually associated with higher atmospheric pressure levels at ground level. This is in line with the expansion of the Azores Anticyclone detected during the twentieth century (Cresswell-Clay et al., 2022)	Low pressures at ground level, as well as very low pressures at 500 HPa (about 550 m), independently of the ground level pressure, are related to the highest rainfall events on the peninsula. Summer low pressures in North Africa determine heat and dust waves over the Iberian Peninsula and the Canary Islands
Strong winds; hurricanes	Winds linked to cyclones and explosive cyclogenesis in Spain show increasing trends since 1971. The projections for strong winds point to an increasing trend at national level, especially in autumn. Both have greater effects in the east and north of the peninsula (Herrera, 2019)	Strong winds are associated with deep low pressure systems and their effects are particularly noticeable along the coasts, destroying infrastructure and flooding coastal plains
Radiation (UV-UVB)	The higher frequency of droughts increases the intensity of visible, IR and UVR (UVA, UVB and UVC) radiation, highly harmful to human health (Byrne & O'Gorman, 2016) (*miteco.gob.es*, s. f.) (Gibson, 2020). However, from the study of the AEMET climate records, available since 2011 with data from the main cities of Spain, no increasing or decreasing evolution can be seen regarding total solar or ultraviolet radiation (AEMET.es)	A long absence of rainfall is usually coupled with a long absence of clouds (especially, vertically developing clouds, which absorb a large amount of ultraviolet radiation—UVR), less humidity, higher temperatures and solar exposure

(continued)

Table 1 (continued)

Environmental factors	Trend	Additional details
Anthropogenic pollutants	The pollution control trend is clear in Spain, with a generalized decrease in emissions (MITECO, 2022). In the case of $PM_{2.5}$, all Spanish provinces have decreased their concentrations, exception for Castellón and Madrid; only Madrid exceeds the limits for NO_2 and Avilés (Asturias) for PM_{10}. However, ground-level O_3 exceeds the limits in most of the peninsula, except for the north, with single exceedance thresholds in all regions, except Cantabria. There is no clear trend, or no data, for the remaining pollutants	All urban pollutants have some toxicity that results in harm to human health; they are especially dangerous during highly stable weather (Echevarría et al., 2021)
Other measured events for which health effects are not directly assessable or data on their evolution are not available		
Hydric resources	Water resources are in clear decline in the southern Spain (especially during the summer), with a surplus in the northern only during the winter (Amblar et al., 2017). This decline is both quantitative and qualitative, due to the combination of droughts and torrential rains mentioned above. There is a close relationship between higher frequency of droughts and lower availability of water resources; hence, their trends are similar	The decline of water resources causes damage to drinking water supply infrastructures, implying that local administrations must be permanently alert to avoid infectious diseases or others related to poor quality due to saline intrusions in aquifers (García Aróstegui et al., 2003; Girardi et al., 2016; McMichael & WHO, 2003). Diseases that were thought to have been eradicated are reappearing due to high water temperatures, facilitating microbial proliferation (Chen et al., 2012; López Alonso et al., 2016; Rose et al., 2000)

(continued)

Table 1 (continued)

Environmental factors	Trend	Additional details
Coastal losses and coastal phenomena	The average trend of mean sea level rise in Spain has been estimated at 2.5 mm/year; the trend to suffer storms has increased along the entire northern Atlantic coast of Spain (including Galicia), the north of the Canary Islands and, to a lesser extent, in the northwest of Balearic Islands, the coast from Barcelona to Valencia, and the Alboran Sea (Medina et al., 2004)	The sea level rise implies a generalized retreat of the coastline, while, paradoxically, the voracity of real estate in neighbouring areas continues unabated, exposing the population to economic and health risks (Hallegatte et al., 2013; Jiang et al., 2015; McInnes et al., 2011; Méndez, 2011; Puertos del Estado, 2022)
Endangered species and habitats	Pollinator species are in clear decline due to intensive agricultural management, the use of pesticides, environmental pollution and invasive alien species, among other causes (Capdevila et al., 2011; Ceballos et al., 2015; Karp et al., 2013). There are no detailed studies about the evolution of endangered species and their territorial distribution	About 2.4% of the wild species present in Spain are included in some category of threat according to the IUCN criteria, increasing the number of endangered species and reducing in biodiversity. However, 36.2% of the terrestrial and 12.3% of the marine territory are protected (Aymerich et al., 2021). In addition, 50% of European animal species live in Spain, 5% of the world's species (National Geographic, 2020). There is an increased frequency of movements to new habitats, favouring zoonosis of external origin. As is known, this is one of the hypotheses for the emergence of the Covid19 pandemic (El Zowalaty & Järhult, 2020; Shaw, 2018)

Modified from Echevarría et al. (2021)

Table 2 shows in the columns on the left side the related natural environmental factors, while on the right side, it shows the anthropogenic pollutants, defined as elements or compounds which proceed directly from human activities; they are potentially harmful for human beings, animals, plants and ecosystems; in addition, they can accumulate in water, soil or the atmosphere, with different decomposition times.

In view of these relationships, we summarize the state of the question in Spain, expressing the impact factors, the prevalence and/or incidence and evolution of the diseases (where information is available) and its mortality (if applicable) linked to each speciality, as can be noted in Figs. 1, 2 and 3. Figure 1 shows the percentage evolution of mortality in each medical specialty, according to data from the Spanish Ministry of Health, based on the database of mortality by disease. Thus, based on the year 2000, the evolution of mortality by specialty is compared in Fig. 1 with the evolution of the Spain's population and global mortality. Figure 2 compares the incidence of reported mortality in each medical specialty since 2000. And finally,

Table 2 Cross-reference of variables related to climate change and medical specialties which have to deal with pathologies related to these variables

Medical specialties	Climate change-related variables																							
	Natural environmental variables													Anthropogenic pollutants										
	a	b	c	d	e	f	g	h	i	j	k	UV	UVB	Volatile compounds						Organic compounds and volatile organic compounds		Inorg. comp. and non-volatile	
														CO–CO_2	NOx	SO_2	O_3	PM_{10}	$PM_{2,5}$	Organic compounds	α-amylase	EDC	Metals
1	×													×						×			
2	×													×					×		×		×
3	×	×	×	×															×		×		
4	×	×		×	×						×			×	×	×	×	×	×				
5	×							×				×	×	×									
6						×													×	×		×	×
7	×						×		×			×					×	×					
8			×		×									×					×				
9	×	×			×			×	×					×	×		×	×		×		×	×
10	×	×																×	×				
11	×														×	×	×	×	×	×		×	×
12	×			×	×				×					×									
13	×	×	×					×	×	×	×	×	×	×				×	×				
14								×				×		×	×			×	×	×		×	×
15						×				×				×	×	×	×	×	×	×			
16												×		×				×	×		×		×

(continued)

Table 2 (continued)

Medical specialties	Climate change-related variables																						
	Natural environmental variables													Anthropogenic pollutants									
	a	b	c	d	e	f	g	h	i	j	k	UV	UVB	Volatile compounds						Organic compounds and volatile organic compounds		Inorg. comp. and non-volatile	
														CO–CO_2	NOx	SO_2	O_3	PM_{10}	$PM_{2.5}$	Organic compounds	α-amylase	EDC	Metals
17	×			×	×	×					×	×											
18	×			×	×						×												
19	×	×		×	×	×		×							×	×	×	×	×				
20	×					×								×	×			×	×				
21	×	×										×	×	×			×	×	×	×	×		×
22	×	×		×	×			×															

1. Allergology; 2. Andrology; 3. Digestive system; 4. Cardiology; 5. Dermatology; 6. Endocrinology/nutrition; 7. Epidemiology; 8. Haematology; 9. Immunology; 10. Nephrology; 11. Neurology; 12. Obstetrics/gynaecology; 13. Ophthalmology; 14. Oncology; 15. Otorhinolaryngology; 16. Paediatrics; 17. Clinical psychology; 18. Psychiatry; 19. Respiratory/pneumology; 20. Rheumatology. 21. Occupational medicine. 22. Emergency medicine

a = Δ Temperatures; b = Heat waves; c = Δ Humidity; d = Flooding; e = Droughts; f = Forest burnings; g = Electric storms; h = Dust storms; i = Pouring rain; j = (High) Pressure; k = Strong winds; hurricanes. EDC: Endocrine disruptor compounds; Metals include arsenic, cadmium, silica, aluminium, beryllium, zirconium, mercury, manganese, copper, lead

Own elaboration

Table 3 Preventive/corrective/palliative measure that can be taken as disease sinks

Global/ regional measures	GHG reduction Increase research on the relationship between climate change and disease, especially in children, but also in lung diseases, immunology, rheumatic or any other disease of non-filial etiology Increase research on Endocrine Disrupting Chemicals (EDCs) Adaptation of the health national systems to a healthier environment International solidarity, with increased investments in adaptation and resilience Adequate territorial management to avoid the effects of flooding and alterations to wildlife, which can cause zoonosis Exhaustive control of biocides, cosmetics and food additives Correct labelling of EDC products Conservation and recovery of ecosystems Systematic vaccinations for preventable pathologies
Local measures	Control of pollution, improving air quality Real-time alerts on pollution peaks (high stability), dust storms, heat waves Control of the food chain to avoid contamination by mycotoxins, bacteria, microalgae and protozoa Treatment suitable for the purification of wastewater, bathing and drinking water Reduce the carbon footprint with the "4R": reduce (waste), reuse, recycle, recover Adequate territorial management to avoid the effects of flooding and alterations to wildlife, which can cause zoonosis Reduction of urban albedo (e.g. increasing green spaces and/or changing built-up areas) Training in occupational risk prevention related to climate and environment Information to the population to achieve a psychosocial adaptation Conservation and recovery of ecosystems
Individual measures	Avoid exposure to environmental pollutants Self-protection against high temperatures, heat waves and cold waves Frequent hydration Self-protection against UVR and light intensity (sunscreen, hat, sunglasses), especially in child population, and reduce the unprotected photo-exposure time Acquisition of healthy habits (i.e. valuing the benefits of Mediterranean diet, more physical activity, non-toxic habits, hygiene) Reduce the carbon footprint with the "4R": reduce (waste), reuse, recycle, recover Self-protection with particle filters in the home Avoid stigmatizing the population requiring specialized help (i.e. psychiatric) Self-protection against biomass and hydrocarbon cookers, which increase exposure to $PM_{2.5}$

Fig. 3 compares the demand for medical consultations by specialty since 2000, which would be equivalent to morbidity; in this figure, data are expressed in logarithmic scale for comparability. In addition, the possible preventive, palliative or curative solutions will be highlighted (Table 3). It is necessary to clarify beforehand that mortality in Spain currently stands at 9.3‰ for men and 8.7‰ for women (INE, 2022b). To avoid an excessively long chapter, in the supplementary material (Suppl. Mat.) the factors which affect each of the diseases of each medical specialty have been specifically developed, as well as some fundamental bibliographic references that support the cause-effect relationship.

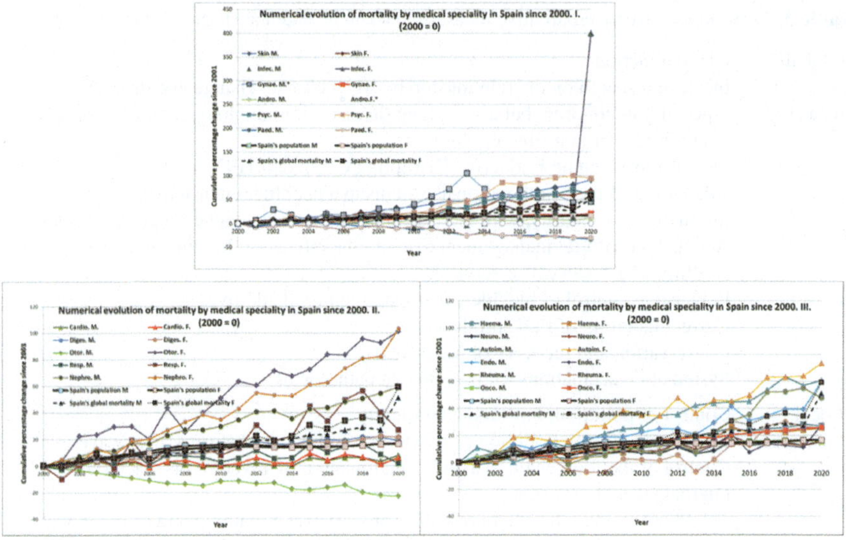

Fig. 1 Numerical evolution of mortality in medical specialties dealing with climate change-related diseases

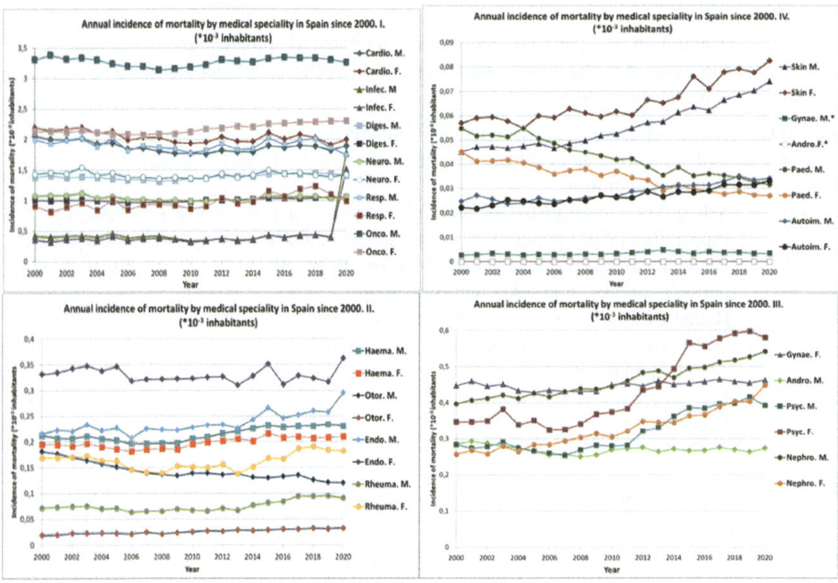

Fig. 2 Incidence of climate change-related mortality trends by medical speciality

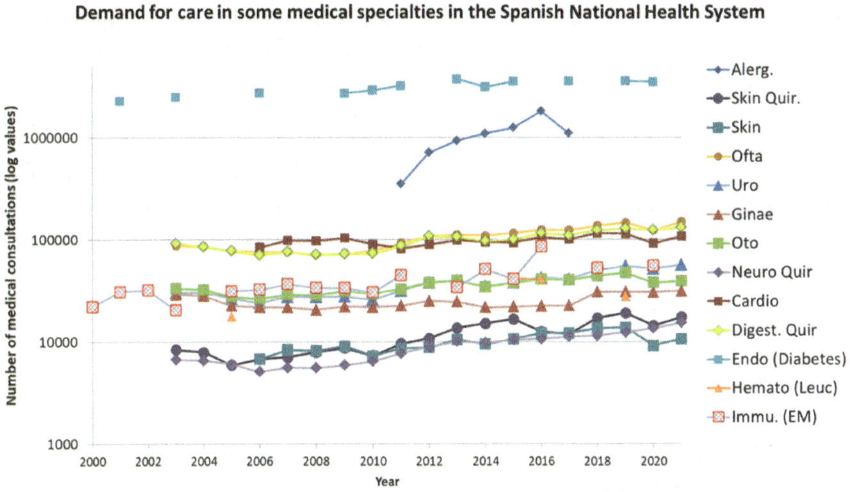

Fig. 3 Evolution of the demand for care in some medical specialties in the Spanish National Health System (logarithmic scale data)

3 Medical Specialties, Climate Change and Evolution in Spain

In this section, we have proceeded to analyze the various sources of information available at the national level on the incidence or prevalence of the different pathologies related to climate change, based on data aggregated by medical specialities. The sources used are diverse, coming in most cases from the different websites of public health institutions (Ministry of Health, National Health System, Carlos III Institute, National Epidemiology Centre), or from public (INE) or private (statista.com) statistics. Occasionally, use has been made of data published by medical societies. Since we have focused especially on mortality, the data usually come from hospital exitus data. In specialties without relevant mortality (allergology, ophthalmology, clinical psychology), the data come from hospital discharges and medical speciality centres.

In general terms, data usually refer to absolute values (incidence) or put in relation to the population (prevalence) and are mentioned as such when appropriate. Usually highest values of incidence and prevalence are coincident, being greater the higher the population (and, with it, the higher the level of urbanization).

3.1 Allergology

(a) Impact factors: pollution, pollen, changes in temperature patterns, food selection and processing.

(b) Evolution in Spain: allergic diseases (including asthma and rhinoconjunctivitis) have increased in recent decades. From 2005 to 2015, rhinitis consultations accounted for 62% of allergology consultations in Spain. The region with the highest number of people suffering from chronic allergies in people over 15 years of age was Madrid, followed by Andalusia and Catalonia. In most communities, women were the most affected, especially in municipalities with more than 400,000 inhabitants, where chronic allergies reached to affect in 2017 by 18.66% of this population (58.6% of them were women) (Colás & Cubero, 2018; INE, 2022e; SEFAC-SEAIC, 2017; Soler, 2022; Statista, 2020).

3.2 *Andrology*

(a) Incident factors: high temperatures, stress, cadmium, pesticides, herbicides, excess pollution, endocrine disruptors.
(b) Evolution in Spain: infertility affects 17% of Spanish couples of reproductive ages. Between 2011 and 2019, infertility multiplied by 8.1. Regarding mortality, male cancers (prostate and testicular) have decreased from 2000 with respect to the population increase. Mortality from prostate cancer and male genital diseases is highest in Andalusia, while mortality from prostate hyperplasia is the highest in Catalonia and testicular cancer in Valencia. About 1.3% of all deaths are due to this cause (Conceptum Fertilitat, 2021; ISCIII, 2021; Statista, 2021d).

3.3 *Digestive System*

(a) Impact factors: infections due to high temperatures, heat waves, high humidity or torrential rains followed by floods; $PM_{2.5}$ and food selection and processing.
(b) Evolution in Spain: in 2020, 13.8% of male hospital discharges and 10.3% of female hospital discharges were due to digestive diseases. Colorectal cancer has been growing at 2.6% per year in men and 0.8% per year in women continuously since 1975, being more frequent in northern Spain. In 2019, mortality due to diseases of the digestive system reached 6.1% of affected patients, increasing from 2000 to 2020 by 11.1% in men and 23.9% in women (prevalence of 0.49‰ and 0.42‰, respectively). About 11.8% of all deaths are due to this cause (Centro Nacional de Epidemiología, 2005; INE, 2021a; ISCIII, 2021; Marín Sillué et al., 2021; SNS, 2020).

3.4 Cardiology

(a) Impact factors: pollution, $PM_{2,5}$, NO_2, SO_2, CO, O_3, PM_{10}, high temperatures, heat and cold waves, hurricanes, droughts and floods.
(b) Evolution in Spain: depending on climate change scenarios, mortality from cardiovascular diseases (CVD) could increase by 10.2%; in fact, it has been estimated that 30% of deaths from CVD disease are related to environmental pollution. From 2006 to 2021, the number of specialist cardiology consultations increased by 28%, accounting for 15% of male and 10.7 of female hospital discharges in 2020. Murcia shows the highest hospitalization rates for CVD. The Canary Islands and Murcia have the highest values for hypertensive disease. Catalonia has the highest values of hospitalization for congestive heart failure. The highest number of deaths from various heart diseases occurs in Andalusia. However, the number of CVD deaths grew from 2000 to 2020 by 7.8% and 6.1% in men and women, which is lower than population growth (16.6 and 16.5%), implying a decreasing prevalence (1.9 and 2‰). About 18.8% of all deaths are due to this cause (Álvarez de Arcaya & Pérez, 2021; González López-Valcárcel et al., 2022; INE, 2021a, 2022a; Ministerio de Sanidad, 2022; SNS, 2020).

3.5 Dermatology

(a) Impact factors: increased temperatures, dust storms, UVR-UVB and pollution.
(b) Evolution in Spain: from 2006 to 2021, the specialized consultations in dermatology increased by 57%, accounting for 1.1% of hospital discharges in men and 1% for women in 2020 (although part of this increase may have been due to dermo-aesthetic interventions). Between 2000 and 2020, the number of deaths from skin and subcutaneous tissue (SST) diseases increased by 114% in men and 95.9% in women; melanomas, in the same period, increased by 67.4% (men) and by 37.7% (women); both SST, melanomas and skin cancer had the highest frequency in men in Andalucia, but melanomas had the highest frequency in Catalonia in women. In general, mortality from skin diseases increased by 91.6% (men) and 69% (women), with a prevalence of 0.06 and 0.07‰, respectively (0.75 of all deaths) (González López-Valcárcel et al., 2022; INE, 2021a, 2022a; ISCIII, 2021; SNS, 2020; Statista, 2021c).

3.6 Endocrinology/Nutrition

(a) Impact factors: endocrine disruptors (EDC), organic compounds, $PM_{2,5}$, forest fire and arsenic.

(b) Evolution in Spain: Galicia and the Canary Islands registered the highest percentage of diabetic population. The Canary Islands and Valencia summarized the highest percentage of population with high cholesterol. Endocrine, nutritional and metabolic diseases accounted for 1.7% of male and 2% of female hospital discharges in 2020. The mortality by diabetes mellitus has fluctuated since 2000; in contrast, other endocrine and metabolic diseases have increased steadily, 2.43-fold from 2000 to 2020. The number of deaths from endocrine and metabolic diseases grew by 59.9% (men) and 27.6% (women), with a prevalence of 0.24 and 0.33‰ (3.2% of all deaths) (INE, 2021b; ISCIII, 2021; Kumar et al., 2020; Ministerio de Sanidad, 2020a; SNS, 2020).

3.7 *Infectious and Parasitic Diseases/epidemiology*

(a) Impact factors: temperatures, pouring rain, UVR, storms, ozone, PM_{10}, polluted water and inadequate interventions in the territory (e.g. deforestation, inappropriate crops, over-irrigation, urbanization on inappropriate sites).
(b) Evolution in Spain: in 2020, there were a total of eight deaths in Seville due to an outbreak of West Nile virus. With the exception of Covid-19, the upward trend in infectious diseases has been limited since 2000 to intestinal diseases and other infectious and parasitic diseases (unspecified). Numerous infectious diseases (see Suppl. Material) are most abundant in central, northern and northeastern Spain, with a tendency towards summer increases. Other infectious diseases lack a spatial pattern. In 2020, infectious and parasitic diseases accounted for 6.4% (men) and 4.9 (women) hospital discharges. In 2019, mortality from these diseases reached 12.1% of affected patients; nevertheless, when Covid-19 is included, values of 400.4% (men) and 397.4% (women), respectively, are reached. It is striking that mortality (without Covid-19) has grown in women above the average population growth. The average mortality rate reached 0.46 and 0.43‰ (with Covid-19) or 0.34 and 0.37‰ (without Covid-19). About 15.8% out of all deaths are due to this cause (3.6% without Covid-19; ISCIII, 2020, 2021; INE, 2021a, 2022a; Ojeda et al., 2020).

3.8 *Haematology*

(a) Impact factors: pollution (especially, by traffic), dry or very wet periods, excessive heat and benzene (organic compound).
(b) Evolution in Spain: in 2019, mortality due to haematological diseases reached 4.38% of affected patients. In 2020, 0.9% of hospital discharges were due to haematological diseases. Blood diseases and leukaemia reached their highest mortality in Madrid. Haematological diseases followed a continuous upward until 2019 (a 64% increase), with an increased mortality by 27 (men) and 25.7% (women) until 2020, and a significant increase in non-Hodgkin's lymphoma. The

average mortality from this cause was 0.21‰ in men and 0.2‰ in women (2.1% of all deaths in Spain) (INE, 2021a; ISCIII, 2020, 2021; Statista, 2021b).

3.9 *Immunology*

(a) Impact factors: high temperatures, heat waves, sunburn, droughts, atmospheric dust, pollution, organic pollutants (benzene), CO_2, UVR, NO_2, O_3, PM_{10}, paints, glazes, inks, pottery or ceramics, organic (pollen, dust storms) and inorganic compounds (silica, talc, silicon, fibreglass) and heavy metals (beryllium, aluminium, titanium, zirconium).

(b) Evolution in Spain: the most frequent immunological diseases are rheumatoid arthritis (see below, Rheumatology) and psoriasis (prevalence of 6.4 and 2.7%, respectively), with the highest values in Cantabria. Immunological diseases, including Multiple Sclerosis (MS) and Amyotrophic Lateral Sclerosis (ALS), reach their highest mortality in Madrid (men) and Valencia (women). In Southern and Eastern Spain, dust storms occur more than 20 times a year, especially in summer, favouring sarcoidosis, a disease affecting 0.11% of the Spanish population (51,000 inhabitants). Despite therapeutic advances, mortality from all immunological diseases increased from 2000 to 2020 by 60.7% (men) and 73.9% (women), although their prevalence is low (0.029 and 0.027‰, respectively) and it reaches a 0.32% out of all deaths (AEMET, 2018; García López et al., 2022; ISCIII, 2020, 2021; Puig et al., 2019; UNIMID, 2019).

3.10 *Nephrology*

(a) Impact factors: high temperatures, heat waves, intense heat or cold, PM_{10} and $PM_{2.5}$.

(b) Evolution in Spain: one in seven adults suffers from chronic kidney disease. Combining all nephrological and urological disorders, mortality increased by 59 (men) and 103.2% (women), with kidney cancer being particularly prevalent. The highest mortality due to these diseases, including cancer, shows the highest values in Andalusia and Catalonia. The average mortality from 2000 to 2020 accounted for 0.46 (men) and 0.33‰ (women), reaching a 4.75% out of all deaths (Gorostidi et al., 2018; ISCIII, 2020, 2021; Ministerio de Sanidad, 2022).

3.11 *Neurology*

(a) Impact factors: high temperatures, pollution, PM_{10}, $PM_{2.5}$, mercury, arsenic, manganese, lead, pesticides, persistent organic pollutants, EDC, biotoxins, NO, CO and O_3, toxins, malnutrition and stress.

(b) Evolution in Spain: between 2.6% (in men) and 2.3% (in women) of hospital discharges were due to neurological diseases in 2020. Given the sharp reduction in cerebrovascular disease, the growth of neurological diseases as a whole has been lower than that of the population, i.e. they have fallen slightly, although there are several neurological diseases which doubled in incidence from 2000 to 2020. Mortality from the most of neurological diseases is the highest in Catalonia (both men and women). This cause implies 12% of total deaths in Spain (INE, 2021a; ISCIII, 2021).

3.12 Obstetrics/Gynaecology

(a) Impact factors: high temperatures, GEIS and pollution, extreme events, droughts, urban habitat, water pollution.
(b) Evolution in Spain: 17.9% out of hospital discharges of women in 2020 were due to complications of pregnancy, childbirth and puerperium; even so, mortality is decreasing in this field, falling by 28.6% from 2000 to 2020. Mortality from breast cancer has risen slightly (15.8%), less than that of the Spanish population in women from 2000 to 2020, although it has increased in men (46.3%). Mortality due to obstetric or gynaecological diseases did increase by 20.7% from 2000 and 2020. The average prevalence was 0.45‰ (2.3% of all deaths). Madrid and Catalonia register the highest mortality due to obstetric complications. Catalonia and Andalusia have the highest values (similar) in mortality due to pregnancy, childbirth and puerperium. According to the national registry of the Spanish Fertility Society, 9.5% of babies born in Spain in 2019 were born thanks to assisted reproduction techniques. Infertility affects 17% of Spanish couples of reproductive age (Andrés, 2022; Conceptum Fertilitat, 2021; INE, 2021a; ISCIII, 2021; Statista, 2021d).

3.13 Ophthalmology

(a) Impact factors: high temperatures, summer, rising minimum temperatures, heat waves, extreme thermal amplitude, high humidity, high pressure, heavy rainfall, floods, drought, sun exposure, high luminosity, higher radiation and albedo, high altitude, blue light, UVR, UVB, evaporation, wind, pollutants, urban pollution, tobacco, $PM_{2.5}$, dust, pollens, solvents and various maternal exposures, factors that exacerbate infectious diseases and contaminated groundwater.
(b) Evolution in Spain: the number of cataract operations increased by 50% from 2004 to 2013. The risk of rhegmatogenous retinal detachment has increased by 17.3% due to the increase in heat waves and high summer temperatures. UVR with high pollution led to an increased risk in age-related Macular Degeneration (AMD). From 2006 to 2016, AMD cases doubled compared to the expected increase. Glaucoma multiplied its incidence from 2011 to 2016 by

3.77 times. The 30-year projections for eye diseases show a projected increase in retinal detachment, AMD, glaucoma and cataracts. Given the strong incidence of sunstroke and high temperatures on eye diseases, these are more prevalent in southern communities (Echevarría et al., 2021; Fernández-Araque et al., 2017).

3.14 *Oncology*

(a) Impact factors: cadmium, arsenic and nickel, as well as benzene and benzopyrene, PM ($_{10}$ and $_{2.5}$), solar exposure, UVR, pollution, endocrine disruptors, environmental toxins, infectious agents, gases and particles in solid or liquid suspension and dust storms.

(b) Evolution in Spain: Andalusia shows the highest absolute mortality values in eleven types of cancer in both sexes, ten types in men and four in women. Catalonia registers the highest absolute mortality values in four types of cancer in both sexes, four in men and ten in women. Moreover, Madrid obtained the highest absolute mortality values in non-Hodgkin's lymphoma in women and Valencia in testicular cancer. The deadliest cancer continues to be lung cancer, followed by colorectal cancer. Overall, cancer mortality prevalence has increased by 15.2% in men (below the population increase), but by 26% in women (9.4% more than the population increase), with lung cancer (men) and breast cancer (women) standing out. About 26.7% of all deaths are due to this cause (ISCIII, 2021; SEOM, 2021).

3.15 *Otorhinolaryngology*

(a) Impact factors: O_3, PM_{10} and $_{2.5}$, CO, SO_2, NO_2, high atmospheric pressure, benzopyrene and forest fires.

(b) Evolution in Spain: the leading fatal diseases in otorhinolaryngology (ORL) are cancer of mouth, pharynx and larynx. In both cases, mortality is mainly in men, in whom it is decreasing (− 7 and − 38.9%, respectively, from 2000 to 2020) but is increasing in female (100 and 110.9%, respectively). Mortality from these diseases reaches its highest values in Andalusia (both sexes), Spain being, according to 2016 data, the country in the world with the highest incidence of larynx cancer ($18 * 10^{-5}$). From 2008 to 2013, the cumulative incidence of morbidity due to malignant otitis externa was $1.3 * 10^{-6}$, especially affecting men; 74.6% of patients were diabetic. About 0.7% of all deaths are due to ORL diseases (Diario Enfermero, 2016; Guerrero-Espejo et al., 2017; ISCIII, 2021).

3.16 Paediatrics

(a) Impact factors: heat, droughts, floods, UVR, pollution, GHGs, contaminated food and water, $PM_{2.5}$, exposure to chemicals (lead, mercury, cadmium, pesticides, benzene, dioxins, phthalates, bisphenol, excessive fluoridation), waste, contamination resulting from armed conflicts, famine caused by droughts or floods, and psychological stress.
(b) Evolution in Spain: Spain is in the medium–low risk zone for children (117 out of 163, with 1 being the highest risk). Thus, paediatric mortality is not only low in our country (0.3% of all deaths) but has been declining sharply for at least the last 20 years (from 5.26 to 3.06‰). However, the birth rate has also been drastically reduced over the same period, having gone from 10.08‰ to 7.1‰. Currently, the main cause of infant death is due to various congenital anomalies, although the fastest growing fatal disease is chromosomal anomalies. Andalusia has the highest number of deaths due to both pathologies in both sexes (FMC, 2021; INE, 2022c, 2022d; ISCIII, 2021; SNS, 2020).

3.17 Clinical Psychology

(a) Impact factors: drought, floods, hurricanes, psychological stress and climate change itself as a threat concept.
(b) Evolution in Spain: counting the people who required help in 2020, 58.6% went to a psychologist. A total of 43.7% of the people who visited a mental health clinic did so for anxiety disorders. However, psychological care is scarce in our country, where there are six clinical psychologists for every 10^5 inhabitants (Carpio, 2021; Gosálvez, 2021).

3.18 Psychiatry

(a) Impact factors: floods, droughts, hurricanes, high temperatures, extreme weather and urban "heat island".
(b) Evolution in Spain: in 2020, there were 2.1 million people suffering from depression, 5.25% of the population over 15 years of age. Of these, 230,000 suffered from major depression. The number of cases in women is almost twice than men. Castilla-León is the community with the highest prevalence of depression. Suicides grew in the male population from 2000 to 2020 below the national growth, but in the female population, they grew by 25.7%; even so, suicides affected almost three times as many men as women, with Andalusia having the highest absolute values in number of suicides. Overall, mortality from various psychiatric diseases increased from 2000 to 2020 by 61.2% in men and 94.8% in women, with an average prevalence of 0.3 and 0.4‰, respectively (4.7% of

all deaths), highlighting the strong weight of dementia (Carpio, 2021; ISCIII, 2021).

3.19 *Respiratory/Pneumology*

(a) Impact factors: pollution, dust, fires, PM_{10} and $_{2.5}$, O_3, NO_2, SO_2, droughts and high temperatures.
(b) Evolution in Spain: respiratory diseases are the most abundant, accounting for 15 (male) and 10.9% (female) hospital discharges in 2020. Deaths due to Chronic Obstructive Pulmonary Disease (COPD), respiratory failure and other respiratory diseases are more abundant in the most populated communities (Andalusia, Catalonia and Madrid). In contrast, deaths due to acute respiratory infections are more common in Madrid. Morbidity due to COPD is the highest in Murcia. Overall, mortality from respiratory diseases in Spain grew by 1.73 (men) and 27% (women), probably due to an increasing in women smoking habits, with a prevalence of 1.91‰ in men and 0.98‰ in women. About 13.1% of all deaths are due to this cause (INE, 2021a; ISCIII, 2020, 2021; Ministerio de Sanidad, 2020a; SNS, 2020).

3.20 *Rheumatology*

(a) Impact factors: high temperatures, heat waves, forest fires, CO, NO, NO_2, NO_x, PM_{10}, $PM_{2.5}$, O_3 and proximity to roads.
(b) Evolution in Spain: 25% of the population suffers from some rheumatic disease. The number of cases of rheumatoid arthritis is multiplied by 4.5 from 2011 to 2019 (346,300 cases in 2019). The highest frequency occurs in the peninsular interior, in women and in urban areas. Crohn's disease has a high prevalence, 0.39%, i.e. $225/10^5$ inhabitants. About 0.26% of the Spanish population over 20 years of age suffers from Ankylosing Spondylitis. Overall, mortality due to rheumatic diseases has increased from 2000 to 2020 by 47.9% in men and 25.6% in women (who suffer twice as many deaths as men from this cause), implying a 1.3% of all deaths (García de Yébenes & Loza, 2018; Heras, 2022; ISCIII, 2021; MSD Salud, 2019; Statista, 2021a).

3.21 *Occupational Medicine*

(a) Impact factors: increase in temperatures, heat waves, UVR, extreme weather, vector-borne diseases and expanded habitats, pollution, changes in the built environment, aerosols, O_3, PM_{10} and $_{2.5}$, α-amylase (in bakers), metals and ecological anxiety.

(b) Evolution in Spain: occupational accidents followed a very low evolution from 2000 to 2020, increasing less than the Spanish population; however, in 2021 they increased drastically, equalling their growth to that of the national population. Castilla la Mancha and Navarra were the communities with the highest incidence in 2020 (more than 3000 accidents at work in each case). Mortality due to occupational accidents accounted for 0.05% of total accidents in 2021, when in 2000 it accounted for 0.2%. All these reveal a recent increase in morbidity, but a decrease in mortality (MITES, 2022; SNS, 2021b).

3.22 *Emergency Medicine*

(a) Impact factors: extreme heat, floods, droughts, dust storms, forest fires, pollution (all factors related to climate change).
(b) Evolution in Spain: The number of emergencies attended in 2020 in hospitals of the national health system (SNS) was $17.3 * 10^6$, i.e. 365.3 per 1000 inhabitants. Ceuta, Melilla and Cantabria are, in that order, the regions with the highest frequentation, although the highest hospitalization values occur in Castilla-León and Asturias. Between 2010 and 2019, emergency care increased by 20.74% (above national growth), falling sharply in 2020 as a result of the blockage of the health system by Covid-19 (Ministerio de Sanidad, 2020b; SNS, 2021a, 2021c).

4 Challenges and Conclusion

There are several medical specialities that deal with diseases whose prevalence in morbidity and mortality has increased above the average for the Spain's population between 2000 and 2020, despite the numerous advances in treatments that have been applied so far in the twenty-first century.

The climate change outlook for Spain (Amblar et al., 2017) foresees average increases in maximum temperatures of 3 °C above current average values, although they could reach up to 6.4 °C at minimum GHG emission control values. In summer, these values could reach 8.4 °C above current values. Continental Spain (the interior of the Peninsula) would reach the maximum values. Likewise, minimum temperatures could increase by up to 5.2 °C in the most emissive scenario, rising especially in summer (up to 7 °C above current values). Exceptional values of daily maximum temperatures have gone from being singular to being common (it is frequent to reach the 90% percentile in daily maximum temperatures , 24–50% more frequent) and warm nights (40–60%), as well as the number of days with heat waves (15–50 days more than usual). In contrast, frost tends to decrease by around 10%.

With such data, infectious diseases, recently exacerbated by the impact of Covid-19, had already been continuously increasing their mortality above the national average since 2015. Given the strong correlation between high temperatures and infectious diseases, the risk of a continued increase in such pathologies is evident. In addition, something similar happens with autoimmune, digestive, gynaecological, haematological, nephrological and psychiatric diseases (all exacerbated because of high temperatures) and the morbidity of allergic diseases (due to phenological changes resulting from high temperatures).

Precipitation tends to fall, with reductions of up to 58% compared to current volumes. This loss is expected to be particularly marked in the Canary Islands, followed by Murcia and Andalusia, precisely the three regions with the lowest water resources. It is in these communities where, in addition, dry periods tend to increase (up to 21 days more). The number of days of precipitation tends to decrease, especially in the north of the peninsula (up to 32 fewer days of rain).

A reduction of between 2 and 8% in average cloudiness is forecast, falling especially in winter in the southern half of the peninsula.

Furthermore, actual evapotranspiration, as a consequence of the decrease in rainfall, also tends to decrease despite the thermal increase (which would imply higher values of potential evapotranspiration), reaching decreases of between 20 and 40% in the south and east of the peninsula.

On the other hand, average wind speed tends to remain stable or even decrease, except in Andalusia and the Ebro valley, where it tends to increase. Something similar happens with the maximum speed.

Finally, extreme events tend to increase in frequency. Thus, rainfall tends to be more intense, especially in the interior of the peninsula, although there is no clear trend at the national level as a whole.

With these prospects, the effectiveness of solar radiation tends to increase, so that the risk of mortality from dermatological and autoimmune diseases, as well as morbidity from ophthalmological diseases, tends to become higher. Increased dryness favours forest fires, which in turn increase the risk of mortality from endocrine and rheumatological diseases. Similarly, torrential rains followed by floods again increase mortality from digestive diseases.

It is necessary to reiterate that infectious, dermatological, nephrological, psychiatric, rheumatological, autoimmune, endocrine, gynaecological, haematological and digestive diseases have seen a steady increase in mortality in recent years in Spain. Something similar is happening with allergological and ophthalmological morbidity.

It remains to be clarified to what extent the increase in mortality from breast cancer in men and from otorhinolaryngological and respiratory diseases in women is due to the particular effects of climate change or to other causes.

There is no doubt that although not all these increases are due to environmental factors (for example, the genetic inheritance or the toxic or dietary habits are not attributable to environmental factors), it is clear that the increases are even contrary to the investment in new human health advances in our country. Therefore, it is plausible that these increases are largely attributable to the climate change, already known from scientific research carried out by medical specialists in various fields.

Among the 17 specialities analyzed that deal with fatal pathologies, 9 show the highest values in women and 8 in men. However, the average increases in mortality for all of them together show 51.1% in men and 59.1% in women. It is significant that the SNS (National Health System) data show increases of 32.5 and 43.3%, respectively, for all causes of death (including accidents and Covid-19).

This reveals that: (1) mortality is being higher in women and (2) specialities with fatal pathologies related to climate change not only show the same pattern (higher mortality in women) but are increasing to a greater extent than the average cause of mortality in Spain.

For this reason, it is necessary to generate disease sinks, that is, immediate global, local and individual actions to reverse the trends of increased morbidity and mortality due to climate change factors. The population needs to know that they can intervene in a simple way and with numerous options in the generation of these sinks. By doing so, it will be possible to maintain and even increase human well-being and life expectancy at birth.

References

AEMET. (2018). *Informe anual, 2017*. AEMET-Ministerio de Agricultura, Pesca y Alimentación. https://www.aemet.es/documentos/es/conocenos/a_que_nos_dedicamos/informes/InformeAnualAEMET_2017_web.pdf

AEMET. (2019). *Climatología de descargas eléctricas y de días de tormenta en España*. https://www.aemet.es/documentos/es/conocermas/recursos_en_linea/publicaciones_y_estudios/publicaciones/Climatologia_de_descargas_electricas/Climatologia_de_descargas_electricas.pdf

AEMET. (2022). *Olas de calor en España desde 1975*. https://www.aemet.es/documentos/es/conocermas/recursos_en_linea/publicaciones_y_estudios/estudios/Olas_calor/Olas_Calor_ActualizacionOctubre2022.pdf

Aemetblog. (2019, enero 26). *La evolución del clima en España desde 1980*. Aemetblog. https://aemetblog.es/2019/01/26/la-evolucion-del-clima-en-espana-desde-1980/

Alonso Capitán, M., & Vázquez Torres, M. C. (2014). *Impactos del Cambio Climático en la Salud*. Ministerio de Sanidad, Servicios Sociales e Igualdad. https://www.sanidad.gob.es/ciudadanos/saludAmbLaboral/docs/CCCompleto.pdf

Álvarez de Arcaya, A., & Pérez, F. (2021, noviembre 23). *El 90% de internistas valora como «amenaza importante» al cambio climático*. Redacción Médica. https://www.redaccionmedica.com/secciones/medicina-interna/el-90-de-internistas-valora-como-amenaza-importante-al-cambio-climatico-4096

Amblar, P., Casado Calle, M. J., Pastor Saavedra, A., Ramos Calzado, P., & Rodríguez Camino, E. (2017). *GUÍA DE ESCENARIOS REGIONALIZADOS DE CAMBIO CLIMÁTICO SOBRE ESPAÑA A PARTIR DE LOS RESULTADOS DEL IPCC-AR5*. Ministerio de Agricultura, Pesca, Alimentación y Medio Ambiente. https://www.miteco.gob.es/es/cambio-climatico/temas/impactos-vulnerabilidad-y-adaptacion/guia_escenarios_ar5_2017_tcm30-485945.pdf

Andrés, M. de. (2022, junio 25). *Aumento de la infertilidad: "Aún hay muy poco apoyo social a una enfermedad como la nuestra"*. La Razón. https://www.larazon.es/sociedad/20220626/qemreqzqyzhnlayh2zskzhg6ri.html

Aymerich, M., Ruiz Franco, B., & Moreno Gutiérrez, C. (2021). *Informe sobre el estado del patrimonio natural y de la biodiversidad en España a 2020*. Ministerio para la Transición Ecológica y el Reto Demográfico (MITECO). https://www.miteco.gob.es/es/biodiversidad/temas/inventarios-nacionales/iepnb20_sexenal_tcm30-527047.pdf

Byrne, M. P., & O'Gorman, P. A. (2016). Understanding decreases in land relative humidity with global warming: Conceptual model and GCM simulations. *Journal of Climate, 29*(24), 9045–9061. https://doi.org/10.1175/JCLI-D-16-0351.1

Capdevila, L., Zilletti, B., & Suárez Álvarez, V. Á. (2011). *Plataforma Nacional de Adaptación al Cambio Climático | Cambio Climático y Especies Exóticas Invasoras en España: Diagnóstico Preliminar y Bases de Conocimiento Sobre Impactos y Vulnerabilidad (2011).* https://www.adaptecca.es/recursos/buscador/cambio-climatico-y-especies-exoticas-invasoras-en-espana-diagnostico-preliminar-y

Carpio, J. A. (2021, diciembre 15). *Los datos de la salud mental en España.* RTVE.es. https://www.rtve.es/noticias/20211215/datos-salud-mental-espana/2238590.shtml

CC.OO. (2019). *Evolución de las emisiones de gases de efecto invernadero en España (1990–2018).* Secretaría Confederal de Medio Ambiente y Movilidad de CC.OO. https://www.ccoo.es/94c96567fa3b77d183b8a4c638e5a1fd000001.pdf

Ceballos, G., Ehrlich, P. R., Barnosky, A. D., García, A., Pringle, R. M., & Palmer, T. M. (2015). Accelerated modern human–induced species losses: Entering the sixth mass extinction. *Science Advances, 1*(5), e1400253. https://doi.org/10.1126/sciadv.1400253

CEDEX. (2017). *EVALUACIÓN DEL IMPACTO DEL CAMBIO CLIMÁTICO EN LOS RECURSOS HÍDRICOS Y SEQUÍAS EN ESPAÑA.* (N.º 42-415-0-001; p. 320). Centro de Estudios Hidrográficos (CEDEX). http://www.cedex.es/NR/rdonlyres/3B08CCC1-C252-4AC0-BAF7-1BC27266534B/145732/2017_07_424150001_Evaluaci%C3%B3n_cambio_clim%C3%A1tico_recu.pdf

Centro Nacional de Epidemiología (España). (2005). *La situación del cáncer en España.* Ministerio de Sanidad y Consumo, Centro de Publicaciones. https://www.sanidad.gob.es/en/ciudadanos/enfLesiones/enfNoTransmisibles/docs/situacionCancer.pdf

Chen, M.-J., Lin, C.-Y., Wu, Y.-T., Wu, P.-C., Lung, S.-C., & Su, H.-J. (2012). Effects of extreme precipitation to the distribution of infectious diseases in Taiwan, 1994–2008. *PLoS ONE, 7*(6). https://doi.org/10.1371/journal.pone.0034651

Colás, C., & Cubero, J. L. (2018). *Perfil de los pacientes en las consultas de alergología de Aragón: Evolución y comparación con el resto de España.* Alergoaragón. http://alergoaragon.org/2018/0002.pdf

Conceptum Fertilitat. (2021, junio 6). *La infertilidad aumenta en España con los años.* Conceptum Fertilitat. https://www.conceptum.es/la-infertilidad-aumenta-en-espana-con-los-anos/

Cresswell-Clay, N., Ummenhofer, C., Thatcher, D. L., Wanamaker, A. D., Denninston, R. F., Asmerom, Y., Polyak, V. J. (2022). Twentieth-century Azores High expansion unprecedented in the past 1,200 years. *Nature Geoscience, 15, 548–553* (2022). https://www.nature.com/articles/s41561-022-00971-w

de las Heras, A. (2022, marzo 29). *El 25% de la población adulta en España sufre una enfermedad reumática.* Inforeuma. https://inforeuma.com/el-25-de-la-poblacion-adulta-en-espana-sufre-una-enfermedad-reumatica/

Diario Enfermero. (2016, octubre 20). España es el país con mayor incidencia de cáncer de laringe de todo el mundo. *Noticias de enfermería y salud.* https://diarioenfermero.es/espana-pais-mayor-incidencia-cancer-laringe-mundo/

Díaz, J., Follos, F., Linares, C. (2020). Cambio climático y Salud. Foro Transiciones, 35 pp. https://forotransiciones.org/wp-content/uploads/sites/51/2020/11/Tiempo-transiciones_cambioclimatico_salud.pdf

Echevarría, L., Senciales-González, J. M., Medialdea-Hurtado, M. E., & Rodrigo-Comino, J. (2021). Impact of climate change on eye diseases and associated economical costs. *International Journal of Environmental Research and Public Health, 18*(13), 7197. https://doi.org/10.3390/ijerph18137197

EL MUNDO. (2022, agosto 1). *La ola de calor de julio cuadruplica el exceso de mortalidad atribuible a las altas temperaturas de años anteriores.* ELMUNDO. https://www.elmundo.es/ciencia-y-salud/salud/2022/08/01/62e81c77fc6c8351778b45cb.html

El Zowalaty, M. E., & Järhult, J. D. (2020). From SARS to COVID-19: A previously unknown SARS—related coronavirus (SARS-CoV-2) of pandemic potential infecting humans—Call for a One Health approach. *One Health, 9*, 100124. https://doi.org/10.1016/j.onehlt.2020.100124

Fagan, M. (2017). Who's afraid of the ecological apocalypse? Climate change and the production of the ethical subject. *The British Journal of Politics and International Relations, 19*(2), 225–244. https://doi.org/10.1177/1369148116687534

Fernández-Araque, A., Giaquinta-Aranda, A., Laudo-Pardos, C., Rojo-Aragüés, A.-A., Fernández-Araque, A., Giaquinta-Aranda, A., Laudo-Pardos, C., & Rojo-Aragüés, A.-A. (2017). Los antioxidantes en el proceso de patologías oculares. *Nutrición Hospitalaria, 34*(2), 469–478. https://doi.org/10.20960/nh.420

FMC. (2021, agosto 23). *Últimas noticias*. Formación Médica Continuada en Atención Primaria. http://www.fmc.es/es-noticia-mil-millones-ninos-viven-paises-31731

Galacho, F. B., Ruiz Sinoga, J. D., & Senciales González, J. M. (2021). ANÁLISIS DE LOS CAMBIOS EN EL USO DEL SUELO AGRARIO Y SU RELACIÓN CON LA EVOLUCIÓN DE LAS CONDICIONES CLIMÁTICAS EN EL ÁMBITO MEDITERRÁNEO ANDALUZ DURANTE LOS ÚLTIMOS 30 AÑOS (1991–2021). En *Libro de resúmenes de los trabajos del XXVII Congreso de la Asociación Española de Geografía* (Asociación Española de Geografía, pp. 89–90).

García Aróstegui, J. L., Cruz Sanjulián, J. J., & Hidalgo Estévez, M. C. (2003). CONTROL DE LA INTRUSIÓN MARINA Y MODELIZACIÓN DEL ACUÍFERO DE VÉLEZ (MÁLAGA, ESPAÑA). *TECNOLOGÍA DE LA INTRUSIÓN DE AGUA DE MAR EN ACUÍFEROS COSTEROS: PAÍSES MEDITERRÁNEOS*, 261–270.

García de Yébenes, M. J., & Loza, E. (2018). Artritis reumatoide: Epidemiología e impacto sociosanitario. *Reumatología Clínica, 14*, 3–6.

García López, F. J., García-Merino, A., Alcalde-Cabero, E., & de Pedro-Cuesta, J. (2022). Incidencia y prevalencia de la esclerosis múltiple en España. Una revisión sistemática. *Neurología*, S0213485322000342. https://doi.org/10.1016/j.nrl.2022.02.006

Gibson, J. H. (2020). *UVB radiation: Definition and characteristics*. Colorado State University. http://uvb.nrel.colostate.edu/UVB/publications/uvb_primer.pdf

Girardi, R., Pinheiro, A., Garbossa, L. H. P., Torres, É., Girardi, R., Pinheiro, A., Garbossa, L. H. P., & Torres, É. (2016). Water quality change of rivers during rainy events in a watershed with different land uses in Southern Brazil. *RBRH, 21*(3), 514–524. https://doi.org/10.1590/2318-0331.011615179

González Gaudiano, É. J. (2012). La representación social del cambio climático: Una revisión internacional. *Revista Mexicana De Investigación Educativa, 17*(55), 1035–1062.

González López-Valcárcel, B., Oliva, J., & Jiménez, J. L. (2022, enero 14). Las listas de espera sanitarias en España (y II): Las consultas a especialistas. *Nada es Gratis*. https://nadaesgratis.es/beatriz-gonzalez-lopez-valcarcel/las-listas-de-espera-sanitarias-en-espana-y-ii-las-consultas-a-especialistas

Gorostidi, M., Sánchez-Martínez, M., Ruilope, L. M., Graciani, A., de la Cruz, J. J., Santamaría, R., del Pino, M. D., Guallar-Castillón, P., de Álvaro, F., Rodríguez-Artalejo, F., & Banegas, J. R. (2018). Prevalencia de enfermedad renal crónica en España: Impacto de la acumulación de factores de riesgo cardiovascular. *Nefrología, 38*(6), 606–615. https://doi.org/10.1016/j.nefro.2018.04.004

Gosálvez, P. (2021, noviembre 14). *España, en terapia*. El País. https://elpais.com/sociedad/2021-11-14/espana-una-semana-en-terapia.html

Gregori, J. (2019, septiembre 12). 300 muertos en España por inundaciones en 20 años. *cadena SER*. https://cadenaser.com/ser/2019/09/12/ciencia/1568304832_760718.html

Guerrero-Espejo, A., Valenciano-Moreno, I., Ramírez-Llorens, R., & Pérez-Monteagudo, P. (2017). Otitis externa maligna en España. *Acta Otorrinolaringológica Española, 68*(1), 23–28. https://doi.org/10.1016/j.otorri.2016.02.010

Hallegatte, S., Green, C., Nicholls, R. J., & Corfee-Morlot, J. (2013). Future flood losses in major coastal cities. *Nature Climate Change, 3*(9), 802–806. https://doi.org/10.1038/nclimate1979

Herrera, S. (2019). *Proyecciones regionales de Cambio Climático para vientos extremos en España para el s.XXI: Caracterización de valores de retorno y frecuencia de configuraciones atmosféricas de peligro* (31 pp.). Universidad de Cantabria. https://adaptecca.es/sites/default/files/u50/2018_uc_proyecciones_regionales_cc_vientos_extremos.pdf

IEEE. (2021). *Las inundaciones en España y en Europa: un problema que sube de nivel. Segunda parte.* Instituto Español de Estudios Estratégicos. https://www.ieee.es/Galerias/fichero/docs_opinion/2021/DIEEEO130_2021_JONGOM_Inundaciones.pdf

INE. (2021a). *Altas hospitalarias en España, 2020.* https://www.ine.es/ss/Satellite?L=es_ES&c=INESeccion_C&cid=1259926692529&p=%5C&pagename=ProductosYServicios%2FPYSLayout&param1=PYSDetalle&param3=1259924822888

INE. (2021b). *Defunciones por causas (lista reducida) por sexo y grupos de edad. Diabetes mellitus y otras enfermedades endocrinas, nutricionales y metabólicas.* INE. https://www.ine.es/jaxiT3/Datos.htm?t=7947#!tabs-grafico

INE. (2022a). *Defunciones por causas (lista reducida) por sexo y grupos de edad(7947).* INE. https://www.ine.es/jaxiT3/Datos.htm?t=7947#!tabs-grafico

INE. (2022b). *Tablas de mortalidad por año, provincia, sexo, edad y funciones.* INE. https://www.ine.es/jaxiT3/Tabla.htm?t=27155

INE. (2022c). *Tasa de Mortalidad de los menores de 5 años según sexo.* INE. https://www.ine.es/jaxiT3/Datos.htm?t=29474

INE. (2022d). *Tasa de Natalidad por comunidad autónoma, según nacionalidad (española/extranjera) de la madre(1433).* INE. https://www.ine.es/jaxiT3/Datos.htm?t=1433

INE. (2022e). Problemas o enfermedades crónicas o de larga evolución padecidas en los últimos 12 meses y diagnosticadas por un médico según sexo y comunidad autónoma. Población de 15 y más años. https://www.ine.es/jaxiT3/Datos.htm?tpx=47145

IPCC. (1992). *Climate change 1992: The supplementary report to the IPCC Scientific assessment—IPCC.* https://www.ipcc.ch/report/climate-change-1992-the-supplementary-report-to-the-ipcc-scientific-assessment/

ISCIII. (2020). *Resultados de la Vigilancia Epidemiológica de las enfermedades transmisibles. Informe anual. Años 2017–2018* (p. 191).

ISCIII. (2021). *Tendencia temporal de la mortalidad.* http://raziel.cne.isciii.es/grafs/fTend.php

Jiang, C., Shaw, K. S., Upperman, C. R., Blythe, D., Mitchell, C., Murtugudde, R., Sapkota, A. R., & Sapkota, A. (2015). Climate change, extreme events and increased risk of salmonellosis in Maryland, USA: Evidence for coastal vulnerability. *Environment International, 83*, 58–62. https://doi.org/10.1016/j.envint.2015.06.006

Johnston, F. H., Henderson, S. B., Chen, Y., Randerson, J. T., Marlier, M., DeFries, R. S., Kinney, P., Bowman, D. M. J. S., & Brauer, M. (2012). Estimated global mortality attributable to smoke from landscape fires. *Environmental Health Perspectives, 120*(5), 695–701. https://doi.org/10.1289/ehp.1104422

Karp, D. S., Moeller, H. V., & Frishkoff, L. O. (2013). Nonrandom extinction patterns can modulate pest control service decline. *Ecological Applications: A Publication of the Ecological Society of America, 23*(4), 840–849. https://doi.org/10.1890/12-0937.1

Kindelán, C. (2013). *PERCEPCIÓN, INFORMACIÓN Y COMUNICACIÓN DEL CAMBIO CLIMÁTICO. CONOCIMIENTO EN ESTUDIANTES UNIVERSITARIOS* [Universidad de Las Palmas de Gran Canaria. Departamento de Didácticas Especiales]. https://accedacris.ulpgc.es/bitstream/10553/10738/2/0686834_00000_0000.pdf

Kumar, M., Sarma, D. K., Shubham, S., Kumawat, M., Verma, V., Prakash, A., & Tiwari, R. (2020). Environmental endocrine-disrupting chemical exposure: Role in non-communicable diseases. *Frontiers in Public Health, 8*. https://doi.org/10.3389/fpubh.2020.553850

Littell, J. S., Peterson, D. L., Riley, K. L., Liu, Y., & Luce, C. H. (2016). A review of the relationships between drought and forest fire in the United States. *Global Change Biology, 22*(7), 2353–2369. https://doi.org/10.1111/gcb.13275

López Alonso, B., Beltrán Rosel, A., Villaverde-Royo, M. V., & López Alonso, I. (2016). Malaria por Plasmodium falciparum en residente en España sin antecedente de viaje reciente a zonas

endémicas. *Medicina De Familia. SEMERGEN, 42*(6), e71–e72. https://doi.org/10.1016/j.semerg.2015.07.014
Manojkumar, N., & Srimuruganandam, B. (2019). Assessment of gaseous emissions and radiative forcing in Indian forest fires. *International Journal of Environmental Studies, 76*(4), 541–557. https://doi.org/10.1080/00207233.2019.1596387
Marín Sillué, S., Daschner, Á., Morales Navas, F. J., Rubio Armendáriz, C., Ruiz Leal, M. J., & Burdaspal Pérez, P. (2021). *Informe del Comité Científico de la Agencia Española de Seguridad Alimentaria y Nutrición (AESAN) en relación a los efectos del cambio climático sobre la presencia de micotoxinas en los alimentos* (p. 41). AESAN. https://www.aesan.gob.es/AECOSAN/docs/documentos/seguridad_alimentaria/evaluacion_riesgos/informes_comite/MICOTOXINAS_CAMBIO_CLIMATICO.pdf
McInnes, K. L., Erwin, T. A., & Bathols, J. M. (2011). Global Climate Model projected changes in 10 m wind speed and direction due to anthropogenic climate change. *Atmospheric Science Letters, 12*(4), 325–333. https://doi.org/10.1002/asl.341
McMichael, A. J., & WHO (Eds.). (2003). *Climate change and human health: Risks and responses.* World Health Organization.
Medina, R., Losada, Í. J., Méndez, F. J., Olabarrieta, M., Liste, M., Ménendez, M., Tomás, A., Abascal, A. J., Agudelo, P., & Guanche, R. (2004). *IMPACTOS EN LA COSTA ESPAÑOLA POR EFECTO DEL CAMBIO CLIMÁTICO. FASE III. ESTRATEGIAS FRENTE AL CAMBIO CLIMÁTICO EN LA COSTA.* Universidad de Cantabria. https://www.miteco.gob.es/es/cambio-climatico/temas/impactos-vulnerabilidad-y-adaptacion/fase3_costas_tcm30-178538.pdf
Meira, P. Á. (2009). *PRINCIPIOS Y RECOMENDACIONES BÁSICAS PARA ENFOCAR LA COMUNICACIÓN DEL CAMBIO CLIMÁTICO* (p. 10) [Artículos de Opinión]. Ministerio de Medio Ambiente y Medio Rural y Marino. https://www.miteco.gob.es/es/ceneam/articulos-de-opinion/2009_12meira_tcm30-163499.pdf
Méndez, M. (2011). *Análisis preliminar de la vulnerabilidad de la costa de Andalucía a la potencial subida del nivel del mar asociada al cambio climático.* Junta de Andalucía. Consejería de Medio Ambiente. https://www.juntadeandalucia.es/medioambiente/portal_web/web/temas_ambientales/clima/actuaciones_cambio_climatico/adaptacion/vulnerabilidad_impactos_medidas/vulnerabilidad_costas.pdf
Mestre, I., Casado, M. J., & Rodríguez, E. (2019). *Tendencias observadas y proyecciones de cambio climático sobre España.* MITECO. https://www.miteco.gob.es/es/cambio-climatico/temas/impactos-vulnerabilidad-y-adaptacion/cap2-tendenciasobservadasyproyeccionesdecambioclimaticosobreespana_tcm30-70204.pdf
Ministerio de Sanidad. (2020a). *Ministerio de Sanidad—Portal Estadístico del SNS—Principales problemas crónicos de salud, porcentaje de población de 15 y más años que padece determinados problemas crónicos, registrado en atención primaria, por comunidad autónoma.* https://www.sanidad.gob.es/estadEstudios/sanidadDatos/tablas/tabla5.htm
Ministerio de Sanidad. (2020b). *Número de atenciones en urgencias hospitalarias.* https://icmbd.sanidad.gob.es/icmbd/indicadorAction.do?method=list
Ministerio de Sanidad. (2022). *Portal Estadístico del SNS—Informes Anuales del Sistema Nacional de Salud. Años anteriores.* https://www.sanidad.gob.es/estadEstudios/estadisticas/sisInfSanSNS/tablasEstadisticas/InfAnSNSAnosanteriores.htm
Ministerio de Sanidad, Consumo y Bienestar Social. (2022). *Indicadores clave del Sistema Nacional de Salud.* http://inclasns.msssi.es/main.html. http://inclasns.msssi.es/main.html
MITECO. (2022). *Histórico de la calidad del aire en España.* Calidad y evaluación ambiental. https://www.miteco.gob.es/es/calidad-y-evaluacion-ambiental/temas/atmosfera-y-calidad-del-aire/calidad-del-aire/evaluacion-datos/datos/Historico_calidad_aire.aspx
Miteco.gob.es. (s. f.). Cambio climático. Recuperado 14 de abril de 2020, de. https://www.miteco.gob.es/es/cambio-climatico/temas/default.aspx
MITES. (2022). *Estadística de Accidentes de Trabajo del año 2021.* https://www.mites.gob.es/es/estadisticas/monograficas_anuales/EAT/2021/index.htm

Montañez, A. I. (2021, marzo 8). Los políticos que niegan (o han negado) el cambio climático. *Verde y Azul*. https://verdeyazul.diarioinformacion.com/los-politicos-que-niegan-o-han-negado-el-cambio-climatico.html

Moya, A. S., Arredondo, R. E., & Yuli Posadas, R. Á. (2017). DETERMINACIÓN DE LA PRESENCIA DE PARTÍCULAS (PM10) EN PERÚ PRODUCIDAS POR QUEMA DE BIOMASA CON AYUDA DE MODELOS NUMÉRICOS. *Revista Internacional de Contaminación Ambiental, 33*(1), 99–108. https://doi.org/10.20937/RICA.2017.33.01.09

MSD Salud. (2019). *Enfermedad de Crohn. Síntomas y Causas*. https://www.msdsalud.es/cuidar-en/msdsalud.es/cuidar-en/enfermedad-crohn.html

National Geographic. (2020, octubre 14). *Animales en peligro de extinción en España*. www.nationalgeographic.com.es. https://www.nationalgeographic.com.es/naturaleza/animales-peligro-extincion-espana_14180

Ojeda, E., Vila Cordero, B., Sastre, M., & Amillategui Dos Santos, R. (2020). *Informe epidemiológico sobre la situación de Herpes Zóster en España, 1998–2018*. Instituto de Salud Carlos III. https://www.isciii.es/QueHacemos/Servicios/VigilanciaSaludPublicaRENAVE/EnfermedadesTransmisibles/Documents/archivos%20A-Z/HERPES%20ZOSTER/Informe_HZ_Espa%C3%B1a_1998-2018.pdf

Pörtner, H.-O., Roberts, D. C., Tignor, M. M. B., Poloczanska, E. S., Mintenbeck, K., Alegría, A., Craig, M., Langsdorf, S., Löschke, S., Möller, V., Okem, A., & Rama, B. (Eds.). (2022). *Climate change 2022: Impacts, adaptation and vulnerability. Contribution of working group II to the sixth assessment report of the intergovernmental panel on climate change.*

Puertos del Estado. (2022). *REDMAR*. https://www.puertos.es/es-es. https://www.puertos.es/es-es

Puig, L., Marsal, S., Escobar, C., Hinojosa, J., Palau, J., Arraiza, A., Casado, P., Codesido, M., Pascual, C., & Saldaña, R. (2019). LA PREVALENCIA DE DIEZ ENFERMEDADES INFLAMATORIAS INMUNOMEDIADAS (IMID) EN ESPAÑA. *Rev Esp Salud Pública., 93*, 14.

Rodrigo, J., & Senciales, J. M. (2021). A regional geography approach to understanding the environmental changes as a consequence of the COVID-19 Lockdown in highly populated Spanish cities. *Applied Sciences, 11*(7), 2912. https://doi.org/10.3390/app11072912

Rose, J. B., Daeschner, S., Easterling, D. R., Curriero, F. C., Lele, S., & Patz, J. A. (2000). Climate and waterborne disease outbreaks. *Journal—AWWA, 92*(9), 77–87. https://doi.org/10.1002/j.1551-8833.2000.tb09006.x

RTVE. (2022a). Cambio climático. Las muertes atribuidas al calor en España triplican la media de los últimos cinco años. https://www.rtve.es/noticias/20220907/muertes-atribuidas-calor-espana/2397831.shtml#:~:text=Muertes%20atribuibles%20al%20calor%3A%204.700%20fallecidos%20en%202022

RTVE. (2022b). https://www.rtve.es/noticias/20220818/incendios-asi-ardido-espana-siglo/2396782.shtml#:~:text=Las%20cifras%20se%20mueven%20en,quemadas%20desde%20el%20a%C3%B1o%202000.&text=Datos%20provisionales%20para%20el%20periodo%202016%2D21%20y%20parciales%20para%202022

Sanz, M. J., & Galán, E. (2020). *IMPACTOS Y RIESGOS DERIVADOS DEL CAMBIO CLIMÁTICO EN ESPAÑA*. Ministerio para la transición ecológica y el reto demográfico. https://www.miteco.gob.es/es/cambio-climatico/temas/impactos-vulnerabilidad-y-adaptacion/informeimpactosriesgosccespana_tcm30-518210.pdf

SEFAC-SEAIC. (2017). *Informe Alergológica 2015*. SEFAC-SEAIC. https://www.sefac.org/sites/default/files/sefac2010/NP_Alergologica.pdf

SEOM. (2021). *Las cifras del cáncer en España, 2021*. Sociedad Española de Oncología Médica. https://seom.org/images/Cifras_del_cancer_en_Espnaha_2021.pdf

Shaw, A. (2018, agosto 28). *Habitat destruction*. National Geographic Kids. https://kids.nationalgeographic.com/explore/science/habitat-destruction/

Sillero, J. A., Pérez-González, M. E., Martínez-Murillo, J. F., & Ruiz-Sinoga, J. D. (2020). Factors affecting eco-geomorphological dynamics in two contrasting Mediterranean environments. *Geomorphology, 352*, 106996. https://doi.org/10.1016/j.geomorph.2019.106996

SNS. (2020). *Estadísticas / Estudios—Sistema de Información Sanitaria del SNS*. https://icmbd.sanidad.gob.es/icmbd/indicadorAction.do?method=list

SNS. (2021a). *Consulta Interactiva del SNS. Actividad Urgencias*. https://pestadistico.inteligenciadegestion.sanidad.gob.es/publicoSNS/C/siae/siae/hospitales/actividad-asistencial/actividad-urgencias

SNS. (2021b). *Indicadores clave del Sistema Nacional de Salud. Accidentes de Trabajo*. http://inclasns.msssi.es/main.html

SNS. (2021c). *Urgencias atendidas en hospitales del Sistema Nacional de Salud (SNS), frecuentación por 1.000 habitantes y porcentaje de urgencias ingresadas sobre el total de urgencias atendidas según comunidad autónoma*. https://www.sanidad.gob.es/estadEstudios/sanidadDatos/tablas/tabla27.htm

Soler, L. N. (2022, mayo 17). *El mapa de la alergia en España: Así es el nivel de alerta por polen en cada zona*. Newtral. https://www.newtral.es/nivel-del-polen/20220517/

Statista. (2020). *Reacciones alérgicas: Número de casos en España 2011–2017*. Statista. https://es.statista.com/estadisticas/991660/numero-de-casos-de-alergia-o-reacciones-alergicas-no-especificadas-en-espana/

Statista. (2021a). *Artritis reumatoide: Número de casos en España 2011–2019*. Statista. https://es.statista.com/estadisticas/1037781/numero-de-casos-de-artritis-reumatoide-en-espana/

Statista. (2021b). *Defunciones por enfermedades de la sangre y órganos hematopoyéticos España*. Statista. https://es.statista.com/estadisticas/810245/muertes-por-enfermedades-de-la-sangre-y-organos-hematopoyeticos-en-espana/

Statista. (2021c). *Enfermedades de la piel: Fallecimientos por género España 2005–2018*. Statista. https://es.statista.com/estadisticas/594965/numero-de-muertes-por-enfermedades-de-la-piel-y-el-tejido-subcutaneo-por-genero-en-espana/

Statista. (2021d). *Infertilidad: Número de casos en España*. Statista. https://es.statista.com/estadisticas/1046501/numero-de-casos-de-infertilidad-en-espana/

UNIMID. (2019). *Asociación de personas con enfermedades crónicas inflamatorias inmunomediadas*. UNIMID. http://www.unimid.es/

United Nations. (1972). *United Nations Conference on the Human Environment, Stockholm 1972*. United Nations; United Nations. https://www.un.org/en/conferences/environment/stockholm1972

United Nations. (1995). *REPORT OF THE CONFERENCE OF THE PARTIES ON ITS FIRST SESSION, HELD AT BERLIN FROM 28 MARCH TO 7 APRIL 1995*. https://unfccc.int/resource/docs/cop1/07a01.pdf

United Nations. (2022). *Stockholm+50*. Stockholm+50. http://www.stockholm50.global/node/2

WHO. (2018). *Ceguera y discapacidad visual* (spanish version). https://www.who.int/es/news-room/fact-sheets/detail/blindness-and-visual-impairment

Climate Change and Environmental Infectious Diseases in Russia: Case Studies in Temperate and Arctic Climate

Svetlana Malkhazova, Fedor Korennoy, and Dmitry Orlov

Abstract This chapter is dedicated to the experience of medico-geographical analysis of certain climate-related diseases' spread in Russia at the end of the 20th century and the beginning of the 21st century using the case of Tularemia and Anthrax. The role of climate change as a trigger factor causing the advancing spread of diseases has been analyzed. The potential change in ranges due to predicted climate warming was studied according to climate model INM-CM5.0. A series of maps was compiled to identify the territories prone to suitability changes for the infection foci for the period up to 2100. It was determined that regions with temperate and arctic climate may become vulnerable to the emergence of climate-related diseases in the course of environmental changes.

Keywords Russia · Climate change · Infectious diseases · Environmental diseases · Medico-geographical modeling · Tularemia · Anthrax

1 Introduction

Climate change could be the biggest health threat that humanity facing. It is expected that between 2030 and 2050 climate change will cause an estimated 250,000 additional deaths per year from heat stress, malnutrition, and infectious diseases (who.int).

Climate change affects the human health in many ways, causing deaths and diseases. It happens because of more and more extreme weather events, such as heat waves, extreme rainfall and flooding, disruption of food systems, increased infectious diseases, including vector-borne diseases. Global warming expands the ranges of disease carriers and vectors, such as mosquitoes, ticks, and fleas, resulting

S. Malkhazova (✉) · D. Orlov
Faculty of Geography, Lomonosov Moscow State University, Moscow, Russian Federation
e-mail: sveta_geo@mail.ru

F. Korennoy
FGBI Federal Center for Animal Health (FGBI ARRIAH), Vladimir, Russian Federation

R. Akhtar (ed.), *Climate Change and Human Health Scenarios*, Global Perspectives on Health Geography, https://doi.org/10.1007/978-3-031-38878-1_18

the wider spread of emerging and re-emerging infections (Carlson et al., 2022; Georgiades et al., 2022; Malkhazova et al., 2018; Mora et al., 2022; who.int).

Tularemia and anthrax are especially dangerous natural focal infections. Their pathogens are among the most pathogenic microorganisms for humans. Due to their high virulence, the multiplicity of transmission routes and the severity of the disease patterns, they are seen as potential agents of biological weapon and bioterrorism. Both infections are widespread in Russia's territory. Tularemia cases in humans are recorded annually, but for anthrax they emerge sporadically in a way of outbreaks.

Tularemia and anthrax are the climate-dependent infectious diseases, i.e., their spread is influenced by various climatic factors (Evengard et al., 2021; Ma et al., 2020; Revich et al., 2012; Ryden et al., 2009). Thus, the currently observed climate warming may lead to changes in the areas of environmental infections, including tularemia and anthrax (Ezhova et al., 2021; Kutz et al., 2005; Pecl et al., 2017).

The persistence of tularemia and anthrax within the natural foci, regular cases of the disease in human population, their manifestation on new territories, all that indicate the need to strengthen epidemiological surveillance and implement territorially differentiated measures to protect the population.

The natural focality of both diseases suggests that their epidemic and epizootic manifestations are confined to areas with a certain set of environmental factors, including climatic characteristics. That's why global warming and climate change will likely result in changes of the boundaries and structure of the disease ranges.

The aim of the study is (1) to identify the relationship between the persistency of tularemia and anthrax natural foci and the set of natural factors in the Arctic and temperate climate zones of Russia and (2) to model the potential and projected infections' ranges affected by global climate change for the period up to 2100 year.

2 Materials and Methods

2.1 Study Area

The model regions in this study are (Fig. 1):

1. For tularemia—the European territory of Russia (ETR).
2. For anthrax—the Arctic Zone of Russia (AZR).

2.2 Disease Data

Locations of isolating the tularemia pathogens ($N = 2333$ points for period 1970–2000) were determined according to the data of Antiplague Center of Rospotrebnadzor.

Fig. 1 Model regions of the research

The location of burial sites of cattle died from anthrax ($N = 154$ points for period 1882–2015) was retrieved from official publications of the Ministry of Agriculture of the Russian Federation (The register of livestock burial sites ..., 2012).

2.3 Analytical Method

To create maps of the current potential and projected ranges of both diseases, the method of ecological niche modeling was used. This method enables identifying territories favorable for the existence of pathogen based on a combination of physico-geographical factors, as well as to predict the change of this factors using model climate indicators. The most developed and documented modeling approach currently is the maximum entropy optimization model (MaxEnt) (Phillips et al., 2006).

Modeling the ranges of tularemia and anthrax using this approach is based on two main assumptions: (1) the points of isolation of tularemia pathogens from environmental objects, the locations of historic anthrax outbreaks, and burials of dead animals can be seen as indicators of the suitability of the local physico-geographical conditions for the existence of following pathogens: *Francisella tularensis* and *Bacillus anthracis, respectively*; (2) identification of places with a similar combination of physico-geographical conditions (including those based on the predictive data on climate change) may indicate the potential danger of the emergence of new or activation of existing natural foci of tularemia and anthrax.

The points of isolation of tularemia and anthrax pathogens are considered as "presence locations" for these diseases.

Climate data from the Worldclim.org data portal (Fick & Hijmans, 2017) on the current (1970–2000) and projected climate at the end of the twenty-first century (2081–2100) were used as explanatory variables. A set of 19 bioclimatic factors BIO_1—BIO_19 was used (Hijmans et al., 2005). The climate model INM-CM5.0 (Volodin & Gritsun, 2018) was chosen as the most detailed reflection of the situation on the territories of the ETR and the AZR. The model is presented in two climate scenarios: mild—SSP1-2.6 and hard—SSP5-8.5. When modeling the range of tularemia, both climate scenarios were used. When modeling the range of anthrax, after analyzing the results, it was decided to use only the severe climate scenario.

Based on the ecological characteristics of the anthrax pathogen, a wider set of natural factors was considered, including in addition to climatic the following: landscape indicators (such as soil type and pH), altitude, and type of vegetation cover.

To reduce the multicolinearity of variables in each model, a preliminary test was carried out using the usdm analysis package in the R software environment (Naimi et al., 2014). Per the test results, the variables with high cross-correlation (VIF > 10) were excluded. The remaining indicators were used to model the disease ranges (Table 1).

The predictive power of the Maxent model was assessed using the AUC score, which estimates the likelihood that a randomly selected actual place of presence will be categorized by the model as "suitable." The relative contribution of variables to the model was determined using the Jackknife test, which consists of: (1) sequential exclusion of each of the variables from the model and comparison of predictive ability with and without this variable; (2) sequential modeling using only one variable and comparing predictive power with the full model.

The result of the model application was creation of "suitability maps," demonstrating the suitability of the study area for existence of the disease pathogens (potential range) based on complex of the analyzed environmental factors. Suitability maps were created for current and projected climates. To visualize the changes in the range, comparative maps were additionally created by calculating the difference between the "forecast" and "current" values and normalizing this difference by standard deviation of the Maxent model for the current climate.

3 Results and Discussion

3.1 *Tularemia*

Modeling of the current potential range of tularemia using the maximum entropy method showed a good predictive ability of the model (AUC = 0.845 ± 0.003) and revealed the largest contribution of the following variables: "bio_8—the average temperature of the wettest quarter" and "bio_9—the average temperature of the driest

Table 1 Environmental factors used as explanatory variables in an ecological niche model after removing highly correlated variables

Variable name	Variable description	Units	Data type	Model in which the variable is used
Alt	Altitude	Meters	Continuous	Anthrax
Bio_1	Annual mean temperature	°C × 10	Continuous	Anthrax
Bio_2	Mean diurnal range (Mean of monthly (max temp–min temp))	°C × 10	Continuous	Anthrax & tularemia
Bio_3	Isothermality (Mean diurnal range/Temp. annual range) (× 100)		Continuous	Tularemia
Bio_5	Max temperature of warmest month	°C × 10	Continuous	Anthrax
Bio_6	Min temperature of coldest month	°C × 10	Continuous	Anthrax
Bio_8	Mean temperature of wettest quarter	°C × 10	Continuous	Tularemia
Bio_9	Mean temperature of driest quarter	°C × 10	Continuous	Tularemia
Bio_12	Annual precipitation	Millimeters	Continuous	Anthrax
Bio_14	Precipitation of driest month	Millimeters	Continuous	Tularemia
Bio_16	Precipitation of wettest quarter	Millimeters	Continuous	Tularemia
Bio_18	Precipitation of warmest quarter	Millimeters	Continuous	Anthrax & tularemia
Bio_19	Precipitation of coldest quarter	Millimeters	Continuous	Anthrax
MGVF	Maximum green vegetation fraction	Proportion	Continuous	Anthrax
Vegetation	Vegetation type	Vegetation categories (see Table 4)	Categorical	Anthrax
Soils	Soil type	Soil categories (see Table 5)	Categorical	Anthrax
Soil pH	Soil pH at zero depth	pH × 10	Continuous	Anthrax

quarter." Metrics for assessing the relative contribution of variables (Table 2) and relative significance according to the Jackknife test results (Fig. 2) are given below.

In terms of the modern climate (1970–2000), the most favorable for tularemia foci functioning are the central and southern parts of the ETR (from the foothills of the North Caucasus in the south to Lake Ladoga in the north, from the Kaliningrad region to the Middle Volga in the east) (Fig. 3).

Table 2 Contribution of variables to the Maxent model of Tularemia in the European territory of Russia

Variable	Percent contribution	Permutation importance
bio_9	36.4	17
bio_8	34.1	42.2
bio_3	18.3	17.7
bio_2	7.9	12.3
bio_18	1.7	8.4
bio_16	1.4	2
bio_14	0.2	0.5

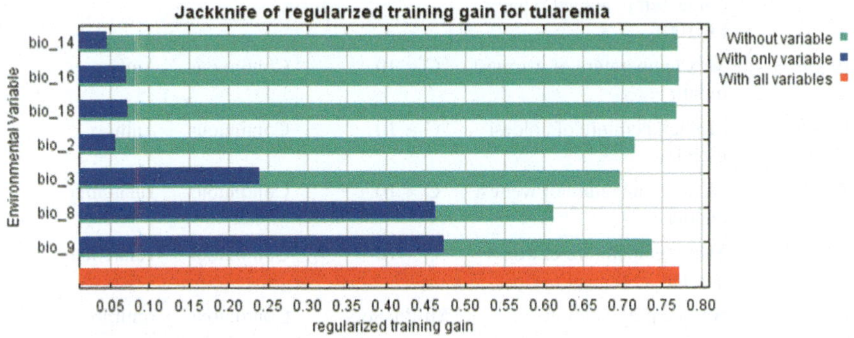

Fig. 2 Significance of variables in modeling the current range of tularemia, assessed using the Jackknife test. The blue bars show the individual contribution of each variable (the longer the bar, the greater the contribution); green columns show a drop in the predictive ability of the model when this variable is excluded (the shorter the column, the more significant information is lost when this variable is excluded)

Under the conditions of the predicted climate (2081–2100, model INM-CM5.0), in the case of a mild scenario SSP1-2.6 (Figs. 4 and 5), the suitability for the functioning of tularemia foci will increase in the northwest (Murmansk region and the Republic of Karelia), in the north (Arkhangelsk region, Komi Republic), southeast of the ETR (Volga region) and will decrease in the center and in the south. Under the severe scenario SSP5-8.5 (Figs. 6 and 7), the suitability for functioning of tularemia foci will decrease in the center of the ETR, Ciscaucasia, on the Kanin Peninsula and part of the Kola Peninsula, but will increase markedly in a number of southern regions.

As the analysis showed, the epidemiological danger will increase in the north of the ETR—an activation of the floodplain-swamp natural foci of tularemia in river valleys, as well as foci of tundra and forest types is expected. The suitability of the territory of the Volga region will increase, possibly due to the increase in moisture and a coresponding change in the ranges of the main pathogen carriers and vectors.

There will be some reduction in the potential epidemiological danger in the center of the ETR. The suitability for tularemia foci functioning will decrease in most of the south of the ETR, apparently due to the increase in dry period, which is unfavorable

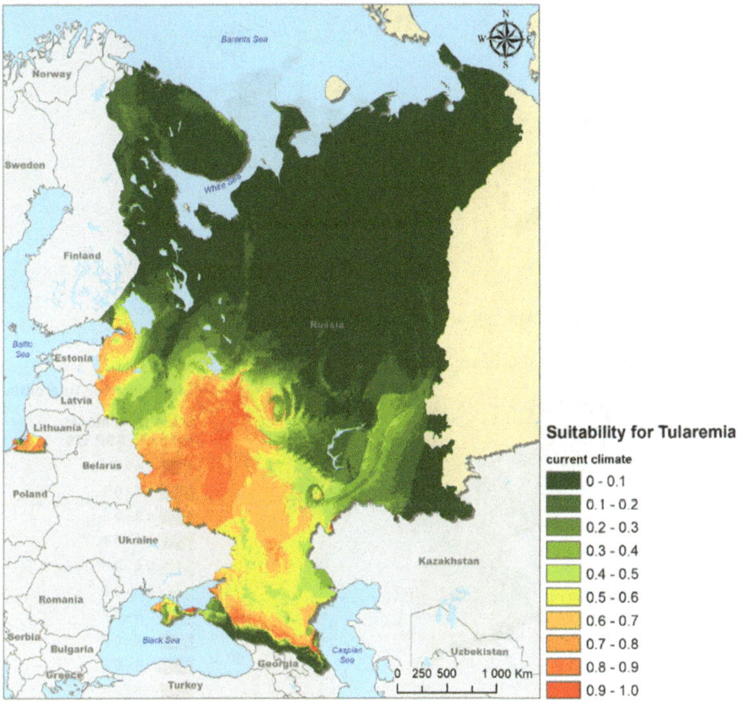

Fig. 3 Suitability of the European territory of Russia (ETR) for the functioning of tularemia natural foci in the current climate

for the members of the parasitic system. Thus, if the global warming scenario persists, the potential epizootic and epidemic danger will increase mostly in the north of the EPR (Republic of Karelia, Murmansk and Arkhangelsk regions, Komi Republic).

3.2 Anthrax

Modeling of the current potential and projected ranges using the maximum entropy method also showed a good predictive ability of the model (AUC = 0.969 ± 0.004). The resulting suitability maps characterizing the favourability to survival of anthrax spores in these physico-geographical conditions are shown in Fig. 8 (for modern climatic conditions) and Fig. 9 (for projected climate). The map in Fig. 10 shows the places of statistically significant increase and decrease of suitability.

In modern climatic conditions, the range of anthrax within the boundaries of the whole Arctic Zone of Russia is limited by the places of registration of anthrax cases in the past. Regions with an increased likelihood of outbreaks include: the southern parts of the Arkhangelsk region and the Republic of Karelia, the north of the Murmansk region, southern regions of the Krasnoyarsk Territory, territory of the

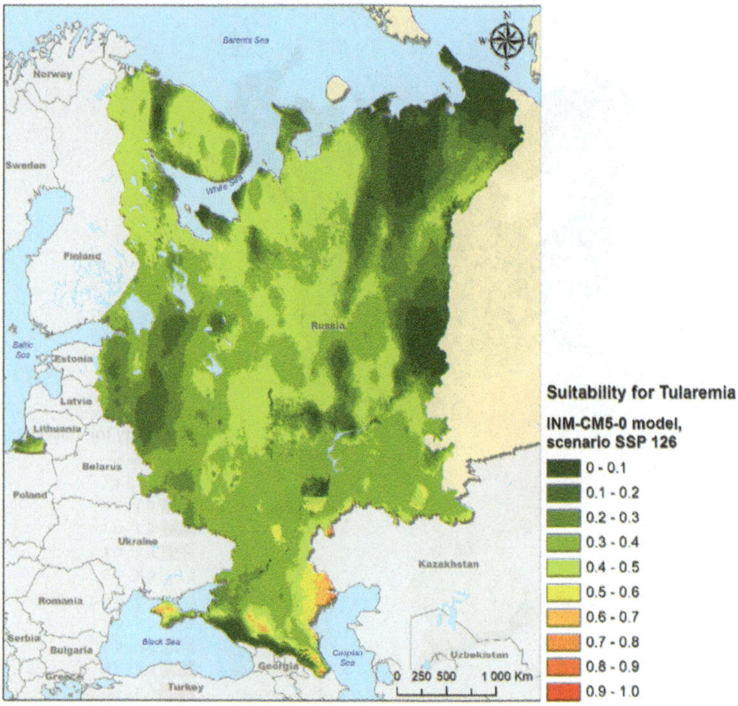

Fig. 4 Suitability of the ETR for the functioning of tularemia natural foci in the projected climate (2081–2100) (INM-CM5.0 model, SSP1-2.6 scenario)

Republic of Sakha (Yakutia) along the Lena River, and in the area of the Kolyma lowland (Fig. 8).

Under the conditions of predicted climate change according to the most severe scenario, by 2100 the potential range of Anthrax will significantly expand, demonstrating an increase in the suitability to disease emergence in almost the entire Russian Arctic. We can especially note a dramatic increase in suitability (from zero to almost maximum) on the Yamal Peninsula, in the coastal areas of the Yamalo-Nenets Autonomous District, the Republic of Sakha (Yakutia), and the Chukotka Autonomous District. The increase in suitability is mainly due to an increase in the average annual air temperature and the maximum average daily air temperature, which can lead to the continuous permafrost thawing and release of conserved soil foci of anthrax (Table 3).

The graph in Fig. 11 clearly demonstrates the relative contribution of each of the variables to the model, estimated with Jackknife test. According to the results of analysis, the variables "soil type" and "soil pH" have the greatest value for the model. Exclusion of these variables from the model leads to greatest drop in its predictivity. This result emphasizes the well-known confinement of anthrax soil foci to certain soil units.

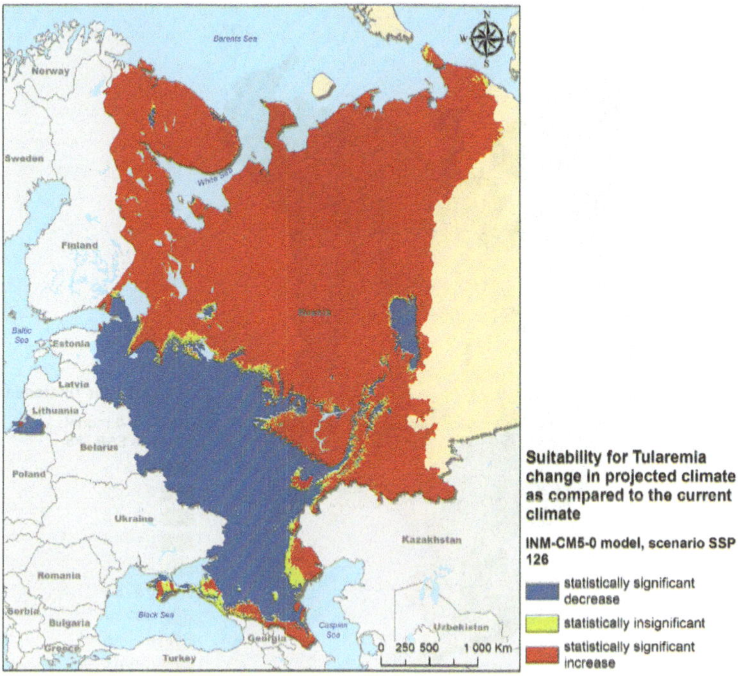

Fig. 5 Change in the suitability of the ETR for the functioning of tularemia natural foci under projected climate (INM-CM5.0 model, scenario SSP1-2.6) compared to the current climate (the difference in suitability is weighted by standard deviation)

4 Preventive measures

Government agencies of the Russian Federation pay much attention to climate change and its impact on public health. The main document at present is the climate doctrine of the Russian Federation—a document that is a system of views on the purpose, principles, content, and ways of implementing the unified state policy of the Russian Federation within the country and in the international arena on issues related to climate change and its consequences. Given the strategic guidelines of the Russian Federation, the Doctrine is the basis for the formation and implementation of climate policy. The document was approved by Decree of the President of the Russian Federation dated December 17, 2009 No. 861-rp.

Much attention is paid to this problem by various ministries and departments that issue constantly updated regulations.

In accordance with the Decree of the Government of the Russian Federation of December 25, 2019 No. 3183-p, the Ministry of Economic Development of Russia collects information on the implementation of measures in the field of climate change adaptation carried out by federal executive authorities and the highest executive authorities of the constituent entities of the Russian Federation and also conducts

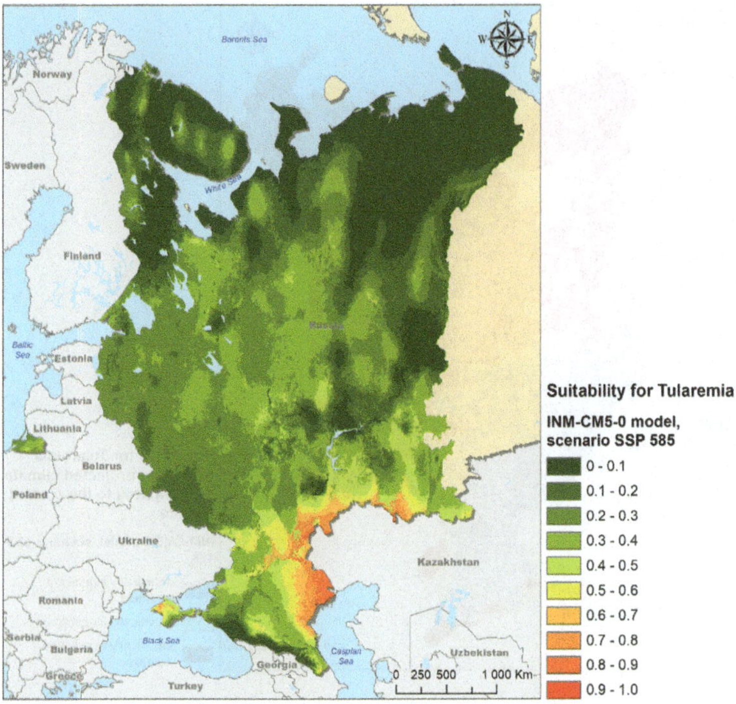

Fig. 6 Suitability of the ETR for the functioning of natural foci of tularemia under projected climate (2081–2100) (INM-CM5.0 model, SSP5-8.5 scenario)

development of a draft national action plan for the second stage of adaptation to climate change for the period up to 2025.

The Ministry of Economic Development of the Russian Federation exercises state administration in the field of limiting greenhouse gas emissions in terms of the functions of developing state policy and legal regulation in the field of limiting greenhouse gas emissions.

The main document of strategic planning in this area is the Strategy for Social and Economic Development of the Russian Federation with Low Greenhouse Gas Emissions until 2050, approved by Decree of the Government of the Russian Federation dated October 29, 2021 No. 3052-r.

The basis for the legal regulation of greenhouse gas emissions was laid down by the Federal Law of July 2, 2021 No. 296-FZ "On Limiting Greenhouse Gas Emissions". In addition, the subjects of the federation have the right to introduce experimental regulation on their territory in accordance with the Federal Law of March 6, 2022 No. 34-FZ "On Conducting an Experiment to Limit Greenhouse Gas Emissions in Certain Subjects of the Russian Federation".

Order of the Ministry of Emergency Situations of Russia dated 10/19/2021 No. 706 "On approval of the sectoral plan for adaptation to climate change in the field of civil

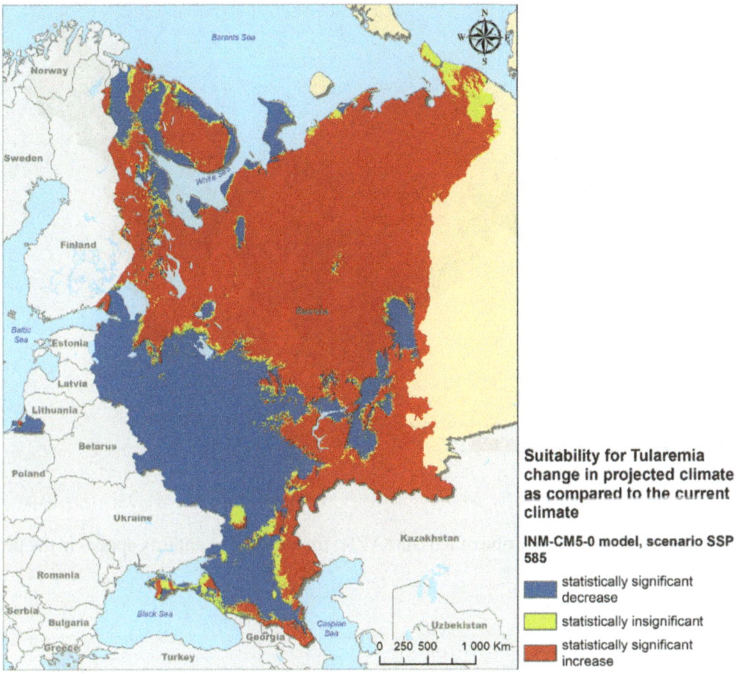

Fig. 7 Change in the suitability of the ETR for the functioning of tularemia natural foci under projected climate (INM-CM5.0 model, SSP5-8.5 scenario) compared to the current climate (fitness difference weighted by standard deviation)

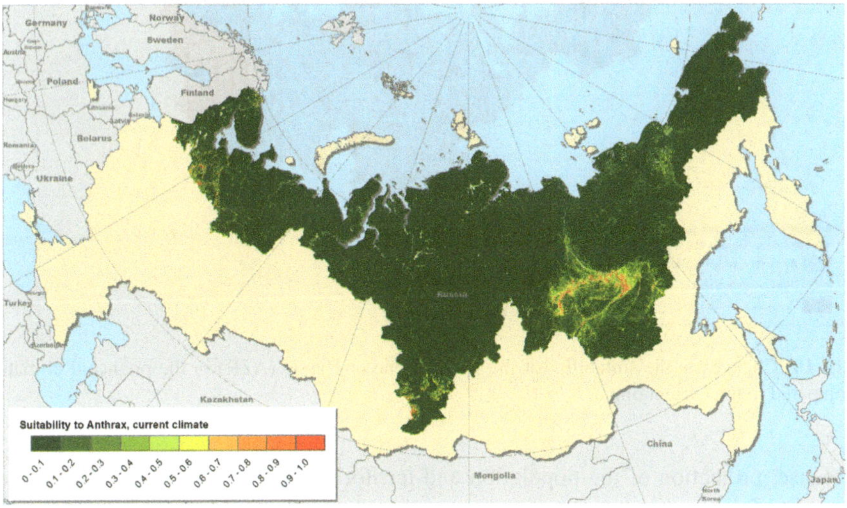

Fig. 8 Suitability of the Arctic Zone of Russia (AZR) for survival of anthrax spores in the current climate

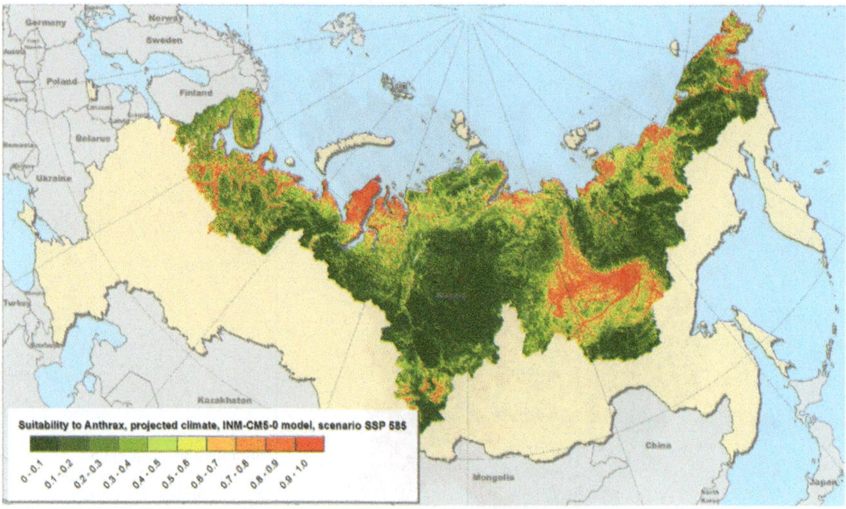

Fig. 9 Suitability of the Arctic Zone of Russia (AZR) for survival of anthrax spores in the projected climate

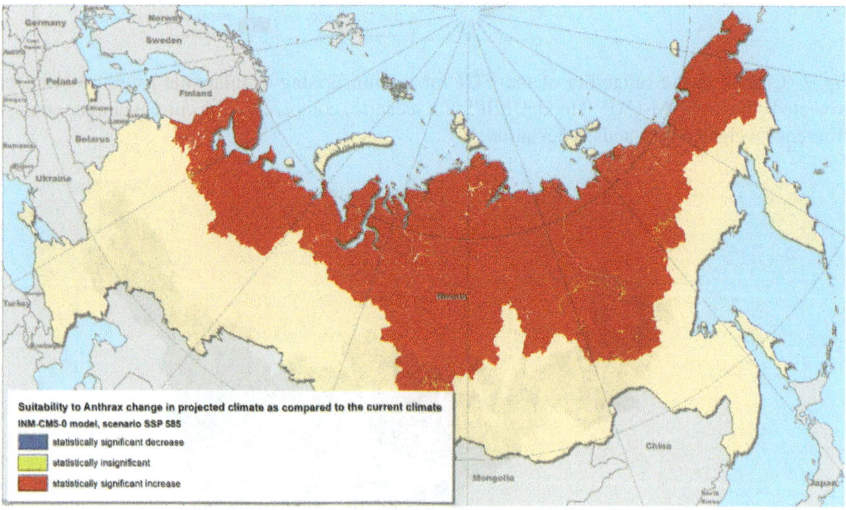

Fig. 10 Changes in the suitability of the Arctic Zone of Russia (AZR) in the predicted climate compared to the current one

defense, protection of the population and territories from natural and technogenic emergencies."

The Ministry of Health of the Russian Federation, on the basis of scientific research, develops sanitary rules and norms for the prevention of climate-related infections.

Table 3 Contribution of variables to the Maxent model of Anthrax in the Arctic zone of Russia

Variable	Percent contribution	Permutation importance
bio_5	31.7	23.6
soil	22.6	7.9
vegetation	16.5	2.3
soil_ph_sl1	12.9	26.6
bio_6	6.5	6.1
alt	5.8	4.8
bio_1	1.4	4.1
bio_2	1	1.7
bio_19	0.7	3.7
bio_18	0.5	5.3
bio_12	0.3	13.9

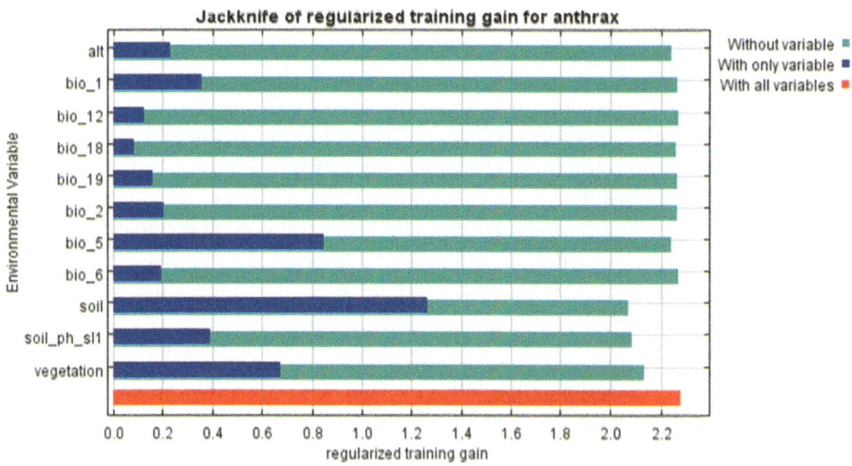

Fig. 11 Significance of variables in the modeling of anthrax range, assessed using the Jackknife test. The blue bars show the individual contribution of each variable (the longer the bar, the greater the contribution); green columns show the drop in the predictive ability of the model when this variable is excluded (the shorter the column, the more significant information is lost when this variable is excluded)

The Rospotrebnadzor has developed methodological recommendations (mr 2.1.10.0273–22), which assesses the accumulated environmental damage to the health of citizens and their life expectancy, in connection with the state of the environment and living conditions of the population. The guidelines were approved by the Chief State Sanitary Doctor of the Russian Federation on January 20, 2022.

According to these documents, the main measures to protect the population of the Russian Federation from the impact of climate change are:

(1) development of a methodology for assessing the impact of climate change on the health of people;
(2) strengthening the epidemiological service, considering the forecasts of the epidemiological situation development;
(3) development and implementation of a communication campaign for targeted informing of the people about impact of climate risk factors on their health;
(4) development of interdepartmental cooperation with the meteorological service, bodies of social protection organizations, and bodies and organizations of rescue services at the local, regional, and federal levels;
(5) improving the energy reliability and energy efficiency of buildings and premises of medical organizations;
(6) repair and reconstruction of buildings;
(7) equipping buildings with modern ventilation and air conditioning systems.

5 Conclusion

Thus, the study allows us to draw the following conclusions.

(1) The method of modeling ecological niches based on physico-geographical factors (including climatic factors) is a good way for assessing the degree of suitability of a territory for the functioning of natural foci. This method is also good to use for mapping the current and predicted potential ranges of natural infections, which is well demonstrated on tularemia and anthrax.
(2) In the current climate, the most epidemiologically dangerous territory for tularemia is the territory of Central Russia. A great influence of the following factors on the spread of tularemia was revealed: average air temperature of a driest quarter, vegetation types, land cover, and soil types. The spread of anthrax is mostly influenced by such factors as type and pH of soil, average annual air temperature, and maximum average daily temperature.
(3) If appropriate preventive and healthcare measures will not be taken, the predicted climate warming will lead to an increase of suitability of Russian territory for the functioning of natural foci of both infections, especially in the northern regions. With the expected climate change for the period up to 2100, the epidemic danger of tularemia will increase in the Arkhangelsk and Murmansk regions, the Republic of Karelia, and the Komi Republic. The potential range of anthrax can also expand significantly, spreading mostly within the Arctic zone of Russia.

6 Appendices

See Tables 4 and 5.

Table 4 Land cover types (adopted from Egorov et al., 2018)

Category number	Land cover type
0, 30	No data
1	Dark evergreen needleleaf forest
2	Light evergreen needleleaf forest
3	Broadleaf forest
4	Deciduous needleleaf forest
5	Evergreen needleleaf shrubs
6, 7, 21, 22	Permanent wetlands
8	Grasslands
9	Broadleaf shrubs
10	Coniferous mixed forests
11	Mixed forests
12	Deciduous mixed forests
13, 19	Open ground and rock outcrops
14	Steppe
15	Coastal vegetation
16	Shrubby tundra
17	Grassy tundra
18	Shrub tundra
20	Water
23	Open deciduous needleleaf forest
24	Fresh burns
31	Urban and built-up areas
32	Snow and ice
33	Arable land

Table 5 Soil units presented in the Anthrax study area* (adopted from The Unified Soil Register of the Russian Federation)

Soil ID	Soil description
25	Podzolics, mainly shallow podzolics
26	Podzolics, mainly rather shallow podzolics
29	Podzolics (without subdivision)
30	Podzolics with the second bleached horizon
42	Sod-podzolics (without subdivision)
50	Sod-podzolics illuvial-ferrugenous
56	Podzols illuvial-ferrugenous (podzols illuvial low-humic)
57	Podzols humic-illuvial
58	Podzols illuvial-humic-ferrugenous (without subdivision)
59	Podzols dry-peaty
62	Podzols gleyic peaty and peat
77	Pales typical
78	Pales podzolized
79	Pales calcareous
80	Pales solodic
81	Gray-pales
83	Sod-calcareouses
102	Light-gray forest
104	Dark-gray forest
115	Pine forest sands
117	Chernozems leached
119	Chernozems ordinary
120	Chernozems southern
125	Chernozems leached glossic
136	Meadow-chernozemics
170	Peaty and peat boggy
178	Meadows
187	Alluvials acid
188	Alluvials saturated
192	Alluvials swamp meadow
222	Taiga gley peaty-muck and soil of spots dystric, including saline
248	Tundra gley peaty and peat and peaty and peat boggy

* only the soil units with the relative contribution to the Maxent model > 0.1 are presented
Available at URL: https://egrpr.esoil.ru/content/1sem.html

Acknowledgements This research was performed according to the Russian Science Foundation project No. 21-47-00016 and Development program of the Interdisciplinary Scientific and Educational School of Lomonosov Moscow State University «Future Planet and Global Environmental Change».

References

Carlson, C. J., Albery, G. F., Merow, C., Trisos, C. H., Zipfel, C. M., Eskew, E. A., Olival, K. J., Ross, N., & Bansal, S. (2022). Climate change increases cross-species viral transmission risk. *Nature*, *607*, 555–562. https://doi.org/10.1038/s41586-022-04788-w

Egorov, V. A., Bartalev, S. A., Kolbudaev, P. A., Plotnikov, D. E., & Khvostikov, S. A. (2018). Land cover map of Russia derived from Proba-V satellite data. *Sovremennye Problemy Distantsionnogo Zondirovaniya Zemli iz Kosmosa*, *15*, 282–286. https://doi.org/10.21046/2070-7401-2018-15-2-282-286

Evengard, B., Destouni, G., Kalantari, Z., Albihn, A., Björkman, C., Bylund, H., Jenkins, E., Koch, A., Kukarenko, N., Leibovici, D., Lemmityinen, J., Menshakova, M., Mulvad, G., Nilsson, L. M., Omazic, A., Pshenichnaya, N., Quegan, S., Rautio, A., Revich, B., Rydén, P., Sjöstedt, A., Tokarevich, N., Thierfelder, T., & Orlov, D. (2021). Healthy ecosystems for human and animal health: Science diplomacy for responsible development in the Arctic. *Polar Record*, *57*(e39), 1–7. https://doi.org/10.1017/S0032247421000589

Ezhova, E., Orlov, D., Suhonen, E., Kaverin, D., Mahura, A., Gennadinik, V., Kukkonen, I., Drozdov, D., Lappalainen, H. K., Melnikov, V., Petäjä, T., Kerminen, V.-M., Zilitinkevich, S., Malkhazova, S. M., Christensen, T. R., & Kulmala, M. (2021). Climatic factors influencing the anthrax outbreak of 2016 in Siberia, Russia. *EcoHealth*. https://doi.org/10.1007/s10393-021-01549-5

Fick, S. E., & Hijmans, R. J. (2017). WorldClim 2: New 1-km spatial resolution climate surfaces for global land areas. *International Journal of Climatology*, *37*(12), 4302–4315.

Georgiades, P., Ezhova, E., Räty, M., Orlov, D., Kulmala, M., Lelieveld, J., Malkhazova, S., Erguler, K., & Petäjä, T. (2022). The impact of climatic factors on tick-related hospital visits and borreliosis incidence rates in European Russia. *PLoS ONE*, *17*(7), 20. https://doi.org/10.1371/journal.pone.0269846

Hijmans, R. J., Cameron, S. E., Parra, J. L., Jones, P. G., & Jarvis A. (2005). Very high resolution interpolated climate surfaces for global land areas. *International Journal of Climatology*, *25*, 1965–1978. https://doi.org/10.1002/JOC.1276

https://www.who.int/news-room/fact-sheets/detail/climate-change-and-health

Kutz, S. J., Hoberg, E. P., Polley, L., & Jenkins, E. J. (2005). Global warming is changing the dynamics of Arctic host-parasite systems. *Proceedings Biological Sciences*, 272(1581), 2571–2576. https://doi.org/10.1098/rspb.2005.3285

Ma, Y., Vigouroux, G., Kalantari, Z., Goldenberg, R., & Destouni, G. (2020). Implications of projected hydroclimatic change for tularemia outbreaks in high-risk areas across Sweden. *International Journal Environmental Research and Public Health*, *17*, 6786. https://doi.org/10.3390/ijerph17186786

Malkhazova, S. M., Mironova, V. A., Orlov, D. S., & Adishcheva, O. S. (2018). Influence of climatic factor on naturally determined diseases in a regional context. *Geography, Environment, Sustainability*, *11*(1), 157–170. https://doi.org/10.24057/2071-9388-2018-11-1-157-170

Mora, C., McKenzie, T., Gaw, I. M., Dean, J. M., von Hammerstein, H., Knudson, T. A., Setter, R. O., Smith, C. Z., Webster, K. M., Patz, J. A., Franklin, E. C. (2022). Over half of known human pathogenic diseases can be aggravated by climate change. *Nature Climate Change*, *12*, 869–875. https://doi.org/10.1038/s41558-022-01426-1

Naimi, B., Hamm, N. A. S., Groen, T. A., Skidmore, A. K., & Toxopeus, A. G. (2014). Where is positional uncertainty a problem for species distribution modelling. *Ecography, 37*, 191–203. https://doi.org/10.1111/j.1600-0587.2013.00205.x

Pecl, G. T., Araújo, M. B., Bell, J. D., Blanchard, J., Bonebrake, T. C., Chen, I. C., Clark, T. D., & Williams, S. E. (2017). Biodiversity redistribution under climate change: Impacts on ecosystems and human well-being. *Science, 355*(6332). https://doi.org/10.1126/science.aai9214

Phillips, S. J., Anderson, R. P., & Schapire, R. E. (2006). Maximum entropy modeling of species geographic distributions. *Ecological Modelling, 190*, 231–259. https://doi.org/10.1016/J.ECOLMODEL.2005.03.026

Revich, B., Tokarevich, N., & Parkinson, A. J. (2012). Climate change and zoonotic infections in the Russian Arctic. *International Journal of Circumpolar Health, 71*, 18792. https://doi.org/10.3402/ijch.v71i0.18792

Ryden, P., Sjöstedt, A., & Johansson, A. (2009). Effects of climate change on tularaemia disease activity in Sweden. *Global Health Action, 2*, 2063. https://doi.org/10.3402/gha.v2i0.2063

The register of livestock burial sites (including anthrax-related) in the territory of the Russian Federation. (2012). Ministry of Agriculture of the Russian Federation. In: Surgucheva, L. M., Borovoy, V. N., & Yaremenko, N. A. (Eds.) Rosinformagrotech. ISBN 978-5-7367-0920-5 (In Russian).

Volodin, E., & Gritsun, A. (2018). Simulation of observed climate changes in 1850–2014 with climate model INM-CM5. *Earth System Dynamics. 9*, 1235–1242. https://doi.org/10.5194/esd-9-1235-2018

Africa

Responding to Climate Change in the Health Sector, Kenya

Andrew K. Githeko

Abstract Kenya has been impacted by the effects of climate change that include epidemics, geographic range expansion of climate-sensitive diseases, droughts and floods. These diseases cause a high health burden. The public health system has put in place intervention measures. Malaria control relies on insecticides and drugs. Rift Valley Fever is managed by vaccinating livestock while dengue and chikungunya are managed using insecticides and larval source management. Victims of drought and floods depend on humanitarian assistance. Water-borne infections can be reduced using safe drinking water sources. A shift from rain-fed to irrigated crops is expected to reduce food insecurity. These adaptation projects require heavy financial investments from the government development budgets.

Keywords Climate · Change · Kenya · Adaptation · Response · Health

1 Introduction

Climate change is caused by an increase in greenhouse gases such as carbon dioxide, methane nitrous oxide, and even water vapor in the atmosphere. The interaction of the atmosphere and oceans causes climate variability which occurs on regional scales. Such phenomenon includes El Niño's, La Niña, Atlantic Ocean oscillations, and the Indian Ocean oscillation that affect rainfall and local temperatures. While the near-surface changes in the global mean temperatures above the pre-industrial level (1850) are low, (0.18 °C/decade between 1981–2022 ₩ (IPCC, 2022)), changes caused by climate variability can be relatively high (1–5 °C/episode) causing more obvious impacts.

Climate variability can cause severe flooding, droughts, and wildfires, which have a direct impact on human health and well-being. It has recently been recognized that

A. K. Githeko (✉)
Climate and Human Health Research Unit Centre for Global Health Research, Kenya Medical Research Institute, Nairobi, Kenya
e-mail: agitheko@gmail.com

R. Akhtar (ed.), *Climate Change and Human Health Scenarios*, Global Perspectives on Health Geography, https://doi.org/10.1007/978-3-031-38878-1_19

persistent extreme weather events have resulted in adverse mental health events ranging from anxiety to suicide (IPCC, 2022).

1.1 Climate-Sensitive Diseases in Kenya

In the last five decades, a number of disease epidemics associated with climate variability have occurred in Kenya. Among the most dramatic were highland malaria, Rift Valley fever (RVF), dengue, cholera, and diarrheal diseases, severe flooding and droughts have had adverse effects through injuries and food shortages. Moreover, such events have affected livestock and cash crop production and thus affected the country's economic backbone and livelihoods.

Besides epidemics and outbreaks, some diseases have expanded their old geographical ranges. For example, prior to the 1990s, there was no report of locally transmitted malaria in the highlands of Central Kenya. However, malaria spread to this region following a change in the mean annual temperature from 17.5 °C to 18 °C and above (Githeko, 2009). In the 1980s, malaria epidemics in the western Kenya highlands were reported in only three districts. However, by the end of the 1990s, epidemics were reported in 15 districts (Githeko & Ndegwa, 2001), and this was associated with positive increase in mean annual temperature anomalies and rainfall above 200 mm/month (Fig. 1).

Cholera outbreaks and epidemics have been associated with flooding following anomalous rainfall (Okaka & Odhiambo, 2018). Examples of such epidemics occurred during the 1982 and 1997–1998 El Niño events. Weather-driven cholera epidemics have largely been confined to the Lake Victoria Basin and the Coastal strip. Moreover, in 2015 when heavy flooding occurred at the beginning of the year, cholera outbreaks were reported in 60% of the counties in Kenya (Githuku et al., 2017). Climate change is expected to increase the frequency of flooding and subsequently the risk of cholera outbreaks and epidemics.

Dengue and chikungunya outbreaks were largely confined to the coastal strip (Fig. 2). In recent years, the two vector-borne diseases have spread to the north eastern and the western parts of the country (Obonyo et al., 2018; Shah et al., 2020; Vu et al., 2017). A study undertaken on four sites in Kenya indicated an association between flooding, anomalously increased temperature, and the abundance of *Aedes aegypti* the dengue virus vector and this increased the risk of dengue transmission (Nosrat et al., 2021).

Chikungunya and dengue viruses are transmitted by the same vector, *Ae. Aegypti,* and the two viruses have a similar distribution in Kenya. However, in arid northern Kenya, *Ae. aegypti* breeds in abundance in open water containers including ground water tanks (Konongoi et al., 2018). This observation suggests that chikungunya and dengue virus transmission can occur during dry spells when people store water in containers within the homestead. Climate change is associated with increased droughts which could increase the storage of water in homesteads.

Rift Valley fever affects both livestock and human beings, and it is associated with extensive flooding. The virus is transmitted by *Aedes*, *Culex*, *Anopheles,* and

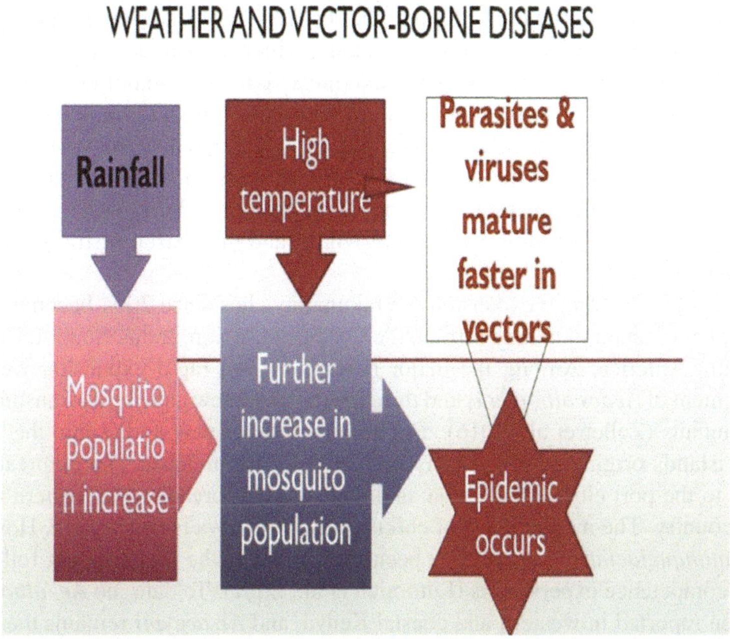

Fig. 1 Rainfall increases mosquito breeding habitats, and their population, while an increase in temperature increases the rate of immature mosquito development and also reduces the time it takes viruses and malaria parasites to become infectious in female mosquitoes

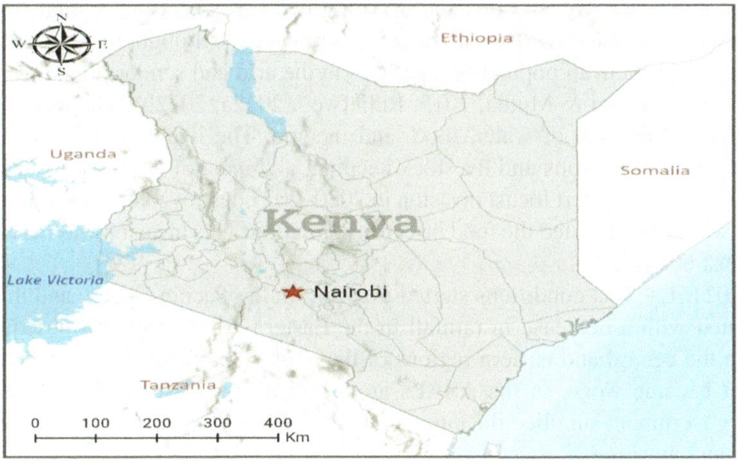

Fig. 2 A map of Kenya

Mansonia mosquito species. In Kenya, the diseases have traditionally occurred in the eastern and north eastern parts of the country which is characterized by arid lands where livestock keeping is the major economic activity. The major epidemics have occurred during El Niño episodes and during positive Indian Ocean dipole events. In 2006–2007, a large outbreak occurred during which the disease was reported in new areas to the west on the original foci of transmission. While in the past, the disease occurred in five districts, it had now spread to 29 of the 49 districts including districts in central, Rift Valley, eastern province, and the coastal districts (Munyua et al., 2010).

The geographic range expansion of chikungunya has since 2004 become a truly global phenomenon including Africa, the Caribbean region, India, South East Asia, and Latin America. Among the major reasons for the rapid expansion were the involvement of *Aedes albopictus* and the appearance of new and highly transmissible viral mutants (Zeller et al., 2016). The 2004 outbreaks that spread into the Indian Ocean islands originated in Lamu Island off the Kenyan coast. The outbreak then moved to the port city of Mombasa, and thereafter, it spread to the western region of the country. The major vector of chikungunya in Kenya is *Ae. aegypti*. However, *Culex quinquefaciatus* has recently been implicated in the transmission following vector competence experiments (Lutomiah et al., 2021). To date, no *Ae. albopictus* has been reported in western and coastal Kenya, and *Ae aegypti* remains the major vector (Ndenga et al., 2017). It is likely that evolving chikungunya gene variants may be contributing to the disease outbreaks.

Frequent and intense droughts continue to impact food shortages, hunger, and malnutrition, especially in the arid and semiarid north and north eastern Kenya. Arid and semiarid land (ASAL) comprises 83% of Kenya's total land surface and available evidence indicates that ASAL is increasing. Following the drought in 2022, 20% (3.1 million) of the Kenyan population that lives in the arid and semiarid land were in a food crisis (Mwangi & Mutua, 2015; Reliefweb, 2022a, 2022b). The drought was characterized by lack of water, food, and income. The livelihood in this area is dependent on food crops and livestock keeping, and most of them were decimated by the drought. A desert locust invasion in 2020–2021 destroyed crops and livestock fodder which exacerbated the food insecurity situation. This invasion was the biggest in the last 60 years.

In 2021, La Niña conditions started evolving in the Pacific Ocean, and this was associated with a decrease in rainfall in the Eastern Africa region. Subsequently, rains in the central and eastern regions of the country performed very poorly. The drought became worse in the ASALs areas and many livestock died. While the Kenya government supplied the human population with food aid, many people had insufficient nutrition.

Flooding events such as those caused by El Niño and the Indian Ocean dipole are frequently followed by Rift Valley fever outbreaks, e.g., 2097–8 and 2006–7 and livestock deaths (Anyamba et al., 2001). Such events result in severe economic decline in the affected areas of north eastern Kenya and subsequent under-nutrition and malnutrition. This condition leads to physical and mental stunting in young children. The frequency of floods has been increasing in Kenya since 2001 (Reliefweb, 2022a, 2022b).

The El Niño Southern Oscillation (ENSO) a weather system that evolves the Pacific Ocean has a global impact. The phenomenon is driven by changes in sea surface temperature anomalies, upwelling of cold ocean water in the Pacific Ocean, and changes in the direction of the trade winds. ENSO affects flooding, and droughts and its impacts are closely related to health particularly in Kenya. It has been speculated that global warming would increase the frequency and intensity of ENSO. Recent findings (Wang et al., 2019) indicate that the severity of El Niños has increased since 1982. There is evidence from this study that warmer weather has increased the intensity of El Niños. This seems to coincide with more intense flooding in Kenya and similar La Nina-driven droughts. This trend implies that a greater effort will be required to mitigate the negative health impacts in the near and the distant future.

2 Sensitivity of Diseases to Climate Change and Variability

The three major components of climate that affect disease transmission are temperature, rainfall, and humidity. Major vector-borne and waterborne diseases have seasonality in their prevalence an indication that they respond to weather and climate variability (Githeko, 2021). Cholera is a water and food-borne disease that occurs in endemic and epidemic forms. Transmission of cholera is dose-dependent, and it takes high concentration of the bacteria in water and food to infect exposed people. The factors that affect the rate of development of the bacteria in the environment include extrinsic temperature, pH, and in some cases salinity. This microbe develops faster at higher temperatures (30 °C). Thus, the level of water contamination with the bacteria increases during hot weather conditions, e.g., during El Niño events that occurred in Kenya in 1982–1983, 1897–1898, and 2006–2007. Such events can have regional effects as observed in the Great Lakes region and the Kenyan coastal strip. Food-borne cholera outbreaks are more localized.

Malaria, RVF, dengue, and chikungunya are mosquito-borne infections. Mosquitoes, parasites, and viruses development is temperature dependent. The mosquito population is a function of water and breeding habitat availability (Fig. 1). Malaria parasite development in the mosquito vector is exponentially dependent on temperature such that small changes in temperate can have a large effect on the extrinsic incubation period of the sporogonic development time. Such temperature changes can expand the geographic range of malaria transmission, especially in the temperature-limited fringe areas of transmission. In the highlands of western and central Kenya, malaria transmission was limited by the 18 °C isotherm at an altitude of approximately 1600–1700 m above sea level. Warming has pushed this isotherm to about 1700 asl, thus expanding the range of malaria transmission. During El Niño events when the temperature is elevated anomalously, the range of transmission is increased further to human populations that have had no prior immunity to malaria, and this results in the occurrence of the severe forms of the disease that is associated with high morbidity and mortality.

Rainfall increases temporary vector habitats that increase the vector population. The rate of development of aquatic stages of the vector is a function of water temperature such that at warmer temperatures mosquito larvae take a shorter time to develop thus increasing the number of mosquito generations per breeding season. The rains lasts for a longer period stabilizes mosquito breeding habitats making them last longer and becoming more productive.

On the contrary during a drought, people tend to store domestic water in open containers close to their houses. Such water provides good breeding habitats for *Aedes aegypti* mosquitoes that are vectors of dengue and chikungunya viruses. Outbreaks of these infections are not uncommon in arid and semiarid north eastern Kenya.

3 The History of Vector-Borne Disease Control in Kenya

3.1 Malaria

Among the vector-borne diseases, malaria outbreaks have the longest history dating back to the 1930s (Mudhune et al., 2011). During those days, the disease was endemic at the coast and in the warm lowlands. Malaria outbreaks were recorded in Nairobi in the 1930s, during a usually warm period (Ebi et al., 2006; Githeko & Shiff, 2005). At that time, malaria control was only undertaken in important urban and commercial areas using Paris Green larviciding and infection treatment using quinine. The discovery of DDT and chloroquine in the 1940s brought hope to malaria control. The discovery of dieldrin, an indoor residual spray insecticide in the late 1940s demonstrated that malaria vector control was feasible. Beyond financial and logistical constraints, the vector developed resistance to insecticides and in the early 1980s *Plasmodium falciparum*, the main malaria parasite developed resistance to chloroquine (Spencer et al., 1987) and in the late 1980s, resistance to pyrimethamine/sulfadoxine appeared (Keuter et al., 1992). At that time, there was no vector control and malaria was managed by the use of drugs. Climate variability-driven malaria epidemics were blamed on drug resistance (Shanks et al., 2005). However, it was argued that while climate variability initiated the epidemics, drug resistance only failed to stop the epidemics.

In early 2000, a great effort was undertaken to discover new antimalarial drugs, and it was at this time that the Chinese drug artemisinin was accepted as potential replacement therapy for chloroquine and pyrimethamine/sulfadoxine (Fansidar). However, the artemisinin-derived monotherapies failed soon after their introduction as replacement therapies (Newton et al., 2011). Artemisinin combination therapies were then recommended as malaria parasites were less likely to develop resistance to this class of drugs.

Later in that decade, trials on the use of pyrethroid-treated nets were undertaken to determine if vector control could reduce malaria-caused morbidity and mortality (Hawley et al., 2003; Nevill et al., 1996). The trials indicated that the insecticide-treated nets (LLIN) reduced the malaria burden in school children, and this led to the introduction of long-lasting insecticide-treated nets with better efficacy. Mass distribution of these nets through the universal LLIN policy significantly reduced malaria morbidity and mortality in the endemic areas. A study carried out in an area of endemic malaria at the Kenyan coast indicated a 5.4-fold reduction in children's hospitalization following the LLIN interventions (O'Meara et al., 2008).

Malaria vectors are notorious for developing resistance to insecticides. After the rollout of the LLINS, the major malaria vectors in Kenya, *Anopheles gambiae, An. arabiensis,* and *An. funestus* developed resistance to pyrethroids (Kawada et al., 2011) and thus reducing the efficacy of the LLIN. In 2017, the World Health Organization (WHO) recommended the combination of pyrethroids with a synergist, piperonyl butoxide (PBO) to restore the efficacy of LLIN and malaria control. The trials carried out to compare the LLIN-PBO and LLINs indicated that LLIN-PBO restored the entomological and parasitological efficacy of the nets (Gleave et al., 2021; Minakawa et al., 2021). The use of effective antimalarial drugs and mass distribution has been used by the health system in Kenya as an intervention against malaria, thus allowing the human population to adapt to living in highly endemic areas of the country. Effective use of this adaptation may also lead to elimination of malaria in areas of low transmission.

Research into the relationship between malaria epidemics in the highland and climate variability led to the development of a weather-based malaria epidemic early warning system (Githeko & Ndegwa, 2001; Githeko et al., 2014). The system detects anomalous temperature and rainfall that are associated with the evolution of malaria epidemics with a lead time of 1–3 months enabling the health authorities to enhance malaria control in vulnerable areas.

Limited application of indoor residual spraying (IRS) using organophosphate insecticides has been implemented in some highly endemic areas to contain malaria transmission, particularly where pyrethroid resistance has been detected (Abong'o et al., 2020). While several trials have been undertaken on larvicides to control malaria vectors in highly endemic areas of western Kenya, none of them have had sufficient impact to qualify for inclusion into malaria control programs (Fillinger et al., 2009; Kahindi et al., 2018).

The ultimate tools for the control of infectious diseases are vaccines. It has been shown for a long time that humans develop parasitological and clinical immunity to malaria infections (Høgh, 1996), thus preventing severe disease and its subsequent complications. Unlike drugs and insecticides, vaccines do not have the problem of resistance.

In the last several decades, scientists in Kenya and elsewhere have made a great effort to develop and test effective vaccines against *Plasmodium falciparum* parasites infections. A vaccine providing partial immunity was launched (Greenwood & Targett, 2011). The RTSS vaccine has been licensed and phase III trials undertaken. However, the vaccine provide only about 40% protection in children, and it requires repeat doses. There is hope that this vaccine will be further developed and could potentially eliminate malaria infections. The World Health Organization has set a target of 75% efficacy for malaria vaccines.

Kenya has been at the forefront of testing new and novel antimalarial arsenal to fight against malaria. This is an important strategy for adaptation aganists an ancient scourge. Climate change unfortunately creates better conditions for the rate of larval development of malaria vectors, the rate of the extrinsic phase of the parasite development, and the transmission of the parasite to humans. It is therefore vital that research into the efficacy of the existing tools, and the development of new tools is supported to reduce or eliminate transmission, morbidity, and mortality. Malaria is now considered a threat to global health and that requires global efforts to control it. Subsequently, international public and private partnerships, financial, and technical support are required to sustain the war against malaria.

While the Kenyan Ministry of health undertakes malaria control through vector control, diagnosis, and early treatment, the Kenya Medical Research Institute and its collaborators undertake epidemiological surveillance, monitoring the efficacy of drugs and insecticides, and undertake trials on vaccines and new control tools. The health system requires enhanced funding to combat the influence of climate change on malaria.

3.2 Dengue and Chikungunya

The history of dengue and chikungunya in Kenya and elsewhere in Africa is not well understood. However, the presence of the major vector *Aedes aegypti* has long been established. The epidemiology of dengue and chikungunya infections has not been well established. It has been suggested that in areas of endemic malaria, dengue infections may have been clinically diagnosed as malaria leading to a significant underestimation of the true burden of the disease. Furthermore, there is a school of thought that people of African descent are less susceptible to severe forms of dengue (Gainor et al., 2022).

Until the recent outbreaks of dengue along the Kenya coast (Lutomiah et al., 2016) and the north eastern region, there was little if any vector control activity. Currently, dengue is not treatable with drugs. However, symptoms are treated until the patient fully recovers. Prevention of dengue and chikungunya is based on vector control. Despite the fact that dengue and chikungunya are the fastest spreading vector-borne diseases worldwide, Kenya does not have a vector control program for the two diseases. Because the major vector feeds on people early morning and at dusk, thus insecticide-treated nets may not provide protection against the vectors.

Aedes aegypti is adapted to the urban environment where it breeds in a variety of water containers including septic tanks and gutters. The control of this vector has been based on larval control by environmental sanitation and larviciding. The adults are controlled by fogging with insecticides. Public health education is critical to both dengue prevention and case management. The health system should have in place a dengue control policies and programs. However, research teams have undertaken some epidemiological and entomological investigations at the Kenyan coastal, western and north eastern regions where little if any control activity is underway. It is now evident that dengue transmission exists in both endemic and epidemic forms in the country. Furthermore, epidemics have been associated with extreme climate events and which are predicted to increase and intensify under climate change scenarios. It is therefore critical that adaptation measures are put in place to prevent health disasters caused by both dengue and chikungunya.

3.3 Rift Valley Fever

Rift Valley fever, a mosquito-borne disease, affects both livestock and human beings. Livestock contributes 12% of the national gross domestic product in Kenya and provides food and livelihood especially to pastoralists living in arid and semiarid area. The disease is closely associated with flooding after heavy rains that are associated with the El Niño Southern Oscillation (ENSO) and the Indian Ocean dipole (IOD). Studies have been carried out to map the RVF endemic and epidemic hotshots (Sang et al., 2017) that would be targets for vaccine deployment.

A vaccine against RVF has been developed for livestock administration, however, no vaccine has been licensed for human use (Mosomtai et al., 2016; Wright et al., 2019). The largest loss of livestock is caused by epidemics following flooding events. Using satellite-derived meteorological and vegetation data, a model for early RVF epidemic warning system was developed (Anyamba et al., 2009). Ideally, the model should be able to identify areas at risk for early vaccination to prevent epidemic evolution.

3.4 Waterborne Diseases

Water, sanitation, and hygiene are critical to prevention of waterborne diseases. These diseases include cholera, typhoid, amebiasis, giardiasis, and hepatitis. Water contamination provides a medium for pathogen transmission to human populations. Children < 5 years of age are particularly susceptible to diarrheal diseases and suffer the highest rates of morbidity and mortality. Ideally, all households should have access to safe piped or community drinking water supplies. However, this provision has been restricted by low economic development, population growth, water availability, and frequent droughts and floods.

According to UNICEF Kenya, 59% of the Kenyan population had access to safe drinking water in 2020. Since the year 2000, the number of households with access to safe drinking water has increased by 12%. However, the urban population has better access to safe drinking water (85%), whereas slightly less than half of the rural population has access to safe drinking water (Marshall, 2011). Kenya's Vision 2030 which is part of the Millennium Development Goals envages access to safe drinking water for all its inhabitants. This would entail the construction of dams, reservoirs, and boreholes. In 2002, the Kenyan government established the National Water Resources Management Authority (WARMA) whose mandate among other things was to address the adaptation of the water sector in response to climate change.

According to WARMA, Kenya has 5541 boreholes, but unfortunately, 60% of them are not functional. Major rehabilitation work is required to bring them back to service.

Kenya has several dams built for multipurpose that includes safe drinking water, irrigation, livestock water, industrial water, and water for power generation. The country has water scarcity and thus the need to construct dams as water reservoirs. Kenya intends to construct 57 new mega dams in order to increase water availability from the current 60% of the population to 80% (CitizenDigital, 2018).

3.5 *Food Security and Nutrition*

According to the Food and Agriculture Organization (FAO), food security exists when all people, at all times, have physical and economic access to sufficient, safe, and nutritious food that meets their dietary needs and food preferences for an active and healthy life. By the year 2018, the arable land in Kenya was 10.1% of the total land mass, and this had been increasing by 1.2% per annum (https://knoema.com/). Climate change is expected to reduce the proportion of arable land and its productivity. The government had come up with a number of strategies to increase food production and food security. Among these strategies is to increase productivity, access to imported foods, expansion of irrigation, increased use of technology, use of genetically modified disease-resistant crops, drought-resistant crops, and credit to farmers. Among the strategies for improving food security is to increase the application of synthetic fertilizers. Reports indicate that the government of Kenya intended to increase the use of fertilizers from 31 kg/heater to 59 kg/heater by 2015 (Oseko & Dienya, 2015).

Rain-fed agriculture will not provide sufficient food and cash crop as demand for consumption and export increases. It is therefore critically important to increase crops under irrigation. The solution to sustainable agriculture lies in the construction of dams to collect and store water. Besides the large multipurpose dams, increasing localized access to water for communities to support their livelihoods by harvesting surface run-off to a cumulative volume of 125,000,000 m^3 by 2023 is planned. Many of the mega dams are expected to increase irrigated acreage and crop yield, e.g., the Thiba Dam will double the rice acreage by an additional 22,000 hectares of irrigated rice. The government plans to put 700,000 acres under irrigation to increase food production.

3.6 *Development of Climate Change Policy in the East African Community Health Sector*

Kenya is part of the East African region, which also includes: Tanzania, Uganda, Rwanda, and Burundi. The region shares a common climate, and the five member states have formed a regional economic block; The East Africa Community (EAC). The EAC secretariat was tasked to develop a climate change policy and strategies to address the adverse impacts of climate change in the region and harness any potential opportunities posed by climate change in the context of the principle of sustainable development. The climate change policy includes adaptation, resilience, and mitigation. The impacts of adaptation are expected to yield results earlier than mitigation, and thus, emphasis should be placed on adaptation. Kenya is currently estimating the cost of approved strategic action implementation of adaptation and resilience to address the expected increased risk due to climate change (WHO, 2015).

3.7 Non-Governmental Partners Addressing Climate Change

The Inter-governmental Authority on Drought and Development (IGADD) which was originally established in 1986 in Nairobi, Kenya, to predict droughts in nine countries in eastern Africa evolved to become IGAD Climate Prediction and Application Centre in 2003, and it is now a World Metrological Organization's (WMO) accredited center of excellence in regional climate prediction and application in a region that has been impacted by climate change and extreme weather events. ICPAC predicts seasonal climate outlooks and collaborates with various sectors in the application and implications of its forecasts. Such information provides early warning for disease outbreaks and other disasters.

3.8 Kenya Red Cross

The Kenya Red Cross Society (KRCS) is a leading humanitarian organization sustainably promoting the well-being, health, and resilience of communities. Its mission is to support the most vulnerable communities facing health and nutrition challenges particularly during emergence such as climate-related crisis (https://www.redcross.or.ke/). The KRCS has taken a lead role with the support of the Kenya government and international partners in executing its mission. As climate change disasters increase in frequency and intensity.

4 Conclusions

Climate change is expected to accelerate in magnitude and frequency. This will increase the geographic range of disease and intensify food insecurity and water deficit. The government has made progress in implementing the climate change policy through the Vision 2030 sustainable development goals. Significant challenges in adaptation remain due to development of insecticide resistance and potential drug resistance, under-nutrition, and malnutrition.

To address these challenges, it is critical to invest in research, so that new tools can be developed and tested for disease control. More effort need to be put on the development and testing of vaccines to protect human and livestock from infections.

It is critical that the ministry of health (MoH) works closely with ministries of agriculture and water and irrigation in addressing population health. In addition, MOH will need to closely collaborate with partners such as IGAD's Climate Prediction and Application Centre (IPAC) in the prediction of adverse weather events. Furthermore, MoH should strengthen support for Kenya Red Cross Society to address humanitarian assistance during climate-related disasters. The Kenya Medical Research Institute will need to develop strategic approaches to address epidemiological surveillance, pathological trends, and monitoring efficacy of existing tools for disease control.

This work was supported by the Director General, Kenya Medical Research Institute. The author previously worked with Prof. Rais Khtar in the Working Group II in IPCC Third Assessment Report.

References

Abong'o, B., Gimnig, J. E., Torr, S. J., Longman, B., Omoke, D., Muchoki, M., Ter Kuile, F., Ochomo, E., Munga, S., Njagi, K., & Samuels, A. M. (2020). Impact of indoor residual spraying with pirimiphos-methyl (Actellic 300CS) on entomological indicators of transmission and malaria case burden in Migori County, western Kenya. *Scientific Reports, 10*, 1–14.

Anyamba, A., Chretien, J.-P., Small, J., Tucker, C. J., Formenty, P. B., Richardson, J. H., Britch, S. C., Schnabel, D. C., Erickson, R. L., & Linthicum, K. J. (2009). Prediction of a Rift Valley fever outbreak. *Proceedings of the National Academy of Sciences, 106*, 955–959.

Anyamba, A., Linthicum, K. J., & Tucker, C. J. (2001). Climate-disease connections: Rift Valley fever in Kenya. *Cadernos De Saude Publica, 17*, S133–S140.

CitizenDigital. (2018). Kenyan Gov't To Construct 57 Dams Across The Country. EAC, 2022. EAC Climate Change Policy Framework.

Ebi, K. L., Smith, J. B., & Burton, I. (2006). *Integration of public health with adaptation to climate change: Lessons learned and new directions*. CRC Press.

Fillinger, U., Ndenga, B., Githeko, A., & Lindsay, S. W. (2009). Integrated malaria vector control with microbial larvicides and insecticide-treated nets in western Kenya: A controlled trial. *Bulletin of the World Health Organization, 87*, 655–665.

Gainor, E. M., Harris, E., & LaBeaud, A. D. (2022). Uncovering the burden of dengue in Africa: Considerations on magnitude, misdiagnosis, and ancestry. *Viruses, 14*, 233.

Githeko, A. K. (2009). *Malaria and climate change: Special feature, Commonwealth health ministers' update 2009*. Pro-Brook Publishing, GB.

Githeko, A. K. (2021). Health related vulnerabilities and enabling institutions to facilitate responses to climate change in East Africa. *East Africa Science, 3*, 1–18.

Githeko, A. K., & Ndegwa, W. (2001). Predicting malaria epidemics in the Kenyan highlands using climate data: A tool for decision makers. *Global Change and Human Health, 2*, 54–63.

Githeko, A. K., Ogallo, L., Lemnge, M., Okia, M., & Ototo, E. N. (2014). Development and validation of climate and ecosystem-based early malaria epidemic prediction models in East Africa. *Malaria Journal, 13*, 1–11.

Githeko, A. K., & Shiff, C. (2005). The history of malaria control in Africa: Lessons learned and future perspectives. In: *Integration of public health with adaptation to climate change: Lessons learned and new directions*. Francis and Taylor.

Githuku, J. N., Boru, W. G., Hall, C. D., Gura, Z., Oyugi, E., Kishimba, R. S., Semali, I., Farhat, G. N., & Park, M. M. (2017). Cholera outbreak in Homa Bay County, Kenya, 2015. *The Pan African Medical Journal, 27*.

Gleave, K., Lissenden, N., Chaplin, M., Choi, L., & Ranson, H. (2021). Piperonyl butoxide (PBO) combined with pyrethroids in insecticide-treated nets to prevent malaria in Africa. *Cochrane Database of Systematic Reviews*.

Greenwood, B., & Targett, G. (2011). Malaria vaccines and the new malaria agenda. *Clinical Microbiology and Infection, 17*, 1600–1607.

Hawley, W. A., Phillips-Howard, P. A., ter Kuile, F. O., Terlouw, D. J., Vulule, J. M., Ombok, M., Nahlen, B. L., Gimnig, J. E., Kariuki, S. K., Hightower, A. W., & Kolczak, M. S. (2003). Community-wide effects of permethrin-treated bed nets on child mortality and malaria morbidity in western Kenya. *The American Journal of Tropical Medicine and Hygiene, 68*, 121–127.

Høgh, B. (1996). Clinical and parasitological studies on immunity to Plasmodium falciparum malaria in children. *Scandinavian Journal of Infectious Diseases. Supplementum, 102*, 1–53.

IPCC. (2022). Climate change 2022: Impacts, adaptation and vulnerability, IPCC Sixth Assessment Report.

Kahindi, S. C., Muriu, S., Derua, Y. A., Wang, X., Zhou, G., Lee, M.-C., Mwangangi, J., Atieli, H., Githeko, A. K., & Yan, G. (2018). Efficacy and persistence of long-lasting microbial larvicides against malaria vectors in western Kenya highlands. *Parasites & Vectors, 11*, 1–10.

Kawada, H., Dida, G. O., Ohashi, K., Komagata, O., Kasai, S., Tomita, T., Sonye, G., Maekawa, Y., Mwatele, C., Mwandawiro, C., & Njenga, S. M. (2011). Multimodal pyrethroid resistance in malaria vectors, Anopheles gambiae ss, Anopheles arabiensis, and Anopheles funestus ss in western Kenya. *PloS one, 6*, e22574.

Keuter, M., Sanders, J., Ronday, M., Veltkamp, S., Kamsteeg, H., Schouten, E., Khalumi, G., Ngwawe, W., Wetsteyn, J. C., & Brandling-Bennett, A. (1992). Parasitological, clinical and haematological response of children with Plasmodium falciparum to 4-aminoquinolines and to pyrimethamine-sulfadoxine with quinine in western Kenya. *Tropical and Geographical Medicine, 44*, 1–8.

Konongoi, S. L., Nyunja, A., Ofula, V., Owaka, S., Koka, H., Koskei, E., Eyase, F., Langat, D., Mancuso, J., Sang, R., & Lutomiah, J. (2018). Human and entomologic investigations of chikungunya outbreak in Mandera, Northeastern Kenya, 2016. *PloS one, 13*, e0205058.

Lutomiah, J., Barrera, R., Makio, A., Mutisya, J., Koka, H., Owaka, S., Koskei, E., Nyunja, A., Eyase, F., Sang, R., & Coldren, R. (2016). Dengue outbreak in Mombasa city, Kenya, 2013–2014: entomologic investigations. *PLoS Neglected Tropical Diseases, 10*, e0004981.

Lutomiah, J., Mulwa, F., Mutisya, J., Koskei, E., Langat, S., Nyunja, A., Koka, H., Konongoi, S., Chepkorir, E., Owaka, S., & Ofula, V. (2021). Probable contribution of Culex quinquefasciatus mosquitoes to the circulation of chikungunya virus during an outbreak in Mombasa County, Kenya, 2017–2018. *Parasites & Vectors, 14*, 1–14.

Marshall, S. (2011). The water crisis in Kenya: Causes, effects and solutions. *Global Majority E-Journal, 2*, 31–45.

Minakawa, N., Kongere, J.O., Sonye, G.O., Lutiali, P.A., Awuor, B., Kawada, H., Isozumi, R., & Futami, K. (2021). Long-lasting insecticidal nets incorporating piperonyl butoxide reduce the risk of malaria in children in Western Kenya: A cluster randomized controlled trial. *The American Journal of Tropical Medicine and Hygiene, 105*, 461.

Mosomtai, G., Evander, M., Sandström, P., Ahlm, C., Sang, R., Hassan, O. A., Affognon, H., & Landmann, T. (2016). Association of ecological factors with Rift Valley fever occurrence and mapping of risk zones in Kenya. *International Journal of Infectious Diseases, 46*, 49–55.

Mudhune, S. A., Okiro, E. A., Noor, A. M., Zurovac, D., Juma, E., Ochola, S. A., & Snow, R. W. (2011). The clinical burden of malaria in Nairobi: A historical review and contemporary audit. *Malaria Journal, 10*, 1–10.

Munyua, P., Murithi, R. M., Wainwright, S., Githinji, J., Hightower, A., Mutonga, D., Macharia, J., Ithondeka, P. M., Musaa, J., Bloland, P., & Breiman, R. F. (2010). Rift Valley fever outbreak in livestock in Kenya, 2006–2007. *The American Journal of Tropical Medicine and Hygiene, 83*, 58.

Mwangi, K. K., & Mutua, F. (2015). Modeling Kenya's vulnerability to climate change–A multifactor approach. *International Journal of Science and Research, 6*, 12–19.

Ndenga, B. A., Mutuku, F. M., Ngugi, H. N., Mbakaya, J. O., Aswani, P., Musunzaji, P. S., Vulule, J., Mukoko, D., Kitron, U., & LaBeaud, A. D. (2017). Characteristics of Aedes aegypti adult mosquitoes in rural and urban areas of western and coastal Kenya. *PloS one, 12*, e0189971.

Nevill, C. G., Some, E. S., Mung'ala, V. O., Mutemi, W., New, L., Marsh, K., Lengeler, C., & Snow, R. W. (1996). Insecticide-treated bednets reduce mortality and severe morbidity from malaria among children on the Kenyan coast. *Tropical Medicine & International Health: TM & IH, 1*, 139–146.

Newton, P. N., Green, M. D., Mildenhall, D. C., Plançon, A., Nettey, H., Nyadong, L., Hostetler, D. M., Swamidoss, I., Harris, G. A., Timmermans, A. E., & Powell, K. (2011). Poor quality vital anti-malarials in Africa—an urgent neglected public health priority. *Malaria Journal, 10*, 1–22.

Nosrat, C., Altamirano, J., Anyamba, A., Caldwell, J. M., Damoah, R., Mutuku, F., Ndenga, B., & LaBeaud, A. D. (2021). Impact of recent climate extremes on mosquito-borne disease transmission in Kenya. *PLoS Neglected Tropical Diseases, 15*, e0009182.

O'Meara, W. P., Bejon, P., Mwangi, T. W., Okiro, E. A., Peshu, N., Snow, R. W., Newton, C. R., & Marsh, K. (2008). Effect of a fall in malaria transmission on morbidity and mortality in Kilifi, Kenya. *Lancet (London, England), 372*, 1555–1562.

Obonyo, M., Fidhow, A., & Ofula, V. (2018). Investigation of laboratory confirmed Dengue outbreak in North-eastern Kenya, 2011. *PLoS ONE, 13*, e0198556.

Okaka, F. O., & Odhiambo, B. (2018). Relationship between flooding and out break of infectious diseasesin Kenya: A review of the literature. *Journal of Environmental and Public Health 2018.*

Oseko, E., & Dienya, T. (2015). Fertilizer consumption and fertilizer use by crop (FUBC) in Kenya. Africafertilizer.org.

Reliefweb. (2022a). Kenya ASAL: IPC Acute Food Insecurity and Acute Malnutrition Analysis (February–June 2022a) Issued in March 2022a.

Reliefweb. (2022b). Kenya: Floods—DREF Operation n° MDRKE047—Final Report.

Sang, R., Lutomiah, J., Said, M., Makio, A., Koka, H., Koskei, E., Nyunja, A., Owaka, S., Matoke-Muhia, D., Lindahl, J., & Bukachi, S. (2017). Effects of irrigation and rainfall on the population dynamics of Rift Valley fever and other arbovirus mosquito vectors in the epidemic-prone Tana River County, Kenya. *Journal of Medical Entomology, 54*, 460–470.

Shah, M. M., Ndenga, B. A., Mutuku, F. M., Vu, D. M., Grossi-Soyster, E. N., Okuta, V., Ronga, C. O., Chebii, P. K., Maina, P., Bosire, C. M., & Jembe, Z. (2020). High dengue burden and circulation of 4 virus serotypes among children with undifferentiated fever, Kenya, 2014–2017. *Emerging Infectious Diseases, 26*.

Shanks, G. D., Hay, S. I., Omumbo, J. A., & Snow, R. W. (2005). Malaria in Kenya's western highlands. *Emerging Infectious Diseases, 11*, 1425.

Spencer, H. C., Kaseje, D. C., Brandling-Bennett, A., Oloo, A. J., Churchill, F. C., & Koech, D. K. (1987). Changing response to chloroquine of Plasmodium falciparum in Saradidi, Kenya, from 1981 to 1984. *Annals of Tropical Medicine & Parasitology, 81*, 98–104.

Vu, D. M., Mutai, N., Heath, C. J., Vulule, J. M., Mutuku, F. M., Ndenga, B. A., & LaBeaud, A. D. (2017). Unrecognized dengue virus infections in children, western Kenya, 2014–2015. *Emerging Infectious Diseases, 23*, 1915.

Wang, B., Luo, X., Yang, Y.-M., Sun, W., Cane, M. A., Cai, W., Yeh, S. W., & Liu, J. (2019). Historical change of El Niño properties sheds light on future changes of extreme El Niño. *Proceedings of the National Academy of Sciences, 116*, 22512–22517.

WHO. (2015). *Climate and health country profile 2015: Kenya.* World Health Organization.

Wright, D., Kortekaas, J., Bowden, T. A., & Warimwe, G. M. (2019). Rift Valley fever: Biology and epidemiology. *Journal of General Virology, 100*, 1187–1199.

Zeller, H., Van Bortel, W., & Sudre, B. (2016). Chikungunya: Its history in Africa and Asia and its spread to new regions in 2013–2014. *The Journal of Infectious Diseases, 214*, S436–S440.

Climate Change Impacts, Adaptation and Mitigation Strategies in Tanzania

Calvin Sindato and Leonard E. G. Mboera

Abstract The United Republic of Tanzania is among the countries that has experienced impacts of climate change including epidemics of climate-sensitive infectious diseases, food and nutrition insecurity. Others include damage to infrastructure caused by flooding and land slides resulting to human injuries, deaths and displacement, and high cost to restore the damaged infrastructure. The climate-sensitive diseases that have occurred in Tanzania include dengue, chikungunya, malaria, Rift Valley fever, leptospirosis, cholera and Human African Trypanosomiasis. The country has recently developed a National Climate Change Strategic Plan to provides a set of interventions on adaptation and mitigation, which are expected to strengthen country's resilience to the impacts of climate change and contribute to the global efforts of reducing greenhouse gas emissions. In addition, the country has developed the Health-National Adaptation Plan to climate change to guide the towards a health system that is more resilient to climate change. However, the efficiency of the operationalization of these strategies are not sufficiently known.

Keywords Infectious diseases · Chikungunya · Rift valley fever · Health national adaptation plan · Tanzania · Nutrition insecurity

1 Background

The United Republic of Tanzania is located at the East Coast of Africa between latitudes 1° South and 12° South and between longitudes 29° East and 41° East. Tanzania shares the borders with Kenya and Uganda to the North, Rwanda, Burundi, Democratic Republic of Congo and Zambia to the West, and Malawi and Mozambique to

---

C. Sindato (✉)
National Institute for Medical Research, Tabora Research Centre, P.O. Box 482, Tabora, Tanzania
e-mail: calvin.kimaro@nimr.or.tz

L. E. G. Mboera (Deceased)
SACIDS Foundation for One Health, Morogoro, Tanzania

R. Akhtar (ed.), *Climate Change and Human Health Scenarios*, Global Perspectives on Health Geography, https://doi.org/10.1007/978-3-031-38878-1_20

the South. The country has a total area of 945,087 square kilometres and a population of 61.7 million people (https://www.nbs.go.tz/index.php/en/census-surveys) with a life expectancy at birth of 64 years (MoH, 2019). The country is dominated by large central plateaus covered with grasslands, plains and rolling hills. There are number of highland belts including Mount Kilimanjaro (5895 m above mean sea-level), and Mount Meru (4566 m). Tanzania's topographical diversity gives rise to four distinct climatic zones; namely the coastal area and immediate hinterland; the central plateau; and the high moist lakes regions. The weather seasons are well defined. From December to March, it is hot and comparatively dry. In the northern part of the country, the heavy rains fall in April and May, while the short rains in October–December (Borhara et al., 2020). The southern regions have only one rainy season, from November to April.

Over the past six decades, Tanzania has experienced several direct and indirect effects of climate change. This has included epidemics of climate-sensitive infectious diseases, food and nutrition insecurity. Other climate change effects include variability in the agriculture, transportation, hydropower generation, deaths of humans, animals and plants, injuries and human displacement (Kangalawe, 2011; Lalika et al., 2015; Mboera et al., 2011). A number of climate-associated infectious disease epidemics, including dengue, chikungunya, malaria, plague, Rift Valley fever, leptospirosis, cholera and Human African Trypanosomiasis have been reported in Tanzania (Hide, 1999; Van den Bossche et al., 2010; Mboera et al., 2011; Sindato et al., 2014). Moreover, the knowledge of climate change among communities of Tanzania is low (Mayala et al., 2011). To mitigate climate change, Tanzania has established a National Climate Change Steering Committee and National Climate Change Technical Committee. Nonetheless, specific climate change associated human health issues are inadequately addressed. The objective of this chapter is to understand the pattern of climate change impacts, adaptation and mitigation strategies in Tanzania.

Tanzania experiences bimodal and unimodal rainfall pattern in different parts of the country. The coastal areas are warm and generally wet experiencing an average of 1000 mm/year of rainfall and daily maximum temperatures ranging between 29 and 32 °C. The central parts of the country experiences a mean annual rainfall of less than 500 mm/year and average daily maximum temperatures, ranging from 27 to 31 °C. The Lake Victoria areas experience a rainfall of 700 mm/year (Tanzania climate action report, 2016).

In terms of rainfall, an annual increase by 10% is expected by 2100, and temperature is projected to increase by 1.5–4.5 °C by the 2090s. Climate projections indicate that northern and southern parts of the country would experience an increase in rainfall ranging from 5 to 45% and that most parts of the country might experience a decrease in rainfall of 10–15% (Mwandosya et al., 1998). During the past 30 years, there has been an increase in temperature, droughts have been recurrent, while water levels in Lakes Victoria, Tanganyika, Rukwa and Babati have dropped significantly. These accelerating impacts threaten the lives and health of most of the human and animal populations. Tanzania has also experience rise in sea levels. Islands of Maziwi in Pangani and Fungu la Nyani in Rufiji districts are already submerged due to rise in sea level. Predictions indicate that Zanzibar and Mafia islands are likely to disappear under water by 2100 following rise in sea level caused by melting of polar ice. Mount Kilimanjaro, the highest mountain in Africa, is losing its 11,700 year old glacial top at an astounding rate. About 80% of glaciers on the mountain have been lost since 1912. It is expected that within the next 10–20 years, the summit will be bare (Thompson et al., 2007).

Natural disasters such as landslides, droughts and floods are frequent in Tanzania. A decade ago, heavy rains accompanied with strong winds have left thousands of people displaced and without food in Muleba, Kilosa and Same districts (Daily News, November 25, 2009). Kilosa district experienced floods from end of December 2009 extending for several weeks in January 2010. In December 2011, Dar es Salaam, the largest city in Tanzania, experienced the heaviest rains and resulting into deaths of 29 people and displacement of 29 people (https://allafrica.com/stories). In October 2009, Ngorongoro, Longido and Monduli districts lost between 3000 and 4000 cattle due to drought (http://www.citizenjournalismafrica.org/), and this has an impact of food security. In a study in Tanzania, heavy rainfall has been reported to have a strong association with child 0–4 years mortality. The same study has reported that monthly average temperature had a stronger association with death in all ages while mortality increased with falling monthly temperature (Mrema et al., 2012).

2 Climate Change and Infectious Diseases

Predicted changes in climate and climate impacts will have direct and indirect impacts on human, animal and plant health. Warming is predicted to increase the incidence of vector-borne and waterborne diseases, and non-communicable diseases through changes in pathogen and vector development rates, migration of the vectors and host populations, changes in transmission dynamics and host susceptibility to infection (Gubler et al., 2001; Patz et al., 2000). The increased frequency of droughts and flooding is in turn likely to increase the frequency and magnitude of epidemics of waterborne diseases such as typhoid and cholera, as well as to influence the incidence of vector-borne diseases (Colwell, 1996; Pascual et al., 2000).

2.1 Vector-Borne Diseases

Malaria: Malaria is among the most important vector-borne disease causing high morbidity and mortality in Tanzania. The endemicity and pattern of malaria transmission between and within the districts depend on many factors including climate and topography. Over many decades, malaria has remained a common disease in low-altitude rural areas of Tanzania (Clyde, 1967). Malaria is endemic in Tanzania, with about 95% of the population at high risk. The disease contributes to about a quarter of all outpatient attendances; accounts for about a third of all hospital admissions and is the leading cause of hospital mortality (Mboera et al., 2018). The disease accounts for 35% of the total years' potential life lost in Tanzania (Rumisha et al., 2020). Though the whole population of the country is at risk for malaria, transmission varies between and within regions (Rumisha et al., 2019). This variation is attributed to both climatic and non-climatic factors (Chirebvu et al., 2016; Rumisha et al., 2019).

A number of districts in Tanzania are prone to malaria epidemics, with high mortality during the rainy season. During the past two decades, malaria epidemics have been reported from highland areas of Muheza, Babati, Hanang, Mbulu, Ngorongoro, Mpwapwa, Muleba, Sumbawanga and Lushoto (Mboera, 2004; Mboera & Kitua, 2001). Factors influencing malaria epidemics in Tanzania include climate variability and land-use that are likely to have some effects on vectors of malaria in these areas. Matola et al. (1987) showed that vegetation clearing that occurred in the East Usambara Mountains played a significant role in the increase of malaria transmission in the area. The occurrence and increase of malaria incidence in the Western and Eastern Usambara Mountains were a result of ecological changes. On the other hand, malaria epidemics in Mbulu and Muleba district followed drought periods support and was associated with food in security (Clyde, 1967; Garay, 1998). The risk of mortality from malaria during the Muleba epidemic was observed to be seven times higher in malnourished than well-nourished children (Garay, 1998).

Dengue: Dengue is caused by dengue virus (DENV) and transmitted by Aedes mosquitoes. It was first reported in the southern coastal areas of Tanzania during the 15th Century by Spanish sailors. Dengue outbreaks were later reported between 1823 and 1870 on the Islands of Zanzibar (Amarasinghe et al. 2011). During the past decade, dengue outbreaks in Tanzania have been reported with increased incidence (Mboera et al., 2016; Vairo et al., 2016; Okada et al., 2019). In addition, seroprevalence studies have reported dengue infection prevalence in several district of Tanzania with rates of 6.4–50.6% (Chipwaza et al., 2014; Kajeguka et al., 2016; Mboera et al., 2016; Mwanyika et al., 2021; Vairo et al., 2014, 2016; Ward et al., 2017). The DENV-1, DENV-2, DENV-3, and DENV-4 serotypes have been reported in Tanzania (Mboera et al., 2016; Vairo et al., 2016). Despite the frequent dengue outbreaks in the country, no effective Aedes vector control measures have been introduced.

Chikungunya: Chikungunya, which literary means a "disease that bends up the joint", is caused by chikungunya virus (CHIKV) and transmitted by Aedes mosquitoes. It was first reported from Newala and Masasi districts of southern Tanzania in an outbreak in 1952–1953 (Robinson, 1955). In Africa, CHIKV infections have been reported in 29 countries (CDC, 2019), with recent outbreaks reported from Angola, Cameroon, Central African Republic, Comoros, The Republic of Congo, The Democratic Republic of Congo, Kenya, Gabon, Guinea, Malawi, Mozambique, Reunion, Nigeria, Seychelles, South Africa, Tanzania and Uganda (Nsoesie et al., 2016; Proesmans et al., 2019; Wahid et al., 2017).

In Tanzania, chikungunya is underreported and go unnoticed despite the fact that seroprevalence studies have indicated the presence of the infection in some districts of Tanzania (Chipwaza et al., 2014; Kajeguka et al., 2016; Kinimi et al., 2018; Mwanyika et al., 2021). Moreover, there are only a few studies that have established chikungunya vectors and transmission indices in Tanzania. Patrick et al. (2018) reported an overall CHIKV infection rate of 30% in *Aedes aegypti* in Karagwe, Kyerwa and Mbeya. An outbreak of the disease has been reported in Tanga, Tanzania affecting four individuals (https://www.thecitizen.co.tz/news/).

Rift Valley fever: Rift Valley fever (RVF) is caused by Rift Valley fever virus in the family Phenuiviridae. It is primarily transmitted by Aedes mosquitoes and via direct contact through infected animal products or contaminated foods or aborted foetuses. The disease is characterized by outbreaks that follow heavy rains and the consequent emergence of large numbers of Aedes and Culex mosquitoes, the later vector been regarded as an outbreak amplifier. The mosquito itself appears to be the reservoir and, since rain is needed to hatch the eggs and heavy rain only occurs every few years, this phenomenon may explain the long periods between epidemics.

RVF epidemics in Tanzania have been recorded to occur in 1956, 1978–1979, 1997–1998, and 2006–2007 (Sindato et al,. 2014). All the past outbreaks were reported in animals before cases could be detected in humans. In January 2007, an outbreak of Rift Valley fever was detected among humans starting from northern Tanzania districts, but later spreading southwards and westwards to affect other parts of the country. Ten regions of the country were affected, and a total of 511 suspect and 186 confirmed RVF cases were reported. All RVF cases were located in the north-central and southern regions of the country. Affected regions were Manyara, Tanga, Dodoma, Morogoro, Dar es Salaam, Coast, Iringa, Mwanza and Singida. Several recent studies have reported RVF in human to be prevalent in a number of districts/regions of Tanzania including Rufiji (3.0%), Ilala (1.8%) Sengerema (1.4%), Buhigwe (%), Kalambo (2.0%), Kinondoni (3.0%), Moshi (2.6%), Mvomero (1.0–19.4%) and Ukerewe (Kumalija et al., 2021; Rugarabamu et al., 2021). A recent study in Ilala by Rufiji and Sengerema has indicated that there is one human case of RVF for every five cases of domestic ruminants; for every nine cases of cattle; and for every three cases of sheep (Sindato et al., 2022). This means that human risk of RVFV is a function of the disease occurrence in animals, and the transmission dynamics depends both on the affected animal species and human-animal contact structure (Sindato et al., 2014, 2022).

Schistosomiasis. Schistosoma mansoni and *S. haematobium* infections are prevalent in Tanzania. Tanzania ranks second among countries with the highest burden of schistosomiasis in Africa (Mazigo et al., 2012). However, the endemicity of the disease in Tanzania varies between the regions. The northern, north-western, central-southern are highly endemic for the two species of Schistosoma (Mazigo et al., 2012). The life cycle of the parasite is complex involving snails, water and human beings. The cycle is susceptible to environmental change, especially in water-associated stages. Available statistics indicate that climate change may alter the geographical distribution of schistosomiasis. This is because climate affects the suitability of fresh-water bodies for hosting the snail population and the parasites. The snail hosts of schistosomes differ for each of the major species of parasite. A major determinant of schistosome distribution is the distribution of the snail host. Snail populations are dependent on temperature, water and water currents. Studies have established that ambient temperature is an important limiting factor of the survival of snails and of the shedding of cercariae (Iijima & Sugiura, 1962).

Leptospirosis: Leptospirosis is one of the important but neglected zoonotic diseases that is also not diagnosed routinely in Tanzania. Humans get leptospirosis through contact with an environment contaminated with *Leptospira* bacteria from reservoir hosts and cats. Prevalence of leptospirosis in cattle in pastoral areas of Tanzania has been reported (Machang'u et al., 1997). A study among patients with fever admitted at a hospital in northern Tanzania reported a leptospirosis prevalence of 9% (Biggs et al., 2011). In July 2022, 20 cases of leptospirosis and 3 deaths were reported from Ruangwa District of southern Tanzania (unpublished). Both Sokoine, Lora, Kenya, Canicola, Habdomadis and Pomona serovars are prevalent in Tanzania (Ahmed et al., 2006; Mgode et al., 2006). *Mastomys natalensis* is responsible for most of the Leptospira transmission in Tanzania.

2.2 Waterborne Diseases

Cholera and other diarrhoeal diseases: Both flooding and unusually low levels of water can result in water contamination and contribute to higher morbidity and mortality rates cholera and other diarrhoeal diseases (Hashizume, 2008). Most often, warmer surface temperatures increase the abundance of phytoplankton, which supports a large population of zooplankton, which serves as a reservoir for cholera bacteria. The combination of higher temperatures, prolonged droughts and floods coupled with scarce water resources and poor sanitation make countries in Sub-Saharan Africa vulnerable to cholera outbreaks and other waterborne diarrhoeal diseases. During the 1997–98 El Niño, a rise in sea surface temperature coupled with excessive flooding was the two important factors in cholera epidemics in Djibouti, Somalia, Kenya, Tanzania and Mozambique. In a study in the Lake Victoria basin in Tanzania, high incidences of cholera coincided with high flow peaks and high temperatures before and during El Niño years (Wandiga et al., 2006).

Reports from the Health Management Information system of the Ministry of Health (unpublished) indicate that in a 10-year period of 2007–2016, a total of 34,544 cumulative cases of cholera were reported in Tanzania. Of these, 515 died translating to overall case fatality rate of 1.5. The spatial distribution of the cholera cases in Tanzania varied by regions. Most cases were reported persistently in the coastal-central-northern parts of the country. Overall, there was an increase in the number of regions which reported outbreaks from 11 in 2007 to 23 regions in 2016. A previous study by Taylor (2009) has also indicated that the five coastal regions of Tanzania are among the top-eight regions by number of cases of cholera per capita.

2.3 Zoonoses

Anthrax: Climate change has also been associated with increase in the prevalence distribution of anthrax. A study in Kenya suggests an increasing effect of rainfall and annual temperature range with the probability of anthrax distribution prediction at specific ranges (Otieno et al., 2021). In Tanzania and elsewhere, temperature and rainfall trends, seasonality and extremes have been found to determine anthrax outbreak distribution (Blackburn et al., 2017; Mwakapeje et al., 2018; Walsh et al., 2018). An increase in rainfall has been reported to influence anthrax outbreaks by exposing buried spores to the surface or increase run-offs that collect and concentrate spores in spot area (Dragon & Rennie, 1995).

Human African Trypanosomiasis: Human Africa Trypanosomiasis (HAT, also known as sleeping sickness) is a zoonotic disease caused by trypanosomes and transmitted by tsetse flies (Brun et al., 2010). The disease is found mainly in Sub-Saharan Africa in two forms depending on the subspecies of the parasite involved (WHO, 2013). *Trypanosoma brucei gambiense*, the chronic form of the diseases with invasion of the central nervous system, is found in west and central Africa while *Trypanosoma brucei rhodesiense,* the acute form of the disease with parasites invading the central nervous system a few months after initial infection, is found in eastern and southern Africa (Brun et al., 2010; Kennedy, 2004; Welburn et al., 2001). The geographic separation of the two forms of the disease coincides with the Rift Valley, with *rhodesiense* present at the east, while *gambiense* is found to the west of the Valley. The rhodesiense HAT has more epidemic potential in humans than the gambiense HAT, and it remains fatal if left untreated (Hide, 1999).

HAT is one of the vector-borne infectious diseases that is expected to respond to climate change and has been predicted to increase in incidence and/or expand their geographical range owing to predicted climate changes (Wildlife Conservation Society, 2008). Although the disease is restricted to Sub-Saharan Africa (Migchelsen et al., 2011), given increased movements and interactions between humans, animals and their environment in the changing climatic conditions, the two forms of the disease are expected to present a geographical overlap (Picozzi et al., 2005) and the disease (especially the rhodesiense form) can spread outside of the endemic foci to non-endemic areas of the world (Gao et al., 2020; Norman et al., 2015; Simarro et al., 2012). Although the total number of reported cases of rhodesiense HAT has shown a gradual descending trend over nearly three decades, a recent study found that almost all of the new cases have been reported in Uganda and Tanzania, and the number of cases in both countries accounted for over 80% of all cases from 1990 to 2007 (Gao et al., 2020).

In Tanzania, tsetse flies occur in over 65% of rangeland savannah ecosystems (Malele, 2020), exposing about four million people in rural communities to the risk of HAT. Their distribution is influenced by climatic conditions particularly temperature (Nnko et al., 2021). The past data indicate that that between 1996 and 2006, 2748 cases were reported in the country, with more than three quarter of the cases been reported from the western parts of the country where the active foci have remained persistent for over the past 80 years (Simarro et al., 2008). Recent data indicate that in 2017, three cases of the disease were reported in Tanzania (Gao et al., 2020), which may not be a true representation of the situation in the absence of active surveillance.

Toxoplasmosis: Toxoplasmosis is a parasitic zoonotic disease caused by a coccidian intracellular protozoan parasite, Toxoplasma gondii. It is of worldwide distribution with at least one-third of human population infected (Flegr et al., 2014; Furtado et al., 2011). The infection may result to cause abortion, mental retardation, blindness and encephalitis. Although the literature describing the effect of climate on the risk of toxoplasmosis, few studies have reported a positive correlation between high rainfall and the risk of occurrence of the disease (Boada-Robayo et al., 2022; Jones & Dubey, 2010; Rudzinski et al., 2013). Higher humid and warm soil conditions have been reported to support oocyst survival than under dry conditions (Lélu et al., 2012). Humans are infected mainly by ingesting food or water contaminated with oocysts shed by cats or by eating undercooked or raw meat containing tissue cysts (Dawson, 2005; Dubey, 2004).

T. gondii prevalence in humans is associated with changing environmental conditions as well as anthropogenic factors. Globally, the seroprevalence of this disease varies between 1 and 100% (Alzaheb, 2018). Higher seroprevalence is in areas where stray cats are plentiful and consumption of raw or under-cooked meat is common (Tenter et al., 2002). In Africa, toxoplasmosis has been reported in humans, domestics and wild animals. Human toxoplasmosis prevalence reports from previous studies were 6.7% in Korea, 12.3% in China, 23.9% in Nigeria, 58.4% in Tunisia, 21% in Mali, 83.5% in Madagascar and 46% in Tanzania although most studies have targeted pregnant women and pastoralists (Gashout et al., 2016; Paul et al. 2018). In a previous study, a high prevalence of toxoplasmosis (52.2%) was reported among livestock keepers in Tanga, north-eastern Tanzania (Swai & Schoonman, 2009). Another study in Mwanza in northern Tanzania has shown that 30.9% of pregnant women were infected with Toxoplasma (Mwambe et al. 2013), whereas in Dar es Salaam, a prevalence of 35% has been reported (Doehring et al., 1995). Evidence from a recent study in Tanzania suggests that *T. gondii* infection accounts for 0.08% of all total hospital deaths and is associated with a number of co-morbidities including human immunodeficiency virus/acquired immunodeficiency syndrome, cryptococcosis and pneumocyts pneumonia (Mboera et al., 2019). Despite evidence of *T. gondii* parasites in human and animals, it has remained among the least prioritized disease in Tanzania healthcare system most likely due to limited diagnostic capacity and lack of awareness.

3 Non-communicable Diseases

The rise in temperature has been associated with incidences of severe heat waves. Such conditions are likely to bring heightened risk to human survival and increase the incidence of some non-communicable diseases (NCDs) including malnutrition, mental disorders, injuries, respiratory diseases, cardiovascular diseases and some cancers (Friel et al., 2011). Evidences from the Global Burden of Disease indicate that the burden of NCD related to climate progressively increased in the period 1990–2019. Most of the health risks depend on climatic influences on global food security, on the trend of infectious diseases, on the integrity of the defences against natural disasters and on the adverse consequences of socioeconomic status, altered social cohesion, migrations and conflicts. NCD and the recently emerged COVID-19 pandemic are linked with environmental harms and with the progressive rise in global temperature (Di Ciaula et al., 2021).

The negative effects of climate change are amplified by the interaction between NCD and socioeconomic factors (McMichael 2013). Although there are no studies that link climate change and NCDs in Tanzania, 27.3% of hospital mortality are due to NCD and injuries with cardiovascular-diseases (31.9%), cancers (18.6%) and chronic respiratory diseases (18.4%) accounting for the largest proportions (Mboera et al., 2017; 2022).

4 Food and Nutrition Security

Increase in food insecurity and malnutrition is among the most important impacts of climate change. Higher temperatures, declining rainfall and water scarcity and floods in Tanzania are impacting negatively on food production resulting in food insecurity. Decreased agricultural productivity in the coming years could lead to hunger and famine in some communities severely affected by climate change. The effects of drought on health include deaths, malnutrition and communicable disease. Malnutrition increases the risk both of acquiring and of dying from an infectious disease. Drought diminishes dietary diversity and reduces overall food consumption, and may therefore lead to micronutrient deficiencies. This would in turn increase illness and death of the most vulnerable groups such as children and women (Shongwe, 2009).

Climate change exacerbate the risks of malnutrition through crop destruction, livestock reductions, critical infrastructure and key community asset damage. Already very high malnutrition exceeding a prevalence of 40% have been reported from nine regions of Dodoma, Geita, Iringa, Kagera, Katavi, Kigoma and Ruvuma (MoH, 2018). Moreover, the regions with the highest number of stunted children and highest prevalence of chronic malnutrition are Dodoma, Kagera, Mbeya and Mwanza.

5 Health Systems

Climate change is among major threats to public health. Climate change is likely to increase the frequency, intensity or duration of extreme weather conditions which increases risks for vulnerable populations and communities in areas exposed to natural hazards. Climate change severely affects some of the most fundamental prerequisites for good health, i.e., sufficient food, clean air and water, adequate shelter and freedom from diseases (WHO, 2009). All these will have negative implications for the achievement of the health-related Sustainable Development Goals of zero hunger and good health.

Most often, extra pressure is placed on health care services by increased demands resulting from weather-related natural hazards such as floods. Electric power outages are common in Tanzania as a result of extreme of drought. Climate change can increase the number of extreme weather events which can damage buildings, roads and other infrastructure. This causes trauma for people having to relocate, as occurred following the rains in Same and Kilosa districts during a decade ago. Health infrastructures are designed for a specific climate. Health risks can arise when any one of these systems fails or becomes compromised—as they may in a changing climate.

In Tanzania, 54% (urban = 79%; rural = 45%) of the households have access to improved water supplies (Taylor, 2009). According to Mimura (2021) since climate change is associated with more extreme precipitation events and rising sea levels, cities will experience more severe and more frequent flooding. Climate change can result in damage to sanitation infrastructure resulting in the spread of disease or threatening a community's ability to maintain its economy, geographic location or cultural-tradition leading to mental stress.

6 Climate Change Adaptation and Mitigation

Adaptation to climate change has been a survival strategy that needs to be well-refined over time for different communities. Innovation, knowledge and coping strategies often exist within the fabric of social structures at the community level (Yamin et al., 2005). Supporting community-led adaptation to climate change means putting communities' centre-stage in determining which vulnerabilities are addressed, how and when they are addressed. Recognizing the adverse impacts of natural disasters and calamities such as floods, droughts, landslides, insect pests and disease epidemics to the well-being of its citizen, Tanzania has recently developed a National Climate Change Strategic Plan to provides a set of interventions on adaptation and mitigation, which are expected to strengthen country's resilience to the impacts of climate change and contribute to the global efforts of reducing GHGs emissions.

Moreover, the government has recently developed Health-National Adaptation Plan (HNAP) to climate change in Tanzania. HNAP aims to guide the country towards a health system that is more resilient to climate change and a sustainable and future for the people of Tanzania. The plan is guided by the following key strategic objectives: (i) reduce vulnerability to the impacts of climate change, by building adaptative capacity and resilience in the health sector; (ii) facilitate the integration of climate change adaptation in a coherent manner, into relevant new and existing policies, programmes and activities within the health sector; (iii) Guide health practitioners on the need to develop and operationalize a climate sensitive early warning systems for disease outbreaks; (iv) Advocate for the mobilization and allocation of resources for adaptation to climate change in the health sector; and (v) Facilitate the integration of health priorities into the National Adaptation Plan and support the implementation process (MoH, 2018). Despite the plan to be in place for the past four year, none of the objectives has been effectively address.

7 Conclusion

Climate change is real and has impacts on health, food and nutrition security. Many adaptation and mitigation strategies need to be strengthened in areas such as intersectoral coordination, good governance, human resources, institutional structures, public finance and natural resource management. It is expected such strategies strengthen the resilience of the country health system, communities and households to the effects of climate change. The two national strategies to mitigate climate change are expected to provide a good starting point for addressing adaptation requirements in the context of adequate food and good health. It is important that adaptation efforts will steps towards mainstreaming climate change issues into all national, sub-national and sectoral planning processes. For advocacy and empowerment of communities, it is important that they are involved and participate in assessments and feed in their knowledge to provide useful climate-health related information. Early warning systems for various infection diseases need to be developed and used to monitor any impending outbreak.

References

Ahmed, N., Devi, S. M., Valverde Mde, L., Vijayachari, P., Machang'u, R. S., Ellis, W. A., & Hartskeerl, R. A. (2006). Multilocus sequence typing method for identification and genotypic classification of pathogenic *Leptospira* species. *Annals of Clinical Microbiology and Antimicrobials, 5*, 28.

Alzaheb, R. (2018). Seroprevalence of *Toxoplasma gondii* and its associated risk factors among women of reproductive age in Saudi Arabia: A systematic review and meta-analysis. *International Journal of Women Health, 10*, 537–544.

Amarasinghe, A., Kuritsk, J. N., Letson, G. W., & Margolis, H. S. (2011). Dengue virus infection in Africa. *Emerging Infectious Diseases, 17*(8), 1349–1354. https://doi.org/10.3201/eid1708.101515

Biggs, H. M., Bui, D. M., Galloway, R. L., Stoddard, R. A., Shadomy, S. V., Morrissey, A. B., et al. (2011). Leptospirosis among hospitalized febrile patients in northern Tanzania. *American Journal of Tropical Medicine and Hygiene, 85*, 275.

Blackburn, J. K., Matakarimov, S., Kozhokeeva, S., Tagaeva, Z., Bell, L. K., Kracalik, I. T., et al. (2017). Modeling the ecological niche of Bacillus anthracis to map anthrax risk in Kyrgyzstan. *American Journal of Tropical Medicine and Hygiene, 96*, 550–556.

Boada-Robayo, L., Cruz-Reyes, D. L., Cifuentes-González, C., Rojas-Carabali, W., Vargas-Largo, Á. P., de-la-Torre, A. (2022). Exploring the association between precipitation and population cases of ocular toxoplasmosis in Colombia. *PLoS Neglected Tropical Diseases, 16*(10), e0010742. https://doi.org/10.1371/journal.pntd.0010742

Brun, R., Blum, J., Chappuis, F., & Burri, C. (2010). Human African trypanosomiasis. *Lancet, 375*(9709), 148–159.

Chipwaza, B., Mugasa, J. P., Selemani, M., Amuri, M., Mosha, F., Ngatunga, S. D., Gwakisa, P. S. (2014). Dengue and Chikungunya fever among viral diseases in outpatient febrile children in Kilosa District Hospital, Tanzania. *PLoS Neglëcted Tropical Diseases, 8*(11).

Chirebvu, E., Chimbari, M. J., Ngwenya, B. N., & Sartorius, B. (2016). Clinical malaria transmission trends and its association with climatic variables in Tubu village, Botswana: A retrospective analysis. *PLoS ONE, 11*, e0139843.

Colwell, R. R. (1996). Global climate and infectious disease: The cholera paradigm. *Science, 274*, 2025–2031. https://doi.org/10.1126/science.274.5295.2025

Clyde, D. F. (1967). *Malaria in Tanzania.* Oxford University Press.

Dawson, D. (2005). Foodborne protozoan parasites. *International Journal of Food Microbiology, 103*(2), 207–227. https://doi.org/10.1016/j.ijfoodmicro.2004.12.032

Dubey, J. P. (2004). Toxoplasmosis, a waterborne zoonosis. *Veterinary Parasitology, 126*(1–2), 57–72. https://doi.org/10.1016/j.vetpar.2004.09.005

Dragon, D. C., & Rennie, R. P. (1995). The ecology of anthrax spores. *Canadian Veterinary Journal, 36*(5).

Flegr, J., Prandota, J., Sovičková, M., & Israili, Z. H. (2014). Toxoplasmosis—A global threat. Correlation of latent toxoplasmosis with specific disease burden in a set of 88 countries. *PLoS ONE, 9*(3), 90–203.

Friel, S., Bowen, K., Campbell-Lendrum, D., Frumkin, H., McMichael, A. J., & Rasanathan, K. (2011). Climate change, noncommunicable diseases, and development: The relationships and common policy opportunities. *Annual Review of Public Health, 32*, 133–147. https://doi.org/10.1146/annurev-publhealth-071910-140612

Furtado, J., Smith, J. R., Belfort, R., Gattey, D., & Winthrop, K. L. (2011). Toxoplasmosis: A global threat. *Journal of Global Infectious Diseases, 3*(1), 281–284.

Gao, J., Qian, Z., Hide, G., Lai, D., Lun, Z., & Wu, Z. (2020). Human African trypanosomiasis: The current situation in endemic regions and the risks for non-endemic regions from imported cases. *Parasitology, 147*(9), 922–931. https://doi.org/10.1017/S0031182020000645

Garay, J. (1998). Epidemiological survey and situation analysis: Malaria epidemic in Nshamba Division, Muleba District, Tanzania, 47 pp. MSF Spain.

Gashout, A., Amro, A., Erhuma, M., Al-Dwibe, H., Elmaihub, E., Bebba, H., Nattah, N., & Abudher, A. (2016). Molecular diagnosis of *Toxoplasma gondii* infection in Libya. *BMC Infectious Diseases, 16*, 157.

Gubler, D. J., Reiter, P., Ebi, K. L., Yap, W., Nasci, R., & Patz, J. A. (2001). Climate variability and change in the United States: Potential impacts on vector-and rodent-borne diseases. *Environmental Health Perspective, 109*, 223–233. https://doi.org/10.2307/3435012(10.2307/3435012)

Hashizume, M. (2008). The effect of rainfall on the incidence of cholera in Bangladesh. *Epidemiology, 19*, 103–110.

Hide, G. (1999). History of sleeping sickness in East Africa. *Clinical Microbiology Reviews, 12*(1), 112–125.

Iijima, T., & Sugiura, S. (1962). Studies on the temperature as a limiting factor for the survival of *Oncomelania nosophora*, the vector snail of *Schistosoma japonicum* in Japan. *Japanese Journal of Medical Sciences and Biology, 15*, 221–228.

Jones, J. L., & Dubey, J. P. (2010). Waterborne toxoplasmosis—Recent developments. *Experimental Parasitology, 124*, 10–25. pmid:19324041.

Kajeguka, D. C., Kaaya, R. D., Mwakalinga, S., Ndossi, R., Ndaro, A., Chilongola, J. O., et al. (2016). Prevalence of dengue and chikungunya virus infections in north-eastern Tanzania: A cross sectional study among participants presenting with malaria-like symptoms. *BMC Infectious Diseases, 16*(1), 1–9.

Kangalawe, R. (2011). Climate change impacts, local knowledge and coping strategies in the great Ruaha River catchment area, Tanzania. *Natural Resources, 2*(4), 212–223.

Kennedy, P. G. E. (2004). Human African trypansomiais of the CNS: Current issues and challenges. *Journal of Clinical Investigation, 113*(4), 496–504.

Kinimi, E., Shayo, M. J., Patrick, B. N., Angwenyi, S. O., Kasanga, C. J., Weyer, J., et al. (2018). Evidence of chikungunya virus infection among febrile patients seeking healthcare in selected districts of Tanzania. *Infection Ecology & Epidemiology, 8*(1), 1553460.

Kumalija, M. S., Chilongola, J. O., Budodo, R. M., Horumpende, P. G., Mkumbaye, S. I., Vianney, J., Mwakapuja, R. S., & Mmbaga, B. T. (2021). Detection of Rift Valley fever virus inter-epidemic activity in Kilimanjaro Region. *North Eastern Tanzania. Global Health Action, 14*, 1. https://doi.org/10.1080/16549716.2021.1957554

Lalika, M. C. S., et al. (2015). Understanding watershed dynamics and impacts of climate change and variability in the Pangani River Basin, Tanzania. *Ecohydrology and Hydrobiology, 15*(1), 26–38.

Lélu, M., Villena, I., Dardé, M.-L., Aubert, D., Geers, R., Dupuis, E., Marnef, E., Poulle, L., Gotteland, C., Dumètre, A., & Gilot-Fromont, E. (2012). Quantitative estimation of the viability of *Toxoplasma gondii* oocysts in soil. *Applied and Environmental Microbiology*. https://doi.org/10.1128/AEM.00246-12

Machang'u, R. S., Mgode, G., & Mpanduji, D. (1997). Leptospirosis in animals and humans in selected areas of Tanzania. *Belgian Journal of Zoology, 127*, 97–104.

Matola, Y. G., White, G. B., & Magayuka, S. A. (1987). The changed pattern of malaria endemicity and transmission at Amani in the eastern Usambara mountains, north-eastern Tanzania. *Journal of Tropical Medicine and Hygiene, 90*, 127–134.

Mazigo, H. D., Nuwaha, F., Kinung'hi, S. M., Morona, D., Pinot de Moira, A., Wilson, S., et al. (2012). Epidemiology and control of human schistosomiasis in Tanzania. *Parasites & Vectors, 5*, 274. https://doi.org/10.1186/1756-3305-5-274

Mboera, L. E. G., & Kitua, A. Y. (2001). Malaria epidemics in Tanzania: An overview. *African Journal of Health Sciences, 8*, 14–18.

Mboera, L. E. G. (2004). Environmental and socio-economic determinants of malaria epidemics in the highlands of Tanzania. *Tanzania Health Research Bulletin, 6*, 11–17.

Mboera, L. E. G., Kishamawe, C., Kimario, E., & Rumisha, S. F. (2019). Mortality pattern of toxoplasmosis and its comorbidities in Tanzania: A 10-year retrospective hospital-based survey. *Frontiers in Public Health, 7*, 25.

Mboera, L. E. G., Kishamawe, C., Rumisha, S. F., Chiduo, M. G., Kimario, E., & Bwana, V. M. (2022). Patterns and trends of in-hospital mortality due to non-communicable diseases and injuries in Tanzania, 2006–2015. *PLoS Global Health* (in press).

Mboera, L. E. G., Mayala, B. K., Kweka, E. J., & Mazigo, H. D. (2011). Impact of climate change on human health and health systems in Tanzania: A review. *Tanzania Journal of Health Research, 13*(5), 407–426.

Mboera, L. E. G., Mweya, C. N., Rumisha, S. F., Tungu, P. K., Stanley, G., Makange, M. R., et al. (2016). The risk of dengue virus transmission in Dar es Salaam, Tanzania during an epidemic period of 2014. *PLoS Neglected Tropical Diseases, 10*(1).

Mboera, L. E. G., Rumisha, S. F., Kumalija, C. J., Chiduo, M. G., Mangu, C. D., Matemba, L. E., & Mhehe, E. (2017). *Hospital Mortality and Causes of Death in Tanzania*. National Institute for Medical Research, Dar es Salaam, Tanzania and Ministry of Health, Community Development, Gender, Elderly and Children, Dar es Salaam, Tanzania. ISBN 978-9987-9143-6-4.

Mgode, G. F., Machang'u, R. S., Goris, M. G., Engelbert, M., Sondij, S., & Hartskeerl, R. A. (2006). New *Leptospira* serovar Sokoine of serogroup Icterohaemorrhagiae from cattle in Tanzania. *International Journal of Systematics and Evolutionary Microbiology, 56*, 593–597.

Migchelsen, S. J., Buscher, P., Hoepelman, A. I., Schallig, H. D., & Adams, E. R. (2011). Human African trypanosomiasis: A review of non-endemic cases in the past 20 years. *International Journal of Infectious Diseases, 15*, e517–e524.

Mimura, N. (2021). Rising seas and subsiding cities. *Nature Climate Change, 11*, 296–297. https://doi.org/10.1038/s41558-021-01009-6

MoH. (2019). Analytical report to inform the Tanzania health sector strategic plan IV, 2015/2016–2019/2020. Ministry of Health, Gender, Community Development, Elderly and Children, United Republic of Tanzania, September 2019.

Mrema, S., Shamte, A., Selemani, M., & Masanja, H. (2012). The influence of weather on mortality in rural Tanzania: A time-series analysis 1999–2010. *Global Health Action, 5*. https://doi.org/10.3402/gha.v5i0.19068

Mwakapeje, E. R., Høgset, S., Fyumagwa, R., Nonga, H. E., Mdegela, R. H., & Skjerve, E. (2018). Anthrax outbreaks in the humans-livestock and wildlife interface areas of Northern Tanzania: A retrospective record review 2006–2016. *BMC Public Health, 18*, 106.

Mwambe, B., Mshana, S. E., Kidenya, B. R., Massinde, A. N., Mazigo, H. D., Michael, D., Majinge, C., & Groß, U., (2013a). Sero-prevalence and factors associated with *Toxoplasma gondii* infection among pregnant women attending antenatal care in Mwanza, Tanzania. *Parasites & Vectors, 6*, 222.

Mwandosya, M. J., Nyenzi, B. S., & Luhanga, M. L. (1998). *The Assessment of Vulnerability and Adaptation to Climate Change Impacts in Tanzania*. Dar-es-Salaam. Centre for Energy, Environment, Science and Technology, Dar es Salaam, Tanzania.

Mwanyika, G. O., Sindato, C., Rugarabamu, S., Rumisha, S. F., Karimuribo, E. D., et al. (2021). Seroprevalence and associated risk factors of chikungunya, dengue and Zika in Tanzania. *International Journal of Infectious Diseases, 111*, 271–280.

Nnko, H. J., Gwakisa, P. S., Ngonyoka, A., Sindato, C., & Estes, A. B. (2021). Potential impacts of climate change on geographical distribution of three primary vectors of African Trypanosomiasis in Tanzania's Maasai Steppe: *G. m. morsitans, G. pallidipes and G. swynnertoni. PLoS Neglected Tropical Disease, 15*(2), e0009081. https://doi.org/10.1371/journal.pntd.0009081

Norman, F. F., Monge-Maillo, B., Martínez-Pérez, Á., Perez-Molina, J. A., & López-Vélez, R. (2015). Parasitic infections in travelers and immigrants: Part I protozoa. *Future Microbiology, 10*, 69–86.

Okada, K., Morita, R., Egawa, K., Hirai, Y., Kaida, A., Shirano, M., et al. (2019). Dengue virus type 1 infection in traveler returning from Tanzania to Japan. *Emerging Infectious Diseases, 25*(9), 1782–1784.

Otieno, F. T., Gachohi, J., Gikuma-Njuru, P., Kariuki, P., Oyas, H., Canfield, S. A., Bett, B., Njenga, M. K., & Blackburn, J. K. (2021). Modeling the potential future distribution of anthrax outbreaks under multiple climate change scenarios for Kenya. International *Journal of Environmental Research and Public Health, 18*(8). https://doi.org/10.3390/ijerph18084176

Pascual, M., Rodo, X., Ellner, S. P., Colwell, R., & Bouma, M. J. (2000). Cholera dynamics and El Niño-southern oscillation. *Science, 289*, 1766–1769. https://doi.org/10.1126/science.289.5485.1766

Patz, J. A., Graczyk, T. K., Geller, N., & Vittor, A. Y. (2000). Effects of environmental change on emerging parasitic diseases. *International Journal of Parasitology, 30*, 1395–1405. https://doi.org/10.1016/S0020-7519(00)00141-7(10.1016/S0020-7519(00)00141-7)

Paul, E., Kiwelu, I., Mmbaga, B., Nazareth, R., Sabuni, E., Maro, A., et al. (2018a) *Toxoplasma gondii* seroprevalence among pregnant women attending antenatal clinic in Northern Tanzania. *Tropical Medicine and Health, 46*, 39.

Picozzi, K., Fèvre, E. M., Odiit, M., Carrington, M., Eisler, M. C., Maudlin, I., Welburn, S. C. (2005). Sleeping sickness in Uganda: a thin line between two fatal diseases. *BMJ*, *331*(7527), 1238–1241. https://doi.org/10.1136/bmj.331.7527.1238

Rugarabamu, S., Mwanyika, G., Rumisha, S. F., Sindato, C., Lim, H.-Y., Misinzo, G., et al. (2021). Seroprevalence of selected zoonotic viral haemorrhagic fevers and associated risk factors in Tanzania. *International Journal of Infectious Diseases, 109*, 174–181.

Rudzinski, M., Meyer, A., Khoury, M., & Couto, C. (2013). Is reactivation of toxoplasmic retinochoroiditis associated to increased annual rainfall? *Parasite 20*, 44. pmid:24225023.

Rumisha, S. F., Shayo, E. H., & Mboera, L. E. G. (2019). Spatio-temporal prevalence of malaria and anaemia in relation to agro-ecosystems in Mvomero District, Tanzania. *Malaria Journal, 18*, 228. https://doi.org/10.1186/s12936-019-2859-y

Simarro, P. P., Jannin, J., & Cattand, P. (2008). Eliminating human African Trypanosomiasis: Where do we stand and what comes next. *PLoS Medicine, 5*, e55.

Simarro, P. P., Franco, J. R., Cecchi, G., Paone, M., Diarra, A., Ruiz Postigo, J. A., et al. (2012). Human African Trypanosomiasis in non-endemic countries (2000–2010). *Journal of Travel Medicine, 19*, 44–53.

Shongwe, S. V. (2009). The impact of climate change on health on the East, Central and Southern African (ECSA) region. *Commonwealth Health Ministers' Update.*

Sindato, C., Karimuribo, E. D., Pfeiffer, D. U., Mboera, L. E. G., Kivaria, F., Dautu, G., Bernard, B., & Paweska, J. T. (2014). Spatial and temporal pattern of Rift Valley fever outbreaks in Tanzania; 1930 to 2007. *PLoS ONE, 9*(2), e8889. https://doi.org/10.1371/journal.pone.0088897

Sindato, C., Karimuribo, E. D., Vairo, F., Misinzo, G., Rweyemamu, M. M., Abdel Hamid, M. M. et al. (2022). Rift Valley fever seropositivity in humans and domestic ruminants and associated risk factors in Sengerema, Ilala and Rufiji districts, Tanzania. *International Journal of Infectious Diseases*, S1201–9712(22)00403–00409. https://doi.org/10.1016/j.ijid.2022.07.012

Swai, E. S., & Schoonman, L. (2009). Seroprevalence of *Toxoplasma gondii* infection amongst residents of Tanga district in northeast Tanzania. *Tanzania Journal of Health Research, 11*, 205–209.

Taylor, B., (2009). Situation analysis of women, children and the water, sanitation and hygiene sector in Tanzania. UNICEF.

Thompson, L., Mosley-Thompson, E., & Mark, B. (2007). *Core Project: Low latitude glacier retreat: Evidence of accelerating climate change and impacts on local to regional water resources*. CWC Funded Project.

Vairo, F., Mboera, L. E. G., de Nardo, P., Oriyo, N. M., Meschi, S., Rumisha, S. F., et al. (2016). Clinical, virologic, and epidemiologic characteristics of dengue outbreak, Dar es Salaam, Tanzania, 2014. *Emerging and Infectious Diseases, 22*(5), 895–899. https://doi.org/10.3201/eid2205.151462

Van den Bossche, P., Stephane, L. R., Hendrickx, G., Bouyer, J. (2010). A changing environment and the epidemiology of tsetse-transmitted livestock trypanosomiasis. *Trends in Parasitology, 26*(5), 236–242.

Walsh, M. G., de Smalen, A. W., & Mor, S. M. (2018). Climatic influence on anthrax suitability in warming northern latitudes. *Science Reports, 8*, 9269.

Wandiga, S. O., Opondo, M., Kathuri, J., Olago, D., Apindi, E., Olaka, L. et al. (2006), *Climate change induced vulnerability to malaria and cholera in the lake Victoria Region: Final report for project no. AF 91*. Assessments of Impacts and Adaptations to Climate Change, International START Secretariat, Washington, DC. www.aiaccproject.org

Ward, T., Samuel, M., Maoz, D., Runge-Ranzinger, S., Boyce, R., Toledo, J., et al. (2017). Dengue data and surveillance in Tanzania: A systematic literature review. *Tropical Medicine and International Health, 22*(8), 960–970.

Welburn, S. C., Fèvre, E. M., Coleman, P. G., Odiit, M., & Maudlin, I. (2001). Sleeping sickness: A tale of two diseases. *Trends in Parasitology, 17*, 19–24.

Wildlife Conservation Society. (2008). 'Deadly Dozen' reports diseases worsened by climate change. Wildlife Conservation Society. http://www.sciencedaily.com/releases/2008/10/081007073928.htm

WHO. (2009). *Protecting Health from Climate Change. Connecting Science, Policy and People.* World Health Organization. http://www.who.int.globalchange_en

WHO. (2013). Control and surveillance of human African Trypanosomiasis. report of a WHO expert committee. WHO Technical Report Series 984. Geneva, Switzerland: World Health Organization.

Yamin, F., Rahman, A., & Huq, S. (2005). Vulnerability, adaptation and climate disasters: A conceptual overview. *IDS Bulletin, 36*, 1–14.

El Niño, Rainfall and Temperature Patterns Influence Perinatal Mortality in South Africa: Health Services Preparedness and Resilience in a Changing Climate

Natalie D. Benschop, Geldine Chironda-Chikanya, Saloshni Naidoo, Nkosana Jafta, Lisa F. Ramsay, and Rajen N. Naidoo

Abstract Southern Africa bears disproportionate consequences of the changing climate. The El Niño Southern Oscillation causes phases of extreme weather events, leading to flooding and extended periods of droughts in different regions of the subcontinent. Our data provides evidence that the extreme El Niño event of 2014–16 increased the risk for perinatal infant mortality in the Northern Cape and North West provinces of South Africa. The maternal health services profile of the Northern Cape and KwaZulu-Natal suggests a compromised health system with limited resilience to respond to the climate crises. The National Department of Health lacks adequate policies and strategies to ensure systems are able to meet the maternal and child health requirements in the context of climate change.

Keywords Pregnancy outcomes · Perinatal mortality · Climate · Aridity · El Niño

1 Introduction

Southern Africa is particularly vulnerable to climate change. It is geographically located in a warm and dry region, with meteorology dominated by subtropical high pressure. The region is projected to become warmer and drier. The mean annual

N. D. Benschop
Discipline of Statistics, School of Mathematics, Statistics and Computer Science, University of KwaZulu-Natal, Durban, South Africa

G. Chironda-Chikanya · N. Jafta · L. F. Ramsay · R. N. Naidoo (✉)
Discipline of Occupational and Environmental Health, School of Nursing and Public Health, University of KwaZulu-Natal, Durban, South Africa
e-mail: NAIDOON@ukzn.ac.za

S. Naidoo
Discipline of Public Health Medicine, School of Nursing and Public Health, University of KwaZulu-Natal, Durban, South Africa

R. Akhtar (ed.), *Climate Change and Human Health Scenarios*, Global Perspectives on Health Geography, https://doi.org/10.1007/978-3-031-38878-1_21

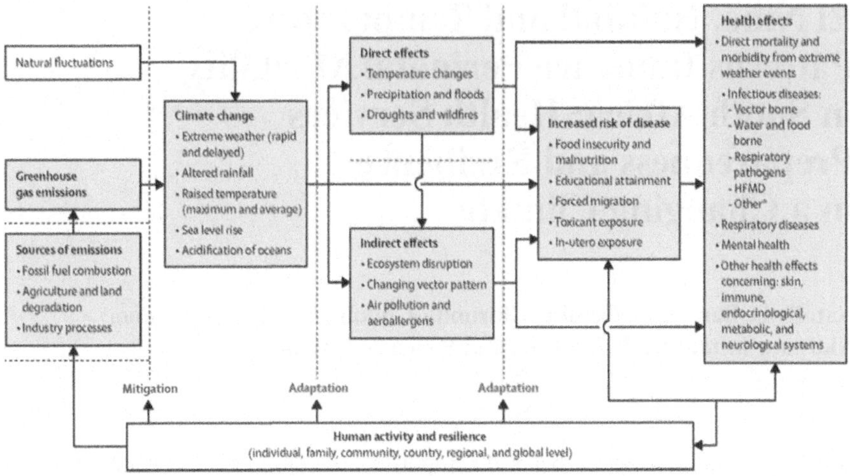

Fig. 1 Climate change and child health: an expanded framework (from Helldén et al., 2021, e167)

temperature has increased by at least 1 °C during the last 50 years (1.5 times the global average) with warming in the interior of southern Africa occurring at about twice the global average rate (Engelbrecht et al., 2015).

Climate change has direct impacts on the social and environmental determinants of health, which include clean air, safe drinking water, sufficient food and secure shelter (Fig. 1). Variations in rainfall patterns, rising sea levels, droughts, floods, also have indirect detrimental effects on human health (Anderko et al., 2020). The interaction of climate variables and human health has implications for the achievement of the Sustainable Development Goals. The region's institutions and finances face multiple developmental demands, reducing the technical and financial capacity for current and future climate change adaptation.

Climate adaptation requires the establishment of resilience such that society and systems can absorb shocks and adjust in the face of disturbance (Folke et al., 2010). Resilience refers to the capacity to absorb, predict, accommodate or recover from the effects of natural hazards in an efficient way through restoration, improvement or preservation of its crucial basic structures and functions through risk management (UNISDR, 2009).

1.1 The Implications of Climate Change for Southern Africa

Based on the reports from the Intergovernmental Panel on Climate Change (Fig. 2; Trisos et al., 2022), a global warming level (GWL) of 1.5 °C, 2 °C and 3 °C above pre-industrial levels will result in southern Africa to be on average 1.2 °C, 2.3 °C and 3.3 °C warmer than the 1994–2005 average respectively. Additionally, the annual number of heatwaves is projected to increase by between 2–4 (GWL 1.5 °C), 4–8 (GWL 2 °C) and 8–12 (GWL 3 °C).

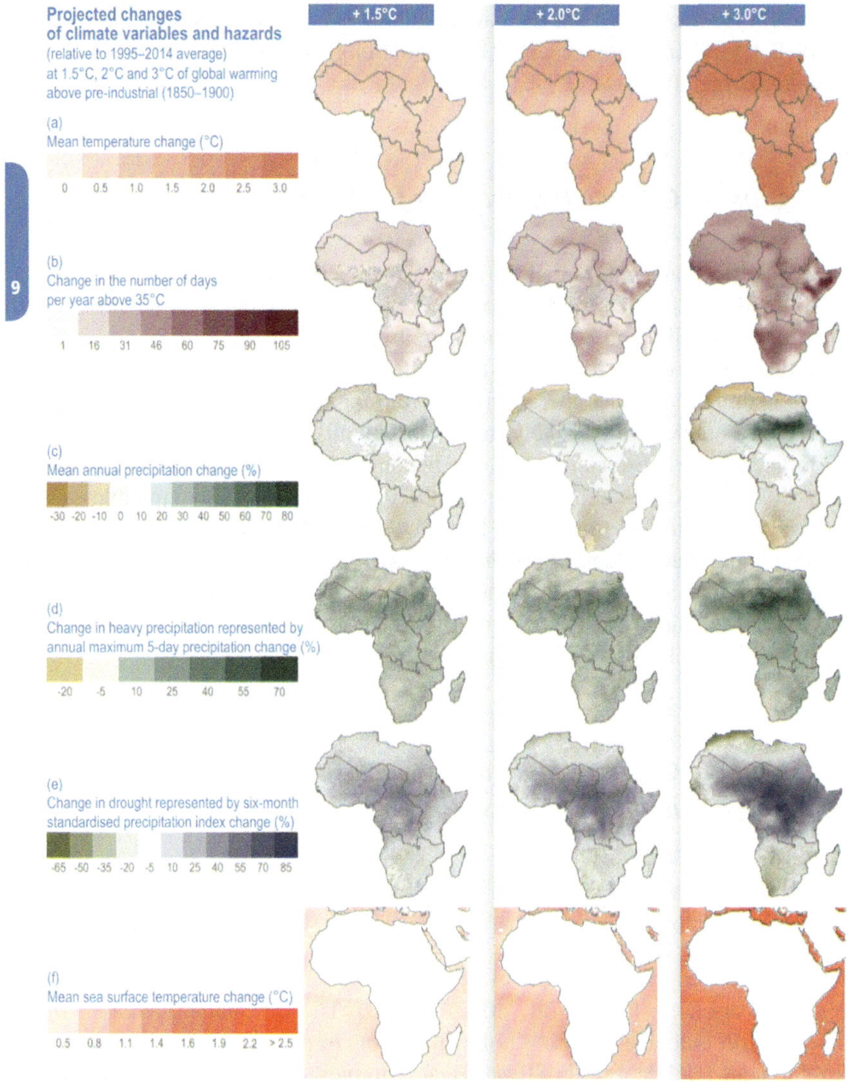

Fig. 2 Projected changes in climate variables and hazards (from the Trisos et al. 2022, Fig. 9.16 on p. 1324)

Mean annual rainfall in the summer rainfall region of southern Africa is projected to decrease by 10–20%, accompanied by an increase in the number of consecutive dry days during the rainy season under a high emissions scenario (RCP8.5[1]). Increases

[1] RCP 8.5 refers to the concentration of carbon that delivers global warming at an average of 8.5 W/m^2 across the planet. The RCP 8.5 pathway delivers a temperature increase of about 4.3 °C by 2100, relative to pre-industrial temperatures.

in drought frequency and duration are projected over large parts of southern Africa at GWL 1.5 °C and unprecedented extreme droughts (compared to the 1981–2010 period) emerge at GWL 2 °C.

1.2 Climate Change Implications for Health

Between 1.5 °C and 2 °C global warming, negative impacts are projected to become widespread and severe with reduced food production, reduced economic growth, increased inequality and poverty, biodiversity loss, increased human morbidity and mortality (Trisos et al., 2022). The IPCC SR1.5 pointed out two major agricultural risks for southern Africa under drastically warmer and drier futures: the maize crop, the region's staple food, is likely to be substantially reduced under 3 °C of global warming, and the livestock industry may become unviable at 3 °C of global warming.

In Southern Africa, children born in 2020, under a 1.5 °C-compatible scenario will be exposed to 3–4 times more heatwaves as well as storms and floods in their lifetimes compared to people born in 1960. As a result, loss of life, injury and damage to infrastructure also increases. The capacity to perform manual labour outdoors decreases dramatically as the occurrence of heat waves increases. Human mortality increases, particularly in urban areas with inadequate housing.

1.3 Climate Change Implications for Maternal Health and Perinatal Outcomes

In South Africa, climate change has major implications for diverse vulnerable groups (Chersich et al., 2018). Pregnant women and newborns are vulnerable due to their physiologic and immunologic status, and experience disproportionate adverse outcomes related to the impacts of heat, increased air pollution and the spread of infectious diseases (Bátiz et al., 2022) (Fig. 3). These result in increased risk of adverse birth outcomes, leading to health effects in neonates and subsequent child development (Helldén et al., 2021). Heat extremes lead to dehydration and renal failure, causing reduced amniotic fluid which affects foetal growth and development (Anderko et al., 2020). Rising body temperature can trigger an inflammatory response in pregnant women reducing placental blood flow and foetal oxygen supply (Anderko et al., 2020). Newborns are sensitive to extreme temperatures due to limited capacity in temperature regulation (Chersich et al., 2018).

Increased air pollutant exposure via the placenta impairs organogenesis and causes adverse birth effects (Bátiz et al., 2022; Ha, 2022; Inoue et al., 2020). In South Africa, pregnant mothers within the coal and industry zones are exposed to poor air quality which leads to neonatal respiratory disease, adverse perinatal outcomes, and higher premature mortality (Langerman & Pauw, 2018).

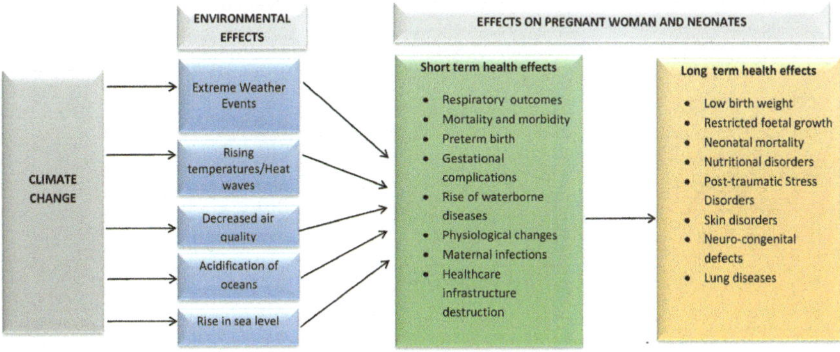

Fig. 3 Climate change, maternal health and perinatal outcomes

The re-emergence and geospatial redistribution of pathogens, directly or indirectly due to climate change, affects the reproductive health of humans (Choudhari, 2022). Extreme weather events in western and coastal regions of South Africa (Chersich et al., 2018; Opoku et al., 2021) introduce disease-carrying organisms into drinking water supplies and recreational waters, causing gastrointestinal and other infections in vulnerable populations such as pregnant women (Segal & Giudice, 2022). Due to their immune system changes, pregnant women are especially vulnerable to vector-borne disease, including Lyme Disease, Dengue and Zika virus, leading to adverse birth outcomes (Ha, 2022).

The nutritional status of communities was impacted by the 2012–2013 drought in South Africa, through reduced food quality, production, transportation, availability and safety (Gomez-Zavaglia et al., 2020). These disruptions influence food access, compromising the pregnancy and the foetal development. Precipitation extremes and drought are associated with child undernutrition and stunting through household food insecurity (Drysdale et al., 2021). Trauma from extreme weather events, heat waves and natural disasters can cause psychological and emotional stress for pregnant and postpartum women (Clark & Zolnikov, 2020). These climatic change events disrupt support networks, behavioural health services and treatment access and consequently pregnant women's ability to cope (Rothschild & Haase, 2022).

1.4 The El Nino Southern Oscillation (ENSO)

El Niño and La Niña are opposite phases of the global ENSO climate pattern associated with teleconnected changes that occur in both the ocean and the atmosphere. The Southern Oscillation describes the irregular periodic variation in winds over the

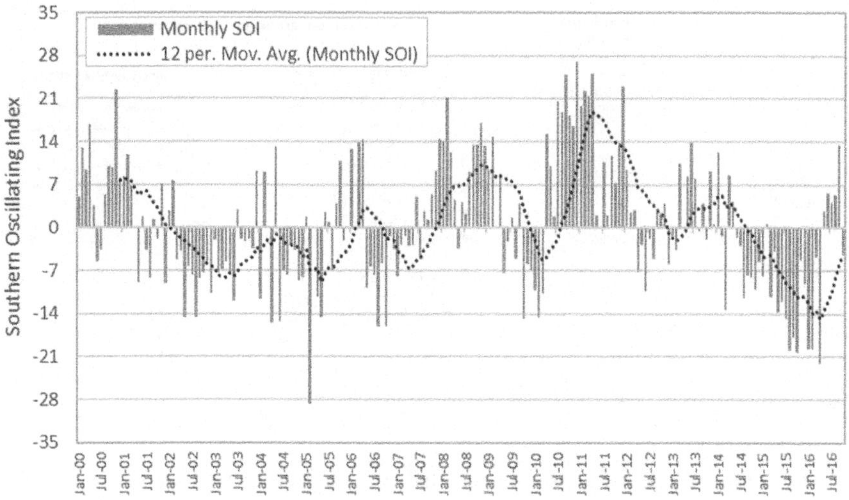

Fig. 4 Southern oscillating index from January 2000 to October 2016

tropical central and eastern Pacific Ocean. The Southern Oscillating Index[2] (SOI) is a standardized metric used to measure ENSO. An SOI below negative seven is indicative of an El Niño episode, whereas values greater than seven are usually associated with a La Niña episode. Figure 4 illustrates the observed values of the SOI between January 2000 and October 2016. The El Niño phases that prevailed during 1982–1983, 1997–1998 and 2014–2016 are regarded as being among the strongest on record (USCPC, 2022). The strength of the 2014–2016 El Niño event contrasts with the preceding weaker El Niño events and added to the global warming trend, with 2014 and 2015 being two of the hottest years on record until that point (Australian Bureau of Meteorology, 2016).

The extreme 2014–2016 El Niño caused severe drought in Southern Africa with Lesotho, Zimbabwe and Swaziland declaring national states of emergency, and the Southern African Development Community (SADC) declaring a regional drought disaster in March 2016 (Riscura, 2016). South Africa also experienced its worst drought in more than thirty years and was forced to import grains in 2016 (Walton, 2016). The El Niño event finally subsided in May 2016.

[2] The Southern Oscillating Index (SOI) compares surface air pressure anomalies at Darwin, Australia, to pressure anomalies at Tahiti. The extent of departure from average conditions indicates the strength of La Niña (positive SOI) or El Niño (negative SOI) conditions.

2 Case Study: Changing Climate and Perinatal Infant Mortality in South Africa, 2000–2016

We investigated the indirect effect of changes in the atmospheric component of the global ENSO climate pattern—the Southern Oscillation—on changes in the rate of perinatal infant mortality in South Africa based on a sample of births that took place between January 2000 and October 2016. It is well established that the El Niño phase of ENSO correlates with increased aridity over much of southern Africa (Tyson & Preston-Whyte, 2000). Aridity is defined as the extent '*to which a climate lacks effective, life-promoting moisture*' (American Meteorological Society, 2022).

Over the study timeframe (January 2000–October 2016), weaker El Niño events were observed in 2002–2003, 2004–2005, 2006–2007, 2009–2010 (USCPC, 2022), before the occurrence of the major El Niño event of 2014–2016. Using meteorological data prepared at a cluster location level (Mayala et al., 2018) for the Demographic and Health Surveys (DHS) Program, we explored how changes in SOI correlate with the variation in the amount of rainfall, temperature and aridity of South Africa between 2000, 2005, 2010 and 2015.

2.1 *Meteorological Analysis*

The annual South African weather estimates (Table 1) are based on an average of the meteorological data for the cluster locations shown in Fig. 5. South Africa experienced on average a drier (and warmer) climate as a result of the 2014–2016 El Niño event, in contrast to the more moderate La Niña influenced conditions that prevailed in 2005. In 2015, total rainfall was 25% lower than in 2010. Temperatures were also more extreme in 2015, with widened diurnal temperature ranges.

The aridity index (AI) (Table 1), the ratio of annual precipitation to annual potential evapotranspiration (Mayala et al., 2018), can identify the development of severe hydrological imbalance caused by a sustained duration of unusually dry weather (drought) (American Meteorological Society, 2022).

Table 1 Average annual weather estimates for South Africa versus average annual SOI

South Africa	2000	2005	2010	2015
Average rainfall (mm)	726.4	558.4	623.1	466.4
Average maximum temperature (°C)	24.0	26.0	25.9	26.4
Average minimum temperature (°C)	10.9	11.5	11.5	11.3
Average diurnal temperature range (°C)	13.1	14.5	14.5	15.1
Average aridity index	22.2	11.2	14.2	8.9
Average SOI (global)	7.8	– 3.6	9.8	– 11.2

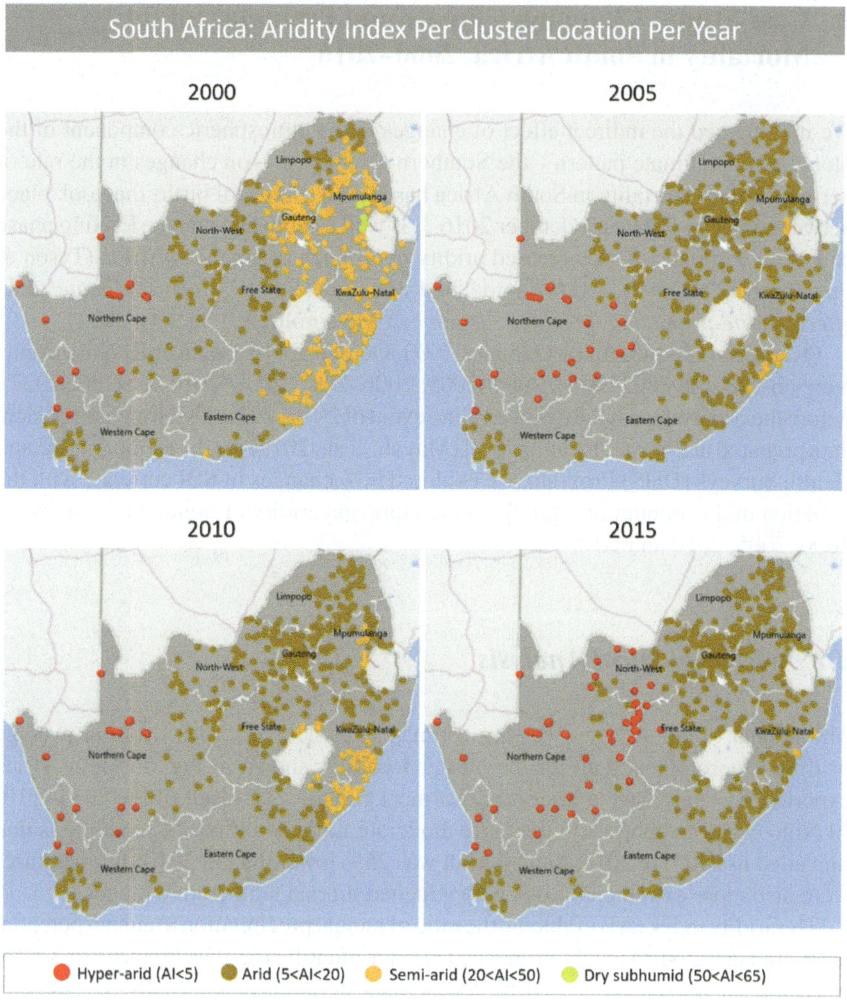

Fig. 5 Aridity index per location for discretized annual periods

The more arid climate observed for South Africa in 2015 (a lower AI) was due to lower rainfall in most provinces (Fig. 6). The usually water-scarce Northern Cape (NC) became predominantly "hyper-arid" in 2015—according to the UNCCD (2022) classification—along with several locations in the North West (NW) province (Fig. 6). Rainfall in these regions was 45% lower as compared to 2010, and temperatures also became notably more extreme in these areas, exceeding an average diurnal range of 17 °C (Fig. 6 and Table 2).

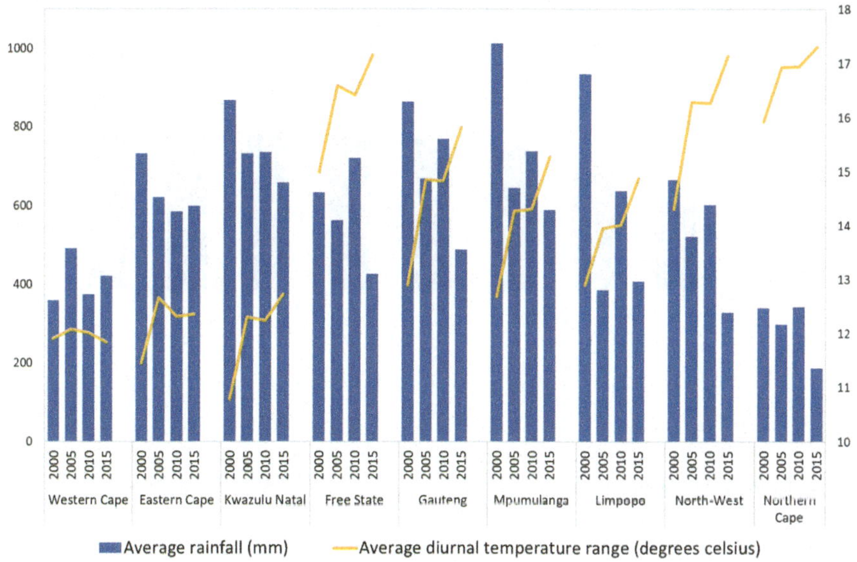

Fig. 6 Average annual weather estimates by province for 2000, 2005, 2010, 2015

Table 2 Aridity index by province and average SOI for 2000, 2005, 2010 and 2015

Aridity Index	2000	2005	2010	2015
Western Cape	9.40	10.4	8.7	8.1
Eastern Cape	22.5	14.9	16.0	12.3
KwaZulu-Natal	31.1	16.7	22.5	13.1
Free State	18.9	10.3	16.4	7.1
Gauteng	26.2	11.0	14.0	8.1
Mpumalanga	33.4	12.5	14.8	10.8
Limpopo	21.8	10.0	12.0	8.5
North-West	20.1	9.0	12.0	6.7
Northern Cape	7.3	5.0	7.2	4.2
Average SOI (global)	7.8	− 3.6	9.8	− 11.2

2.2 *Analysis of Perinatal Infant Mortality in South Africa*

The perinatal health data used in this case study were collected from a sample of mothers aged 15–49 during the South Africa Demographic and Health Survey (SADHS) of 2016. The main aim of SADHS 2016 was to obtain estimates on basic demographic and health care indicators, including childhood mortality and maternal health (National Department of Health (NDoH), 2019). Detailed interviews were

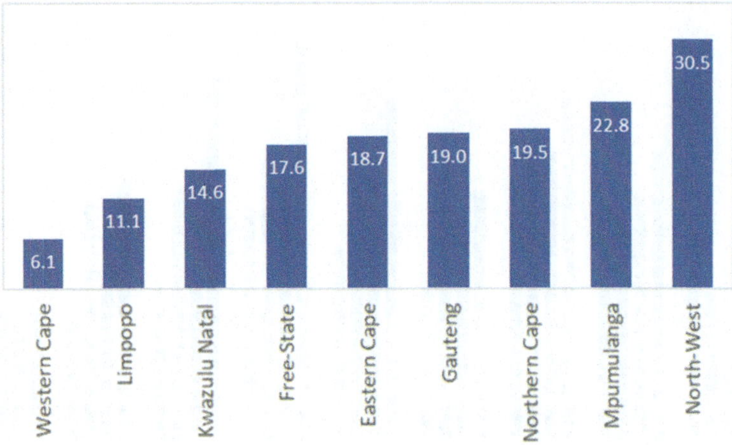

Fig. 7 Estimated perinatal mortality rate by province (January 2000–October 2016)

conducted between June and November 2016, selected from throughout the country spanning both urban ($n = 430$) and rural ($n = 279$) areas of South Africa.[3]

Although a total sample of $n = 6124$ mothers were interviewed during the survey, the case study analysis was restricted to a sample of $n = 5402$ mothers of verified age, who had given birth to at least one child in a singleton pregnancy post the year 1999. The sample consisted of 9777 single births between January 2000–October 2016 on known dates within a known province and area type (rural or urban) of South Africa. Multiple births were excluded as an inherent risk factor for perinatal death (Moura et al., 2014), and births prior to 2000 because of the likelihood of recall bias.

Based on this sample, the perinatal death (death of an infant within the first 7 days of birth) rate for South Africa in the review period was estimated at 17.9 per 1000 births. An overall increasing and cyclical trend in the perinatal death rate for South Africa is observed based on 12-month moving average (Fig. 8). The rate of perinatal infant death varies by province and is highest in the NW (Fig. 7) for the period under study.

2.3 *Assessing the Impact of Increased Aridity on Perinatal Mortality*

There was a low negative correlation between the estimated 12-month moving perinatal mortality rate and the 12-month moving average SOI - statistically significant for a 3–9 month lagged SOI effect (Fig. 9).

[3] The sampling procedure and questionnaires is described in the SADHS 2016 report and is available online at Stats SA website: www.statssa.gov.za.

Fig. 8 Estimated perinatal death rate per 1000 births in South Africa (12-month moving average)

Fig. 9 Correlation between perinatal death rate per 1000 births in South Africa, and lagged Southern Oscillating Index (SOI), based on 12-month moving average estimates

To model the probability of a birth resulting in perinatal infant death, multivariate logistic regression was performed in which the average 12-month moving SOI at the time of birth (i.e. in the month of birth) was used as the main explanatory variable, adjusting for maternal age and smoking during pregnancy, along with province of birth and area type, as covariates that may influence this relationship. The SOI variable was treated as binary to indicate whether or not El Niño conditions had likely been prevailing before and during pregnancy, based on a cut-off value of -7. Regression models included a main model, as well as province specific models.

Province of birth was found to be a significant factor associated with birth outcome while controlling for all other variables, with the odds of a birth resulting in perinatal death being significantly higher if the birth took place in the NW, as opposed to the Western Cape (p-value $= 0.004$), Limpopo (p-value $= 0.01$) or KwaZulu-Natal

(KZN) (p-value = 0.03). Type of region of birth (rural versus urban) was not found to be significantly associated with birth outcome, nor was SOI, while maternal age (> 35 years) and smoking were significant predictors.

Provincial specific models produced varying results. While controlling for other explanatory variables, the odds of a birth in NC resulting in perinatal death were determined to be 2.8 (95% CI: 1.1–7.2) times higher when the average 12-month SOI at the time of birth was < -7. SOI was the most significant factor associated with perinatal death in NC, after maternal antenatal smoking. Evidence of a similar association was also found between SOI and birth outcomes in the NW. No significant association was found in any of the other provinces of South Africa. In KZN, the modelling results indicated with low confidence (p-value = 0.25) that prevailing El Niño conditions may be associated with *lower* odds of perinatal death. This can be explained by KwaZulu-Natal's wetter climate which is more susceptible to flooding during *non*-El Niño periods, and less susceptible to aridity during El Niño events (unlike NW and NC), and possibly, as described below, better maternal health services.

3 The Maternal Health Profile and Service Provision: Health Services Preparedness and Resilience

Health services delivery takes place at local government level in South Africa. To investigate health services preparedness and resilience in response to the ENSO phenomenon, based on the analysis above, a municipality from NC and KZN provinces were selected for comparison with respect to maternal health indicators and service provision.

3.1 Characteristics of KwaZulu-Natal and Northern Cape Provinces

Of the total South African population of 60,466,705, KZN and NC contribute approximately 19.6% (11,503,917) and 2.1% (1,240,254) respectively (StatsSA, 2020). NC is characterised by dryness, fluctuating temperatures and lower annual rainfall (50–400 mm). Frances Baard District (FBD) municipality in NC has a population of 438,904, while the eThekwini Metropolitan Municipality (EM) in KZN has a population of 3,442,361. EM is a substantially better resourced municipality with a gross domestic product of €24billion as compared to FBD with €1.9billion.

Across both municipalities, more than 80% of the population have no private health insurance, with 62% below the age of 35 years and 50.4% females in the childbearing age, and with substantial poverty rates of 46.3% (FBD) and 56.6% (EM). This implies a dependency on the public sector maternal services (StatsSA, 2020).

Table 3 Indicators of maternal child health for South Africa, eThekwini Municipality and Francis Baard District

Indicator	RSA	EM	FBD
Antenatal visits coverage (per total number of women aged 15–49 with a live birth in the same period) (%)	83.1	90.8	125.3
Delivery in 10–19 years in facility rate (per total number of deliveries in public health facilities over the same period) (%)	13.2	12.4	15.9
Maternal mortality rate (per 100 000 live birth)	88	91.5	172.8
Perinatal mortality rate (stillbirth and early neonatal mortality per 1000 total births)	30	30	N/A
Early neonatal mortality per 1000 live birth (0–7 days)	9.9	10.1	13.8
Neonatal mortality rate (child death within 28 days per 1000 live births)	12.0	11.6	20.6

3.2 Maternal and Child Health Indicators

South Africa is confronted with substantial maternal, perinatal, neonatal and early neonatal mortality (Massyn et al., 2020; NDoH, 2022a, 2022b) (Table 3). Of particular note is the pregnancy rate among 10-19 year olds —ages for high risk pregnancy and poorer neonatal outcomes. Across almost all of the health indicators shown (Table 3), FBD in particular fares far worse than the national average.

3.3 Maternal Health Service Provision

Maternal and child health services is an essential component of primary healthcare (PHC), provided at no cost in South Africa (Ramavhoya et al., 2022). The maternal PHC services include weekday antenatal care, child health, family planning and postnatal clinics, and 24-h comprehensive service with an obstetric unit run by midwives (NDoH, 2016). The district hospitals receive referrals for antenatal care for high-risk women, and provide 24-h labour and delivery service including caesarean delivery, postnatal and postoperative care. The PHC approach enables uninsured maternal citizens to access antenatal care within a referral network at district, regional and tertiary hospitals (Massyn et al., 2020).

However, service delivery and quality varies. Only 50.9% and 24.1% of facilities achieved the 'Ideal Clinic' accreditation in EM and FBD, respectively (defined as a clinic with infrastructure, staff, medicines and supplies, administrative processes, adequate bulk supplies, use of clinical policies, protocols and guidelines and partner and stakeholder support, to ensure the provision of quality health services to the community) (Hunter et al., 2017; Massyn et al., 2020).

A benchmark of one district hospital per 100,000 population is recommended (Sambo & Kirigia, 2014), however, within EM this indicator is 0.11/100,000. The

South African maternal health guidelines recommends a minimum of eight antenatal consultations with a skilled healthcare professional during pregnancy (NDoH, 2016). However, 94% of women who gave birth in the five years preceding the SADHS had received only a single consultation.

The World Health Organisation (WHO) (2016) standards recommends approximately 297 skilled healthcare professionals per 100,000 population at PHC level, however, EM has 178.6 per 100,000. In NC province, the majority of community health centres do not operate 24 h due to lack of staff and adequate healthcare infrastructure (NCOP, 2015).

4 Adapting and Building a Resilient Health Sector

4.1 The WHO Framework for Health Sector Resilience

The WHO's framework to guide the health sector's development of a National Adaptation Plan (NAP) is focused on establishing health sector resilience. The framework ensures that management of health risks of climate change is integrated into the overall NAP process, through assessing risks; identifying, prioritising, and implementing adaptation options; and monitoring and evaluating the adaptation process (WHO, 2014). The guide presents ten key components for building resilience in the health system (Fig. 10).

These ten key components can be grouped as (i) information, (ii) foundation; and (iii) risk management. The foundation component is driven through public health services and systems, governance and collaboration and capacity development (WHO, 2014). This is echoed by agencies such as the Network of African Science Academies (NASAC), who urge that, in adapting and building resilience to climate change in the health sector, it is critical to strengthen health systems, improve infrastructure, provide access to services, and retain and upskill personnel in the sector (NASAC, 2015).

4.2 Health Systems Resilience

Health systems resilience is the ability of health systems to plan for the effects of climate change, minimize the negative consequences of such disruptions, recover as quickly as possible and adapt by learning lessons from the experience to become even better performing and more prepared (WHO, 2014). Maternal, perinatal and neonatal health care outcomes require well-coordinated and responsive healthcare systems with an adequate health workforce and resourced facilities (NDoH, 2021). With over 80% of the population lacking health insurance cover in EM and FBD, the number dependent on public health services is substantial. The maternal services

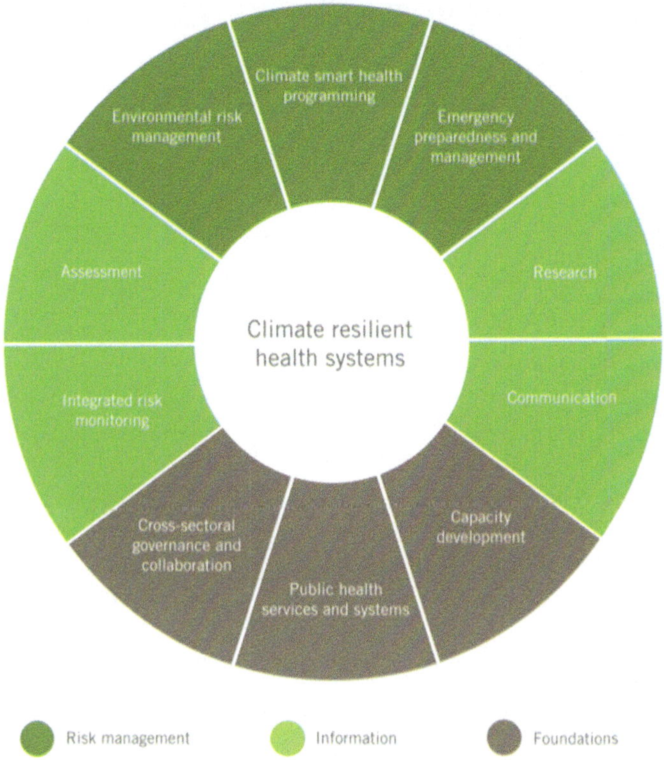

Fig. 10 Ten key components for building climate resilience. (Adapted from the WHO Operational framework for building climate-resilient health systems) (WHO, 2014, p. 15)

within the PHC systems of EM and FBD are already inadequately resourced in terms of infrastructure and human resources—rendering them incapable of responding to extreme weather events or climate emergencies in general.

In EM, in April 2022, at least 23 hospitals and 34 clinics and community health centres were damaged in major floods at a cost of R187m, according to the National Minister of Health (NDoH, 2022a, 2022b). However, the social contexts within which these health services are located further compromise service delivery in times of emergencies and disasters. Disruption of roads, water and electrical supply result in poorly or non-functioning services, as is evident from the 2022 flood crisis in eThekwini (Keipopele, 2022). The inability to rapidly restore services within the health sector specifically, or within the community generally, is reflected by the duration that affected communities spend in make-shift homes or relying upon short-term mobile clinic facilities (Matangira, 2022).

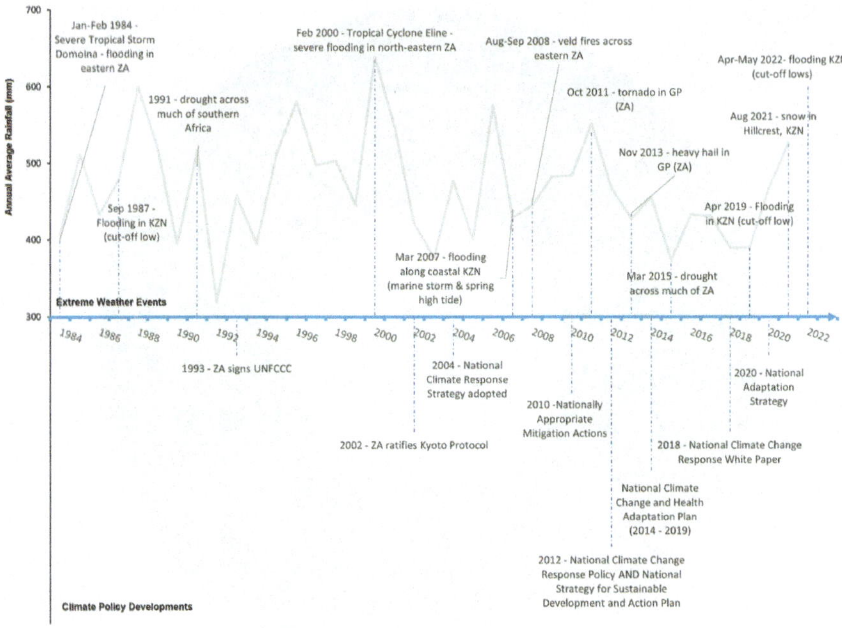

Fig. 11 Average rainfall in South Africa, extreme weather events and climate policy interventions, 1984–2022 (rainfall data from World Bank, 2023)

4.3 South Africa Governance and Management Policies in Response to Climate Change

South Africa has been part of the international community implementing strategies for addressing climate change, from mitigating to building resilience of the communities. A variety of policies and strategic documents have been produced to guide the response and building of resilience to climate change (Fig. 11).

4.4 The South African National Adaptation Strategy and the Health Sector

Prior to 2018, few of the South African policies defined how the health sector must adapt or mitigate the effects of climate change. The 2019 National Climate Change Adaption Strategy (DFFE, 2019) has health as a lead sector, and among the key response actions are (i) to prepare for the climate impacts and (ii) to respond at the time of climate change impact or disaster. Within this Strategy, there are specific expectations of the Health sector (Table 4).

Table 4 The health sector expectations as per the 2018 national adaptation strategy

Governance and policy
• Create a national structure which convenes regularly to plan and implement processes that addresses adaptation through an inter-sectoral forum
• Discuss climate change and health issues and develop an implementation plan at local levels through local structures, which can forge community partnerships to implement strategies
• Ensure that disaster management plans at all tiers of the government in the health sector address climate change and related health concerns in vulnerable communities
Implementation
• Evaluate and prepare public health intervention for extreme weather events, this including developing special working conditions for vulnerable groups such as outdoor workers
• To equip healthcare facilities to manage climate change-related health impacts through provision of sufficient resources and adequate capacity to manage increased patient-load

5 Discussion

Our data provide growing evidence that the changing climate and weather patterns such as El Niño, contribute to adverse health impacts resulting in an increased perinatal mortality risk. At a national level, the data show a variation in perinatal mortality and in extreme weather events over the period of review. It is clear that certain provinces in South Africa are at greater risk for these outcomes and experience the greatest impacts of the weather phenomena.

The El Niño phases that prevailed during 1982–83, 1997–98, and 2014–16 are regarded as being among the strongest on record. The hyperaridity experience in the NC in the 2015/2016 period resulting from the El Niño conditions has been described as an 'El Niño Shock' (Sazib et al., 2020). The decline in agricultural production by 8.4% in 2015 was attributed to the extreme drought conditions, with productivity 20% below the country's long-term average (Sazib et al., 2020). Our findings suggest that food insecurity and indirect health impacts of increased aridity and increased temperature may have contributed to the contrasting perinatal mortality seen in the EM and FBD municipalities in this period.

However, our review of the maternal health services and perinatal health indicators suggests that these two municipalities have different health service foundations which may contribute to the health outcomes during the extreme weather events, such as El Niño. The poor maternal and perinatal health indicators strongly suggest that health services are compromised, and the factors influencing these include lack of facilities, lack of health professionals and failure of facilities to provide adequate services (not achieving the "ideal clinic" benchmark). Thus, when faced with new threats, particularly related to the environment or extreme weather events, the ability of these services to respond to the crises is non-existent. The lack of this resilience has been objectively reflected in responses of EM to the 2022 flooding disaster.

However, policy and strategic interventions have been implemented by the South Africa Government. The government released a Bill in 2022 that seeks to address a climate change response and a just transition to a low-carbon and climate-resilient economy. The National Climate Response Policy (NCCRP) is central to this policy framework. Unfortunately, this policy framework omits references to building resilience within the health services broadly and in the health system itself. Of note, the NDoH's Strategic Plan 2020/21–2024/25 makes no reference to health services and health systems resilience in the context of the changing climate (NDoH, 2020). The World Bank's annual reports on the national responses to climate change emphasizes that despite the progress made by South Africa, no climate policies are specific to health (World Bank, 2021). The NDoH did release a "National Climate Change and Health Adaptation Plan, 2014–2019" (NDoH, 2014), which provided plans of action, including health service readiness, however, there have been no reports since to indicate the extent of implementation of this plan.

In conclusion, perinatal infant mortality is affected by weather phenomena, particularly aridity influenced by El Niño. Maternal health services in regions most at risk are already compromised, and lack the required resilience to respond to these extreme events. Nationally, the absence of meaningful health systems approaches and the lack of a health policy framework to address maternal and infant health in the changing climate context, undermines the ability of the health sector to develop strong resilience.

6 Recommendations

In line with the WHO (2014) framework of building climate resilient health systems, these recommendations are contextualized at PHC as the first line of service for maternal and child health in South Africa. The recommendations take into consideration multisectoral policy and action, integrated health services and empowered communities (WHO, 2014) to mitigate the effects of climate change and improve the PHC system resilience (Fig. 12).

6.1 Strict Environmental Protection and Regulation

Enforcement of the world standard existing regulations that guide air and emission control of environmental pollutants in South Africa is necessary. This requires local and provincial government to deploy trained staff, well versed in the legislation as agents of enforcement. Engagement between producers of emissions and enforcement agencies must be established to achieve legislated standards.

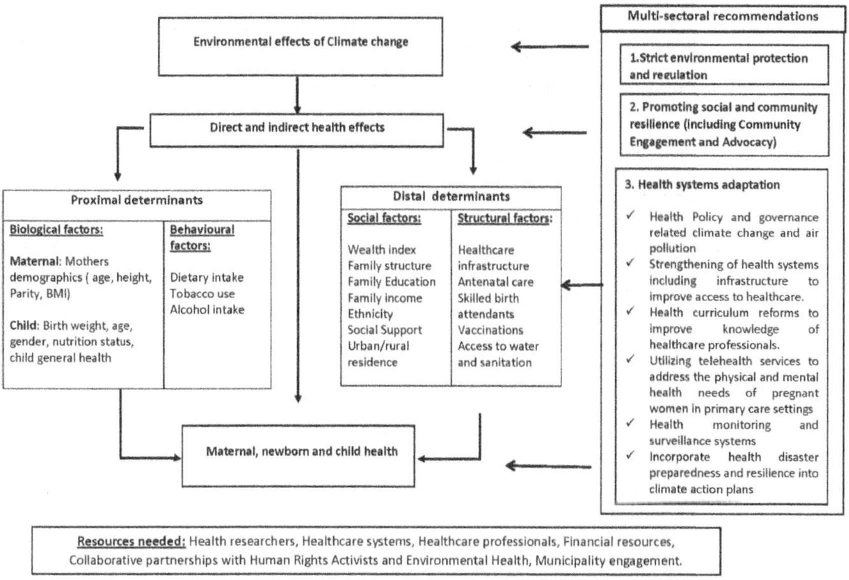

Fig. 12 Recommendations to mitigate effects of climate change in relation to maternal, newborn and child health for South Africa

6.2 Promoting Community Resilience Through Community Engagement and Advocacy

Communities need to be engaged to ensure that they understand the effects of climate change on the health of pregnant mothers and neonates. There is need to scale up public communications, and population awareness and maternal education. Climate and health information should be included in clinic materials and diverse maternal educational programmes, shared through various social media platforms. Pregnant mothers should be made aware of specific interventions such as limiting the amount of time spent outdoors and staying hydrated. Engagement with community-based organisations, religious structures and resident associations need to be strengthened to building advocacy within communities.

6.3 Healthcare Systems Adaptation

South Africa has developed and implemented plans and policies for improving maternal, new-born and child survival and health (NDoH, 2021) in line with the recommendations of the WHO (WHO, 2014) and the Sustainable Development Goals 2030 among others, located within the PHC approach. However, there are no policies that address the impacts of temperature and extreme weather events on perinatal health. Health policies need to implement effective plans for maternal health resilience in the changing climate. Additionally, a national heat vulnerability index for pregnant women in high heat risk areas for healthcare preparedness is necessary.

Strengthening national health systems would significantly improve the responsiveness to health hazards associated with climate change. Improving and modernising existing infrastructure associated with sanitation and water resource management is critical. Health care access is necessary: locating health facilities in close proximity to communities, innovative operational hours and reducing waiting times in health facilities are some options. Equipping facilities for emergency preparedness for extreme weather events must be prioritised.

There is need for innovative approaches, including improved remuneration, towards the training and retention of human resources for health in LMIC countries including South Africa. Human healthcare resource capacitation for maternal, newborn and child is crucial to meet the increasing demands for services. Education curricula for health professionals and continuing professional education need new focus on climate-related health effects.

Surveillance and research into existing and potential diseases due to climate change need to be commissioned and/or supported by national government. The District Health Information System collects information on various health outcomes, however, there is no clear linking of these indicators to outcomes associated with climate change. Open platform data allowing research is necessary to determine the likely associated burden of maternal and negative perinatal outcomes and healthcare system resilience. This will allow policymakers to determine the service needs.

6.4 Agricultural and Irrigation Systems and Policy

Although we have focused on the responses necessary by the health sector to protect maternal and child health, agencies in the agriculture and water conservation sector play critical roles in ensuring optimal health. Our lagged-correlation analysis showed that soil data can be used to predict declines in food production capacity, which can serve as early warning to identify alternate food supply or improve irrigation in at-risk communities.

References

American Meteorological Society. (2022). Aridity. *Glossary of Meteorology*. Available from: https://glossary.ametsoc.org/wiki/aridity. Accessed 19 February 2023.

Anderko, L., et al. (2020). Climate changes reproductive and children's health: A review of risks, exposures, and impacts. *Pediatric Research, 87*(2), 414–419.

Australian Bureau of Meteorology. (2016). El Niño is over, but has left its mark across the world. Available from: http://www.bom.gov.au/climate/updates/articles/a018.shtml. Accessed 4 February 2023.

Bátiz, L. F., et al. (2022). Climate change and preterm birth: A narrative review. *Environmental Advances, 10*, 100316.

Chersich, M. F., et al. (2018). Impacts of climate change on health and wellbeing in South Africa. *International Journal of Environmental Research and Public Health, 15*(9), 1884.

Choudhari, R. H. (2022). Multidimensional impact of climate change on human reproduction and fertility: A medical perspective on changing dynamics (Chap. 83). In *Research anthology on environmental and societal impacts of climate change* (pp. 1672–1709). IGI Global.

Clark, T., & Zolnikov, T. R. (2020). Climate change and mental health. In *The Palgrave handbook of climate resilient societies* (pp. 1–26). Palgrave-Macmillan.

Drysdale, R. E., et al. (2021). Coping through a drought: The association between child nutritional status and household food insecurity in the district of iLembe South Africa. *Public Health Nutrition, 24*(5), 1052–1065.

Engelbrecht, F., et al. (2015). Projections of rapidly rising surface temperatures over Africa under low mitigation. *Environmental Research Letters, 10*, 1–16.

Folke, C., et al. (2010). Resilience thinking: Integrating resilience, adaptability and transformability. *Ecology and Society, 15*(4), 20.

Gomez-Zavaglia, A., et al. (2020). Mitigation of emerging implications of climate change on food production systems. *Food Research International, 134*, 109256.

Ha, S. (2022). The changing climate and pregnancy health. *Current Environmental Health Reports, 9*, 263–275.

Helldén, D., et al. (2021). Climate change and child health: A scoping review and an expanded conceptual framework. *The Lancet Planetary Health, 5*(3), e164–e175.

Hunter, J. R., et al. (2017). The ideal clinic in South Africa: Progress and challenges in implementation. *South African Health Review, 2017*(1), 111–123.

Inoue, K., et al. (2020). Air pollution and adverse pregnancy and birth outcomes: Mediation analysis using metabolomic profiles. *Current Environmental Health Reports, 7*(3), 231–242.

Keipopele, K. (2022). KZN floods: Healthcare services take a battering. *Health e-News: Journalism for Public Health.* Available from: https://health-e.org.za/2022/04/20/kzn-floods-healthcare-services-take-a-battering/. Accessed 8 February 2023.

Langerman, K. E., & Pauw, C. J. (2018). A critical review of health risk assessments of exposure to emissions from coal-fired power stations in South Africa. *Clean Air Journal, 28*(2), 68–79.

Massyn, N., Barron, P., & Day, C. (2020). District health barometer. *Health Systems Trust.* Available from: https://www.hst.org.za/publications/District%20Health%20Barometers/DHB%202019-20%20Publisher%20information.pdf. Accessed 19 February 2023.

Matangira, L. (2022). Over 1,000 children affected by April floods receive play and learning materials. *Unicef South Africa.* Available from: https://www.unicef.org/southafrica/stories/over-1000-children-affected-april-floods-receive-play-and-learning-materials. Accessed 8 February 2023.

Mayala, B., et al. (2018). *The DHS program geospatial covariates datasets manual* (2nd ed.). ICF.

Moura, P. M. S. S., et al. (2014). Risk factors for perinatal death in two different levels of care: A case–control study. *Reproductive Health, 11*(11).

National Council of Provinces (NCOP). (2015). Portfolio committee on health and social services meeting report. Available from: https://pmg.org.za/committee-meeting/21130/. Accessed 20 February 2023.

National Department of Forestry, Fisheries and the Environment (DFFE). (2019). National climate change adaptation strategy: South Africa. Available at: https://www.dffe.gov.za/sites/default/files/docs/nationalclimatechange_adaptationstrategy_ue10november2019.pdf. Accessed 07 February 2023.

National Department of Health (NDoH). (2014). National climate change & health adaptation plan. Available from: https://www.unisdr.org/preventionweb/files/57216_nationalclimatechangeandhealthadapt.pdf. Accessed 20 February 2022.

National Department of Health (NDoH). (2016). Guidelines for maternity care in South Africa: A manual for clinics, community health centres and district hospitals, Republic of South Africa.

Available from: https://www.idealhealthfacility.org.za/App/Document/Download/42. Accessed 8 February 2023.

National Department of Health (NDoH), et al. (2019). South Africa demographic and health survey 2016.

National Department of Health (NDoH). (2020). Health: strategic plan 2020/21–2024/25. Pretoria. 2020. Available at: https://www.health.gov.za/wp-content/uploads/2020/11/depthealthstrategicplanfinal2020-21to2024-25-1.pdf. Accessed 20 February 2023.

National Department of Health (NDoH). (2021). South African maternal, perinatal and neonatal health policy. Available from: https://www.knowledgehub.org.za/system/files/elibdownloads/2021-06/SA%20MPNH%20Policy%2023-6-2021%20signed%20Web%20View%20v2.pdf. Accessed 8 February 2023.

National Department of Health (NDoH). (2022a). Media statement: Health minister constitutes team for KZN floods 19 April 2022. Available at: https://www.health.gov.za/wp-content/uploads/2022/04/Health-Minister-constitutes-Team-for-KZN-floods-19-April-2022.pdf. Accessed 20 February 2023.

National Department of Health (NDoH). (2022b). Annual performance plan, Republic of South Africa. Available from: https://www.health.gov.za/wp-content/uploads/2021/10/Annual-Performance-Plan-2021-2022-1.pdf. Accessed 8 February 2023.

Network of African Science Academies (NASAC). (2015). *Climate change adaptation and resilience in Africa: Recommendations to policymakers*. Network of African Science Academies.

Opoku, S. K., et al. (2021). Climate change and health preparedness in Africa: Analysing trends in six African countries. *International Journal of Environmental Research and Public Health, 18*(9), 4672.

Ramavhoya, T. I., et al. (2022). Developing and prioritising strategies to improve the implementation of maternal healthcare guidelines in South Africa: The nominal group technique. *African Journal of Primary Health Care and Family Medicine, 14*(1), 1–8.

Riscura. (2016). The effect of El Niño on key Southern African countries. Available from: https://www.riscura.com/effect-southern-african-countries/. Last accessed 4 February 2023.

Rothschild, J., & Haase, E. (2022). The mental health of women and climate change: Direct neuropsychiatric impacts and associated psychological concerns. *International Journal of Gynecology and Obstetrics, 160*(20), 405–413.

Sambo, L. G., & Kirigia, J. M. (2014). Investing in health systems for universal health coverage in Africa. *BMC International Health and Human Rights, 14*, 28. https://doi.org/10.1186/s12914-014-0028-5

Sazib, N., et al. (2020). Assessing the impact of ENSO on agriculture over Africa using earth observation data. *Frontiers in Sustainable Food Systems, 4*, 509914.

Segal, T. R., & Giudice, L. C. (2022). Systematic review of climate change effects on reproductive health. *Fertility and Sterility, 118*(2), 215–223.

StatsSA. (2020). Mid-year population estimates 2020. Statistics South Africa. Available from: https://www.statssa.gov.za. Accessed 8 February 2023.

Trisos, C. H., et al. (2022). 'Africa', *climate change 2022: Impacts, adaptation and vulnerability. Contribution of Working Group II to the Sixth Assessment Report of the Intergovernmental Panel on Climate Change* (pp. 1285–1455). Cambridge University Press.

Tyson, P. D., & Preston-Whyte, R. A. (2000). *The Weather and Climate of Southern Africa*. Oxford University Press.

United Nations Convention to Combat Desertification (UNCCD). (2022). Reporting glossary. Available at: https://www2.unccd.int/sites/default/files/inline-files/Glossary_UNCCD-2022-reporting-process_EN.pdf. Accessed 19 February 2023

United Nations Office for Disaster Risk Reduction (UNISDR). (2009). Terminology and disaster risk reduction. Available at: https://www.undrr.org/publication/2009-unisdr-terminology-disaster-risk-reduction. Accessed 19 February 2023.

United States Climate Prediction Center (USCPC). (2022). Historical El Niño/La Niña episodes (1950–present). Available from: https://origin.cpc.ncep.noaa.gov/products/analysis_monitoring/ensostuff/ONI_v5.php. Accessed 4 February 2023.

Walton, B. (2016). El Niño is over but the damage continues. *Circle of Blue*, 19 April. Available from: http://www.circleofblue.org/2016/africa/el-nino-damage-continues. Accessed 4 February 2023.

World Bank. (2021). Climate risk country profile: South Africa. Washington DC, USA. Available from: https://climateknowledgeportal.worldbank.org/sites/default/files/country-profiles/15932-WB_South%20Africa%20Country%20Profile-WEB.pdf. Accessed 07 February 2023.

World Bank. (2023). Country: South Africa, current climate > climatology [Online] climate change knowledge portal. Available from: https://climateknowledgeportal.worldbank.org/country/south-africa/climate-data-historical. 18 February 2023.

World Health Organization (WHO). (2014). WHO guidance to protect health from climate change through health adaptation planning. Available at: https://www.paho.org/en/documents/who-guidance-protect-health-climate-change-through-health-adaptation-planning-2014. Last accessed 07 February 2023.

World Health Organisation (WHO). (2016). Health workforce requirements for universal health coverage and the sustainable development goals. *Human Resources for Health Observer*, 17.

Americas

Climate Change, Mental Health, and Substance Use—USA

Olaniyi Olayinka and Brook Alemu

Abstract There is an accumulating body of evidence showing the human health impacts of climate change. Although research on physical human effects predominates, recent studies show a significant mental health impact of climate-and-weather-related events. Awareness of climate-related mental health issues has important ramifications for the implementation of national healthcare policies. This is more so in the United States (the third most populous country and a leading contributor to global greenhouse gas emission), where over one in ten adults live with severe mental illness and/or a substance use disorder (SUD). In the 2016 *Climate and Health Assessment* report, the U.S. Global Change Research Program (USGCRP) noted (with very high confidence in some instances) that exposure to weather and climate extremes increases the risk of trauma-and-stressor-related disorders, anxiety, depression and substance use, among other psychological issues. Given that mental and SUDs are the leading cause of years lived with disability globally, preventing the exacerbation of this reality by a changing climate is crucial. In this chapter, we introduce the reader to established mental health consequences of climate change with focus on the United States.

Keywords Mental health · Substance use disorders · Climate change · Health care · Policy

O. Olayinka (✉)
University of Texas Health Science Center at Houston, Houston, TX 77021, USA
e-mail: olaniyi.o.olayinka@uth.tmc.edu

B. Alemu
Health Sciences Program, School of Health Sciences, Western Carolina University, Cullowhee, NC 28723, USA
e-mail: balemu@email.wcu.edu

R. Akhtar (ed.), *Climate Change and Human Health Scenarios*, Global Perspectives on Health Geography, https://doi.org/10.1007/978-3-031-38878-1_22

1 Introduction

There is an accumulating body of scientific evidence showing the significant human health impact of climate change (Kim, 2016; McMichael et al., 2006; Watts et al., 2015). This is particularly so for mental health and substance use disorders, both of which are leading causes of disability globally (Lopez & Murray, 1998; Whiteford et al., 2013). In the 2016 *Climate and Health Assessment* report, the US Global Change Research Program (USGCRP) noted (with very high confidence in some instances) that exposure to weather and climate extremes increases the risk of trauma- and stressor-related disorders, anxiety, depression, and substance use, among other psychological problems (Dodgen et al., 2016). Unfortunately, research on the mental health impact of the changing climate is sparse (Lawrance et al., 2021).

In countries with a high burden of psychiatric and substance use disorders, awareness of specific health threats of climate change is vital in developing effective public health policies (Watts et al., 2015; Collaborators, 2022). This includes the USA—*the third most populous country and a leading contributor to global greenhouse gas emission*, with over one in ten adults living with a severe mental illness and/or a substance use disorder (SUD) (Boden et al., 2017; Substance Abuse and Mental Health Services Administration, 2020). As a result, strong policy statements on the public health threats of climate change, including on mental health, have been adopted by leading academic, public health, and clinical associations in the USA (Ursano et al., 2017). In this chapter, we introduce the reader to established impact of climate change on mental health and substance use, with a special focus on the USA.

2 Climate Change and Mental Health Problems

The consequences of a warming climate are being observed today, and with far reaching effects on the health of Americans (Dodgen et al., 2016; Melillo et al., 2014; Upperman et al., 2017; Ursano et al., 2017). The mental health effects of climate change range in severity from mild to major mental health problems and have short- and long-term implications (Cianconi et al., 2020). According to current evidence, extreme heat and precipitation—which have become far more frequent across the USA in the past few decades—increase the risks associated with wildfires, drought, and flooding (Duffy & Tebaldi, 2012; Melillo et al., 2014). The mental health risks of these extreme weather events can be characterized as direct and indirect (Costello et al., 2009; Hayes et al., 2018; Portier et al., 2013). For example, flood victims that were rescued in the aftermath of Hurricane Sandy might have developed Posttraumatic stress disorder (PTSD) (*a direct effect*), while close relatives of the flood victims might themselves have developed the same disorder (*indirect effect*). Berry and colleagues share a similar approach in their review of the pathophysiology of climate-related health problems (Berry et al., 2010). That is, acute/subacute weather events may impact mental health directly and indirectly. Despite the sparse literature on the subject, we discuss in this chapter the most common mental health problems that have been described in the disaster literature.

2.1 Trauma- and Stressor-Related Disorders

Extreme weather events are often a source of trauma to humans and may cause trauma- and stressor-related disorders (TSRD). According to DSM-5, TSRD include PTSD, acute stress disorder, and adjustment disorders. The most described TSRD in the disaster literature is PTSD (Berry et al., 2010; Dodgen et al., 2016; Parker et al., 2016); hence, it is the focus of the next few paragraphs.

Specifically, PTSD results from 'exposure to actual or threatened death, serious injury, or sexual violence,' with trauma-associated symptoms lasting for a period of one month or more. The disorder affects approximately four to six percent of Americans and constitutes a huge economic burden (Davis et al., 2022; Goldstein et al., 2016; Kessler et al., 2005). A 2018 study estimated the national cost of the disorder at over US$230 billion, with approximately 80% of the financial burden borne by US civilians (Davis et al., 2022). Unfortunately, the future public health cost of PTSD is likely to worsen, as exposure to climate-related events increases. In fact, findings from the 2012 to 2013 National Epidemiologic Survey on Alcohol and Related Conditions-III (NESARC-III) show that approximately one in ten Americans were exposed to natural disaster (Goldstein et al., 2016).

Extreme weather events are associated with significant physical and psychological trauma, which invariably increases risk for PTSD, as has been described by several epidemiologic studies (Galea et al., 2005; Goldstein et al., 2016). This is particularly of public health concern given that vulnerable populations/communities (e.g., children, the elderly, pregnant women, racial/ethnic minorities, and socially disadvantaged individuals) are more prone to the negative impacts of disasters. Of note, disparities in the impact of severe weather events occur even when disaster exposures are similar (Alexander et al., 2017; Clayton, 2021; Davidson et al., 2013; Raker et al., 2020; Steinglass & Gerrity, 1990). For instance, a 2013 study of the mental health impact of Hurricane Ike on a sample of 1249 adults in Galveston and Chambers counties found a higher report of PTSD and depression among African Americans, compared with White and Latino Americans (Davidson et al., 2013). It is worth mentioning that when all traumatic exposures were considered, the prevalence of PTSD in the NESARC-III study was higher among Native American women (Goldstein et al., 2016).

In terms of factors that increase the risk for PTSD after exposure to a disaster, the Resilience in Survivors of Katrina (RISK) study (*a longitudinal, mixed-methods study of the health effects of disaster on a sample of economically disadvantaged college students in New Orleans, Louisiana*) shows that 'bereavement, fear for one's life and uncertainty about the safety of loved one' (Raker et al., 2020). The risk study revealed other useful information. For example, the risk of PTSD appears to be highest during the immediate period (up to one year according to some studies) following a disaster, decreasing modestly over time (Raker et al., 2019). Furthermore, compared with subject of other race and those without a history of mental illness, subjects who identified as black and with a pre-disaster history of psychological distress (PD) had approximately three and five times the odds of developing PTSD, respectively. Pre-disaster mental health functioning, however, appears to explain the association between race and the risk for PTSD after a disaster (Alexander et al., 2017). Specific types of disaster also appear to confer a higher risk for PTSD, compared with other disaster types. For instance, severe anxiety and other trauma-associated symptoms have been reported in those exposed to disasters that occur acutely—such as floods, wildfires, and heat waves (Berry et al., 2010).

Given the high risk for PTSD in vulnerable, trauma-exposed individuals, interventions that mitigate the mental health impact of disasters are crucial. Addressing mental health disparities among Americans is one way to decrease the risk of developing PTSD following a disaster. This includes providing trauma-informed care especially to at-risk populations such as children and socially disadvantaged communities (Williams, 2022).

2.2 *Depression*

Climate-related events such as heat waves, hurricanes, and severe flooding are associated with a range of psychological problems, ranging from mild distress and acute stress to severe anxiety and depression (Cianconi et al., 2020). Depression is characterized by a constellation of psychological and physical symptoms that often emerge following a significant stressor. In the setting of extreme climate events, the associated loss of loved ones and property may cause pervasive feelings of sadness, worthlessness, and helplessness, sleep disturbance, anxiety, and suicidal thoughts.

Contextual factors (including a prior history of depression, adverse childhood experience, and social factors), however, may increase the risk for depression in disaster-exposed individuals (Lowe et al., 2015). For instance, Tracy and colleagues found in their study of 658 Americans exposed to Hurricane Ike in Galveston, Texas, that life time stressors and sociodemographic factors increased the risk for depression post-Hurricane Ike (Tracy et al., 2011). While the past month prevalence of depression in the study participants was approximately 5%, similar studies report a higher prevalence of depression in survivors of natural disasters, especially among vulnerable populations. Among the 475 persons whose homes were damaged by the severe flooding of 2006 in El Paso County, Texas, about 18% reported depression four months after the flooding (Collins et al., 2013). The El Paso study also showed that while factors such as older age and financial problems increased the risk for depression and other mental health problems, being born outside the USA, having access to medical services, and a lack of proficiency in spoken English decreased the risk by 60%, 70%, and 71%, respectively. Interestingly, studies show a similar trend in the general population that is depression is higher among US born subjects, compared with foreign-born individuals (González et al., 2010). In essence, public health decision makers are encouraged to examine factors that are protective of mental problems in other populations for the benefit of Americans.

3 Climate Change and Substance Use Disorders

Substance use disorders (SUD) have significant impacts on people, families, and communities in the USA (Hasin & Grant, 2015; Murray & Lopez, 2013; Reeves et al., 2011; World Health Organization, 2013). In the following paragraphs, we examine SUD in the context of climate change and then assess its burden among different population sub-groups in the USA.

Climate change has been identified as 'the defining issue' for public health in the twenty-first century (Sheehan et al., 2017). Increasing global temperatures due to contemporary human activity are causing major physical, chemical, and ecological changes on the planet. With increasing global surface temperatures, the possibility of more droughts and increased intensity of storms are also exacerbated. Further, rising sea levels expose higher locations not usually affected by climate change (United States Geological Survey, n.d.). From a public health perspective, climate change poses significant threats to human health, safety, and security. Extreme weather events including winter weather, heat waves, floods, drought, dust storms, wildfires, tropical cyclones, hurricanes, and tornadoes cause disaster—defined as a 'situation or event which overwhelms local capacity, necessitating a request to a national or international level for external assistance; an unforeseen and often sudden event that causes great damage, destruction, and human suffering' (Below et al., 2009; Oliver-Smith, 1998). Weather-related disasters initiate a series of subsequent stressors that can last from months to years (Simpson et al., 2011). Psychological symptoms that emerge during the days or weeks following a disaster can take months or years to dissipate. Some disaster victims develop chronic mental health problems such as post-traumatic stress disorder, depression, anxiety, and substance use (Vestal, 2017). These public health problems are expected to strain the US healthcare system as the intensity and frequency of hurricanes, floods, tornadoes, wildfires, earthquakes, and other natural disasters increase in the coming decades (Vestal, 2017).

Although the topics of climate change and SUD have been well documented independently, the literature on the association between the two is increasing but somewhat limited. Schuler et al. conducted a longitudinal study utilizing substance abuse treatment discharge data from New Orleans from the years 2006–2011 (Shuler et al., 2017). The study examined substance users by looking at the prevalence of substance use, treatment characteristics, and demographics. They found 35% of all discharges had a co-occurring psychiatric and substance use disorder (COD). Further, after controlling for race, employment, treatment service setting at discharge, primary substance problem, and the discharge's principal source of referral, patients with COD were 29% less likely to complete treatment as compared to those with no COD. Treatment completion among discharges with a COD has significantly declined from 36.8% in 2006 to 18.7% in 2011. Notable significant trends in homelessness, criminality, and heroin use were also identified among subjects with a COD. The findings suggest that in the event of a natural disaster, substance abuse treatment undergoes various changes and may increase challenges for successful treatment completion for vulnerable populations.

Cepeda et al. also examined the relationship between disaster-related experiences and mental health outcomes among highly disadvantaged Hurricane Katrina evacuees (Cepeda et al., 2010). The study included African Americans with SUD from inner-city New Orleans neighborhoods. They found that disaster-related experiences including negative life changes, disaster exposure, post-disaster stressors, and resource loss had negative relationships with mental health among individuals with SUD. In addition, resource loss had the strongest inverse relationship with mental health and disaster exposure. Disaster-related experiences also had a negative interactive effect on psychological distress and anxiety. Overall, the study showed a nonlinear relationship between disaster-related experiences and mental health. Rather, natural disasters, caused by climate change, are uniquely linked to various mental health outcomes among substance users indicating a heightened psychological vulnerability among those without suitable resources.

3.1 Substance Use Disorder Among Adolescents

Substance use among young people remains prevalent globally and nationally, causing significant physical, social, and economic costs to individuals, families, and society. In the 2019 Monitoring the Future adolescent survey, the investigators found that alcohol, marijuana, and tobacco use remain prevalent among American youths (The National Institute on Drug Abuse, 2020). This, and other adolescent surveys, also report an estimated 50% of those in 9th–12th grade has used cannabis while about two-thirds have used alcohol by their 12th grade (Centers for Disease Control & Prevention, 2020). Unfortunately, current marijuana laws and the high exposure to vaping products appear contribute to the rising prevalence of substance use among adolescents (Borodovsky et al., 2017; Chadi et al., 2019; Yu et al., 2020).

With early substance use initiation, however, comes the significant risk of developing a substance use disorder post-adolescence (Jordan & Andersen, 2017; Morales et al., 2020; Piehler et al., 2012; Schulte & Hser, 2013). For example, individuals exposed to alcohol before the age of 15 years reportedly have a higher prevalence of alcohol use disorder than those who initiated alcohol use after the age of 21 years (Substance Abuse and Mental Health Services Administration, 2014). It is therefore crucial to continually characterize factors associated with substance use among youths as it may inform effective strategies to reduce the untoward impact of drug use in this population. This is in tandem with one of seven national drug priorities set forth by the current US administration which is supporting evidence-based prevention efforts to reduce youth substance use (Office of National Drug Control Policy, 2021).

3.2 *Epidemiology and Trends in Adolescent Substance Use Disorder*

Trends of adolescent use before 2020 were on the rise and had consistently been for years. Reports of adolescent use included that 65,000 or 8.14% of 12–17-year-olds reported using drugs in the last month while 15% reported having ever used drugs in their lifetime (Jones et al., 2012; National Center for Drug Abuse Statistics, 2019). The spectrum of drugs being used by adolescents is also alarming. Among adolescents, 83.08% reported using marijuana in the last month, 12.65% reported using marijuana in the last year, and 36.8% reported using in their lifetime (Jones et al., 2012; National Center for Drug Abuse Statistics, 2019). A total of 0.38% reported using cocaine in the last year and 3.9% reported cocaine use in their lifetime while 0.13% reported using methamphetamine in the last year and 2.1% reported methamphetamine use in their lifetime (Jones et al., 2012; National Center for Drug Abuse Statistics, 2019). It was also reported that up to 0.06% used heroin in the last year and 1.8% reported use of heroin in their lifetime while 2.63% reported using pain relievers (Jones et al., 2012; National Center for Drug Abuse Statistics, 2019). The data also showed that 9.15% of all 12–17-year-olds used alcohol in the last month (National Center for Drug Abuse Statistics, 2019). This percentage was lower than the Youth Risk Behavior Surveillance Survey score which identified that 29.2% of adolescents reported current alcohol use which was defined by 1 drink in the last thirty days (Jones et al., 2012). Of additional concern is that 3.13% of all 12–17-year-olds met the criteria for SUD in the last year and 1.50% of all 12–17-year-olds met the criteria for AUD in the last year (National Center for Drug Abuse Statistics, 2019).

Another concerning trend, and an unfortunate risk associated with substance use, is the potential for an overdose (Compton et al., 2019; National Center for Drug Abuse Statistics, 2019; Nelson et al., 2017). In 2019, there were 4777 overdose deaths of individuals ages 15–24 which was a decrease from 2017 in which 5455 individuals aged 15–24 died from an overdose (National Center for Drug Abuse Statistics, 2019). 2017 was the height of the opioid epidemic and the presence of fentanyl and carfentanil in the drug supply was high across the country (Quinones, 2021). This identifies that there were unprecedented quantities of lethal drugs and increased access to them at that time (Quinones, 2021). To put the 2019 data into some perspective, 10 years prior (2009) the death toll from overdoses for 15–24-year-olds was 3377 (National Center for Drug Abuse Statistics, 2019). In looking back 20 years, the death toll from overdoses for 15–24-year-olds was 1240 (National Center for Drug Abuse Statistics, 2019). The number of overdose deaths has quadrupled in the last 20 years for this population which should cause significant concern for the road ahead if something is not done to change that trajectory (National Center for Drug Abuse Statistics, 2019). The risk-taking behavior of adolescents predisposes them to potential harm, substance use, and unintentional overdoses (Steinberg, 2007). The death of one adolescent from a drug overdose is too many.

3.3 Substance Use Disorder Among Pregnant Women

Pregnant women are among the most affected population groups by substance use disorder (Patrick et al., 2012). Increasing at an alarming rate, opioid use in pregnancy increased fivefold between 2000 and 2012 (Patrick et al., 2015). The 2011 US National Survey on Drug Use and Health determined that 5% of pregnant women, 15–44 years of age, reported using illicit drugs (Substance Abuse and Maternal Health Services Administration, 2012).

Substance dependence in pregnancy complicates the clinical management of an already vulnerable group of patients. Dependence increases the risk of poor maternal and perinatal outcomes (Forray & Foster, 2015; Benningfield et al., 2010; Brown et al., 2018; Whiteman et al., 2014). Women of reproductive age who use and misuse drugs, whether prescribed or illicit, are more likely to have a lower socioeconomic status, family instability, receive inadequate prenatal care, and suffer from alcohol, tobacco, and illicit drug use. In addition to the risks associated with substance dependence, these comorbid conditions further increase the risk of maternal and fetal outcomes including maternal cardiac arrest, maternal blood transfusion, fetal growth restriction, placental abruption, preterm labor, oligohydramnios, fetal demise, and hospital lengths of stay > 7 days (Kaltenbach et al., 1998; Maeda et al., 2014; Wendell, 2013).

4 Conclusion

The negative impact of a warming climate on mental health has been observed globally including in the USA. While literature on the mental health impacts of climate change is understandably challenging and sparse, numerous studies exist that describe the negative impacts on extreme weather events. In general, the mental health effects of climate change has been characterized as (1) direct or indirect, (2) short term or long term, and (3) acute or delayed. The most widely reported mental health sequelae of climate change include PTSD, anxiety, depression, and substance use. These result from the threat to human lives, loss of properties, and displacement of populations that follow extreme weather events (such as floods, hurricanes, and wildfires). It is worth noting that most climate and health studies focus on the direct impacts of climate change, with a significant gap in studies involving pediatric populations, and those related to community-level factors and health disparities. Unfortunately, current evidence shows that early childhood events increase the risk for mental health disorders and substance use in later life which stresses the importance of climate studies that focus on this population.

While addressing risk factors for mental disorders is crucial, limiting global temperature to below 1.5 °C could equally minimize the threat to human health of global warming. Ultimately, as nations engage in collaborative efforts to curb the production of earth-warming pollutants, building the psychological resilience of at-risk populations is equally important and should be encouraged.

References

Alexander, A. C., Ali, J., Mcdevitt-Murphy, M. E., Forde, D. R., Stockton, M., Read, M., & Ward, K. D. (2017). Racial differences in posttraumatic stress disorder vulnerability following Hurricane Katrina among a sample of adult cigarette smokers from New Orleans. *Journal of Racial and Ethnic Health Disparities, 4*, 94–103.

Below, R., Wirtz, A., & Guha-Sapir, D. (2009). Disaster category classification and peril terminology for operational purposes.

Benningfield, M. M., Arria, A. M., Kaltenbach, K., Heil, S. H., Stine, S. M., Coyle, M. G., Fischer, G., Jones, H. E., & Martin, P. R. (2010). Co-occurring psychiatric symptoms are associated with increased psychological, social, and medical impairment in opioid dependent pregnant women. *The American Journal on Addictions, 19*, 416–421.

Berry, H. L., Bowen, K., & Kjellstrom, T. (2010). Climate change and mental health: A causal pathways framework. *International Journal of Public Health, 55*, 123–132.

Boden, T. A., Marland, G., & Andres, R. J. (2017). National CO_2 emissions from fossil-fuel burning, cement manufacture, and gas flaring: 1751–2014. Carbon Dioxide Information Analysis Center, Oak Ridge National Laboratory, US Department of Energy.

Borodovsky, J. T., Lee, D. C., Crosier, B. S., Gabrielli, J. L., Sargent, J. D., & Budney, A. J. (2017). US cannabis legalization and use of vaping and edible products among youth. *Drug and Alcohol Dependence, 177*, 299–306.

Brown, J. D., Goodin, A. J., & Talbert, J. C. (2018). Rural and Appalachian disparities in neonatal abstinence syndrome incidence and access to opioid abuse treatment. *The Journal of Rural Health, 34*, 6–13.

Centers For Disease Control And Prevention. (2020). Teen substance use & risks: National center on birth defects and developmental disabilities, 2020. Retrieved from https://www.cdc.gov

Cepeda, A., Saint Onge, J. M., Kaplan, C., & Valdez, A. (2010). The association between disaster-related experiences and mental health outcomes among drug using African American Hurricane Katrina evacuees. *Community Mental Health Journal, 46*, 612–620.

Chadi, N., Hadland, S. E., & Harris, S. K. (2019). Understanding the implications of the "vaping epidemic" among adolescents and young adults: A call for action. *Substance Abuse, 40*, 7–10.

Cianconi, P., Betrò, S., & Janiri, L. (2020). The impact of climate change on mental health: A systematic descriptive review. *Frontiers in Psychiatry, 11*, 74.

Clayton, S. (2021). Climate change and mental health. *Current Environmental Health Reports, 8*, 1–6.

Collins, T. W., Jimenez, A. M., & Grineski, S. E. (2013). Hispanic health disparities after a flood disaster: results of a population-based survey of individuals experiencing home site damage in El Paso (Texas, USA). *Journal of Immigrant and Minority Health, 15*, 415–426.

Compton, W. M., Jones, C. M., Baldwin, G. T., Harding, F. M., Blanco, C., & Wargo, E. M. (2019). Targeting youth to prevent later substance use disorder: An underutilized response to the US opioid crisis. *American Journal of Public Health, 109*, S185–S189.

Costello, A., Abbas, M., Allen, A., Ball, S., Bell, S., Bellamy, R., Friel, S., Groce, N., Johnson, A., & Kett, M. (2009). Managing the health effects of climate change: Lancet and University College London Institute for Global Health Commission. *The Lancet, 373*, 1693–1733.

Davidson, T. M., Price, M., Mccauley, J. L., & Ruggiero, K. J. (2013). Disaster impact across cultural groups: Comparison of sWhites, African Americans, and Latinos. *American Journal of Community Psychology, 52*, 97–105.

Davis, L. L., Schein, J., Cloutier, M., Gagnon-Sanschagrin, P., Maitland, J., Urganus, A., Guerin, A., Lefebvre, P., & Houle, C. R. (2022). The economic burden of posttraumatic stress disorder in the United States from a societal perspective. *The Journal of Clinical Psychiatry, 83*.

Dodgen, D., Donato, D., Kelly, N., La Greca, A., Morganstein, J., Reser, J., Ruzek, J., Schweitzer, S., Shimamoto, M., & Tart, K. T. (2016). Mental health and well-being (Chapter 8). US Global Change Research Program.

Duffy, P. B., & Tebaldi, C. (2012). Increasing prevalence of extreme summer temperatures in the U.S. *Climatic Change, 111*, 487–495.
Forray, A., & Foster, D. (2015). Substance use in the perinatal period. *Current Psychiatry Reports, 17*, 1–11.
Galea, S., Nandi, A., & Vlahov, D. (2005). The epidemiology of post-traumatic stress disorder after disasters. *Epidemiologic Reviews, 27*, 78–91.
Goldstein, R. B., Smith, S. M., Chou, S. P., Saha, T. D., Jung, J., Zhang, H., Pickering, R. P., Ruan, W. J., Huang, B., & Grant, B. F. (2016). The epidemiology of DSM-5 posttraumatic stress disorder in the United States: Results from the national epidemiologic survey on alcohol and related conditions-III. *Social Psychiatry and Psychiatric Epidemiology, 51*, 1137–1148.
González, H. M., Tarraf, W., Whitfield, K. E., & Vega, W. A. (2010). The epidemiology of major depression and ethnicity in the United States. *Journal of Psychiatric Research, 44*, 1043–1051.
Hasin, D. S., & Grant, B. F. (2015). The national epidemiologic survey on alcohol and related conditions (NESARC) waves 1 and 2: Review and summary of findings. *Social Psychiatry and Psychiatric Epidemiology, 50*, 1609–1640.
Hayes, K., Blashki, G., Wiseman, J., Burke, S., & Reifels, L. (2018). Climate change and mental health: Risks, impacts and priority actions. *International Journal of Mental Health Systems, 12.*
Jones, H. E., Finnegan, L. P., & Kaltenbach, K. (2012). Methadone and buprenorphine for the management of opioid dependence in pregnancy. *Drugs, 72*, 747–757.
Jordan, C. J., & Andersen, S. L. (2017). Sensitive periods of substance abuse: Early risk for the transition to dependence. *Developmental Cognitive Neuroscience, 25*, 29–44.
Kaltenbach, K., Berghella, V., & Finnegan, L. (1998). Opioid dependence during pregnancy: Effects and management. *Obstetrics and Gynecology Clinics of North America, 25*, 139–151.
Kessler, R. C., Chiu, W. T., Demler, O., & Walters, E. E. (2005). Prevalence, severity, and comorbidity of 12-month DSM-IV disorders in the national comorbidity survey replication. *Archives of General Psychiatry, 62*, 617.
Kim, E. J. (2016). The impacts of climate change on human health in the United States: A scientific assessment, by us global change research program. *Journal of the American Planning Association, 82*, 418–419.
Lawrance, E., Thompson, R., Fontana, G., & Jennings, N. (2021). The impact of climate change on mental health and emotional wellbeing: current evidence and implications for policy and practice. Available at: https://www.imperial.ac.uk/grantham/publications/all-publications/the-impact-of-climate-change-on-mentalhealth-and-emotional-wellbeing-current-evidence-and-implications-for-policy-and-practice.php
Lopez, A. D., & Murray, C. C. J. L. (1998). The global burden of disease, 1990–2020. *Nature Medicine, 4*, 1241–1243.
Lowe, S. R., Sampson, L., Gruebner, O., & Galea, S. (2015). Psychological resilience after Hurricane Sandy: The influence of individual- and community-level factors on mental health after a large-scale natural disaster. *PLoS ONE, 10*, e0125761.
Maeda, A., Bateman, B. T., Clancy, C. R., Creanga, A. A., & Leffert, L. R. (2014). Opioid abuse and dependence during pregnancy: Temporal trends and obstetrical outcomes. *Anesthesiology, 121*, 1158–1165.
Mcmichael, A. J., Woodruff, R. E., & Hales, S. (2006). Climate change and human health: Present and future risks. *The Lancet, 367*, 859–869.
Melillo, J. M., Richmond, T., & Yohe, G. (2014). Climate change impacts in the United States. *Third national climate assessment, 52.*
Morales, A. M., Jones, S. A., Kliamovich, D., Harman, G., & Nagel, B. J. (2020). Identifying early risk factors for addiction later in life: A review of prospective longitudinal studies. *Current Addiction Reports, 7*, 89–98.
Murray, C. J., & Lopez, A. D. (2013). Measuring the global burden of disease. *New England Journal of Medicine, 369*, 448–457.
National Center For Drug Abuse Statistics. (2019). Drug abuse statistics. Retrieved from https://drugabusestatistics.org/

Nelson, J., Bundoc-Baronia, R., Comiskey, G., & Mcgovern, T. F. (2017). Facing addiction in America: The surgeon general's report on alcohol, drugs, and health: A commentary. *Alcoholism Treatment Quarterly, 35*, 445–454.

Office of National Drug Control Policy. (2021). The Biden-Harris administration's statement of drug policy priorities for year one. Executive Office of the President. Retrieved from https://www.whitehouse.gov/

Oliver-Smith, A. (1998). Global changes and definitions of disaster. In E. L. Quarantelli (Eds.), What is a disaster. Perspectives on the Question.

Parker, G., Lie, D., Siskind, D. J., Martin-Khan, M., Raphael, B., Crompton, D., & Kisely, S. (2016). Mental health implications for older adults after natural disasters–a systematic review and meta-analysis. *International Psychogeriatrics, 28*, 11–20.

Patrick, S. W., Davis, M. M., Lehmann, C., & Cooper, W. O. (2015). Increasing incidence and geographic distribution of neonatal abstinence syndrome: United States 2009 to 2012. *Journal of Perinatology, 35*, 650–655.

Patrick, S. W., Schumacher, R. E., Benneyworth, B. D., Krans, E. E., Mcallister, J. M., & Davis, M. M. (2012). Neonatal abstinence syndrome and associated health care expenditures: United States, 2000–2009. *JAMA, 307*, 1934–1940.

Piehler, T. F., Véronneau, M.-H., & Dishion, T. J. (2012). Substance use progression from adolescence to early adulthood: Effortful control in the context of friendship influence and early-onset use. *Journal of Abnormal Child Psychology, 40*, 1045–1058.

Portier, C. J., Tart, K. T., Carter, S. R., Dilworth, C. H., Grambsch, A. E., Gohlke, J., Hess, J., Howard, S. N., Luber, G., & Lutz, J. T. (2013). A human health perspective on climate change: A report outlining the research needs on the human health effects of climate change. *Journal of Current Issues in Globalization, 6*, 621.

Quinones, S. (2021). *The least of Us: True tales of America and hope in the time of Fentanyl and Meth*. Bloomsbury Publishing.

Raker, E. J., Arcaya, M. C., Lowe, S. R., Zacher, M., Rhodes, J., & Waters, M. C. (2020). Mitigating health disparities after natural disasters: Lessons from the RISK project: Study examines mitigating health disparities after natural disasters. *Health Affairs, 39*, 2128–2135.

Raker, E. J., Lowe, S. R., Arcaya, M. C., Johnson, S. T., Rhodes, J., & Waters, M. C. (2019). Twelve years later: The long-term mental health consequences of Hurricane Katrina. *Social Science & Medicine, 242*, 112610.

Reeves, W. C., Pratt, L. A., Thompson, W., Ahluwalia, I. B., Dhingra, S. S., Mcknight-Eily, L. R., Harrison, L., D'angelo, D. V., Williams, L., & Morrow, B. (2011). Mental illness surveillance among adults in the United States.

Schulte, M. T., & Hser, Y.-I. (2013). Substance use and associated health conditions throughout the lifespan. *Public Health Reviews, 35*, 1–27.

Sheehan, M. C., Fox, M. A., Kaye, C., & Resnick, B. (2017). Integrating health into local climate response: Lessons from the US CDC climate-ready states and cities initiative. *Environmental Health Perspectives, 125*, 094501.

Shuler, M., Suzuki, S., Podesta, A., Qualls-Hampton, R., & Wallington, S. F. (2017). A post-Hurricane Katrina examination of substance abuse treatment discharges with co-occurring psychiatric and substance use disorders. *Journal of Dual Diagnosis, 13*, 144–156.

Simpson, D. M., Weissbecker, I., & Sephton, S. E. (2011). Extreme weather-related events: Implications for mental health and well-being. *Climate change and human well-being*, 57–78.

Steinberg, L. (2007). Risk taking in adolescence: New perspectives from brain and behavioral science. *Current Directions in Psychological Science, 16*, 55–59.

Steinglass, P., & Gerrity, E. (1990). Natural disasters and post-traumatic stress disorder short-term versus long-term recovery in two disaster-affected communities1. *Journal of Applied Social Psychology, 20*, 1746–1765.

Substance Abuse And Maternal Health Services Administration. (2012). Results from the 2011 national survey on drug use and health: summary of national findings-NSDUH Series H-44, 2012. Retrieved from https://www.samhsa.gov/

Substance Abuse And Mental Health Services Administration. (2014). Results from the 2013 National Survey on Drug Use and Health: summary of national findings, NSDUH Series H-48, HHS Publication No. (SMA) 14-4863. 2014. Retrieved from https://www.samhsa.gov/

Substance Abuse And Mental Health Services Administration. (2020). Key substance use and mental health indicators in the United States: Results from the 2019 National Survey on Drug Use and Health (HHS Publication No. PEP20-07-01-001, NSDUH Series H-55). Center for Behavioral Health Statistics and Quality, Substance Abuse and Mental Health Services Administration. Retrieved from https://www.samhsa.gov/data/

The National Institute on Drug Abuse. (2020). Monitoring the future national survey results on drug use 1975–2019: Overview, key findings on adolescent drug use, 2020. Retrieved from http://www.monitoringthefuture.org//pubs/monographs/mtf-overview2019.pdf

Tracy, M., Norris, F. H., & Galea, S. (2011). Differences in the determinants of posttraumatic stress disorder and depression after a mass traumatic event. *Depression and Anxiety, 28*, 666–675.

United States Geological Survey (n.d.). How can climate change affect natural disasters? Retrieved from https://www.usgs.gov

Upperman, C. R., Parker, J. D., Akinbami, L. J., Jiang, C., He, X., Murtugudde, R., Curriero, F. C., Ziska, L., & Sapkota, A. (2017). Exposure to extreme heat events is associated with increased hay fever prevalence among nationally representative sample of US adults: 1997–2013. *The Journal of Allergy and Clinical Immunology: In Practice, 5*, 435-441.e2.

Ursano, R., Morganstein, J., & Cooper, R. (2017). *American Psychiatric Association Position Statement: Mental Health and Climate Change*. https://doi.org/10.13140/RG.2.2.26846.64325

Vestal, C. (2017). Katrina brain': The invisible long-term toll of megastorms. *The Agenda.*

Watts, N., Adger, W. N., Agnolucci, P., Blackstock, J., Byass, P., Cai, W., Chaytor, S., Colbourn, T., Collins, M., & Cooper, A. (2015). Health and climate change: Policy responses to protect public health. *The Lancet, 386*, 1861–1914.

Wendell, A. D. (2013). Overview and epidemiology of substance abuse in pregnancy. *Clinical Obstetrics and Gynecology, 56*, 91–96.

Whiteford, H. A., Degenhardt, L., Rehm, J., Baxter, A. J., Ferrari, A. J., Erskine, H. E., Charlson, F. J., Norman, R. E., Flaxman, A. D., & Johns, N. (2013). Global burden of disease attributable to mental and substance use disorders: Findings from the Global Burden of Disease Study 2010. *The Lancet, 382*, 1575–1586.

Whiteman, V. E., Salemi, J. L., Mogos, M. F., Cain, M. A., Aliyu, M. H., & Salihu, H. M. (2014). Maternal opioid drug use during pregnancy and its impact on perinatal morbidity, mortality, and the costs of medical care in the United States. *Journal of Pregnancy.*

Williams, Y. (2022). *Trauma-informed psychological supports when natural disasters come.* Springer International Publishing.

World Health Organization. (2013). Mental health action plan 2013–2020. Retrieved from http://www.who.int/mental_health/publications/action_plan/en/

Yu, B., Chen, X., Chen, X., & Yan, H. (2020). Marijuana legalization and historical trends in marijuana use among US residents aged 12–25: Results from the 1979–2016 National Survey on drug use and health. *BMC Public Health, 20*, 1–10.

Change Exposes the Complications of Wildland Fire Full Suppression Policy and Smoke Management in the Sierra Nevada of California, USA

Donald Schweizer, Ricardo Cisneros, and Trent Procter

Abstract Fire and climate change in the Sierra Nevada of California, USA have a complex interaction with human land management and forest ecology. Fire was an important agent of change for the fire prone forests of this landscape. Many species, such as the Giant Sequoia (*Sequoiadendron giganteum*) evolved to take advantage of frequent fire as this natural process sculpted the environment. Native Americans used fire widely for socioeconomic benefit and fuel reduction with moderate intensity fire encouraged to burn across the land. Euro-American settlement brought about an era of suppression that increased fuels and changed the forest composition and structure. But, suppression was and is the simple seeming solution. Even if suppression is not sustainable it will garner support. Historic suppression has currently brought an extreme fuel problem that has manifested into a greater and greater threat of destructive high intensity fire not typical of this ecosystem. Fire policy was and is slow to change due to risk aversion and lack of urgency. This is not in small part from increased smoke impacts as a result of heavy fuel loads and returning fire to the landscape. These emissions were essentially mortgaged to the current age from previous generations. Fire and land management policies collide with air regulatory policy in California because of already heavily anthropogenically polluted air with little to no capacity for an additional emission source. However, fire and the subsequent smoke are inevitable. Public smoke tolerance is low and a significant deterrent to bringing fire back to California wilderness.

D. Schweizer (✉) · R. Cisneros
Health Sciences Research Institute, University of California at Merced, 5200 N. Lake Road, Merced, CA 95343, USA
e-mail: dschweizer@ucmerced.edu

R. Cisneros
e-mail: rcisneros@ucmerced.edu

D. Schweizer
US Forest Service, Pacific Southwest Region, 351 Pacu Lane, Bishop, CA 93514, USA

T. Procter
US Forest Service, 1600 Tollhouse Road, Clovis, CA 93611, USA

R. Akhtar (ed.), *Climate Change and Human Health Scenarios*, Global Perspectives on Health Geography, https://doi.org/10.1007/978-3-031-38878-1_23

Keywords Wildland fire · Air quality · Smoke · Management · Regulation · Policy

1 Fire and Smoke Management History and Policy Review

1.1 Historic Native American Burning and Frequent Fire Use

Throughout California, Indigenous economies depended on widely using both direct and indirect management of fire effectively for food supplies and raw materials (Anderson, 2018, p. 381). Native North American burning was extremely variable from little to no burning to extensive use of fire, which provided a mosaic of frequent fire from natural sources (lightning) that were not suppressed (Vale, 2002, pp. 297–300). Fire and smoke was integrated successfully into landscape management (Levy, 2005, p. 303) and were effectively used as a tool to drive game for hunting or encourage the growth of plants for food, tools, and medicine. In addition to purposely putting fire on the landscape, there was little suppression which created a landscape both healthy and resistant to extreme fire. Natural ignition wildland fires along with tribal burning produced long-term stability of forest biomass (Knight et al., 2022, pp. 4–5). Frequent low intensity human set fires created forest structure in the Sierra Nevada of California (Klimaszewski-Patterson & Mensing, 2016, p. 46). Native American burning practices were stopped with Euro-American settlement. The 1850 Act for the Government and Protection of Indians outlawed intentional burning. The loss of these traditional practices has contributed to the increased fuels and loss of forest resiliency from the suppression era (Kimmerer & Lake, 2001, p. 40). European settlement of the United States altered how fire was used and perceived on the landscape.

1.2 Euro-American Fire Management History

The losses of indigenous burning in California were significant. Initially, with Euro-American settlement in the nineteenth century, there were some who integrated fire into forest management using traditional techniques. As settlement continued, there was less interest in using fire and fire began being equated to evil and destruction. Large fires in the 1800s such as the Peshtigo Fire in Wisconsin of 1871 (burning about 1.2–1.5 million acres) reinforced this idea. It burned after an unusually dry summer and was one of the most damaging fires in American history killing over 1000 people. It was argued at this time that fires threatened commercial timber and that the U.S. should set aside national forests. By 1872 the U.S. Congress established Yellowstone as the first National Park and then began setting aside forest reserves for protection. The United States Forest Service (USFS) was established in 1905 to manage these lands. Light burning (similar to prescribed fire today) was advocated in the early 1900s by people such as Walter Fry, a California USFS Forest Supervisor, Gifford Pinchot, the first Chief of the USFS, and the first Superintendents of Sequoia and Kings Canyon National Parks in California. Their advocacy of including fire faced opposition from people such as Horace Albright, Deputy Director of the National Park Service (NPS), who believed in fire exclusion. By 1910 with one of the driest years in memory, came the "Big Blowup" where multiple fires in northern Idaho, western Montana, and British Columbia Canada burned approximately 3 million acres. The Big Blowup initiated the suppression era of fire management. The USFS embraced a more systematic suppression approach to fire management and the suppression of wildland fire. The "10 am rule," where all fires, when detected, were to be put out at 10 acres or less by 10 am of the next day, was put in place by 1935. World War II encouraged prevention policies and further solidified the full suppression policy of land management.

By the late 1960s and early 1970s, the USFS and NPS began to understand that fire was an ecological process that could be beneficial. Land management agencies began crafting policy that was not full suppression. Certain fires where land managers thought size and intensity would be consistent with the historic normal and where there was no threat to life or property were allowed to burn. These fires were managed with minimal direct suppression and terms such as "let burn," "prescribed natural fire," "wildland fire use," and "fire managed for resource benefit" or "managed fire" were used to define the use of unplanned ignitions to achieve resource benefit. The Federal Wildland Fire Policy was implemented in 2009. This policy was intended to make managing fire consistent across federal agencies and allowed for managed fire for resource benefit on public lands. In 2014, the National Cohesive Wildland Fire Management Strategy shifted fire management policy to both suppress wildfires when needed while allowing others to burn where people and property were not at risk. This policy attempts to safely restore fire resilient landscapes and create fire-adapted communities. Current fire policy nomenclature describes wildland fires as either planned (prescribed) or unplanned and allows land managers to have wildland fires that are not full suppression and can be managed where appropriate for resource benefit.

The policy of fire suppression in California has failed (Busenberg, 2004, p. 154). Current mega fire response is forced to be focused on evacuating citizens, providing point protection of structures, and concentrating on keeping firefighters safe. Strategic containment in a meaningful ecologically beneficial way with any control of emissions is out of reach as massive amounts of smoke are spread over exceptionally large areas. The extreme levels of smoke impacting air quality from these fires did not need to happen and the potential exists to reduce smoke exposure by returning these environmental systems to their normal fire return intervals and fuel loading (Schweizer et al., 2020b, p. 7).

Fire exclusion and climate change increase the complex process of managing fire on the landscape in social and regulatory contexts (Peterson et al., 2022, pp. 241–242). Climate change is expected to increase severity (Flannigan et al., 2000, p. 221) and wildfire activity (Westerling & Bryant, 2008, pp. 244–248). The number of days of extreme fire weather each year will increase (Goss et al., 2020, pp. 11–12). Extreme wind events, higher temperatures, and drought from climate change can be expected to increase fire activity but fuel loading is an important contributor (Keeley and Syphard, 2019, p. 1). Climate change is having an impact on the intensity and length of fire seasons. The fuel build up from the suppression of wildland fire and climate change are combining to increase the area burned in the United States and California (Fig. 1). Delayed onset of autumn rain, record warm summers and autumns, and fuel loading intensify extreme wind driven events in California (Nauslar et al., 2018, pp. 13–14).

It has been shown that fire management can influence fire locally, however, landscape area burned is controlled primarily by a warmer drier climate (Vachula et al., 2019, p. 6). However, the largest shifts in burned area correspond with socioecological change that amplifies or buffers fire changes from climate change. For example, the loss of Native American fire management made fire strongly driven by interannual moisture and fuel variation, with a shift to weaker fire-climate relationships with suppression and Euro-American settlement (Taylor et al., 2016, p. 13684).

Suppression and reduced use of fire in the nineteenth and twentieth centuries increased fuel loading and effectively mortgaged the smoke from these lost fires to future generations. Climate change is amplifying the size and intensity of wildland fires and exacerbating both the ecological impacts and smoke exposure over larger and larger geographic areas. Prescribed fires and ecologically beneficial managed wildfires can mitigate future wildfire severity (Prichard et al., 2021, pp. 20–21) but come with air quality impacts from smoke that are both unpopular to the public and in conflict with federal and state air regulations designed to protect human health.

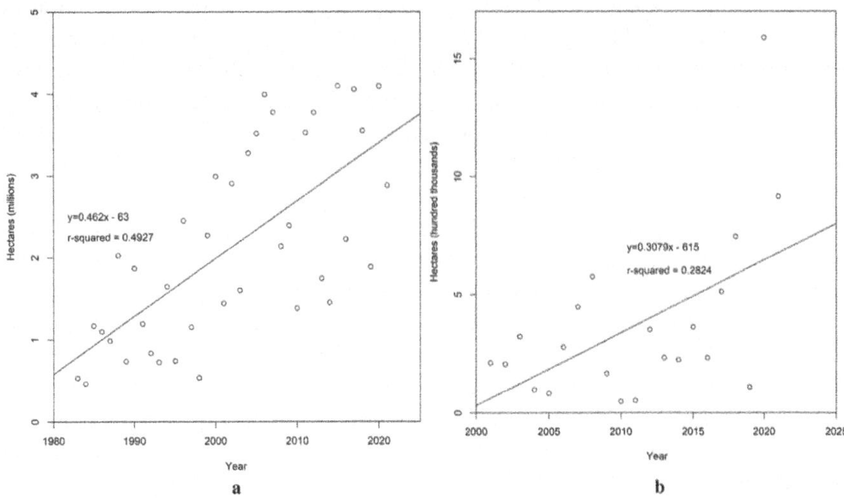

Fig. 1 **a** Millions of hectares of wildland fire in the United States from 1983 to 2021 and **b** Hundred thousands of hectares of wildland fire in California from 2001 to 2022 (data from the National Interagency Fire Center (NIFC) https://www.nifc.gov/ accessed 10/20/2022)

1.3 Federal and State Air Regulations

Fire and smoke both play an important role in California ecosystems. Smoke from fire can also be harmful to human health and welfare, impair visibility, and contribute greenhouse gases. Smoke can thus lead to policy differences between air and land managers' competing missions.

Climate change has a myriad of impacts to human health including climate-sensitive infectious diseases, extreme heat, increased wildfire, and increased drought (Romanello et al., 2021, p. 1653–1654). The deleterious effects of wildland fire smoke on human health are well documented for both the general public (Rice et al., 2021, p. 923) and occupationally for firefighters (Groot et al., 2019, p. 121). While smoke from wildland fire cannot be avoided, strategies have been implemented to mitigate smoke exposure. Public outreach during smoke events is designed to communicate smoke extent and levels for outdoor smoke avoidance (e.g., daily patterns of smoke transport can often allow fire and air managers to recommend avoiding outdoor activities at certain times of the day). Outreach during non-smoke periods recommend being smoke ready through avoidance, the use of masks, installing high-efficiency air conditioner filters, using certified air cleaners, and creating an indoor clean air space (from the California Air Resources Board (CARB) https://ww2.arb.ca.gov/protecting-yourself-wildfire-smoke accessed 1/18/2023).

Air quality in California is managed through federal, state, and local rules and regulations. The Environmental Protection Agency (EPA) is the federal agency that ensures compliance with the U.S. Clean Air Act (CAA). The EPA determines regulatory requirements such as the National Ambient Air Quality Standards (NAAQS) which are used to determine attainment to thresholds of six Criteria Pollutants established under the CAA. The CAA gives individual states the responsibility for meeting federal air quality regulations. Individual states, tribes, or local agencies are also allowed to establish rules that are stricter than federal standards. The EPA oversees state work and enforcement along with setting the federal air quality and emission standards.

The California Air Resources Board (CARB) and local Air Pollution Control Districts (APCDs) are responsible for meeting air quality regulations in California, as well as developing approved State Implementation Plans (SIPs) to achieve federal standards. The California Clean Air Act of 1988 established stricter standards of Criteria Pollutants in California along with establishing ambient air quality standards for other pollutants. CARB sets the state emission standards while the 35 local APCDs regulate stationary sources in their air sheds (Fig. 2).

California air quality is challenging to regulate. Many APCDs are in nonattainment of federal and state standards and wildland fire emissions are increasing into these air sheds (Ahuja and Procter, 2018, p. 442).

Adding to the wildland fire policy and management complexity is a smoke regulatory environment driven by outdated regulations. For instance, natural ignitions managed for resource benefit are considered prescribed fire and therefore under APCD's allocation and approval authority. The last amendments to the Clean Air Act were in 1990, the EPA developed an Interim Air Quality Policy on Wildland and

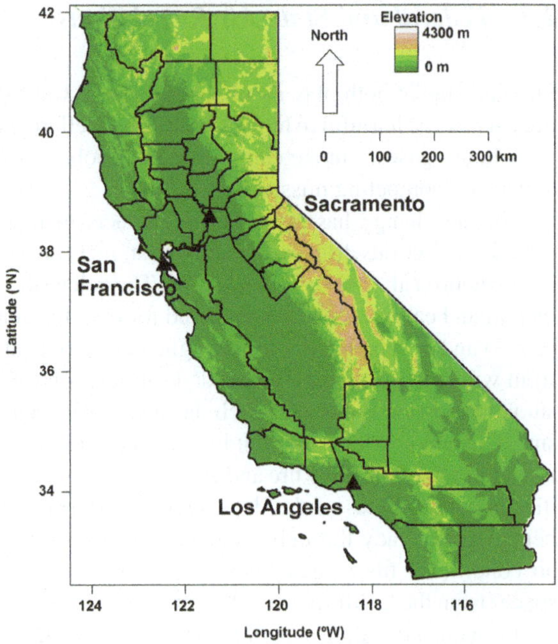

Fig. 2 The 35 California air districts

Prescribed Fires in May of 1998, and the California Smoke Management Guidelines were last revised in 2000. The California Smoke Management Guidelines provide the structure and requirements that the 35 air districts in California use to develop their individual smoke management regulations related to prescribed fire. The California Guidelines were developed in response to the EPA interim policy.

Fortunately, the land and air quality agencies in California have successfully worked together to develop management actions that attempt to recognize the regulatory shortcomings and interpret new policy respective of the outdated regulations. Significant time has been invested over the last 20 years to train staff and develop good relationships around the respective missions of the land and air quality agencies. The legal framework of the federal Clean Air Act, EPA policy, and California Smoke Management Guidelines need revision to catch up with collaborative policy that air and land agencies have developed. Lawmakers need now to provide a thoughtful framework that minimizes legal conflict.

2 The Challenge of Fire and Air Management

It has long been established that frequent low intensity fire is essential to California forests (Kilgore, 1981, p. 81, 1973, p. 496). The need to have regular fire as a required agent of change in fire prone forests has been understood for decades (Kilgore et al., 1979, p. 138–141; Parsons, 1976, p. 97; Parsons & DeBenedetti, 1979, p. 21) and also understood that the era of full suppression needed to end and fire be restored (Keifer et al., 2000, p. 268). A low to moderate severity fire regime was in place pre-European settlement of the American west (Miller & Safford, 2017, p. 75). But, fire management policy had defaulted to suppression for far too long (Adams, 2013, p. 250; Dellasala et al., 2004, p. 984). Prescribed fire and mechanical treatments can reduce the extreme fire behavior and potential for extreme large high intensity fires (Stephens, 1998, p. 21). Climate change impacts can also be reduced using prescribed and wildfires managed for ecological benefit; but mechanical thinning needs to focus on fire management goals rather than the maximization of commercial timber production to obtain the same benefits (Stephens et al., 2020, pp. 358–359).

Land managers meet resistance through an institutional lack of ecological policy support (Boer et al., 2015, p. 920; North et al., 2015, p. 921; Thompson et al., 2015, p. 1280–1281) and human health concerns about air quality from a smoke averse society (Schweizer & Cisneros, 2017, p. 33). The poor air quality in California from anthropogenic sources makes additional smoke impacts an important factor in fire management. The opportunity to forgo these smoke impacts to the future provides an easy reason not to burn and can be appealing to a vocal smoke averse public.

Smoke management is difficult because of this smoke averse public. It is estimated that area burned in just California annually pre-1800 (prior to suppression and Euro-American settlement) is 88% of the United States total area in current "extreme fire" years (Stephens et al., 2007, p. 213). While it has been known for many years these forests need to burn much more, not enough has been done to mitigate the chance of a destructive wildfire occurring. Climate change is only making these avoidable catastrophes more likely (Goss et al., 2020, p. 12) from increased extreme fire behavior fueled by years of suppression. Climate change is increasing the probability that the backlog of smoke left from the generations before us will come with high intensity fire instead of ecologically beneficial fire; increasing smoke, burning communities, type converting forests, and giving reduced, if any, ecological benefit.

3 Smoke Management and Fire Impacts to Air Quality

Smoke has an impact on air quality that is widely studied and documented (Aguilera et al., 2021, p. 3; Sanderfoot et al., 2021, p. 4). Human health impacts and smoke management are a nuanced problem difficult to manage particularly with a skeptical and smoke averse public accustomed to unsustainable smoke free skies of the suppression era. But, managing smoke with prescribed and ecologically beneficial fire can mitigated the extreme impacts from uncontrolled destructive wildfires (Hill et al., 2022, p. 25).

Research into smoke impact to health tradeoffs from large high intensity fire versus prescribed and ecologically beneficial fire is almost non-existent with health outcomes focused on large high intensity fires. Smoke's role in the environment has been understudied. Smoke has long been a part of the western landscape and removing it has had ecological consequences (Kobziar et al., 2018, p. 8). There is a need to increase our understanding of smoke ecology and the implications of altering the smoke cycle through management actions. Changing the timing and intensity of fire may have unintended negative consequences on smoke exposure and human health that can be mitigated with prescribed and ecologically beneficial fire (Schweizer et al., 2019, p. 91).

Suppression has left a backlog of unburned fuels creating a smoke debt from the suppression era of fire. Smoke from wildfire will come to California. Whether it comes in large high intensity fire when suppression fails, prescribed fire, or ecologically beneficial fire is determined by political will and conditions dictating fire behavior; but smoke exposure appears to be minimized with ecologically beneficial wildland fires while prescribed fire can be used to dictate the timing and amount of emissions (Schweizer et al., 2019, p. 93). Controlling emissions through prescribed fire keeps the smoke exposure more local and reduced (Navarro et al., 2018, p. 193). Fire and air quality management policy makers need continued collaboration for a required nuanced policy.

4 Smoke Impacts from Suppression, Ecologically Beneficial Burns, and Prescribed Fire

Currently, the emphasis is to use prescribed fire to return fire to the system. Prescribed fire, similar to suppression, is not the only solution. Even if desired, prescribed fires cannot be completed at a scale to restore fire to the landscape similar to the natural fire regime. There is a time and a place for both suppression and prescribed fire, but they have little benefit in reducing smoke exposure over time when compared to fires at historic size and intensity (Schweizer et al., 2019, p. 87) (Fig. 3).

Prescribed fire takes advantage of burning less at once to control the smoke impacts and it is unclear if this will work at a larger scale (Schweizer et al., 2019, p. 92). Additionally, the health impacts are likely less because homes, cars, and other manmade items are usually not burned. A more holistic approach which includes allowing fire to burn is essential to manage forests with frequent fire (Stephens et al., 2020, p. 359) and reduce the negative health impacts of smoke (Schweizer et al., 2020a, p. 7). Self-regulating ecologically beneficial fires, when safe to do so, should be the policy and regulatory goal for these forests.

5 Fuel Loading and Climate Change Complications

Climate change increases and amplifies extreme weather events and lengthens the fire season (Flannigan et al., 2013, p. 59). In addition to increased extreme weather, human changed ignition patterns, land-use, and suppression, along with forest dieback and fragmentation are also extremely important (Pausas & Keeley, 2021, p. 393). Climate change does not explain all the change in forest structure and

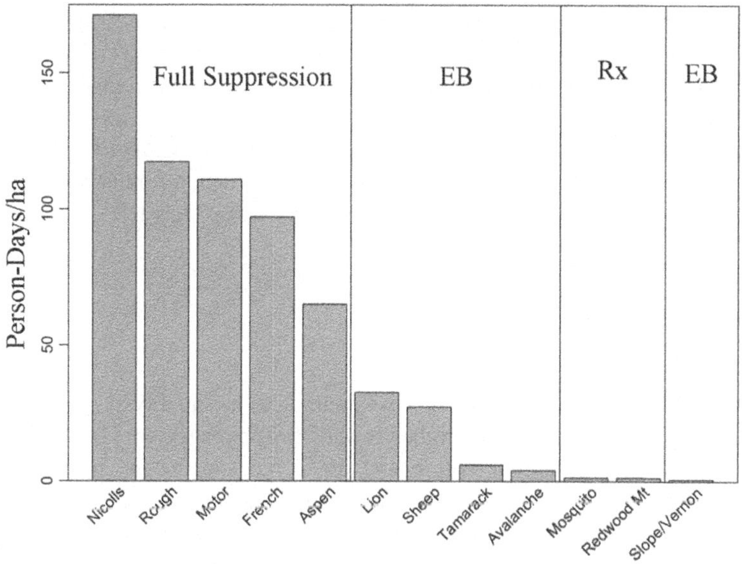

Fig. 3 Population impacts of smoke for select Sierra Nevada fires 2010–2016 in number of person days per hectare burned showing the increased impacts of large high intensity (Full Suppression) to ecologically beneficial fire (EB) to prescribed (Rx) fire

density during the last 3000 years. Tribal burning helped create a stable fire regime unlike the current suppression shaped forest structures and fuel loads (Knight et al., 2022, p. 3). Increased fuel loading from long-term suppression has created conditions where the landscape is more vulnerable to climate change enhanced occurrences of drought and fire (Hagmann et al., 2021, p. 23).

Climate change from anthropogenic emissions is the main driver of increased extreme fire weather (Zhuang et al., 2021, p. 1). This produces an environment where more and larger fires can occur. Human caused increases in temperature since the 1970s have increased fuel dryness approximately doubling western US wildland fire area beyond what is expected from natural variability (Abatzoglou & Williams, 2016, p. 11773). Warmer and drier conditions both make fires harder to put out and help them to spread more quickly and also narrows prescription windows and opportunities to manage lightning fire safely. Additionally, climate change is increasing the range and availability of stressors to forest health. For example, the mountain pine beetle (*Dendroctonus ponderosae*) are developing faster and increasing their range creating epidemics orders of magnitude larger than ever recorded (Mitton & Ferrenberg, 2012, p. E163). Bug kill has created an immense load of dead fuel in California creating additional difficulties to restoring the primary low to moderate intensity burns needed.

Current climate science is raising the alarm for urgent action to mitigate climate change (Armstrong McKay et al., 2022, p. 1). Returning fire in a fire prone ecosystem can help improve forest health and resiliency. Fire management in wilderness areas needs to return the fire and smoke cycle (Cisneros et al., 2018b, p. 123). Fire may be akin to a "keystone species" in these systems (He et al., 2019, p. 2004). Additionally, ecologically beneficial fires in Wilderness and other protected undeveloped forest areas reduce the overall smoke exposure of surrounding communities (Navarro et al., 2018, p. 8). When fires are allowed to burn at historic levels and to self-regulate more of the smoke tends to remain in these remote areas. As counter intuitive as it seems, more frequent low intensity fire is needed to reduce smoke impacts.

6 Concluding Remarks

The interaction of social-political environments and air quality regulations has a crucial impact on community attitudes toward fire management. Public nuisance regulations are embedded in statute and are a mechanism for communities to bring attention to smoke impacts to regulators. In the more recent collaborative environment, air regulators and land management agencies often work together to explore mitigations for complaints including pro-active outreach for prescribed fire. Additionally, public complaints may sometimes go directly political. Community members complaining directly to elected officials can sometimes lead to extreme decisions without the opportunity for thoughtful and effective mitigations.

Smoke footprints typically exceed the fire footprint. How people perceive the impact of smoke can vary widely depending on location and sensitivity. It is important to understand how the impact of smoke is considered by communities with homes and property at risk versus those outside the fire risk area where personal health and economics may be the only, albeit important, consideration. Support for managed fire versus full suppression may vary widely on the spectrum of smoke impacts and should be considered in the development of public education and messaging.

Fire managers need to be allowed to accept greater risk of escaped fires and high intensity fire to increase moderate fire severity to improve forest restoration (Huffman et al., 2017, p. 402). Wildland fire is likely going to increase in severity with climate change while ecosystem services such as carbon sequestration along with species richness are expected to be reduced (Lecina-diaz et al., 2021, p. 1696). Increasing fire to manage forest health does not significantly impact net greenhouse gas emissions or total ecosystem carbon (Volkova et al., 2021, p. 9). This provides an opportunity for forest managers to get ecologically beneficial fire back into the environmental system and not increase emissions impacting climate change while improving forest health and the benefits from a forest ecosystem. This requires the political will of the public, land managers, air regulators, and elected officials.

In the United States, there is a general agreement with the use of prescribed fire (Bowker et al., 2008, pp. 269–272). While there is agreement on using fire, even a minority of smoke averse public can limit the successful management of fires (Cisneros et al., 2018a, p. 680). Land and air managers working together to accomplish increased use of ecologically beneficial fire on the landscape is essential to successful forest management even with, at times, apparent conflict in policy.

Public support is critical and needs to include support and protection for agencies implementing the burning. This should include helping the public understand and accept the risk of a prescribed fire failing either by losing containment or smoke dispersal not going as planned. Messaging that effectively articulates the benefits of restoring fire to the losses incurred when suppression fails are needed for a skeptical public. Continued research is needed to better understand the health and economic cost tradeoffs from the failure to return fire as a natural process. Government and the public need to stop relying on the unrealistic expectation of smoke free skies where fire is an important component to the local ecosystem.

Our attempt to dominate fire through suppression did not work. A focus on relying on prescribed fire only also may not work. Focusing on the immediate predicted air quality impacts may be just deferring the risk, and not best incorporate natural process fire into a fire prone ecosystem.

Smoke managers need to allow more fires whenever it is safe. Of course, extreme high intensity fires particularly where life and property are at risk in the American West need to have a strong suppression response. But, unless fire is returned to the landscape, suppression will continue to fail as climate change exacerbates the harm done from the removal of this natural process during the era of suppression.

Conservation of the fire prone environmental system of our forests is a necessity for public health. While the benefits of a healthy green forest are obvious, the reduction of smoke impacts over time is easily overlooked by a smoke averse public. Outreach and education are needed from every component of our society. Whether it is realtors telling potential homeowners that they are buying in a smoke "flood plain" or it is local government's inspiring citizens to protect their health from smoke by creating a clean room in their house or public expectation of smoke in fire prone areas. Local ordinances to national policy need to reflect the reality of the backlog of smoke we have inherited. Everyone is needed to acknowledge and undertake the work and sacrifice required to dig out of the smoke deficit resulting from the prior policy of smoke avoidance.

More wildland fire smoke is inevitable. Policy and regulations should accelerate alignment with scientific understanding of local fire ecology and adapt to allow for more fire on the landscape, creating healthier forests more resilient to climate change if needed. Currently, we are limiting our tools for smoke management and need to eradicate long held beliefs in suppression within fire and air management, operations, and the public. We can walk obliviously into the smoke headed our way or try to fix failing policy and a system that rewards the most risk adverse.

References

Abatzoglou, J. T., & Williams, A. P. (2016). Impact of anthropogenic climate change on wildfire across western US forests. *Proceedings of the National Academy of Sciences, 113*, 11770–11775.

Adams, M. A. (2013). Mega-fires, tipping points and ecosystem services: Managing forests and woodlands in an uncertain future. *Forest Ecology and Management, 294*, 250–261.

Aguilera, R., Corringham, T., Gershunov, A., & Benmarhnia, T. (2021). Wildfire smoke impacts respiratory health more than fine particles from other sources: Observational evidence from Southern California. *Nature Communications, 12*, 1493.

Ahuja, S., & Procter, T., et al. (2018). Fire, air quality, and greenhouse gases. In J. van Wagtendonk (Ed.), *Fire in California's ecosystems* (pp. 439–458). University of California Press.

Anderson, M. K., et al. (2018). The use of fire by native Americans in California. In J. van Wagtendonk (Ed.), *Fire in California's ecosystems* (pp. 381–397). University of California Press.

Armstrong McKay, D. I., Staal, A., Abrams, J. F., Winkelmann, R., Sakschewski, B., Loriani, S., Fetzer, I., Cornell, S. E., Rockström, J., Lenton, T. M. (2022). Exceeding 1.5 °C global warming could trigger multiple climate tipping points, *Science*, 377.

Boer, M. M., Price, O. F., & Bradstock, R. A. (2015). Wildfires: Weigh policy effectiveness. *Science, 350*, 920.

Bowker, J. M., Lim, S. H., Cordell, H. K., Green, G. T., Rideout-Hanzak, S., & Johnson, C. Y. (2008). Wildland fire, risk, and recovery: Results of a national survey with regional and racial perspectives. *Journal of Forestry, 106*, 268–276.

Busenberg, G. (2004). Wildfire management in the United States: The evolution of a policy failure. *Review of Policy Research, 21*, 145–156.

Cisneros, R., Alcala, E., Schweizer, D., & Burke, N. (2018a). Smoke complaints caused by wildland fire in the southern Sierra Nevada region, California. *International Journal of Wildland Fire, 27*, 677.

Cisneros, R., Schweizer, D., (Lee) Tarnay, L., Navarro, K., Veloz, D., Procter, C. T. (2018b). Climate change, forest fires, and health in California. In R. Akhtar, & C. Palagiano (Eds.), *Climate change and air pollution* (pp. 99–130). Springer International Publishing.

Dellasala, D. A., Williams, J. E., Williams, C. D., & Franklin, J. F. (2004). Beyond smoke and mirrors: A synthesis of fire policy and science. *Conservation Biology, 18*, 976–986.

Flannigan, M., Cantin, A. S., De Groot, W. J., Wotton, M., Newbery, A., & Gowman, L. M. (2013). Global wildland fire season severity in the 21st century. *Forest Ecology and Management, 294*, 54–61.

Flannigan, M., Stocks, B., & Wotton, B. (2000). Climate change and forest fires. *Science of the Total Environment, 262*, 221–229.

Goss, M., Swain, D. L., Abatzoglou, J. T., Sarhadi, A., Kolden, C. A., Williams, A. P., Diffenbaugh, N. S. (2020). Climate change is increasing the likelihood of extreme autumn wildfire conditions across California. *Environmental Research Letters*, 15.

Groot, E., Caturay, A., Khan, Y., & Copes, R. (2019). A systematic review of the health impacts of occupational exposure to wildland fires. *International Journal of Occupational Medicine and Environmental Health, 2*, 121–140.

Hagmann, R. K., Hessburg, P. F., Prichard, S. J., Povak, N. A., Brown, P. M., Fulé, P. Z., Keane, R. E., Knapp, E. E., Lydersen, J. M., Metlen, K. L., Reilly, M. J., Sánchez Meador, A. J., Stephens, S. L., Stevens, J. T., Taylor, A. H., Yocom, L. L., Battaglia, M. A., Churchill, D. J., Daniels, L. D., Falk, D. A., Henson, P., Johnston, J. D., Krawchuk, M. A., Levine, C. R., Meigs, G. W., Merschel, A. G., North, M. P., Safford, H. D., Swetnam, T. W., Waltz, A. E. M. (2021). Evidence for widespread changes in the structure, composition, and fire regimes of western North American forests, *Ecological Applications*, 31.

He, T., Lamont, B. B., & Pausas, J. G. (2019). Fire as a key driver of Earth's biodiversity. *Biological Reviews, 94*, 1983–2010.

Hill, L. A. L., Jaeger, J. M., Smith, A. (2022). *Can prescribed fires mitigate health harm? A Review of air quality and public health implications of wildfire and prescribed fire*. Oakland.

Huffman, D. W., Sánchez Meador, A. J., Stoddard, M. T., Crouse, J. E., & Roccaforte, J. P. (2017). Efficacy of resource objective wildfires for restoration of ponderosa pine (*Pinus ponderosa*) forests in northern Arizona. *Forest Ecology and Management, 389*, 395–403.

Keeley, J. E., Syphard, A. D. (2019). Twenty-first century California, USA, wildfires: Fuel-dominated vs. wind-dominated fires. *Fire Ecology, 15*, 24.

Keifer, M., Stephenson, N. L., Manley, J. (2000). Prescribed fire as the minimum tool for wilderness forest and fire regime restoration: A case study from the Sierra Nevada, California. In D. Cole, S.

McCool, W. Borrie, & J. O'Loughlin (Eds.), *Wilderness science in a time of change conference—volume 5: Wilderness ecosystems, threats, and management: 1999 May 23–37* (pp. 266–269). U.S. Department of Agriculture, Forest Service, Rocky Mountain Research Station, Ogden, UT.

Kilgore, B. (1981). Fire in ecosystem distribution and structure: Western forests and Scrublands. In H. Mooney, T. Bonnicksen, & N. Christensen (Eds.), *Proceedings of the conference: fire regimes and ecosystem properties* (pp. 58–89). USDA Forest Service, General Technical Report WO-GTR-26.

Kilgore, B. M. (1973). The ecological role of fire in Sierran conifer forests. Its application to national park management. *Quaternary Research, 3*, 496–513.

Kilgore, B. M., Taylor, D., Kilgore, B. M., & Taylor, D. (1979). Fire history of a sequoia-mixed conifer forest. *Ecology, 60*, 129–142.

Kimmerer, R. W., & Lake, F. K. (2001). The role of indigenous burning in land management. *Journal of Forestry, 99*, 36–41.

Klimaszewski-Patterson, A., & Mensing, S. A. (2016). Multi-disciplinary approach to identifying native American impacts on Late Holocene forest dynamics in the southern Sierra Nevada range, California, USA. *Anthropocene, 15*, 37–48.

Knight, C. A., Anderson, L., Bunting, M. J., Champagne, M., Clayburn, R. M., Crawford, J. N., Klimaszewski-Patterson, A., Knapp, E. E., Lake, F. K., Mensing, S. A., Wahl, D., Wanket, J., Watts-Tobin, A., Potts, M. D., & Battles, J. J. (2022). Land management explains major trends in forest structure and composition over the last millennium in California's Klamath Mountains. *Proceedings of the National Academy of Sciences, 119*, 1–11.

Kobziar, L. N., Pingree, M. R. A., Larson, H., Dreaden, T. J., Green, S., & Smith, J. A. (2018). Pyroaerobiology: The aerosolization and transport of viable microbial life by wildland fire. *Ecosphere, 9*, e02507.

Lecina-diaz, J., Martınez-vilalta, J., Alvarez, A., Vayreda, J., & Retana, J. (2021). Assessing the risk of losing forest ecosystem services due to wildfires. *Ecosystems, 24*, 1687–1701.

Levy, S. (2005). Rekindling native fires. *BioScience, 55*, 303–308.

Miller, J. D., & Safford, H. D. (2017). Corroborating evidence of a pre-Euro-American low- to moderate-severity fire regime in yellow pine-mixed conifer forests of the Sierra Nevada California, USA. *Fire Ecology, 13*, 58–90.

Mitton, J. B., & Ferrenberg, S. M. (2012). Mountain pine beetle develops an unprecedented summer generation in response to climate warming. *The American Naturalist, 179*, E163–E171.

Nauslar, N., Abatzoglou, J., & Marsh, P. (2018). The 2017 North Bay and Southern California fires: A case study. *Fire, 1*, 18.

Navarro, K., Schweizer, D., Balmes, J., & Cisneros, R. (2018). A review of community smoke exposure from wildfire compared to prescribed fire in the United States. *Atmosphere, 9*, 185.

North, M. P., Stephens, S. L., Collins, B. M., Agee, J. K., Aplet, G., Franklin, J. F., & Fule, P. Z. (2015). Reform forest fire management. *Science, 349*, 1280–1281.

Parsons, D. J. (1976). The role of fire in natural communities: An example from the southern Sierra Nevada, California. *Environmental Conservation, 3*, 91–99.

Parsons, D. J., & DeBenedetti, S. H. (1979). Impact of fire suppression on a mixed-conifer forest. *Forest Ecololgy and Management, 2*, 21–33.

Pausas, J. G., & Keeley, J. E. (2021). Wildfires and global change. *Frontiers in Ecology and the Environment, 19*, 387–395.

Peterson, J. L., Pitrolo, M. C., Schweizer, D. W., Striplin, R. L., Geiser, L. H., Holm, S. M., Hunter, J. D., Croft, J. M., Chappell, L. M., Lahm, P. W., Amezquita, G. E., Brown, T. J., Cisneros, R. G., Connolly, S. J., Halofsky, J. E., Loudermilk, E. L., Navarro, K. M., Nick, A. L., Procter, C. T., ... West, C. D. (2022). *Wildland fire smoke in the United States*. Springer International Publishing.

Prichard, S. J., Hessburg, P. F., Hagmann, R. K., Povak, N. A., Dobrowski, S. Z., Hurteau, M. D., Kane, V. R., Keane, R. E., Kobziar, L. N., Kolden, C. A., North, M., Parks, S. A., Safford, H. D., Stevens, J. T., Yocom, L. L., Churchill, D. J., Gray, R. W., Huffman, D. W., Lake, F.

K., & Khatri-Chhetri, P. (2021). Adapting western North American forests to climate change and wildfires: 10 common questions. *Ecological Applications, 31*, 1–30.

Rice, M. B., Henderson, S. B., Lambert, A. A., Cromar, K. R., Hall, J. A., Cascio, W. E., Smith, P. G., Marsh, B. J., Coefield, S., Balmes, J. R., Kamal, A., Gilmour, M. I., Carlsten, C., Navarro, K. M., Collman, G. W., Rappold, A., Miller, M. D., Stone, S. L., & Costa, D. L. (2021). Respiratory impacts of wildland fire smoke: Future challenges and policy opportunities. *Annals of the American Thoracic Society, 18*, 921–1084.

Romanello, M., McGushin, A., Di Napoli, C., Drummond, P., Hughes, N., Jamart, L., Kennard, H., Lampard, P., Solano Rodriguez, B., Arnell, N., Ayeb-Karlsson, S., Belesova, K., Cai, W., Campbell-Lendrum, D., Capstick, S., Chambers, J., Chu, L., Ciampi, L., Dalin, C., … Hamilton, I. (2021). The 2021 report of the *Lancet* countdown on health and climate change: Code red for a healthy future. *The Lancet, 398*, 1619–1662.

Sanderfoot, O. V., Bassing, S. B., Brusa, J. L., Emmet, R. L., Gillman, S. J., Swift, K., & Gardner, B. (2021). A review of the effects of wildfire smoke on the health and behavior of wildlife. *Environmental Research Letters, 16*, 123003.

Schweizer, D., Cisneros, R., & Navarro, K. (2020a). The effectiveness of adding fire for air quality benefits challenged: A case study of increased fine particulate matter from wilderness fire smoke with more active fire management. *Forest Ecology and Management, 458*, 117761.

Schweizer, D., Nichols, T., Cisneros, R., Navarro, K., & Procter, T. (2020b). Wildland fire, extreme weather and society: Implications of a history of fire suppression in California, USA. *Extreme weather events and human health* (pp. 41–57). Springer International Publishing.

Schweizer, D., Preisler, H. K., & Cisneros, R. (2019). Assessing relative differences in smoke exposure from prescribed, managed, and full suppression wildland fire. *Air Quality, Atmosphere & Health, 12*, 87–95.

Schweizer, D. W., & Cisneros, R. (2017). Forest fire policy: Change conventional thinking of smoke management to prioritize long-term air quality and public health. *Air Quality, Atmosphere & Health, 10*, 33–36.

Stephens, S. L. (1998). Evaluation of the effects of silvicultural and fuels treatments on potential fire behaviour in Sierra Nevada mixed-conifer forests. *Forest Ecology and Management, 105*, 21–35.

Stephens, S. L., Martin, R. E., & Clinton, N. E. (2007). Prehistoric fire area and emissions from California's forests, woodlands, shrublands, and grasslands. *Forest Ecology and Management, 251*, 205–216.

Stephens, S. L., Westerling, A. L., Hurteau, M. D., Peery, M. Z., Schultz, C. A., & Thompson, S. (2020). Fire and climate change: Conserving seasonally dry forests is still possible. *Frontiers in Ecology and the Environment, 18*, 354–360.

Taylor, A. H., Trouet, V., Skinner, C. N., & Stephens, S. (2016). Socioecological transitions trigger fire regime shifts and modulate fire–climate interactions in the Sierra Nevada, USA, 1600–2015 CE. *Proceedings of the National Academy of Sciences, 113*, 13684–13689.

Thompson, M., Dunn, C., & Calkin, D. (2015). Wildfires: Systemic changes required. *Science, 350*, 920–920.

Vachula, R. S., Russell, J. M., & Huang, Y. (2019). Climate exceeded human management as the dominant control of fire at the regional scale in California's Sierra Nevada. *Environmental Research Letters, 14*, 104011.

Vale, T. (Ed.). (2002). *Fire native peoples, and the natural landscape.* Island Press.

Volkova, L., Roxburgh, S. H., & Weston, C. J. (2021). Effects of prescribed fire frequency on wildfire emissions and carbon sequestration in a fire adapted ecosystem using a comprehensive carbon model. *Journal of Environmental Management, 290*, 112673.

Westerling, A. L., & Bryant, B. P. (2008). Climate change and wildfire in California. *Climatic Change, 87*, 231–249.

Zhuang, Y., Fu, R., Santer, B. D., Dickinson, R. E., & Hall, A. (2021). Quantifying contributions of natural variability and anthropogenic forcings on increased fire weather risk over the western United States. *Proceedings of the National Academy of Sciences, 118*, 1–9.

Climate Change, Wildfires, and Health in Canada

Robin Meadows

Abstract While still an active area of research, the increasing average temperature and decreasing atmospheric humidity due to climate change is predicted to increase the frequency and intensity of large wildfires in Canada. Wildfire smoke causes immediate respiratory distress, although there is a noted absence of research into prolonged exposure and long-term health outcomes. Further, evacuation from wildfires causes short-term hardships which leads to long-term mental health outcomes. Options to adapt to wildfires are limited and our capacity to prevent ever worsening wildfires in the future may be overwhelmed.

Keywords Canada · Climate change · Wildfires · Respiratory health · Mental health · Evacuation · Adaptation

1 Wildfires

1.1 What is a Wildfire?

Wildland fires, also called wildfires, are fires that occur in combustible vegetation. The largest wildfires may be difficult to put out, they may spread quickly, and they may burn for long periods, threatening the health of populations through fire and smoke. Depending on the type of vegetation that may burn, there are three main types of wildfires: forest fires (the most common wildfire type in Canada), bushfires or shrub fires, and grassland fires or grass fires. Wildfires are caused predominantly by lightning or by humans (either accidentally or intentionally), although fires have been known to start by other means; for instance, a landslide may result in rocks colliding and causing sparks that ignite nearby vegetation. Not all wildfires are harmful to people; they are in fact a naturally occurring phenomena in some ecosystems, and

R. Meadows (✉)
Calgary, AB, Canada
e-mail: meadows.rfm@gmail.com

R. Akhtar (ed.), *Climate Change and Human Health Scenarios*, Global Perspectives on Health Geography, https://doi.org/10.1007/978-3-031-38878-1_24

there is an ecological need for fire on the landscape, although we must also protect communities.

1.2 What Are Conditions Conducive to a Wildfire?

Drier conditions are more conducive to a fire igniting and spreading. Vegetation may be dry because of high temperatures, low humidity and precipitation, and strong winds. Once a fire starts, the heat of the fire itself can preheat and ignite vegetation (e.g., green vegetation) that was not as dry as the vegetation that initially started on fire (e.g., fallen leaves) and hence can spread farther and grow. The Canadian Forest Fire Weather Index (FWI) is an index that estimates the potential for wildfires in Canada (NRCAN: CFS, n.d.). In the FWI, the first level of assessment considers how weather conditions (i.e., temperature, relative humidity, wind, rain) may affect fuel moisture in the forest floor. The next level of assessment considers the rate at which a fire may spread (based on the previous assessments together with wind speed) and how much fuel is available to burn (NRCAN: CFS, n.d.). The result of these assessments is an index that estimates the potential fire danger in forests in Canada (NRCAN: CFS, n.d.). In other words, "fire danger" can be thought of as antecedent weather conditions determining fuel moisture and "fire behavior" can be thought of as potential fire characteristics such as rate of spread based on type of vegetation and current conditions.

Global warming has been associated with increased moderate risk of wildfire damage (IPPC 2019, p. 17). With the year-round increase in temperature throughout the world, it follows that the potential for conditions to start wildfires will also increase (Field et al. 2007, p. 623) in the form of changes in wildfire season length (Coogan et al., 2019, p. 476–477; Hanes et al., 2019, pp. 264, 267) and in the number of days conducive to wildfires (Jain et al., 2020, pp. 206–210).

2 Wildfire Trends in Canada

2.1 What Are the Observed Trends?

Figures 1 and 2 summarize data collected by the CWFIS Datamart of Natural Resources Canada and the Canadian Forest Service (NRCAN CFS 2022). The datasets include forest fires over all of Canada for the years 1970–2021. The datasets do not include fires in unmanaged forests, but they remain the best nationwide data we have. Forest fire data prior to 1970 were not collected as consistently or comprehensively, which is why this section examines trends from 1970 onwards. Figure 1 shows the number of forest fires in Canada of all sizes and the total area burned. From 1970 to 2021, there has been a general decreasing trend in the total number of

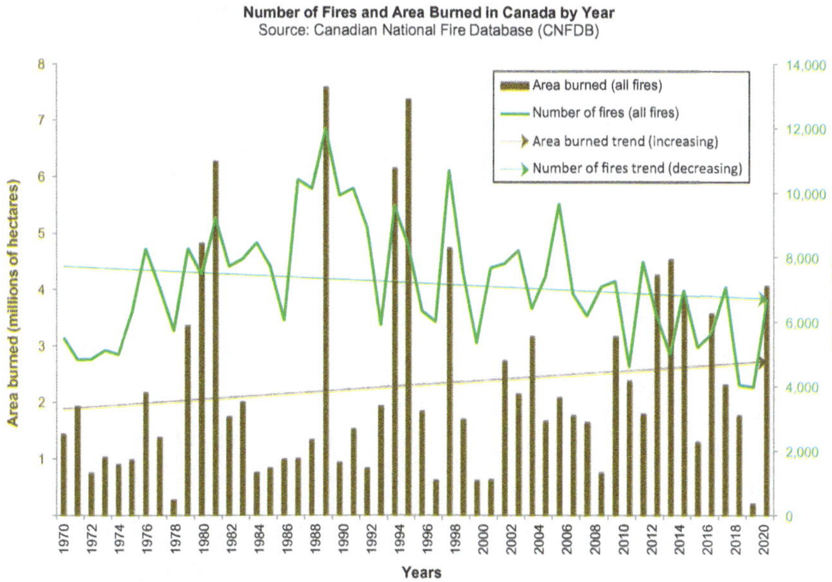

Fig. 1 The number of fires and area burned in Canada by year of fires of all sizes. Figure adapted from the original figure in NFDB_point_stats.xlsx (NRCAN CFS 2022) from data used under Canada's Open Government License

fires while the total area burned has experienced a general increase. Figure 2 shows general increasing trends for both the total number of fires larger than 200 ha as well as total area burned by fires larger than 200 ha. In Canada, forest fires larger than 200 ha account for a small proportion of the total number of all fires; however, they account for around 98% of total area burned. For example, in 2021, 596 of 6709 fires were large fires (i.e., > 200 ha), yet those 596 fires (8.88% of all fires in 2021) accounted for 4.039 million hectares burned (99.01% of all area burned in 2021). From 1970 to 2021, the total number of large fires comprised 4.28% of all fires, which also accounted for 98.77% of all area burned. Thus, the increasing trend of the number of large fires is more important and worrisome than the decreasing trend of all fires (i.e., the combined number of small fires and large fires).

2.2 *What Are Differences in Cause?*

In the period 1980–2015, human-caused fires constituted around 50% of the total number of fires but accounted for under 10% of the total area burned, while lightning-caused fires constituted around 50% of the total number of fires but accounted for over 90% of the total area burned (Hanes et al., 2019, p. 261). Lightning-caused fires have displayed a general increasing trend while human-caused fires have displayed a

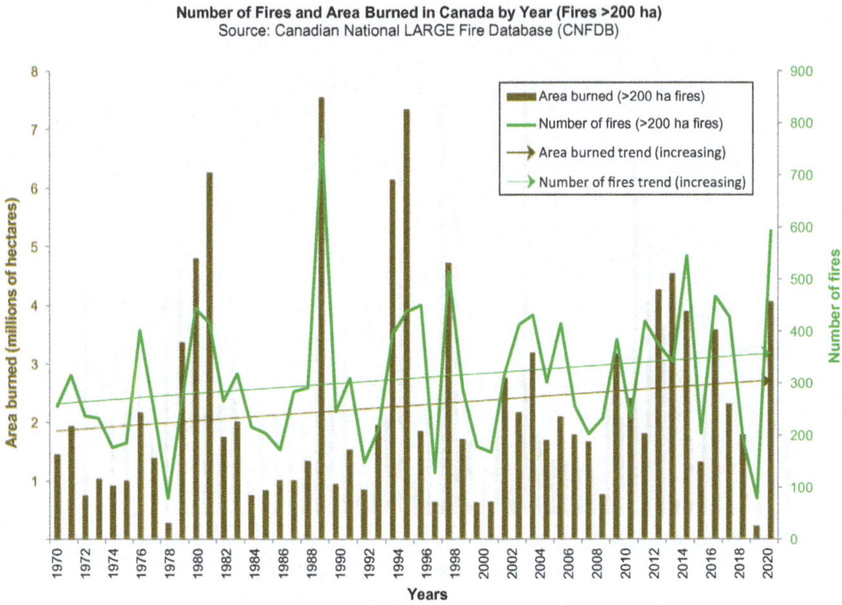

Fig. 2 The number of fires and area burned in Canada by year of fires larger than 200 ha. Figure adapted from the original figure in NFDB_point_stats.xlsx (NRCAN CFS 2022) from data used under Canada's Open Government License

general decreasing trend (Hanes et al., 2019, p. 264); however, the authors note that the increasing numbers of reported lightning-caused fires may be due to the increased capability of detecting lightning-caused fires since 1980 (Hanes et al., 2019, pp. 267). Coogan et al. reported that human-caused fires consisted of 55% of fires of all sizes but consisted of only 48% of fires $\geq$ 2 ha for the period 1959–2018, with a fall to 53% of fires of all sizes and 44% of fires $\geq$ 2 ha for the period 1981–2018 (Coogan et al., 2020, p. 475). For the period 1981–2018, the number of lightning-caused fires peaked from mid-June to mid-August, while human-caused fires peaked around early May (Coogan et al., 2020, p. 475).

2.3 *What Are Changes in Seasonality?*

In Canada, wildfire season typically coincides with the snow-free period of April to October, with the highest fire activity from June to August in some parts of the country; the seasonality of wildfires varies based on the different ecozones throughout Canada. Hanes et al. found that between 1959 and 2015, there was a trend in the fire season length, with fires starting around nine days earlier and lasting for around nineteen days longer in Canada overall (these changes were statistically significant); in the period 1980–2015, most ecozones experienced earlier seasons (five of nine ecozones) and later seasons (five of nine ecozones), while a plurality experienced longer seasons (four ecozones experienced longer seasons, three experienced shorter seasons, two experienced shifts of one day) (not all regions reported statistically significant changes) (Hanes et al., 2019, pp. 264, 267). Similarly, Coogan et al. found that from 1959 to 2018, the fire season had been starting earlier and ending later in some ecozones but not others (Coogan et al., 2020, p. 476). In contrast, for the narrower period of 1981 to 2018, the fire season had been starting later and ending earlier in Canada overall, although some ecozones experienced earlier starts and later ends (Coogan et al., 2020, p. 477). Another consequence of global warming is an increased capacity of the atmosphere to hold moisture. This may lead to changes in regional precipitation, which may create moister conditions in some areas and therefore fewer fires or smaller fires in those areas as well as differences in seasons. Conversely, other areas may see longer periods of drought and therefore experience a greater number of fires and larger fires. For changes in seasons, Coogan et al. suggested that examining fire season length overall throughout Canada may be less important than examining changes in different ecozones as they possess different environmental conditions (Coogan et al., 2020, pp. 482).

2.4 *What Are Predictions of the Future?*

There are numerous studies, based on sophisticated climate models that predict increases in fire weather severity as well as fire activity. Wang et al. predicted an increase in fire size and area burned in Canada overall by 64% and 93%, respectively, by 2080 (Wang et al., 2020, pp. 5–6). Area burned in the period 2046–2065 in Western Canadian forests is predicted to increase, likely due to increased air temperatures and 500 hPa geopotential heights (where 500 hPa geopotential height is important in the prediction of area burned in boreal forests) (Yue et al., 2015, pp. 10034, 10043–10044). In two Western Canadian provinces, Jain et al. predicted an increase in the number of potential spread days (i.e., days where conditions are more conducive to burning) by 2060 and 2090 (Jain et al., 2020, pp. 206–210). In the Westernmost province of British Columbia in the Cordilleran forests, Nitschke and Innes predicted that mean fire size will increase by 2050 and 2080; as well, they predicted that fires less than 1000 ha will occur less and fires greater than 1000 ha will occur more by the 2080s (Nitschke and Innes 2012, pp. 584–585). Yue et al. predicted increased area burned in Central Canada and Southern Canada in the period 2046–2065 (Yue et al., 2015, pp. 10043–10044). In Eastern Canada in the Acadian Forest Region of Nova Scotia, Whitman et al. predicted drying conditions that will lead to increased fire intensity and potential fire weather severity (Whitman et al. 2014, p. 1455).

In the boreal forest of Western Canada, in a comparison of observed 2001–2007 and estimated 2091–2100 conditions, the severity of wildfire weather conditions was predicted to increase under three models and three scenarios set out by de Groot et al., with increases in total fuel consumption rate (de Groot et al., 2013, p. 4).

Gaudreau et al., using 1950–2010 data, predicted an increased mean area burned in the boreal forest of Quebec (Central Canada) by 2100 (Gaudreau et al., 2016, p. 23). Conversely, Yue et al. predicted a decrease in area burned in the period 2046–2065 in some regions of Northern Canada, such as the Taiga Plain, owing to a prediction of wetter conditions, albeit the decreases in area burned were noted by the authors as small and not significant (Yue et al., 2015, pp. 10043–10044). By 2100, in the Waswanipi boreal forest area in Quebec, annual area burned could increase by 7% (le Goff et al., 2009, p. 2375).

In a study of wildfire risk in areas around eleven cities in Canada, including two cities in the northern territories, Gaur et al. estimated that over a 50-year period under global warming of either 2 or 3.5 °C, wildfire seasons will extend (i.e., start earlier and end later) and extreme wildfires will occur more frequently in the interface of wildland and urban areas (Gaur et al., 2020, pp. 7–8), which is a relatively small yet impactful region where wildfires can occur. These estimates were made despite the estimates that total precipitation will also increase (Gaur et al., 2020, p. 6), suggesting that increased precipitation may not always completely counter the effects of increasing temperatures (Flannigan et al., 2016, pp. 61, 63–65; Wang et al., 2020, p. 9).

Wildfire season is projected to increase by up to 35 days (2 °C) or 61 days (3.5 °C) (Gaur et al., 2020, p. 7). Moreover, an increase in head fire intensity (i.e., the energy output of fire at the front or head of a wildfire) is also expected, leading to an increase in the frequency of extreme wildfires (Gaur et al., 2020, p. 8).

3 Health Impacts

3.1 What Are the Impacts of Wildfire Smoke?

Smoke from wildfires consists of a variety of air pollutants, such as fine particulate matter, carbon monoxide, carbon dioxide, nitrogen oxides, sulfur oxides, and volatile organic compounds, that with enough exposure can cause respiratory distress (e.g., shortness of breath, coughing, wheezing, asthma attacks), irritated eyes and nose, headache, chest pains, heart palpitations, and dizziness (CCOHS 2022; ECCC 2021). Exposure to wildfire smoke also increases the risk of mortality of all causes, and of cardiovascular mortality in particular, in addition to increasing the number of respiratory-related visits to hospitals (Barn et al., 2016, p. 1). Fine particulate matter (PM) of 2.5 microns or less (PM2.5) is thought to be of the greatest concern (Barn et al., 2016, p. 1). PM refers to small particles in the air that can be composed of a variety of components such as soot, dust, pollen, organic compounds, and inorganic compounds (EPA 2022). Small particles are able to enter the body through the throat and nose and can damage the heart and lungs (EPA 2022). At highest risk are children, the elderly, pregnant women, people with lung or heart conditions, and people who work outside and therefore have the greatest amount of exposure (ECCC 2021; EPA 2022).

The impacts of wildfire season in Canada were estimated to cost $0.4–$1.8 billion annually for acute health impacts and $4.3–$19 billion annually for chronic health impacts in the period 2013–2018 (Matz et al., 2020, p. 5). As an immediate response to periods of high smoke, the number of hospital visits has been seen to statistically significantly increase for respiratory-related distress but not for cardio-related distress (Howard et al., 2020, p. 5; Henderson et al., 2011, p. 1270). Two research teams examined the impact of smoke on respondents to the 2016 Fort McMurray wildfire incident in Alberta, the most expensive natural disaster in Canada's history (Moitra et al., 2021, p. 1). In a cross-sectional study of Royal Canadian Mounted Police officers (i.e., national police officers) two years after the 2016 wildfires, a marginal association was found between air pollution and reduced residual volume in the lungs but not between air pollution and total lung capacity (Moitra et al., 2021, pp. 6–7). However, this association was only statistically significant within the first 90 days following deployment and not after 90 days (i.e., between 90 days and the two years following the 2016 wildfires). In a case–control study of firefighters two to three years after the 2016 wildfires, firefighters were more likely to develop asthma (Cherry et al., 2021, p. 781). They were also found to have decreased forced expiratory volume and forced vital capacity (i.e., indicators of exhalation strength) (Cherry et al., 2021, p. 782). Moreover, 19.5% of assessed firefighters tested positively in the methacholine challenge test (i.e., a positive result implies reduced lung function) while 21.3% exhibited presence of bronchial wall thickening (i.e., bronchial tube damage) (Cherry et al., 2021, p. 782).

Unfortunately, there has been a lack of long-term studies on exposure to wildfires (Koopmans et al., 2022, p. 13).

3.2 What Are the Impacts of Evacuation?

Evacuation is often a last resort in response to wildfires that have gotten close to communities. An evacuation may be the temporary resettlement of residents fleeing wildfire or it may be permanent resettlement if the community is later destroyed by the fire. People are uprooted from their homes and displaced, potentially losing their homes, their livelihoods, their lives.

In an extensive and comprehensive search, Beverly & Bothwell (2011) compiled and explored estimates of the total number of people in Canada evacuated because of wildfires each year from 1980 to 2007. Over this time period, the median number of evacuees was 3590, the lowest number was 40 (1984), the third highest was approximately 17,000 (2007), the second highest was approximately 28,000 (1990), and the highest was 51,346 (2003) (Beverly & Bothwell, 2011, p. 580).

In 2016, the wildfires in Northern Alberta caused 88,000 people to evacuate (Belleville et al., 2021, p. 2). This 2016 incident involved evacuating nearly twice the number of people evacuated in 2003 from wildfires in all Canada (i.e., 51,346), and it does not include the number of evacuees from wildfires in the rest of Canada in 2016. Of the total number of evacuees in the period 1980–2007, only six years saw more than 10,000 people evacuated (Beverly & Bothwell, 2011, p. 580). With that in mind, more than 10,000 people were evacuated in 2011 from just two wildfire incidents: up to 8650 people in communities in May in Northern Alberta (Botey & Kulig, 2013, p. 1473; Christianson et al., 2019, pp. 12–13) and up to 4476 people in communities in July in Northern Ontario (Asfwa et al. 2019, p. 2). Again, these are isolated numbers that do not include evacuees from the rest of the country for 2011; however, combined they are well above the median number of evacuees of 3590 in the period 1980–2007. An updated estimate of the total number of annual evacuees would allow us to see if there is a trend of increasing numbers of evacuees from the year 1980 onwards.

The Indigenous people of Canada are disproportionately affected by wildfires. In Canada, 12.3% of Canada's population lived within the wildland–human interface (i.e., areas where forest fuels intermingle with urban area or industrial areas) in 2011, including 32.1% of the Indigenous people of Canada who lived on First Nations reserves (Erni et al., 2021, pp. 1362, 1365). Other researchers report that, at 4% of the population, 80% of Indigenous communities in general (as opposed to strictly First Nations reserves) were located "in or adjacent to forest ecosystems that burn frequently"; further, in the period 1980–2007, a third of wildfire evacuee incidents were from Indigenous communities (Asfaw et al., 2019, pp. 1–2). During evacuation, entire families and communities are upended as they transport themselves to host communities, which involves several hardships such as a lack of vehicles or not enough gas; confusion over where to go; a lack of housing and supplies in host communities; discomfort at living in temporary housing and anxiety over not knowing when or if they can return home; racism experienced by Indigenous evacuees in primarily non-Indigenous host communities; the lack of essentials such as money, clothes, toiletries, and medication; and the absence of space available for recreation or for childcare, among many other concerns (Asfaw et al., 2019, pp. 4–7; Christianson et al., 2019, pp. 14–19).

On top of everything else, evacuated communities may have to pay host communities a share of the cost of temporary housing (Christianson et al., 2019, p. 20) while families may have to deal with insurance companies (Botey & Kulig, 2013, p. 1477). Returning evacuees also have to deal with practical problems, such as throwing out things ruined by the fire and smoke, waiting for power and other utilities to be reactivated, and throwing out food that has gone bad (Christianson et al., 2019, pp. 18). As well, evacuees experience mental distress from the negative experiences of the wildfire and the evacuation. Some children have reacted with anxiety, anger, nightmares, and crying even seven months after returning home (Botey & Kulig, 2013, p. 1477). A year after the 2016 wildfires in Northern Alberta, in a phone survey of evacuees of Fort McMurray, 37.7% were found to have clinically significant psychological symptoms, with prevalence estimates of 28.5% for insomnia disorder, 15.4% for post-traumatic stress disorder, 15.0% for major depressive disorder, 14.2% for generalized anxiety disorder, and 7.9% for substance use disorder. (Belleville et al., 2021, pp. 5–6). In the future, we can expect overall worsening mental health in terms of distress from experiencing natural disasters and from anxiety over the prospect of experiencing more natural disasters as the planet warms (Hayes et al., 2019, pp. 3–4).

4 Wildfire Prevention, Mitigation, and Adaptation

4.1 What Are Ways to Respond to Wildfire Smoke?

Staying indoors is a common response to the appearance of wildfire smoke. For example, in a rehabilitation facility in 2020 in Vancouver, sensors recorded PM2.5 levels from August 21st to October 31st of smoke from wildfires in the United States. Within this period, during an 11-day air quality advisory, a rooftop sensor averaged 72.0 $\mu g/m^3$ PM2.5 while the indoor sensors collectively averaged 29.6 $\mu g/m^3$ PM2.5 (Nguyen et al., 2021, p. 9); in other words, there was 41% less particulate matter indoors. As an additional measure, portable air cleaners (i.e., cleaners with high efficiency particulate air (HEPA) filters) can filter PM2.5 from the air (Barn et al., 2016, p. 2). In one study of portable air cleaner use during wildfire events, use of portable air cleaners was associated with lower odds (0.54) of worsening respiratory symptoms; in another study, use of portable air cleaners saw a reduction of PM2.5 levels of 45% indoors, with estimates that greater adoption of portable air cleaners could reduce smoke-related deaths by 30% and hospital admissions by 45% during periods of wildfire smoke (Barn et al., 2016, pp. 3–4). Improved forest fire smoke forecasts would help alert the public about days of poor air quality (Yuchi et al., 2021, pp. 309–316).

4.2 What Are Ways to Prevent, Mitigate, and Adapt to Wildfires?

Not building or living in areas of high wildfire risk (a luxury not everyone may be able to afford) is one way to avoid some of the harm that can be caused by wildfires (e.g., smoke can be carried great distances and can threaten the health of people in communities that are otherwise not at risk of nearby wildfires). Governments in Canada implement fire bans and fire restrictions (https://www.albertaparks.ca/albertaparksca/advisories-public-safety/fire-bans/) as well as promote fire safety with recommendations on ways to upgrade your home and reduce the amount of vegetation in areas around your home (https://firesmartcanada.ca/about-firesmart/). To help slow the spread and intensity of all fires in an area, governments also carry out planned and controlled fires (i.e., prescribed fires), which reduces the amount of vegetation available for fire to burn (https://www2.gov.bc.ca/gov/content/safety/wildfire-status/prevention/vegetation-and-fuel-management/prescribed-burning). Scientists continue to develop new approaches to reduce the amount of wildfire fuel, which will also reduce the risk of wildfires (Coogan et al., 2019, p. 1020; Prichard et al., 2021, pp. 3–30; Yemshanov et al., 2021, pp. 3–26). A nationwide framework to create and update fire-risk maps was suggested (Parisien, 2016, p. 297). It should be noted again that wildfires are a natural part of ecosystems; for this reason, wildfire management often take the approach of "appropriate response" whereby wildfires that put people at risk are fought and put out but non-harmful wildfires are usually, but not always, left to burn.

In a report that examined, among other aspects of wildfires, wildfires by cause from 1978 to 2011, investigators attributed the decreasing number of human-caused fires to a variety of factors, including migration of people from rural to urban areas, fire prevention and public education programs, and improved forest industry guidelines (B. J. Stocks Wildfire Investigations, 2013, p. 28). When it comes to managing wildfires in general, the investigators noted that increasing costs of fire management may imply the presence of more challenging fire conditions (B. J. Stocks Wildfire Investigations 2013, p. 30). Even a small increase in wildfire size and area burned could threaten to overwhelm our current capacity to manage fires (Wang et al., 2020, p. 9). Coogan et al., in a reflective and encompassing examination of wildfire risk in Canada, noted that traditional approaches to wildfire suppression may be reaching their limit and that with predicted increases in wildfire frequency, intensity, and seasons, managing fires will become even more challenging (Coogan et al., 2019, p. 1019).

However, future trends are difficult and complex to predict. For example, climate change could theoretically cause a positive feedback loop whereby dryer, hotter conditions cause more wildfires which cause dryer, hotter conditions which in turn cause more wildfires, et cetera (Bowman et al., 2020, p. 508). We know that different regions experience different changing conditions and that although the frequency of wildfires and the amount of area burned is expected to increase generally in Canada, some regions may stay the same while others in fact experience decreases (Flannigan et al., 2009, p. 501).

Acknowledgements For answering my questions and providing feedback on this book chapter, I would like to thank the fire scientists at Natural Resources Canada (NRCAN) in the Canadian Forest Service (CFS) department. Prior to the final draft, this manuscript was peppered with "personal communication" citations regarding wildfire fundamentals, data, and management before it was decided to remove those citations and instead recognize the assistance of the NRCAN scientists in this acknowledgments section. Thank you. I would also like to thank the anonymous reviewers for their feedback and comments. This chapter was better for it.

Disclaimer

Note that any subject matter simplifications or mistakes that may be present in this summary book chapter are the author's alone and not the fault of the scientists at NRCAN.

References

Asfaw, H. W., Sandy Lake First Nation, McGee, T. K., & Christianson, A. C. (2019, August). A qualitative study exploring barriers and facilitators of effective service delivery for Indigenous wildfire hazard evacuees during their stay in host communities. *International Journal of Disaster Risk Reduction, 41*, 1–10. https://doi.org/10.1016/j.ijdrr.2019.101300

Barn, P. K., Elliot, C. T., Kosatsky, T., Rideout, K., & Henderson, S. B. (2016, November).Portable air cleaners should be at the forefront of the public health response to landscape fire smoke. *Environmental Health, 15*, 1–8. https://doi.org/10.1186/s12940-016-0198-9

Belleville, G., Ouellet M-A., Lebel, J., Ghosh, S., Morin, C. M., Bouchard, S., Guay, S.,Bergeron, N., Campbell, T., & MacMaster, F. P. (2021, May). Psychological symptoms among evacuees from the 2016 Fort McMurray wildfires: a population-based survey one year later. *Frontiers in Public Health, 9*, 1–15. https://doi.org/10.3389/fpubh.2021.655357

Beverly, J. L., & Bothwell, P. (2011, March). Wildfire evacuations in Canada 1980–2007. *Natural Hazards, 59*, 571–596. https://doi.org/10.1007/s11069-011-9777-9

B. J. Stocks Wildfire Investigations. (2013). Evaluating past, current, and future forest fire load trends in Canada. *Canadian Council of Forest Ministers*. https://www.ccfm.org/releases/evaluating-past-current-and-future-forest-fire-load-trends-in-canada/. Accessed on November 2, 2022.

Bowman, D. M. J. S., Kolden, C. A., Abatzoglou, J. T., Johnston, F. H., van der Werf, G. R., & Flannigan, M. (2020, August). Vegetation fires in the Anthropocene. *Nature Reviews Earth & Environment, 1*, 500–515. https://doi.org/10.1038/s43017-020-0085-3

Botey, A. P., & Kulig, J. C. (2013, July). Family functioning following wildfires: Recovering from the 2011 Slave Lake fires. *Journal of Child and Family Studies, 23*, 1471–1483. https://doi.org/10.1007/s10826-013-9802-6

Canadian Centre for Occupational Health and Safety (CCOHS). (2022). Forest fires and wildfire smoke. *Canadian Centre for Occupational Health and Safety*. https://www.ccohs.ca/oshanswers/safety_haz/forest_fires.html. Accessed on November 2, 2022.

Cherry, N., Barrie, J. R., Beach, J., Galarneau, J.-M., Mhonde, T., & Wong, E. (2021, September). Respiratory outcomes of firefighter exposures in the fort McMurray fire. *Journal of Occupational and Environmental Medicine, 63*(9), 779–786. https://doi.org/10.1097/JOM.0000000000002286

Christianson, A. C., McGee, T. K., & Whitefish Lake First Nation 459. (2019, January). Wildfire evacuation experiences of band members of Whitefish Lake First Nation 459, Alberta, Canada. *Natural Hazards, 98*, 9–29. https://doi.org/10.1007/s11069-018-3556-9

Coogan, S. C. P., Cali, X., Jain, P., Flannigan, & M. D. (2020, February). Seasonality and trends in human- and lightning-caused wildfires $\geq$ 2 ha in Canada, 1959–2018. *International Journal of Wildland Fire, 29*, 473–485. https://doi.org/10.1071/WF19129

Coogan, S. C. P., Robinne, F.-N., Jain, P., & Flannigan, M. D. (2019, May). Scientists' warning on wildfire-a Canadian perspective. *NRC Research Press, 49*, 1015–1023. https://doi.org/10.1139/cjfr-2019-0094

de Groot, W. J., Flannigan, M. D., & Cantin, A. S. (2013, April). Climate change impacts on future boreal fire regimes. *Forest Ecology and Management, 294*, 35–44. https://doi.org/10.1016/j.foreco.2012.09.027

Environment and Climate Change Canada (ECCC). (2021, July). Wildfire smoke and your health. *Environment and Climate Change Canada*. https://www.canada.ca/en/environment-climate-change/services/air-quality-health-index/wildfire-smoke/wildfire-smoke-health.html. Accessed on November 2, 2022.

Environmental Protection Agency (EPA). (2022, October). Why wildfire smoke is a health concern. *United States Environmental Protection Agency*. https://www.epa.gov/wildfire-smoke-course/why-wildfire-smoke-health-concern. Accessed on November 2, 2022.

Erni, S., Johnston, L., Boulanger, Y., Manka, F., Bernier, P. Eddy, B., Christianson, A., Swystun, T., & Gauthier, S. (2021, February). Exposure of the Canadian wildland–human interface and population to wildland fire, under current and future climate conditions. *Canadian Journal of Forest Research, 51*, 1357–1367. https://doi.org/10.1139/cjfr-2020-0422

Field, C. B., Mortsch, L. D., Brklacich, M., Forbes, D. L., Kovacs, P., Patz, J. A., Running, S. W., & Scott, M. J. (2007). North America. In M. L. Parry, O. F. Canziani, J. P. Palutikof, P. J. van der Linden, & C.E. Hanson (Eds.), *Climate change 2007: Impacts, adaptation and vulnerability. Contribution of working group ii to the fourth assessment report of the intergovernmental panel on climate change* (pp. 617–652). Cambridge University Press. https://www.ipcc.ch/pdf/assessment-report/ar4/wg2/ar4_wg2_full_report.pdf

Flannigan, M. D., Krawchuk, M. A., de Groot, W. J., Wotton, B. M., & Gowman, L. M. (2009, August). Implications of changing climate for global wildland fire. *International Journal of Wildland Fire, 18*, 483–507. https://doi.org/10.1071/WF08187

Flannigan, M. D., Wotton, B. M., Marshall, G. A., de Groot, W. J., Johnston, J., Jurko, N., & Cantin, A. S. (2016, January). Fuel moisture sensitivity to temperature and precipitation: Climate change implications. *Climatic Change, 134*, 59–71. https://doi.org/10.1007/s10584-015-1521-0

Gaudreau, J., Perez, L., & Drapeau, P. (2016, January). BorealFireSim: A GIS-based cellular automata model of wildfires for the boreal forest of Quebec in a climate change paradigm. *Ecological Informatics, 32*, 12–27. https://doi.org/10.1016/j.ecoinf.2015.12.006

Gaur,A., Benichou, N., Armstrong, M., & Hill, F. (2020, November). Potential future changes in wildfire weather and behavior around 11 Canadian cities. *Urban Climate, 35*, 1–12. https://doi.org/10.1016/j.uclim.2020.100735

Hanes, C. C., Wang, X., Jain, P., Parisien, M.-A., Little, J. M., & Flannigan, M. D. (2019). Fire-regime changes in Canada over the last half century. *Canadian Journal of Forest Research, 49*(3), 256–269. https://doi.org/10.1139/cjfr-2018-0293

Hayes, K., Berry, P., & Ebi, K. L. (2019, May). Factors influencing the mental health consequences of climate change in Canada. *International Journal of Environmental Research and Public Health, 16*, 1–13. https://doi.org/10.3390/ijerph16091583

Henderson, S. B., Brauer, M., MacNab, Y. C., & Kennedy, S. M. (2011, September). Three measures of forest fire smoke exposure and their associations with respiratory and cardiovascular health outcomes in a population-based cohort. *Environmental Health Perspectives, 119*(9), 1266–1271. https://doi.org/10.1289/ehp.1002288

Howard,C., Rose, C., Dodd, W., Kohle, K., Scott, C., Scott, P., Cunsolo, A., & Orbinski, J. (2020, November). SOS! Summer of Smoke: a retrospective cohort study examining the cardiorespiratory impacts of a severe and prolonged wildfire season in Canada's high subarctic. *BMJ Open, 11*, 1–10. https://doi.org/10.1136/bmjopen-2020-037029

Intergovernmental Panel on Climate Change. (2019). Summary for Policymakers. In P.R. Shukla, J. Skea, E. Calvo Buendia, V. Masson-Delmotte, H.-O. Pörtner, D. C. Roberts, P. Zhai, R. Slade, S. Connors, R. van Diemen, M. Ferrat, E. Haughey, S. Luz, S. Neogi, M. Pathak, J. Petzold, J. Portugal Pereira, P. Vyas, E. Huntley, K. Kissick, M. Belkacemi, & J. Malley (Eds.), *Climate change and land: an IPCC special report on climate change, desertification, land degradation, sustainable land management, food security, and greenhouse gas fluxes in terrestrial ecosystems* (pp. 1–36). Cambridge University Press. https://doi.org/10.1017/9781009157988.001

Jain, P., Tye, M. R., Paimazumder, D., & Flannigan, M. (2020, October). Downscaling fire weather extremes from historical and projected climate models. *Climatic Change, 163*, 189–216. https://doi.org/10.1007/s10584-020-02865-5

Koopmans, E., Cornish, K., Fyfe, T. M., Bailey, K., & Pelletier, C. A. (2022, January).Health risks and mitigation strategies from occupational exposure to wildland fire: A scoping review. *Journal of Occupational Medicine and Toxicology, 17*, 1–17. https://doi.org/10.1186/s12995-021-00328-w

le Goff, H., Flannigan, M. D., & Bergeron, Y. (2009). Potential changes in monthly fire risk in eastern Canadian boreal forest under future climate change. *Canadian Journal of Forest Research, 39*, 2369–2380. https://doi.org/10.1139/X09-153

Matz, C. J., Egyed, M., Xi, G., Racine, J., Pavlovic, R., Rittmaster, R., Henderson, S. B., & Stieb, D. M. (2020, November). Health impact analysis of PM2.5 from wildfire smoke in Canada (2013–2015, 2017–2018). *Science of the Total Environment, 725*, 1–9. https://doi.org/10.1016/j.scitotenv.2020.138506

Moitra, S., Tabrizi, A. F., Fathy, D., Kamravaei, S., Miandashti, N., Henderson, L., Khadour, F., Naseem, M. T., Murgia, N. Melenka, L., & Lacy, P. (2021, November). *International Journal of Environmental Research and Public Health, 18*, 1–12. https://doi.org/10.3390/ijerph182211787

Nguyen, P. D. M., Martinussen, N., Mallach, G., Ebrahimi, G., Jones, K., Zimmerman, N., & Henderson, S. B. (2021, September). Using low-cost sensors to assess fine particulate matter-infiltration (pm2.5) during a wildfire smoke episode at a large inpatient healthcare facility. *International Journal of Environmental Research and Public Health*, *18*, 1–17. https://doi.org/10.3390/ijerph18189811

Nitschke, C. R., & Innes, J. L. (2012, June). Potential effect of climate change on observed fire regimes in the Cordilleran forests of South-Central Interior, British Columbia. *Climatic Change*, *116*, 579–591. https://doi.org/10.1007/s10584-012-0522-5

NRCAN Canadian Forest Service. (No date). Background Information: Canadian Forest Fire Weather Index (FWI) System. *Natural Resources Canada*. https://cwfis.cfs.nrcan.gc.ca/background/summary/fwi. Accessed on November 2, 2022a.

NRCAN Canadian Forest Service. (2022). National fire database: NFDB_point_stats.xlsx. https://cwfis.cfs.nrcan.gc.ca/datamart/download/nfdbpnt. Accessed on November 2, 2022.

Parisien, M.-A. (2016, June). Science can map a solution to a fast-burning problem. *Nature*, *534*, 297. https://doi.org/10.1038/534297a

Prichard, S. J., Hessburg, P. F., Hagmann, R. K., Povak, N. A., Dobrowski, S. Z., Hurteau, M. D., Kane, V. R., Keane, R. E., Kobziar, L. N., Kolden, C. A., North, M., Parks, S. A., Safford, H. D., Stevens, J. T., Yocom, L. L., Churchill, D. J., Gray, R. W., Huffman, D. W., Lake, F. K., & Khatri-Chhetri, P. (2021). Adapting western North American forests to climate change and wildfires: 10 common questions. *Ecological Applications*, *31*(8), 1–30. https://doi.org/10.1002/eap.2433

Wang, X., Studens, K., Parisien, M.-A., Taylor, S. W., Candau, J.-N., Boulanger, Y., & Flannigan, M. D. (2020). Projected changes in fire size from daily spread potential in Canada over the 21st century. *Environmental Research Letters*, *15*, 104048. https://doi.org/10.1088/1748-9326/aba10

Whitman,E., Sherren, K., & Rapaport, E. (2014, September). Increasing daily wildfire risk in the Acadian Forest Region of Nova Scotia, Canada, under future climate change. *Regional Environmental Change*, *15*, 1447–1459. https://doi.org/10.1007/s10113-014-0698-5

Yemshanov, D., Liu, N., Thompson, D. K., Parisien, M.-A., Barber, Q. E., Koch, F. H., & Reimer, J. (2021, October). Detecting critical nodes in forest landscape networks to reduce wildfire spread. *PLOS ONE*, *16*, 1–31. https://doi.org/10.1371/journal.pone.0258060

Yuchi,W., Yao, J., McLean, K. E., Stull, R., Pavlovic, R., Davignon, D., Moran, M. D., & Henderson, S. B. (2021, September). Blending forest fire smoke forecasts with observed data can improve their utility for public health applications. *Atmospheric Environment*, *145*, 308–317. https://doi.org/10.1016/j.atmosenv.2016.09.049

Yue, X., Mickley, L. J., Logan, J. A., Hudman, R. C., Martin, M. V., & Yantosca, R. M. (2015, September). Impact of 2050 climate change on North American wildfire: Consequences for ozone air quality. *Atmospheric Chemistry and Physics*, *15*, 10033–10055. https://doi.org/10.5194/acp-15-10033-2015

Climate Change and Human Health in Mexico: Public Health Trends and Government Strategies

María E. Ibarrarán, Gabriela Pérez-Castresana, Romeo A. Saldaña-Vázquez, and Tamara Pérez-García

Abstract Climate change directly affects health through extreme weather events, and indirectly through the effect of these events on the dynamics of pathogens and vector-borne diseases, as well as on the productivity of crops that impact human nutrition. Mexico's geographic location is a relevant factor for exposure to hydrometeorological phenomena, such as cyclones, storms, and floods. This chapter reviews how extreme hydrometeorological events affect ecosystems and therefore morbidity and mortality in Mexico. It also discusses health impacts from the lack of water and food security. This leads to the need of specific public policies for adaptation to climate change. We describe Mexico's institutional framework regarding health and climate change. We discuss whether it has the necessary policies and resources to face this phenomenon. Finally, we make recommendations to make health and general policy climate sensitive.

Keywords North America · Climate-sensitive health policy · Drought · Floods · Food security · Health-related adaptation · Health vulnerability

1 Introduction

The increase in the planet's surface temperature results in the modification of weather patterns expressed in changes in temperature and precipitation. This produces extreme events related to the water cycle, such as floods and droughts. The latter can also result in fires, due to a high accumulation of biomass that can be accidentally or intentionally ignited. These events can be directly related to human health, but they are also indirectly related.

Climate change can cause disease in people by sickening and reducing populations of animals that are part of the food chain (Thornton et al., 2014). Also, by changing the distribution of populations of wild organisms that are important for food production,

M. E. Ibarrarán (✉) · G. Pérez-Castresana · R. A. Saldaña-Vázquez · T. Pérez-García
Xabier Gorostiaga Environmental Research Institute, Universidad Iberoamericana Puebla, San Andrés Cholula, Mexico
e-mail: mariaeugenia.ibarraran@iberopuebla.mx

R. Akhtar (ed.), *Climate Change and Human Health Scenarios*, Global Perspectives on Health Geography, https://doi.org/10.1007/978-3-031-38878-1_25

such as pollinators (Jackson et al., 2022). Finally, climate change can put people's health at risk from increasing populations of species that can be disease vectors, such as mosquitoes (Franklinos et al., 2019). Interactions then go in all directions, between human, animal, and ecosystem health, as already established in the One Health Framework (Stephen, 2022).

Through a indepth literature review, we searched for studies on Mexico, on the effects of climate change on human health, and on policies implemented in the country to face these issues. Thus, the first section of this chapter reviews threats to human health in Mexico because of climate change, focusing on changes in the ecology and health of wild and domestic organisms that are part of our food system, and/or can be reservoirs of zoonotic diseases. The second section describes the main human diseases that result from the impact on ecosystems. Specifically, diarrhea, respiratory illnesses, and vector-borne diseases are addressed and their possible evolution over time, given their association with more extreme climatic conditions. The third section discusses public policy regarding these health effects and how it is expected to evolve. It reviews budget allocation to epidemiological surveillance and analyzes whether it is sufficient to deal with future pandemics. Finally, some conclusions and public policy considerations are drawn.

2 Climate Change and Human Nutrition in Mexico

Climate change will alter the spatial distribution of ideal climate for different species of organisms to flourish, including our food. For example, it will reduce the regions for coffee production in Mexico, especially in the states of Veracruz, Chiapas, and Guerrero (Bunn et al., 2015; Imbach et al., 2017), without increasing the suitable area in other states. To reduce the effects of climate change on this crop, shade production has been suggested to reduce water loss in soil, as well as protect plantations from hurricanes or torrential rains (Jat et al., 2016; Lin et al., 2008; Schroth et al., 2009).

Climate change will decrease abundance and distribution of pollinators in some parts of the country, which will indirectly lead to malnutrition by reducing crop productivity (Sosenski & Domínguez, 2018). Eighty five percent of plant food production in Mexico depends on pollinators (Ashworth et al., 2009). Agricultural products mostly affected by their reduction or uncoupling of pollinator-plant interactions will be coffee and agave (Gómez-Ruiz & Lacher, 2019; Imbach et al., 2017). Unfortunately, no further information is available on the potential variation on pollinator distribution and abundance associated with climate change in Mexico.

Animal health of both wild and domestic organisms, will be affected by climate change (Filho et al., 2022). This will shift the geographic range of pathogens and their vectors to places where they were not previously found. One of the few related studies predict that the vampire bat *Desmodus rotundus'* distribution, one of the vectors of paralytic bovine rabies, will decrease in southeastern Mexico and increase in the highlands and northern Mexico (Zarza et al., 2017). This implies a higher risk of rabies mortality in cattle, with consequences on human nutrition, especially because animal protein is an important source of nutrients for the Mexican population (Tello et al., 2020; Thornton et al., 2014). Livestock will also be threatened by climate change in another front. In years of high drought, production may fall by up to 3% (Murray-Tortarolo & Jaramillo, 2019). Although 3% at the national level may not seem much, there are states that will be more affected since they have higher livestock production, such as Chihuahua, Sonora and Tamaulipas, where meat is mainly produced. This idea coincides with projections of vulnerability to climate change from animal food production in Mexico (Godber & Wall, 2014; Murray-Tortarolo & Jaramillo, 2020).

Extreme droughts related to climate change are among the main factors promoting wildfires in different biomes of the planet (Andela et al., 2017). Smoke from these fires is responsible for 8% of the 3.3 million annual premature human deaths worldwide related to poor air quality (Lelieveld et al., 2015), as well as respiratory diseases such as asthma and chronic obstructed lung disease (Reid et al., 2016). Therefore, it can be expected that in regions with a higher frequency of wildfires, their population is more at risk from respiratory diseases and associated premature deaths.

Mexico is within the regions where about 15% of forests burned in the last 17 years (Andela et al., 2017, Fig. 1). Although the annual percentage is not as high as in other countries, the density of the exposed population can be higher in some cases, causing health problems associated with forest fire pollution. In addition, premature deaths from wildfire smoke can affect ecosystem processes linked to human health and the wild and domestic organisms on which humans depend. Fires can damage soil, and thus agricultural or livestock land, as well as protected natural areas, increasing erosion and a decrease in soil nutrients. It can also cause death of plant tissue, deterioration of wood properties of forest species, and changes in trajectories of plant succession in forests (Castillo et al., 2003). Despite all this, there are no studies in Mexico that have evaluated the effects of forest fires on human and ecosystem health. There are only analyses of emissions related to forest fires (Rodríguez-Trejo et al., 2007). Figure 1 shows tree coverage loss due to forest fires in Mexico. Shading in brown color pixels represents areas of tree cover loss due to fires compared to all other drivers of tree cover loss accumulated from 2001 to 2021. Fires were responsible for 15% of tree cover loss in Mexico between 2001 and 2021.

According to climate change forecasts, the northern, western, and central regions of Mexico will be the most impacted by droughts (Magaña et al., 2012; Moreno & Huber-Sannwald, 2011; Mundo-Molina, 2015; Ramírez Sánchez et al., 2022; Fig. 2). Maize and sorghum are grown in these regions as staple foods for human nutrition, either directly or as feed for cattle and poultry. Approximately 20 million Mexicans are nutritionally dependent on corn. The reduction in areas suitable for corn cultivation is expected to reduce yields due to increased drought. Estimates show that in some places, maize production could drop from 4 to 1 ton/ha (Arce Romero et al., 2020; Murray-Tortarolo et al., 2018; Ramírez Sánchez et al., 2022). This reduction is because most maize in Mexico is rainfed, and thus, the most affected due to less precipitation (Moreno & Huber-Sannwald, 2011; Ureta et al., 2020). Under this scenario, solutions proposed are to increase adaptive capacities of farmers based on efficient water management, adequate use of fertilizers, and preservation of genetic

Fig. 1 Loss of tree coverage, 2001–2021. *Source* Own, with data from Global Forest Watch and Tyukavina et al. (2022)

diversity of domesticated maize which contain drought-resistant genes, and use of rainfall forecasting technology (Donatti et al., 2019; Hellin et al., 2014; Liverman, 1990). Figure 2 shows an example of the distribution of drought in Mexico. The range of yellow and red represents regions of the country with the greatest drought for May 2022.

Other crops that will reduce their productivity due to reduced precipitation will be wheat, coffee, potatoes, and beans (Arce Romero et al., 2020; Torres Castillo et al., 2020). In the case of coffee, low productivity associated with rising temperatures is linked to the increase in fungus infections of *Hemileia vastatrix*, which causes "coffee rust" (Torres Castillo et al., 2020). On the other hand, crops such as soybean and avocado are expected to expand their suitable distribution and productivity in the Yucatan Peninsula and Michoacan, respectively (Arce Romero et al., 2020; Charre-Medellín et al., 2021; Hernandez-Ochoa et al., 2018), due to the increase in temperature and humidity in these areas. Both crops could be an option to maintain the population´s nutritional levels if their production is sustainable. Finally, data for sugarcane production is contradictory, with some authors forecasting low yield scenarios, while others predict increasing yields (Baez-Gonzalez et al., 2018; Santillán-Fernández et al., 2016).

On the other hand, increase in ocean temperature results in precipitation anomalies. "El Niño" phenomenon, which is characterized by an increase in the temperature of the Pacific Ocean, brings with it an increase in intensity of hurricanes and torrential

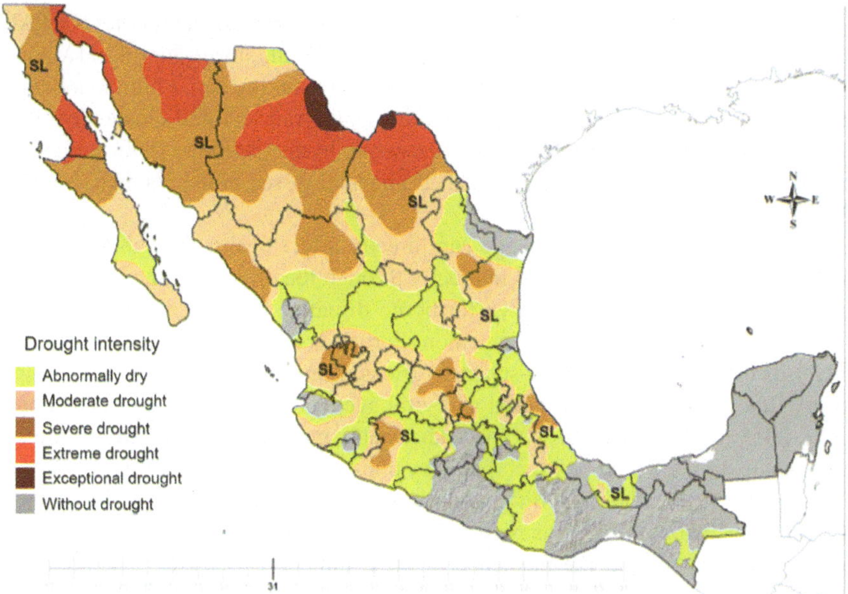

Fig. 2 Distribution of drought in Mexico, May 2022. *Source* CONAGUA (2023)

rains throughout the country. Studies show that the rise in ocean's temperature associated with "El Niño" is positively correlated with zoonotic diseases such as malaria and hantavirus (Rupasinghe et al., 2022). In Mexico, there is no record of the number of zoonotic diseases associated with climatic anomalies. However, there are isolated studies for brucellosis. This is a zoonotic bacterial disease that causes fever, joint pain, and fatigue in humans. It can be acquired through direct contact with infected animals or from the consumption of unpasteurized dairy products (Rodriguez-Morales, 2013).

Another anomalous climatic event is frost caused by a sudden drop in temperature. In the last 100 years, duration and severity of low temperatures have been reduced (Cuervo-Robayo et al., 2020). This may result in benefits to food production and human health. However, frosts have increased in the Baja California and Yucatan peninsulas, where there is low population density and food production is not significant (Cuervo-Robayo et al., 2020).

3 Effects on Human Health

The alteration of the global climate is generating numerous negative consequences on human health. In Mexico, effects are becoming more frequent and health risks are increasing with social vulnerability conditions and severe environmental degradation. In addition, the country's geographic location is a relevant factor for exposure to hydrometeorological phenomena, such as cyclones, storms, and floods, since its territorial limits border oceanic waters on its eastern and western boundaries (CEPAL, 2022; COFEPRIS, 2017; Delgado et al., 2010). In 2013, the storms Ingrid and Manuel simultaneously hit both coasts within a 24-h period resulting in 5.7 billion dollars in damages. In rural areas, extreme temperatures and erratic rainfall drastically affect agricultural productivity, including both crops and livestock. Since 1990, agriculture has accounted for 80% of weather-related financial losses in the country (USAID, 2017).

As stated above, climate change can directly affect health through heat waves and extreme weather events, and indirectly through the effect of climate change on the dynamics of pathogens and vector-borne diseases, as well as on the productivity of crops that impact human nutrition (Pörtner et al., 2022). Among the indirect effects is the increase in climate-sensitive diseases, such as diarrhea and acute respiratory illnesses caused by bacterial and viral agents (Riojas Rodríguez & Hurtado Díaz, 2015; Rodríguez et al., 2006). Additionally, more than half of known human pathogenic diseases may be exacerbated by climate change (Mora et al., 2022).

According to the National Mitigation and Adaptation Strategy to Climate Change, a strong positive correlation is registered for central Mexico between the average annual temperature and the incidence of these infectious diseases. The scenarios projected for 2030 and 2050, considering a temperature increase of 4 °C, show that the incidence of these diseases could increase by up to 200% in some municipalities (SEMARNAT, 2013). However, the impacts of climate change linked to the increase in diarrheal diseases could be greater in other parts of the country, considering the problems of drinking water supply and sanitation. At the national level, only 58% of the population receives water daily at home and has improved basic sanitation, this being the most critical problem in the southern states of Mexico (CONAGUA, 2020).

Climate change is exacerbating and making the existing health problem more complex since the so-called "climate-sensitive" diseases, such as those enclosed within the black circle in Fig. 3, show an increasing trend in Mexico that is linked to social conditions and environmental factors that promote its incidence (SSA, 2021a; WHO, 2022), and the changes that are projected in temperature and environmental conditions linked to climate change in Mexico (Gutiérrez et al., 2021; Pörtner et al., 2022), favor the proliferation and spread of pathogens, as well as the development of chronic respiratory and cardiovascular diseases (Pörtner et al., 2022; WHO, 2015).

Climatic risk factors for respiratory tract infections due to multiple pathogens, such as bacteria, viruses, and fungi, include temperature and humidity extremes, dust storms, extreme precipitation events and increased climate variability (Pörtner et al., 2022). Influenza and pneumonia represent a relevant public health problem in Mexico that worsens at all ages and show a tendency to increase over time (Pörtner et al., 2022). This is shown in Fig. 4.

This situation could worsen in time since, according to Liu et al. (2020), climate change could increase influenza epidemics by 20–50% in highly populated temperate zones by the end of the century due to sudden climatic variations that are now taking place in Autumn, which weakens the human immune system and makes it more susceptible to the flu virus. In Mexico, there have been geographical variations in

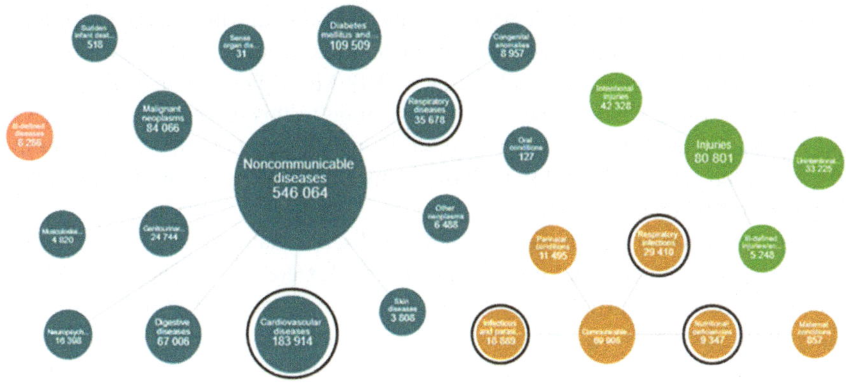

Fig. 3 Distribution of causes of total deaths grouped by category for 2018 in México. *Source* WHO (2022)

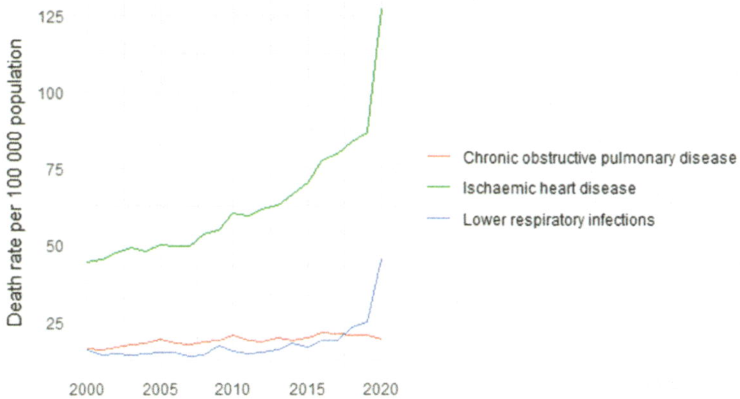

Fig. 4 Mortality trends from different diseases for Mexico. *Source* Own based on WHO (2022)

the incidence of influenza, with higher rates in regions with dry climates and high shifts in weather, as well as in the most polluted cities (de Jesús Coria-Lorenzo et al., 2018).

One of the main risks of climate change is the increase in zoonoses. Zoonoses, as mentioned already, are defined as a disease that is shared by vertebrate animals and humans. Mexico is among the countries with the least research on zoonoses and climate change (Filho et al., 2022), which leaves a high degree of uncertainty to predict possible epidemics related to the expansion of related diseases. Ticks of the *Ixodes scapularis* species, which transmit Lyme disease to humans, will remain stable in their distribution in the northeast of the country (Feria-Arroyo et al., 2014). The flies of the genus *Lutzomia* and the mice of the genus *Neotoma*, vectors and hosts, respectively, of protozoan species of the genus *Leishmania*, which cause Leishmaniasis in humans, will increase their presence in northern Mexico, moving towards the United States and Canada (González et al., 2010; Moo-Llanes et al., 2013). Therefore, regions of the country that were not endemic to this disease (Leishmaniasis) will now suffer from it. The vector bugs of the parasite that causes Chagas disease (*Trypanozoma cruzi*) will modify their distribution towards northern Mexico and the United States (Garza et al., 2014).

Dengue is another animal vector-mediated disease that may become a growing public health problem with climate change. In Mexico, the number of cases has increased in recent years despite vector prevention and control efforts (Navarro et al., 2021). Rodríguez et al. (2006) point out that the increase in dengue cases may be due to the increase in temperature, recording in their study a significant relationship between the number of cases and temperature for a period of 15 years, between 1994 and 2010, in states where there is a high risk of contracting the disease such as Chiapas, Veracruz, Oaxaca, Guerrero, and Colima.

The presence of *Aedes aegypti,* carrier of the dengue virus, and dengue outbreaks have been reported in the coastal states of the Pacific and the Gulf of Mexico, where climatic conditions are favorable for the transmission of this disease and cases of infections have been found in all states, as shown in Fig. 5. The study by Ryan et al. (2019), shows that towards 2080, and in the worst scenario of climate change, *Aedes aegypti* will be able to reproduce on practically the entire planet, and will threaten 7 billion people, 1 billion more than today.

Among other direct effects of climate change on human health are heat waves, which are and will affect the population, particularly those living in the north of the country (Pörtner et al., 2022). In the study by Riojas Rodríguez and Hurtado Díaz (2015), a positive and statistically significant association was found between the

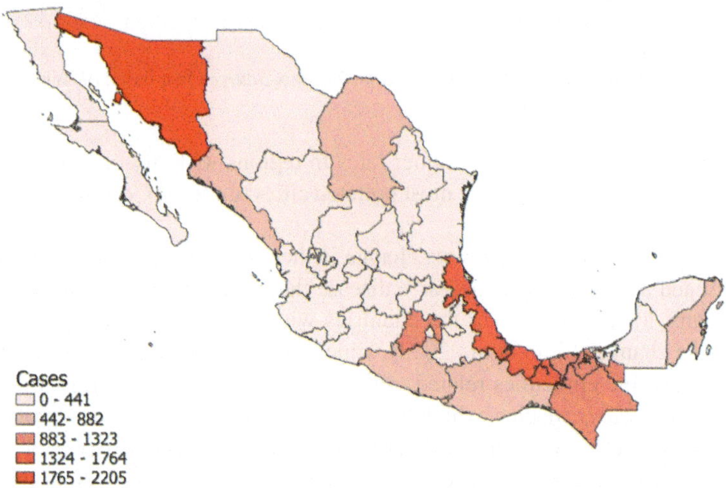

Fig. 5 Cases and incidence of Dengue by State in 2022. *Source* Own based on SSA (2023)

occurrence of heat strokes with ambient maximum and minimum temperatures. The models estimated that by increasing the monthly temperature by 1 °C, mortality from heat stroke increases, on average, by 1.22% in Sonora and 1.35% in Baja California.

Other diseases that are influenced by environmental agents and are among the main causes of death in the country, with increasing trends over time, are cardiovascular and chronic respiratory illnesses. Within respiratory diseases, Chronic Obstructive Pulmonary Disease (COPD) and asthma stands out (Pörtner et al., 2022). An increase in their frequency is expected due to the direct effect of climate change on air quality, with higher concentration of particulate matter and highly toxic ozone. Aeroallergens are also increasing due to this phenomenon, becoming an important health problem for non-communicable outbreaks of asthma, skin and respiratory diseases (Mora et al., 2022). This is especially true in densely populated regions of the country where people are exposed to a huge number of dangerous pollutants of industrial, urban, and agricultural origin. In Mexico, there are more than 40 Regions of Sanitary and Environmental Emergency (RESA), recognized by the federal government, in which various emissions and discharges of all types of pollutants that negatively impact the environment and society are concentrated and overlapping, causing harm to people's health (Barreda Marín, 2020).

On the other hand, cardiovascular diseases are associated with extreme temperatures. Exposure to high temperatures triggers compensatory mechanisms to release heat into the environment, redistributing blood flow and increased heart rate, which could cause heart attacks and ischemia. Exposure to low temperatures causes an increase in blood pressure due to activation of the sympathetic system, dehydration due to increased urination, and vasoconstriction, all triggering effects of cardiovascular events (Peña et al., 2022). According to Pörtner et al., 2022cardiovascular disease mortality could increase by 18.4%, 47.8% and 69.0% in the 2020s, 2050s and 2080s worldwide respectively under RCP4.5.

Finally, another climate change related health effect has to do with the occurrence of extreme hydrometeorological events and the high risk of disasters (Martinez-Austria, 2020). Vulnerabilities arise due to deficient territorial planning and limited access to public health systems, where 20% lack health coverage. Every year in Mexico, an average of five tropical cyclones impact, which can trigger large-scale disasters, especially on the Pacific coasts. The occurrence of these extreme events can cause premature deaths and injuries. Historically, the greatest impacts and the propensity for flooding are concentrated in 17 states that are home to 62% of the population (CONAGUA, 2020).

Thus, climate change is expected to increase health effects in Mexico. It will do so by affecting ecosystems and through them human health. The next section describes the public policies that Mexico has undertaken recently in this front, and how this is expected to evolve.

4 Current Health Policies Related to Climate Change

We first describe Mexico´s institutional framework in the climate change field. According to the General Law on Climate Change (Ley General de Cambio Climático, LGCC), issued in 2012, Mexico has two main planning instruments: the Climate Change National Strategy (Estrategia Nacional de Cambio Climático) and the Special Climate Change Program (Programa Especial de Cambio Climático). Both define concrete actions to address the impacts of climate change on health through four lines of action (Gay y García & Rueda Abad, 2015).

In addition, the Law establishes the creation of an Inter-Ministerial Commission on Climate Change (Comisión Intersecretarial de Cambio Climático, CICC), which has a permanent character, is chaired by the head of the Federal Executive branch, and is integrated by the heads of 14 ministries, including the Ministry of Health. Among the main powers of this Commission is to promote the coordination of actions among the agencies and entities of the public administration on climate change, formulate and implement national mitigation and adaptation policies, as well as their incorporation into the corresponding sectoral programs and actions (SEMARNAT, 2015).

The National Institute of Ecology and Climate Change oversees generating and integrating technical and scientific knowledge for the formulation and evaluation of public policies that lead to the mitigation and adaptation of climate change in the country. In 2006, in collaboration with the National Institute of Public Health, they published a "Diagnostic study on the effects of climate change on human health in Mexico", which examined the potential effects of climate change on health by carrying out an exploratory analysis, that included developing thematic maps showing the behavior of health events over time and the association between these events and climatic variables (Rodríguez et al., 2006).

The Commission for the Protection of Sanitary Risks (COFEPRIS for its acronym in Spanish) is the institution in charge of disseminating knowledge about the climate change process in Mexico, as well as its repercussions on health. One of its actions is to establish the Mexican Network on Climate Change and Health to disseminate, exchange and generate information for decision-making and the development of climate change and health plans at the federal and state levels (Gay y García & Rueda Abad, 2015). In 2016, they carried out diagnoses of health vulnerability in the face of climate change for all states (Garza Galván, 2016), and documented the correlation between the occurrence of diseases and the variation in temperatures, in addition to showing the areas with the greatest health vulnerability. However, many of these diagnoses are no longer accessible for public consultation, only 5 of them are available (Campeche, Oaxaca, Quintana Roo, San Luis Potosí and Tlaxcala).

As for tracking disease nationwide, currently, the 2019–2024 Health Sector Plan includes 5 priority objectives, among which is epidemiological surveillance. Specific strategies are proposed to guide programs and actions, based on scientific information, to facilitate health promotion and prevention, as well as epidemiological control. Nevertheless, health risks derived from climate change found in the Environmental Health Strategy are only mentioned in a very superficial way, namely as "the emerging risks of climate change that affect the health of the population will be monitored to anticipate due attention and foresee measures to reduce their impact" (SSA, 2019b). The brief reference to the issue clearly shows the low priority given by the current government to issues related to the climate crisis, leaving aside recommendations such as those of the Pan American Health Organization (PAHO, 2021), which invites Latin America and the Caribbean to address the problem through three strategic lines of action to ensure a healthy life and promote the welfare of the entire population.

Additional to the provisions of the LGCC, Mexico has a National Epidemiological Surveillance System (SINAVE), which is responsible of epidemiological information for public health decision-making across more than 20 thousand care units throughout the country. SINAVE monitors the incidence of vector-borne diseases and is associated with PAHO and the World Health Organization (WHO), which send epidemiological alerts to health sectors of all countries when health risks are detected. Mexico participates in the dengue working group (GT-Dengue) (SSA, 2020).

The National Center for Preventive Programs and Disease Control (CENAPRECE) is the decentralized body of the Ministry of Health in charge of conducting and implementing programs for the prevention and control of diseases in the population to reduce mortality and morbidity, among them the Rabies and other Zoonoses Program, the Vector-Borne Diseases Program (2017) and the Epidemiological Emergencies and Disasters Program (2015). The first of these programs has identified that half of the territory, where approximately 60% of the population lives, is susceptible to becoming a dengue endemic zone, putting 50 million Mexicans at risk, in areas where there are primary sector activities, but also oil production, industrial and tourist activities. No updates to these programs have been done during the current administration (CENAPRECE, 2017).

The Mexican health system has kept a registry of diseases since 1994 through the periodic notification of infectious diseases. Since that date, the number of diseases monitored has been modified several times (SSA, 2021b). In 2019, 14 infectious diseases of the digestive system, five of the respiratory system, five of malnutrition and 19 of dysplasias and neoplasias were monitored; vector-borne diseases included dengue, chikungunya and zika (SSA, 2019a). In 2020, COVID-19 was added as a respiratory disease (SSA, 2021b).

The budget is crucial for the health sector to carry out its functions, particularly disease monitoring. Between 2010 and 2022 Mexico only spent between 2.5 and 2.9% its GDP on health, half of what best practices indicate internationally. This led to severe constraints in response capabilities of the health sector in general, and to specific pandemics such as COVID-19.

For 2023, spending for the health sector is estimated to be 3% more than what was approved for 2022. However, Mexico's public spending on health remains between 2 and 3% of GDP, below 6% of the OECD average (Cámara de Diputados, 2022).

Of the total health budget, the Ministry of Health gets about 22% of resources of the health sector, and the rest of the budget goes to public funded contributive health systems. From these remaining resources, only a small share is used to prevent and treat sanitary-related risks and perform epidemiological monitoring through two programs. The Health Risk Protection Program consists of carrying out strategies for the analysis of health risks to which the population is exposed and provides and evaluates the safety, efficacy, and accessibility of medicines. The Epidemiological Surveillance Program, on the other hand, contributes to consolidating disease surveillance, prevention and control actions through early identification and risk control, as well as the timely and specific treatment of new cases of infectious and communicable diseases under the responsibility of CENAPRECE, and the General Directorate of Epidemiology. It also evaluates the technical performance of the National Network of Public Health Laboratories; maintains continuous updating of the personnel of SINAVE, and provides diagnostic services nationwide through the National Network of Public Health Laboratories (RNLSP). The federal budget for these two programs and their evolution can be seen in Fig. 6. It shows how they practically have been falling since 2016, and for the case of protection against health risks, the drop has been dramatic since 2019. The small rebound effects was due to the COVID-19 pandemic. It is not expected that this budget will increase significantly in the near future.

In recent years, the budget of the Ministry of Health, which is responsible for these two programs, has been reduced in real terms. The largest drop, of approximately 60%, has been observed in the budget for protection against health risks. Spending on epidemiological surveillance has been reduced by almost 40%.

In 2023, the Federal Expenditure Budget, for the health branch, is 3% higher in real terms than in 2022. The Ministry of Health will allocate resources for the Protection against Sanitary Risks, where it proposed to elaborate environmental health regulation that includes the issue of climate change. Resources will also be increased for the Epidemiological Surveillance Program, which will allow for the expansion of training on infectious and communicable diseases and programmed supervisory visits to the states (SHCP, 2022). Therefore, this meager increase is more to elaborate regulation and training rather than for true surveillance campaigns.

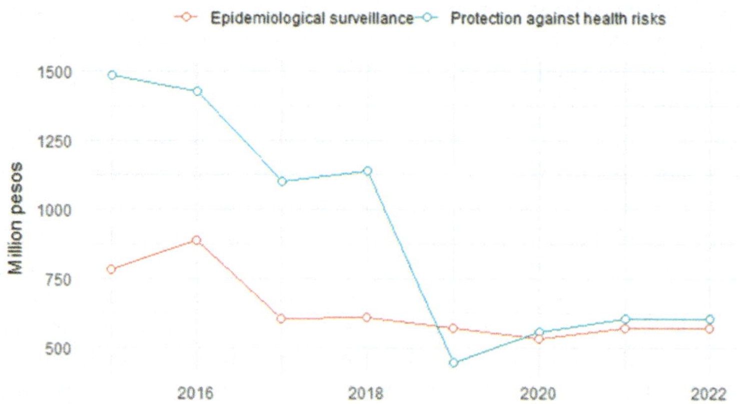

Fig. 6 Federal budget for prevention of sanitary risks and epidemiological surveillance (Million pesos, base year 2020). *Source* CIEP (2022)

5 Conclusions and Public Policy Recommendations

Climate change is already causing various health effects, especially in developing countries such as Mexico. The increase in the frequency of extreme meteorological events, together with the continuous environmental degradation, generates a considerable impact on the quantity and quality of water and food to which the population has access, on the quality of the air they breathe, on the distribution and propagation of vectors. This has induced changes in ecosystems that have negative consequences for human health.

Climate change related health impacts represent a great challenge for both scientists and decision-makers. Scientists have yet to quantify the impacts due to the large number of factors involved. Decision-makers, on the other hand, have to choose actions based on sound scientific knowledge that lead to sensible actions to face these challenges.

Mexico must consider the precariousness of its health sector and the large impacts climate change will have on health. Ignoring this will leave Mexico highly vulnerable to climate change on yet another front. Furthermore, data must be available to the broader public and research should be supported to determine the impacts climate change has been having on both the ecosystems and the population. Transparent access to information is vital for prevention.

Given the information reviewed, the Mexican government is not investing systematically in the health sector or on monitoring of diseases, much less in those related to climate change that could worsen in the face of more frequent and intense extreme weather events expected shortly and chronic conditions of environmental deterioration that will be added over time. The recent budget increases are not enough to make up for the huge cut in resources over the last 8 years.

References

Andela, N., Morton, D. C., Giglio, L., Chen, Y., Van Der Werf, G. R., Kasibhatla, P. S., DeFries, R. S., Collatz, G. J., Hantson, S., Kloster, S., Bachelet, D., Forrest, M., Lasslop, G., Li, F., Mangeon, S., Melton, J. R., Yue, C., & Randerson, J. T. (2017). A human-driven decline in global burned area. *Science, 356*(6345), 1356–1362. https://doi.org/10.1126/SCIENCE.AAL4108/SUPPL_FILE/FIREMIP_DATA.ZIP

Arce Romero, A., Monterroso Rivas, A. I., Gómez Díaz, J. D., Palacios Mendoza, M. Á., Navarro Salas, E. N., López Blanco, J., Conde Álvarez, A. C., Arce Romero, A., Monterroso Rivas, A. I., Gómez Díaz, J. D., Palacios Mendoza, M. Á., Navarro Salas, E. N., López Blanco, J., & Conde Álvarez, A. C. (2020). Crop yield simulations in Mexican agriculture for climate change adaptation. *Atmósfera, 33*(3), 215–231. https://doi.org/10.20937/ATM.52430

Ashworth, L., Quesada, M., Casas, A., Aguilar, R., & Oyama, K. (2009). Pollinator-dependent food production in Mexico. *Biological Conservation, 142*(5), 1050–1057. https://doi.org/10.1016/j.biocon.2009.01.016

Baez-Gonzalez, A. D., Kiniry, J. R., Meki, M. N., Williams, J. R., Alvarez Cilva, M., Ramos Gonzalez, J. L., & Magallanes Estala, A. (2018). Potential impact of future climate change on sugarcane under dryland conditions in Mexico. *Journal of Agronomy and Crop Science, 204*(5), 515–528. https://doi.org/10.1111/jac.12278

Barreda Marín, A. (2020, November 21). Del envenenamiento neoliberal a la reconstrucción socioambiental. *La Jornada Del Campo*, 3. https://www.jornada.com.mx/2020/11/21/delcampo/delcampo158.pdf

Bunn, C., Läderach, P., Ovalle Rivera, O., & Kirschke, D. (2015). A bitter cup: Climate change profile of global production of Arabica and Robusta coffee. *Climatic Change, 129*(1–2), 89–101. https://doi.org/10.1007/S10584-014-1306-X/FIGURES/5

Cámara de Diputados. (2022, October 26). Comisión de Salud emite opinión positiva sobre el presupuesto del Sector de Salud en 2023. *Boletín.* http://comunicacionsocial.diputados.gob.mx/index.php/boletines/comision-de-salud-emite-opinion-positiva-sobre-el-presupuesto-del-sector-de-salud-en-2023

Castillo, M., Pedernera, P., & Peña, E. (2003). Incendios forestales y medio ambiente: una síntesis global. *Rvista Ambiente y Desarrollo de CIPMA, XIX*(3 y 4), 44–53. https://keneamazon.net/Documents/Publications/Virtual-Library/GRFFS/18.pdf

CENAPRECE. (2017). *Dirección del Programa de Enfermedades Transmitidas por Vectores.* Gobierno de México. https://www.gob.mx/salud/cenaprece/acciones-y-programas/direccion-del-programa-de-enfermedades-transmitidas-por-vectores#:~:text=Las. ETV's representan un importante,ambientales que favorecen su transmisión.

CEPAL. (2022). *Panorama Social de América Latina y el Caribe, 2022.* https://repositorio.cepal.org/bitstream/handle/11362/48518/1/S2200947_es.pdf

Charre-Medellín, J. F., Mas, J.-F., Chang-Martínez, L. A., Charre-Medellín, J. F., Mas, J.-F., & Chang-Martínez, L. A. (2021). Potential expansion of Hass avocado cultivation under climate change scenarios threatens Mexican mountain ecosystems. *Crop and Pasture Science, 72*(4), 291–301. https://doi.org/10.1071/CP20458

CIEP. (2022). *Gasto en Salud y Objetivos de Desarrollo Sostenible (ODS).* https://ciep.mx/wp-content/uploads/2022/10/Gasto-en-Salud-y-Objetivos-de-Desarrollo-Sostenible.pdf

COFEPRIS. (2017). *Impactos del cambio climático en la salud*. Gobierno de México. https://www.gob.mx/cofepris/acciones-y-programas/impactos-del-cambio-climatico-en-la-salud

CONAGUA. (2020). *Programa Nacional Hídrico 2020–2024.* https://www.gob.mx/cms/uploads/attachment/file/642632/PNH_2020-2024__ptimo.pdf

CONAGUA. (2023). *Monitor de sequía en México.* CONAGUA. https://smn.conagua.gob.mx/es/climatologia/monitor-de-sequia/monitor-de-sequia-en-mexico

Cuervo-Robayo, A. P., Ureta, C., Gómez-Albores, M. A., Meneses-Mosquera, A. K., Téllez-Valdés, O., & Martínez-Meyer, E. (2020). One hundred years of climate change in Mexico. *PLoS ONE, 15*(7), e0209808–e0209808. https://doi.org/10.1371/journal.pone.0209808

de Jesús Coria-Lorenzo, J., Mirón-Calderón, X., Pla-Esquivel, E., Ortiz-García, E. R., Dávila-Chávez, E., & Mancilla-Bernal, L. E. (2018). La influenza en el Estado de México y su relación climática. De la pandemia a la estacionalidad. *Acta Pediátrica de México, 39*(5), 289–297.

Delgado, G. C., Gay y García, C., Imaz, M., & Martínez, M. A. (2010). *México frente al cambio climático: Retos y oportunidades* (UNAM (ed.)). UNAM.

Donatti, C. I., Harvey, C. A., Martinez-Rodriguez, M. R., Vignola, R., & Rodriguez, C. M. (2019). Vulnerability of smallholder farmers to climate change in Central America and Mexico: Current knowledge and research gaps. *Climate and Development, 11*(3), 264–286. https://doi.org/10.1080/17565529.2018.1442796/SUPPL_FILE/TCLD_A_1442796_SM1457.DOCX

Feria-Arroyo, T. P., Castro-Arellano, I., Gordillo-Perez, G., Cavazos, A. L., Vargas-Sandoval, M., Grover, A., Torres, J., Medina, R. F., de León, A. A. P., & Esteve-Gassent, M. D. (2014). Implications of climate change on the distribution of the tick vector Ixodes scapularis and risk for Lyme disease in the Texas-Mexico transboundary region. *Parasites & Vectors, 7*(1), 1–16.

Filho, W. L., Ternova, L., Parasnis, S. A., Kovaleva, M., & Nagy, G. J. (2022). Climate change and zoonoses: A review of concepts, definitions, and bibliometrics. *International Journal of Environmental Research and Public Health, 19*(2), 893. https://doi.org/10.3390/IJERPH19020893

Franklinos, L. H. V., Jones, K. E., Redding, D. W., & Abubakar, I. (2019). The effect of global change on mosquito-borne disease. *The Lancet Infectious Diseases, 19*(9), e302–e312. https://doi.org/10.1016/S1473-3099(19)30161-6

Garza Galván, S. G. (2016). *Primer Foro Nacional de Adaptación al Cambio Climático. Memorias* (Senado de la Republica (ed.); I). Senado de la Republica. https://www.senado.gob.mx/comisiones/biblioteca/docs/CAMBIO_CLIMATICO_INTERNET.pdf

Garza, M., Feria Arroyo, T. P., Casillas, E. A., Sanchez-Cordero, V., Rivaldi, C. L., & Sarkar, S. (2014). Projected future distributions of vectors of *Trypanosoma cruzi* in North America under Climate Change Scenarios. *PLOS Neglected Tropical Diseases, 8*(5), e2818–e2818. https://doi.org/10.1371/JOURNAL.PNTD.0002818

Gay y García, C., & Rueda Abad, J. C. (2015). *Reporte Mexicano de Cambio Climático GRUPO II Impactos, vulnerabilidad y adaptación*. https://www.researchgate.net/profile/Carlos-Gay-Garcia/publication/286250417_Reporte_Mexicano_de_Cambio_Climatico_Impactos_Vulnerabilidad_y_Adaptacion_Grupo_II/links/566725ac08ae8905db8b95e3/Reporte-Mexicano-de-Cambio-Climatico-Impactos-Vulnerabilidad-y-A

Godber, O. F., & Wall, R. (2014). Livestock and food security: Vulnerability to population growth and climate change. *Global Change Biology, 20*(10), 3092–3102. https://doi.org/10.1111/gcb.12589

Gómez-Ruiz, E. P., & Lacher, T. E. (2019). Climate change, range shifts, and the disruption of a pollinator-plant complex. *Scientific Reports, 9*(1), 1–10. https://doi.org/10.1038/s41598-019-50059-6

González, C., Wang, O., Strutz, S. E., González-Salazar, C., Sánchez-Cordero, V., & Sarkar, S. (2010). Climate change and risk of Leishmaniasis in North America: Predictions from ecological niche models of vector and reservoir species. *PLOS Neglected Tropical Diseases, 4*(1), e585–e585. https://doi.org/10.1371/JOURNAL.PNTD.0000585

Gutiérrez, J. M., Jones, R. G., Narisma, G. T., Alves, L. M., Amjad, M., Gorodetskaya, I. V., Grose, M., Klutse, N. A. B., Krakovska, S., Li, J., Martínez-Castro, D., Mearns, L. O., Mernild, S. H., Ngo-Duc, T., van den Hurk, B., & Yoon, J. H. (2021). Atlas. In B. Z. Masson-Delmotte, V., P. Zhai, A. Pirani, S.L. Connors, C. Péan, S. Berger, N. Caud, Y. Chen, L.Goldfarb, M.I. Gomis, M. Huang, K. Leitzell, E. Lonnoy, J. B. R. Matthews, T. K.Maycock, T. Waterfield, O. Yelekçi, R. Yu (Eds.), *Climate change 2021: The physical science basis. Contribution of working group I to the sixth assessment report of the intergovernmental panel on climate change*. Cambridge University Press. http://interactive-atlas.ipcc.ch/

Hellin, J., Bellon, M. R., & Hearne, S. J. (2014). Maize landraces and adaptation to climate change in Mexico. *Journal of Crop Improvement, 28*(4), 484–501. https://doi.org/10.1080/15427528.2014.921800

Hernandez-Ochoa, I. M., Asseng, S., Kassie, B. T., Xiong, W., Robertson, R., Luz Pequeno, D. N., Sonder, K., Reynolds, M., Babar, M. A., Molero Milan, A., & Hoogenboom, G. (2018). Climate change impact on Mexico wheat production. *Agricultural and Forest Meteorology, 263*, 373–387. https://doi.org/10.1016/J.AGRFORMET.2018.09.008

Imbach, P., Fung, E., Hannah, L., Navarro-Racines, C. E., Roubik, D. W., Ricketts, T. H., Harvey, C. A., Donatti, C. I., Läderach, P., Locatelli, B., & Roehrdanz, P. R. (2017). Coupling of pollination services and coffee suitability under climate change. *Proceedings of the National Academy of Sciences of the United States of America, 114*(39), 10438–10442. https://doi.org/10.1073/PNAS.1617940114/-/DCSUPPLEMENTAL/PNAS.201617940SI.PDF

Jackson, H. M., Johnson, S. A., Morandin, L. A., Richardson, L. L., Guzman, L. M., & M'Gonigle, L. K. (2022). Climate change winners and losers among North American bumblebees. *Biology Letters, 18*(6), 20210551. https://doi.org/10.1098/RSBL.2021.0551

Jat, M. L., Dagar, J. C., Sapkota, T. B., Yadvinder-Singh, Govaerts, B., Ridaura, S. L., Saharawat, Y. S., Sharma, R. K., Tetarwal, J. P., Jat, R. K., Hobbs, H., & Stirling, C. (2016). Climate change and agriculture: Adaptation strategies and mitigation opportunities for food security in South Asia and Latin America. In D. L. Sparks (Ed.), *Advances in agronomy* (vol. 137, pp. 127–235). Academic Press. https://doi.org/10.1016/bs.agron.2015.12.005

Lelieveld, J., Evans, J. S., Fnais, M., Giannadaki, D., & Pozzer, A. (2015). The contribution of outdoor air pollution sources to premature mortality on a global scale. *Nature, 525*(7569), 367–371. https://doi.org/10.1038/nature15371

Lin, B. B., Perfecto, I., & Vandermeer, J. (2008). Synergies between agricultural intensification and climate change could create surprising vulnerabilities for crops. *BioScience, 58*(9), 847–854. https://doi.org/10.1641/B580911

Liu, Q., Tan, Z.-M., Sun, J., Hou, Y., Fu, C., & Wu, Z. (2020). Changing rapid weather variability increases influenza epidemic risk in a warming climate. *Environmental Research Letters, 15*(4), 44004.

Liverman, D. M. (1990). Drought impacts in Mexico: climate, agriculture, technology, and land tenure in Sonora and Puebla. *Annals of the Association of American Geographers, 80*(1), 49–72. https://doi.org/10.1111/j.1467-8306.1990.tb00003.x

Magaña, V., Zermeño, D., & Neri, C. (2012). Climate change scenarios and potential impacts on water availability in northern Mexico. *Climate Research, 51*(2), 171–184. https://doi.org/10.3354/cr01080

Martinez-Austria, P. F. (2020). Climate change and water resources in Mexico. In *Water Resources of Mexico* (pp. 157–175). Springer.

Moo-Llanes, D., Ibarra-Cerdeña, C. N., Rebollar-Téllez, E. A., Ibáñez-Bernal, S., González, C., & Ramsey, J. M. (2013). Current and future niche of North and Central American sand flies (Diptera: Psychodidae) in climate change scenarios. *PLOS Neglected Tropical Diseases, 7*(9), e2421–e2421. https://doi.org/10.1371/JOURNAL.PNTD.0002421

Mora, C., McKenzie, T., Gaw, I. M., Dean, J. M., von Hammerstein, H., Knudson, T. A., Setter, R. O., Smith, C. Z., Webster, K. M., & Patz, J. A. (2022). Over half of known human pathogenic diseases can be aggravated by climate change. *Nature Climate Change, 12*(9), 869–875. https://doi.org/10.1038/s41558-022-01426-1

Moreno, T. A., & Huber-Sannwald, E. (2011). Impacts of drought on agriculture in Northern Mexico. In *Coping with global environmental change, disasters and security* (pp. 875–891). Springer. https://doi.org/10.1007/978-3-642-17776-7_51

Mundo-Molina, M. (2015). Climate change effects on evapotranspiration in Mexico. *American Journal of Climate Change, 04*(02), 163–172. https://doi.org/10.4236/AJCC.2015.42012

Murray-Tortarolo, G. N., & Jaramillo, V. J. (2019). The impact of extreme weather events on livestock populations: The case of the 2011 drought in Mexico. *Climatic Change, 153*(1–2), 79–89. https://doi.org/10.1007/s10584-019-02373-1

Murray-Tortarolo, G. N., & Jaramillo, V. J. (2020). Precipitation extremes in recent decades impact cattle populations at the global and national scales. *Science of the Total Environment, 736*, 139557. https://doi.org/10.1016/j.scitotenv.2020.139557

Murray-Tortarolo, G. N., Jaramillo, V. J., & Larsen, J. (2018). Food security and climate change: The case of rainfed maize production in Mexico. *Agricultural and Forest Meteorology, 253–254*, 124–131. https://doi.org/10.1016/j.agrformet.2018.02.011

Navarro, A. V., Vázquez, E. B., Casas, A. S., Ledezma, J. C. R., & Palafox, J. L. I. (2021). Infección por Dengue, un problema de salud pública en México. *Journal of Negative and No Positive Results, 6*(2), 293–306.

PAHO. (2021). *Agenda para las Américas sobre Salud, Medioambiente y Cambio Climático 2021–2030* (PAHO (ed.); I). PAHO. https://iris.paho.org/bitstream/handle/10665.2/55212/OPSCDECE210004_spa.pdf?sequence=1&isAllowed=y

Peña, R., Valdés, M., & Nazzal, C. (2022). Efecto de las temperaturas extremas en la incidencia de enfermedades cardiovasculares: Revisión bibliográfica 2016–2022. *Revista Chilena De Cardiología, 41*(1), 51–64.

Pörtner, H.-O., Roberts, D. C., Adams, H., Adler, C., Aldunce, P., Ali, E., Begum, R. A., Betts, R., Kerr, R. B., & Biesbroek, R. (2022). *Climate change 2022: impacts, adaptation, and vulnerability. Contribution of working group ii to the sixth assessment report of the intergovernmental panel on climate change*. 3056. https://doi.org/10.1017/9781009325844

Ramírez Sánchez, H. U., Fajardo Montiel, L. A., Ortiz Bañuelos, A. D., & De la Torre Villaseñor, O. (2022). The Agricultural Sector and Climate Change in Mexico. *Journal of Agriculture and Ecology Research International*, 19–44. https://doi.org/10.9734/jaeri/2022/v23i330222

Reid, C. E., Brauer, M., Johnston, F. H., Jerrett, M., Balmes, J. R., & Elliott, C. T. (2016). Critical review of health impacts of wildfire smoke exposure. *Environmental Health Perspectives, 124*(9), 1334–1343. https://doi.org/10.1289/EHP.1409277

Riojas Rodríguez, H., & Hurtado Díaz, M. (2015). *Cambio climático e impactos regionales en la salud humana en México.*

Rodriguez-Morales, A. J. (2013). Climate change, climate variability and brucellosis. *Recent Patents on Anti-Infective Drug Discovery, 8*(1), 4–12. https://doi.org/10.2174/1574891113805290791

Rodríguez-Trejo, D. A., Tchikoué-Maga, H., & Santillán-Pérez, J. (2007). Emisiones contaminantes duranre la temporada 2003 de incendios en México. *Revista Chapingo Serie Ciencias Forestales y Del Ambiente, 13*(1), 33–39.

Rodríguez, H. R., Díaz, M. H., Velandia, J. I., & Grameix, H. V. (2006). Estudio diagnóstico sobre los efectos del cambio climático en la salud humana de la población en México. *INE/INSP, México DF*.

Rupasinghe, R., Chomel, B. B., & Martínez-López, B. (2022). Climate change and zoonoses: A review of the current status, knowledge gaps, and future trends. *Acta Tropica, 226*, 106225. https://doi.org/10.1016/J.ACTATROPICA.2021.106225

Ryan, S. J., Carlson, C. J., Mordecai, E. A., & Johnson, L. R. (2019). Global expansion and redistribution of Aedes-borne virus transmission risk with climate change. *PLoS Neglected Tropical Diseases, 13*(3), e0007213.

Santillán-Fernández, A., Santoyo-Cortés, V. H., García-Chávez, L. R., Covarrubias-Gutiérrez, I., & Merino, A. (2016). Influence of drought and irrigation on sugarcane yields in different agroecoregions in Mexico. *Agricultural Systems, 143*, 126–135. https://doi.org/10.1016/J.AGSY.2015.12.013

Schroth, G., Laderach, P., Dempewolf, J., Philpott, S., Haggar, J., Eakin, H., Castillejos, T., Moreno, J. G., Pinto, L. S., Hernandez, R., Eitzinger, A., & Ramirez-Villegas, J. (2009). Towards a climate change adaptation strategy for coffee communities and ecosystems in the Sierra Madre de Chiapas, Mexico. *Mitigation and Adaptation Strategies for Global Change, 14*(7), 605–625. https://doi.org/10.1007/S11027-009-9186-5/FIGURES/6

SEMARNAT. (2013). *Estrategia Nacional de Cambio Climático. Visión 10–20–40.* https://www.gob.mx/cms/uploads/attachment/file/41978/Estrategia-Nacional-Cambio-Climatico-2013.pdf

SEMARNAT. (2015). *Comisión Intersecretarial de Cambio Climático (CICC).* Secretaría de Medio Ambiente y Recursos Naturales. https://www.gob.mx/semarnat/acciones-y-programas/comision-intersecretarial-de-cambio-climatico-cicc

SHCP. (2022). *Previsiones de Gasto Programable para 2023.* https://www.ppef.hacienda.gob.mx/work/models/8uLX2rB7/PPEF2023/mo2h2PK/docs/exposicion/EM_Capitulo_3.pdf

Sosenski, P., & Domínguez, C. A. (2018). El valor de la polinización y los riesgos que enfrenta como servicio ecosistémico. *Revista Mexicana de Biodiversidad, 89*(3). https://doi.org/10.22201/ib.20078706e.2018.3.2168

SSA. (2019a). *Informe semanal de casos nuevos de enfermedades.* Sistema Nacional de Salud. https://epidemiologia.salud.gob.mx/anuario/suive/Formato_SUIVE_12019.pdf

SSA. (2019b). *Programa Sectorial de Salud 2021–2024.* http://saludsinaloa.gob.mx/wp-content/uploads/2019/transparencia/PROGRAM_SECTORIAL_DE_SALUD_2019_2024.pdf

SSA. (2020). *Sistema Nacional de Vigilancia Epidemiológica.* Dirección de Vigilancia Epidemiológica. https://www.gob.mx/salud/acciones-y-programas/sistema-nacional-de-vigilancia-epidemiologica

SSA. (2021a). *Anuario de Morbilidad 1984–2021a.* Dirección General de Epidemiología. https://epidemiologia.salud.gob.mx/anuario/html/index.html

SSA. (2021b). *Manual de Procedimientos Estandarizados para la Vigilancia Epidemiológica Convencional.* https://epidemiologia.salud.gob.mx/gobmx/salud/documentos/manuales/32_ManualSuive.pdf

SSA. (2023). *Panorama Edpidemiológico de Dengue 2022.* Dirección General de Epidemiología. https://www.gob.mx/salud/documentos/panorama-epidemiologico-de-dengue-2022

Stephen, C. (2022). One Health most often has people as the primary beneficiary. How must One Health policies and practice change to make animal, plant and ecosystem health a primary focus that is influenced by human and environmental factors? *Research Directions: One Health, 1*, e3.

Tello, J., Garcillán, P. P., & Ezcurra, E. (2020). How dietary transition changed land use in Mexico. *Ambio, 49*(10), 1676–1684. https://doi.org/10.1007/s13280-020-01317

Thornton, P. K., Ericksen, P. J., Herrero, M., & Challinor, A. J. (2014). Climate variability and vulnerability to climate change: A review. *Global Change Biology, 20*(11), 3313–3328. https://doi.org/10.1111/gcb.12581

Torres Castillo, N. E., Melchor-Martínez, E. M., Ochoa Sierra, J. S., Ramirez-Mendoza, R. A., Parra-Saldívar, R., & Iqbal, H. M. N. (2020). Impact of climate change and early development of coffee rust—An overview of control strategies to preserve organic cultivars in Mexico. *Science of The Total Environment, 738*, 140225. https://doi.org/10.1016/j.scitotenv.2020.140225

Tyukavina, A., Potapov, P., Hansen, M. C., Pickens, A. H., Stehman, S. V, Turubanova, S., Parker, D., Zalles, V., Lima, A., & Kommareddy, I. (2022). Global trends of forest loss due to fire from 2001 to 2019. *Frontiers in Remote Sensing, 3.*

Ureta, C., González, E. J., Espinosa, A., Trueba, A., Piñeyro-Nelson, A., & Álvarez-Buylla, E. R. (2020). Maize yield in Mexico under climate change. *Agricultural Systems, 177*, 102697. https://doi.org/10.1016/J.AGSY.2019.102697

USAID. (2017). *Climate risk profile Mexico.* https://www.climatelinks.org/sites/default/files/asset/document/2017_USAID_ClimateChange Risk Profile_Mexico.pdf

WHO. (2015). *Climate and Health Country Profile. Mexico.* https://apps.who.int/iris/bitstream/handle/10665/246116/WHO-FWC-PHE-EPE-15.26-eng.pdf

WHO. (2022). *Interactive platform visualizing mortality data.* WHO Mortality Data Base. https://platform.who.int/mortality

Zarza, H., Martínez-Meyer, E., Suzán, G., & Ceballos, G. (2017). Geographic distribution of Desmodus rotundus in Mexico under current and future climate change scenarios: Implications for bovine paralytic rabies infection. *Veterinaria México OA, 4*(3), 10–25. https://doi.org/10.21753/vmoa.4.3.390

Climate Change in Argentina. Implications on Health

Daniel Oscar Lipp

Abstract Climate change, caused by the increase in Greenhouse Gas (GHG) emissions, is inducing significant climatic alterations in Argentina, so it is a priority to deepen the knowledge of its impact, particularly on human health. In this context, the main diseases that cause serious threats to the population are presented. Aspects as relevant as the effects that extreme temperatures expressed as heat waves and cold waves, air or water quality and the possible spread of diseases have on the morbidity and mortality of the population are addressed. It is concluded that the health effects caused by climate change require new strategies to mitigate them, with a multidisciplinary and intersectoral approach, where prevention and health promotion actions are essential in addressing this issue. In addition, the article intends to present a summary of the mitigation and adaptation measures applied in Argentina on the health sector. Both measures are complementary and, although they present different challenges, they converge in the final objective.

Keywords Infectious diseases · Climate change · Heat waves · Climate vulnerability · Dengue · Cholera

1 Introduction

The carbon dioxide content in the atmosphere has increased remarkably since the Industrial Revolution. It is believed that since 1860, the global atmospheric concentration of carbon dioxide went from the pre-industrial value of 280 ppmv to the current level of 380 ppmv (Smith and Smith, 2007, p. 42). Experts estimate that the continuation of this trend would cause the Earth to warm up more than normal and trigger climate change. There is already overwhelming evidence that the annual

D. O. Lipp (✉)
Universidad Católica de Salta, Buenos Aires, Argentina
e-mail: danieloscarlipp@yahoo.com.ar

Universidad de Buenos Aires, Buenos Aires, Argentina

Universidad del Salvador, Buenos Aires, Argentina

R. Akhtar (ed.), *Climate Change and Human Health Scenarios*, Global Perspectives on Health Geography, https://doi.org/10.1007/978-3-031-38878-1_26

global average temperature in this period rose by about 0.7 °C, as global carbon emissions increased. Argentina, for its part, is already suffering some of the many effects of global climate change, such as an increase in the average annual temperature, changes in the rainfall pattern, the rise in the average sea level, the reduction of the cryosphere, and modifications in the patterns of extreme weather events (IPCC, 2013, p. 25). These climatic transformations are causing significant economic, social, environmental, and health effects in Argentina. The most recent report from the Intergovernmental Panel on Climate Change (IPCC, 2022, p. 38) paints a worrying picture: Climate change is already affecting all corners of the world with only 1.1 °C of warming, and far greater impacts are ahead, severe if we fail to halve our greenhouse gas (GHG) emissions in this decade and immediately scale up adaptation efforts. The 2022 IPCC report also details which climate adaptation approaches are most effective and feasible, as well as which groups of people and ecosystems are most vulnerable.

Among the effects of climate change in the country, we find the increase in average temperatures. Between the years 1960 and 2010, the average temperature in the country has increased by about 0.5 °C in the central-northern region of Argentina. The minimum temperatures increased approximately 1 °C, and the maximums decreased almost in the same proportion during that time. The region of the country that has suffered the greatest increase in temperatures compared to other parts of the territory is the Andean region of Patagonia and Cuyo, reaching temperatures above 1 °C in certain areas. On the other hand, summers tend to be longer and winters less harsh. In addition, the occurrence of frosts has decreased, and the frequency of heat waves has increased. Future estimates indicate that warming could reach 0.5 and 1 °C in most of the Argentine territory. These values could even be higher in the extreme northwest of the country. Toward the end of this century, the Pampas and Patagonia region will present a warming of 1.5 °C, the center-north and the Cuyo region 2 °C and the northwest could exceed 2.5 °C (Camilloni, 2018, p. 42). Regarding annual rainfall, between 1961 and 2016, in the central-eastern region of the country, they increased between 10 and 40% and the largest increases occurred in the center of the provinces of Santa Fe, Entre Ríos, and Misiones. Due to the increase in rainfall, there were numerous floods on the banks of the Paraná and Uruguay rivers and also flooding in lowlands that affected numerous provinces. On the contrary, in the northwest region, Cuyo and Patagonia, rainfall has decreased considerably, causing prolonged droughts. Regarding the fluvial systems of the region since the 1980s, the San Juan, Atuel, Negro, Limay, Neuquén, and Colorado rivers show a reduction in their annual flows. These trends indicate the decrease in the masses of water stored in the glaciers present in the high mountains and generate a potential risk of water deficit in these sites. Regarding the rise in sea level in Argentina, it could generate saline intrusions in some areas of the Province of Buenos Aires. These could reach important aquifers that provide fresh water for human consumption (Madanes et al., 2008, p. 25). With these changes, of course, heat-related illnesses, vector-borne, and waterborne infectious diseases will increase. According to the IPCC, higher temperatures will also allow the spread of vector-borne diseases, such as West Nile virus, Lyme disease, and malaria, as well as waterborne diseases such as cholera (IPCC, 2022, p. 38). The objective of this document is the study of some selected events in the field of health and that impact our country in a special way. The impacts of climate change on health can be direct and indirect, understanding as direct effects those that occur as a result of the occurrence of extreme weather events such as cold and heat waves, floods, droughts, and strong winds. The indirect effects of climate on health, on the other hand, are those mediated by alterations in ecological or human systems. This group may include vector- and rodent-borne diseases, diseases transmitted by water or food, associated respiratory diseases to air pollution, malnutrition, and occupational risks modified by changes in climate.

2 Direct Impacts on Health. Waves of Heat and Cold

If we were to give a general classification of these health impacts, according to whether they are directly or indirectly related to the increase or decrease in temperature, heat waves and cold waves would cause major disasters. These would be aggravated by the presence of humidity levels higher than the current ones. Table 1 shows the potential impacts, both direct and indirect, that tend to affect the organism due to climate change.

2.1 Heat Waves

Heat waves are extreme situations. Although there is no universal definition of heat waves, this phenomenon is understood as a prolonged period of unusually hot weather, lasting at least 3 days, generally with an appreciable impact on human and natural systems. In Argentina, the National Meteorological Service defines heat waves as the period in which the maximum and minimum temperatures equal or exceed, for at least 3 consecutive days and simultaneously, the 90th percentile, calculated from daily data for the months of October to March (warm semester in the southern hemisphere) from the period 1961 to 2010 (S.M.N., 2019, p. 26).

Heat waves are extreme events that have increased significantly in recent decades, although the phenomenon has been observed much earlier but not with the frequency and intensity of recent summers, especially in the Southern Hemisphere. In Argentina, due to climate change, heat waves increased in the north and center of the country, as evidenced by the hospitalizations and mortality registered in that period. For example, a recent study reported the high mortality rate due to the heat waves that occurred in the center and north of the country in the summer of 2013–2014. In said summer, three heat wave events were registered: one in December that reached 16 provinces, one in January that affected 14 provinces, and one in February with 4 affected provinces. These heat waves presented characteristics of severity in terms of their duration and number of provinces reached, registering an excess of

Table 1 Climate impacts on health

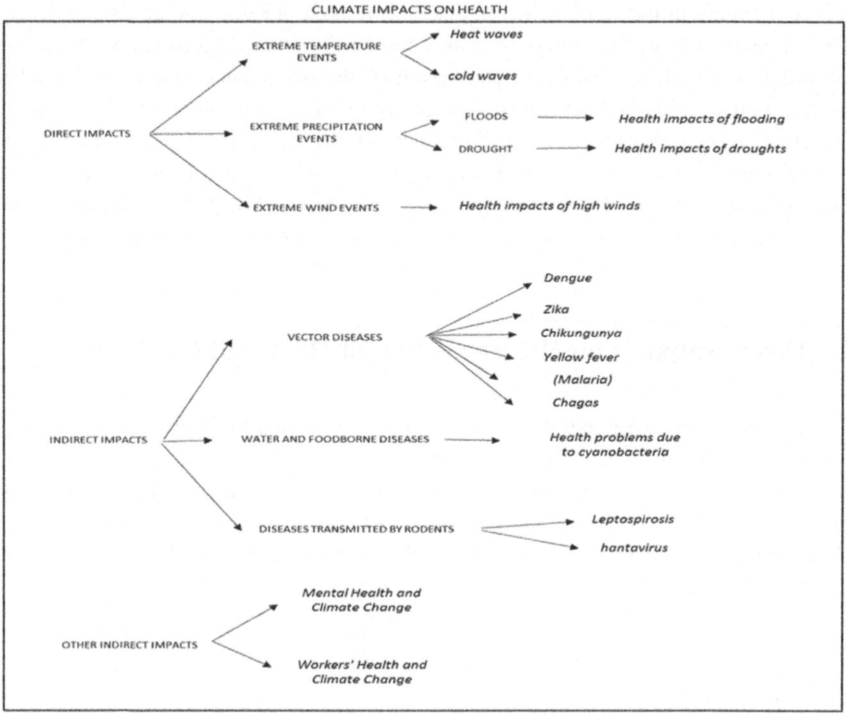

Source Elaboration of the author

1877 deaths with respect to the normal for the period (S.M.N., 2013, p. 58). Figure 1 shows the frequency of heat waves in Argentina during the period 2005–2017.

2.2 *Cold Waves*

However, just like what happens with heat waves, cold waves also increased notably in Argentina. In the south of our country, several locations—such as Río Gallegos, Puerto Santa Cruz, or El Calafate—were affected in July 2022 by intense cold waves, according to the National Meteorological Service (SMN). For example, southern Patagonia on July 4 was affected by an air mass with very low temperatures. In some cities, the maximum temperatures were close to 0 °C, and the minimum temperatures were below − 10 °C. As a result of this, our territory experienced the fifth coldest autumn in the last 62 years (Fig. 2). Although climate change generates extreme heat waves, it is also associated with extreme cold waves. In truth, there is no valid and uniform criterion to define a cold wave (S.M.N., 2018, p. 51). However, when the maximum temperatures are lower than a certain threshold, mortality increases

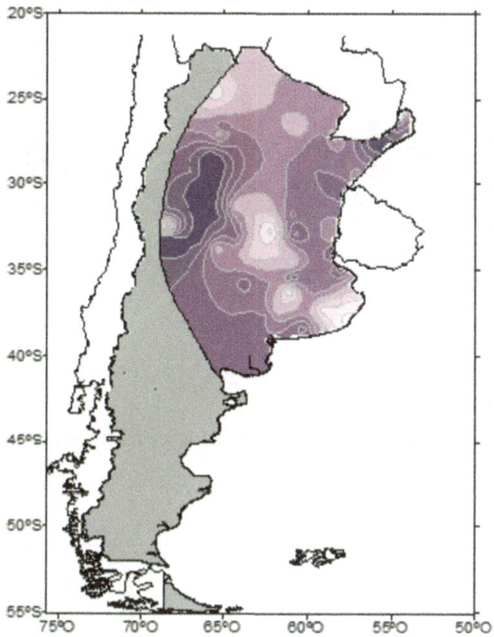

Fig. 1 Frequency of heat waves in Argentina (2005–2017). The gray area is excluded from the health and heat wave early warning system. *Source* National Weather Service

significantly. The National Meteorological Service defines a cold wave as the excessively cold period in which the maximum and minimum temperatures are equal to or lower than the 10th percentile of the cold semester (April–August), for at least three consecutive days and simultaneously. The health impacts of cold snaps include direct effects such as hypothermia and indirect effects such as increased rates of pneumonia, influenza-like illness, and other respiratory illnesses.

Epidemiological evidence suggests that cold waves also generate increases in mortality patterns from ischemic heart disease, cerebrovascular disease, and respiratory disease. In 2007 in Argentina, considered one of the coldest years, cold waves occurred in almost the entire center and north of the country. Figure 3 shows the frequency of cold waves that occurred in Argentina in the period 2005–2017.

3 Indirect Impacts on Health. Malaria

In particular, I will deal here with the numerous pathologies, mostly infectious, that will emerge and re-emerge due to climate change (Table 2), many of them considered eradicated. I will do so with special emphasis, referring above all to developing countries, among which Argentina is included, possibly the most affected by these pathologies.

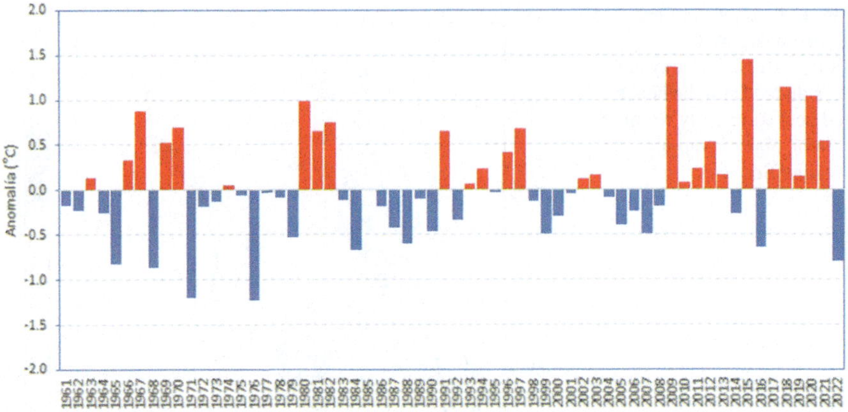

Fig. 2 Average temperature anomaly at the country level. Autumn (1961–2022). *Source* National Weather. https://www.smn.gob.ar/

Fig. 3 Frequency of cold waves in Argentina (2005–2017). The gray area is excluded from monitoring of cold waves due to climatological criteria. *Source* National Weather Service

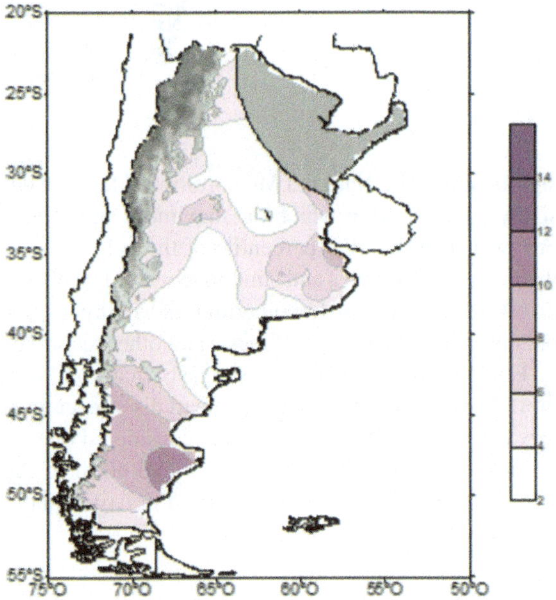

Infectious diseases, both those transmitted by vectors or by water and food, are sensitive to variations in climate change, they can be affected by various factors such as temperature, humidity, rainfall pattern, and wind. These meteorological phenomena certainly influence the reproduction and maturation of vectors, and on the pathogen itself, which requires temperatures and humidity that favor its development, although logically a factor of undoubted significance for the epidemic outbreak

Table 2 Main vector-borne tropical diseases and probability of changes in their distribution as a result of climate change

Disease	Vector	Number of people at risk (millions)	Number of people infected or new cases per year	Current distribution	Probability of distribution modification due to climate change
Malaria	Mosquito	2400	300–500 million	Subtropical tropics	+++
Schistosomiasis	Aquatic snail	600	200 million	Subtropical tropics	++
Lymphatic filariasis	Mosquito	1094	117 million	Subtropical tropics	+
African trypanosomiasis	Tsetse fly	55	250,000–300,000 cases per year	Tropical Africa	+
Dracunculiasis	Crustaceans (copepod)	100	100,000 years	South Asia, Middle East, Central, and West Africa	?
Leishmaniasis	Sandflies	350	12 million infected, 500,000 new cases	Asia, Southern Europe, Africa, America	+
Onchocerciasis	Blackflies	123	17.5 million	Africa, Latin America	++
American trypanosomiasis	Triatomids	100	18–20 million	Central and South America	+
Dengue	Mosquito	2500	50 million	Subtropical tropics	++
Yellow fever	Mosquito	450	< 5000 cases year	South America and tropical Africa	++

+ = probable; ++ = very likely; +++ = highly probable; = it is unknown
Fuente: OMS (1996)

to occur is the unhealthy setting, lack of hygiene, limited access to drinking water supplies, and insufficient medical care offer all cause the disease to spread.

A case that is particular is malaria or paludism, which generates more than a million deaths per year in developing countries. Surely as temperatures rise and rainfall patterns alter, the Anopheles mosquito will expand its habitat to higher latitudes and altitudes considered malaria-free areas. This will increase the number of people susceptible to contracting the disease. On the other hand, as the environmental temperature increases, the *plasmodium* takes less time to develop inside the anopheles and this would increase its transmissibility. We should indicate that most of the vectors of this type of disease are cold-blooded arthropods highly sensitive to environmental temperatures. These increase speed up your metabolism, increase egg production, and the need to feed. On the other hand, the rains have an indirect effect on the longevity of the vector, due to the increase in humidity, which creates a favorable habitat for its development. Malaria transmission is clearly influenced by climate, since malaria transmission does not occur in climates where the mosquito cannot survive. However, in recent years, adaptive changes have been observed in the insect, such as surviving at higher than usual heights, above 2600 m, and at lower temperatures (up to 8 °C), as was recently observed in Bolivia in 2008, where tails were found of *Anopheles pseudopunctipenis*, in Oruro, at about 3710 m of altitude. Of course, climate change is not the only variable that will aggravate the expansion of this pathology, a series of complementary explanatory factors are also postulated, such as the increase in resistance to antimalarials, cessation of vector control measures, deforestation, and human migrations (Brandt et al., 2018, p. 21).

In addition, the risk of getting sick and dying from this disease is closely related to the socioeconomic conditions of the affected country. In Argentina, these conditions, also described as "*social variables*", are low family income, "*suboptimal*" living conditions that imply vector risk, the mother's schooling, overcrowding, inadequate access to health care, and the type of health system, among other social, environmental, and economic variables. It has been proven that this infection slows down social progress in our country, by affecting populations with socioeconomic problems more. In fact, it has been affirmed that malaria constitutes a "*poverty trap*", to the extent that it increases vulnerability and exacerbates poverty in families. For example, the material conditions of the dwelling constitute an important risk factor in Argentina: case–control studies have shown the risk associated with unplastered mat and adobe walls and homeless areas, as well as the proximity of rivers, ditches, and fruit crops. In addition, housing is an indicator of the economic situation of the family. It also constitutes a means to satisfy biological, psychological, and social needs, and is one of the first scenarios on which action can be taken to mitigate the risk of malaria illness and to promote health.

In relation to schooling and income, some studies in our country report economic losses due to malaria and consider this disease as a "*proxy*" of poverty and school retardation. However, in research on malaria, it is not frequent to find studies on its relationship with conditions or lifestyles. In general, there is little experience in the study of the social determinants of health, applied to this disease. Figure 4 shows the malarious areas in Argentina.

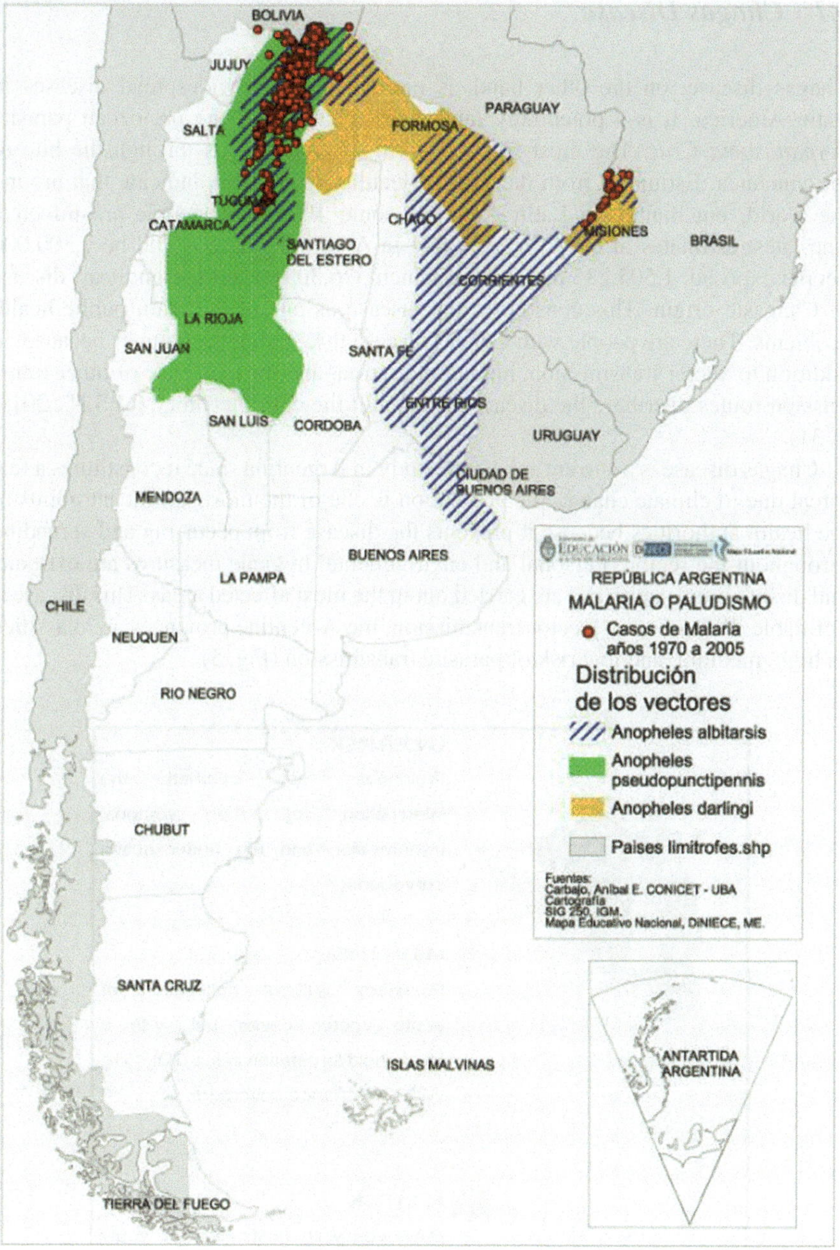

Fig. 4 Malaria areas in Argentine territory. *Source* Ministry of education. Available at: http://www.mapaeducativo.edu.ar/atlas/mapas-de-contexto/salud-yeducacion/malaria/

3.1 Chagas Disease

Chagas disease, on the other hand, is one of the most widespread diseases in Latin America. It is a potentially fatal disease caused by the protozoan parasite *Trypanosoma Cruz*. The most frequent form of contagion is through the bite of the vinchuca. Estimates from the World Health Organization indicate that around the world, but mainly in Latin America, some 10 million people are infected. The latest estimates of cases indicate that in Argentina there would be 7,300,000 people exposed, 1,505,235 infected, of which 376,309 would present heart disease of Chagasic origin. This constitutes the disease as one of the main public health problems. There are people with Chagas disease throughout the country because, in addition to vector transmission, human migration, and the existence of other transmission routes distribute the disease throughout the entire territory (M.S.N., 2019, p. 31).

Chagas disease is a current subject of study in Argentina since it constitutes a real threat due to climate change. Its prevention is one of the most significant points of the health authorities because it prevents the disease from occurring and spreading throughout the region. Personal and environmental hygiene measures are extreme, and disinfection campaigns are carried out in the most affected areas. This disease is notifiable. Regarding its vector transmission, the Argentine provinces are classified as high, medium, and low risk of parasite transmission (Fig. 5).

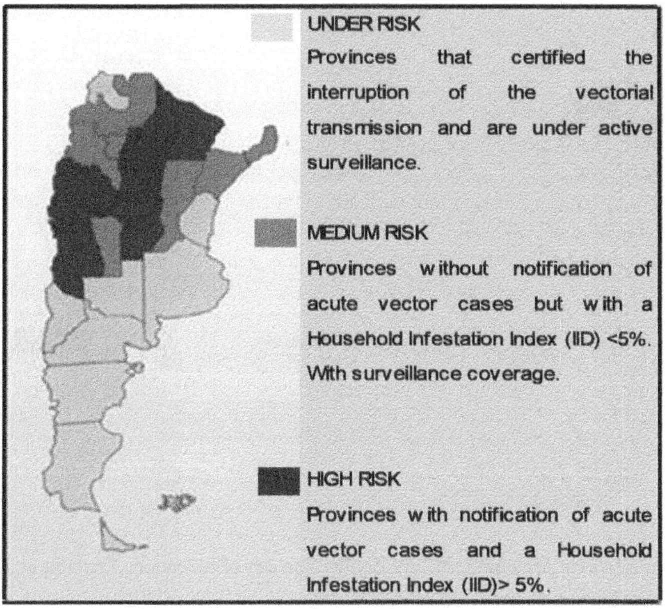

Fig. 5 Mal de Chagas in Argentina. *Source* Argentina. Epidemiology Department. Ministry of Health of the Nation

There are also some so-called risk-free zones due to the magnitude of the number of existing vectors. In which areas of the country does the disease exist? In our country, it is found in desert or semi-desert areas, its appearance being sporadic in more humid regions. The provinces that present a significant home infestation by this vector are Chaco, Catamarca, Formosa, Santiago del Estero, San Juan, and Mendoza, and to a lesser extent Córdoba, Corrientes, La Rioja, Salta, and Tucumán (M.S.N., 2015, p. 21).

3.2 *Dengue*

Another evil is dengue, which will worsen with climate change, a viral disease also transmitted by mosquitoes. The virus is spread by the female *Aedes Aegypti* mosquito or other species of the Aedes genus *(Aedes Albopictus),* widespread in the tropics. These mosquitoes are also vectors of the chikungunya, yellow fever, and Zika viruses, making exposure to them extremely dangerous. Although dengue disease is favored by very hot and humid climatic conditions, the severity of the risk, however, depends locally on social and environmental factors. Severe dengue carries a higher risk of death if not properly treated. This was first identified in the 1950s, during an epidemic in the Philippines and Thailand. Today, it affects most of the countries of Asia and Latin America and has become one of the main causes of hospitalization and death among children and adults in these regions (O.P.S., 2019, p. 54).

The first outbreak of dengue, on the other hand, that affected our country occurred in 1916, causing, especially in the provinces of Corrientes and Entre Ríos, infectious sources of relative importance. The disease had been introduced from Paraguay, which was suffering from a widespread epidemic of classic dengue. Fortunately, it did not appear again until 1998 when a new epidemic outbreak devastated the Chaco region of Salta, with its epicenter in the city of Tartagal, and caused several hundred cases, all of them classic dengue. This time, the disease had been brought from Bolivia to Argentina. Likewise, from December 1999 to May 2000, the Muñiz Hospital in Buenos Aires assisted more than 50 cases of dengue fever imported from Paraguay. Other serotypes also began to circulate on the northwestern border, with reduced epidemic outbreaks until, in 2004, an extended outbreak occurred in some cities of the Chaco of Salta, such as Salvador Maza, Orán, Tartagal, Embarcación Aguaray, and Pichanal, with thousands of cases of dengue. The situation on the northwest border, however, worsened in 2006 due to the floods suffered by Tartagal.

Both in northwestern Argentina and in Iguazú (Misiones), there were also outbreaks of malaria and dengue. In the Salta Chaco in 2006, there were 55 cases of dengue and 96 of malaria and in Iguazú 90 of dengue and 13 of malaria. In this last place, the cases were mostly imported through the transit of people in the so-called "triple border". Due to the epidemic situation in Paraguay, the neighboring Argentine provinces are considered high risk, especially the province of Formosa where the abundance of the Aedes Aegypti mosquito continues to be high. Regarding the Metropolitan Region of Buenos Aires, compared to the scenarios seen before, the

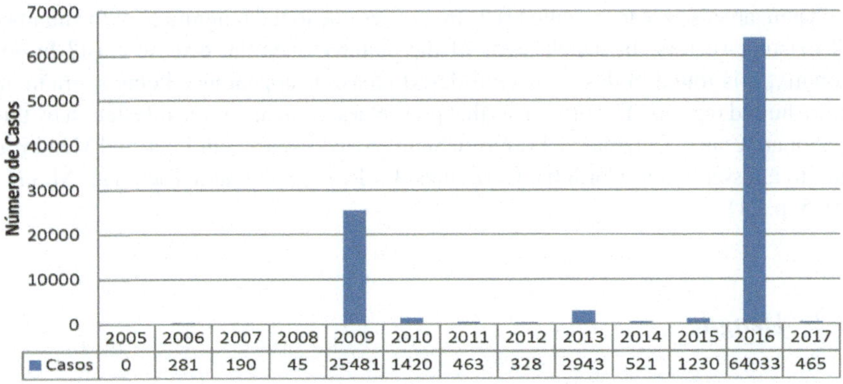

Fig. 6 Number of dengue cases, Argentina (2005–2017). *Source* Autonomous city of Buenos Aires. Ministry of Health of the Nation, 2019. "Climate and health in Argentina: diagnosis of the 2019 situation". Buenos Aires. https://bancos.salud.gob.ar/recurso/clima-y-salud-en-la-argentina-diagnostico-de-situacion-2019

risk is still low. However, global climate change will transform these low-risk areas into regions with autochthonous transmission in a few years (M.S.N., 2014, p. 18).

Dengue currently affects between 50 and 100 million people annually and will increase in developing countries as climate change intensifies. As in the case of malaria, its expansion will occur thanks to progressively warmer and more humid climates. But also promoted by the growing unplanned urbanization and whose cities offer favorable habitats for the development of the larvae of the transmitting mosquito. Such is the situation in Latin America, a region with 77% of the urban population and where a considerable number of people live in highly urbanized communities with deficient sanitary networks. Figure 6 shows the cases of dengue in Argentina from 2005 to 2017.

4 Infectious Diseases Transmitted by Water and Food

Water is a carrier agent for pathogenic microorganisms. Studies have been undertaken in this regard about the toxicity caused by these agents and that can endanger health and life. The most frequent of these pathogens in water are mostly responsible for infections in the intestinal tract such as typhoid and paratyphoid fever, dysentery, and cholera, among others. Modern disinfection techniques have greatly diminished this danger in some countries, yet in others, especially in the developing world, these ailments, such as cholera, are still very common. We should underline that waterborne pathologies are prone to the effects of climate change. For example, the survival and persistence of these microorganisms are directly affected by temperature. It has been proven that the causative agents of acute gastroenteritis multiply faster in warmer

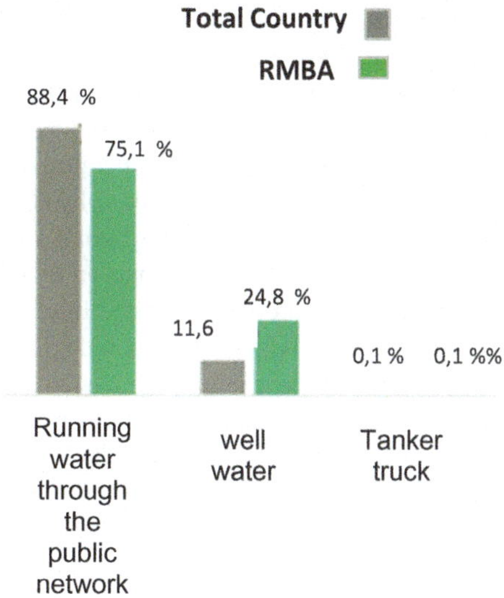

Fig. 7 Different types of access to drinking water in the Argentine Republic. Total of the country and metropolitan region of Buenos Aires (RMBA) (In percentages). *Source* EDSA-Equity Agenda Series (2017–2025)

conditions, and on the other hand, a positive and significant correlation was found between the incidence of cholera and the rise in temperature.

An increase in the frequency of outbreaks and epidemics of diseases such as cholera and typhoid fever is to be expected if water quality deteriorates or deteriorates. These diseases will find a more propitious climate for their expansion with the increase in floods or the scarcity of drinking water.

Only half of Argentines have access to drinking water. According to official estimates, the deficit reaches seven million people and has a greater impact on the most vulnerable areas. In turn, the households that have water inside the dwelling are broken down according to the source of the water, that is, if it comes from the public piped water network, from drilling with a motor pump, from drilling with a manual pump or another source (Fig. 7). The strong growth of cities with between 100,000 and more inhabitants in the metropolitan region not only exacerbated the deficit of network water and sewers in areas with low service coverage, particularly in the Buenos Aires suburbs. Minors are the ones who suffer the most from the drinking water deficit. A 39.9% of those who do not have a guaranteed right to water are boys and girls between the ages of 5 and 11. The regions that require immediate attention are the already historical situations in the north and west of the country, while in the province of Buenos Aires, the needs are concentrated in the suburbs.

4.1 Diarrheal Diseases

Diarrhea remains one of the leading causes of infant death in the world. According to the latest IPCC report for 2022, the increased risk of flooding could cause up to 48,000 additional deaths in children under 15 years of age in 2030 due to diarrhea (IPCC, 2022, p. 36). Temperature increases associated with climate change are expected to greatly increase the global risk of diarrhea by the end of the century. These diseases are more frequent in summer due to the climate that favors the spread of the bacteria that cause them. With high temperatures, the risk of dehydration increases. These infectious diseases are transmitted by the ingestion of infected fecal matter particles through the consumption of contaminated food or water. Evidence was found that the impact of climate change on diarrhea outbreaks could differ depending on the underlying source of infection. In general, diarrhea-causing bacterial infections increased in areas with hot and humid conditions, while viruses became less prevalent in these regions. For example, the risk of *E. coli* infections increased with increasing temperatures, while that of rotavirus became more common in cold periods. In the region of the America, diarrheal diseases are among the five causes of death in all ages in 17 countries. In our country during the year 2000, the notification of diarrhea affected almost 18% of the population under 6 years of age. In some regions, such as the northern region, diarrhea was reported in 25% of patients under 6 years of age, with extreme examples such as Salta and Jujuy, in which the number of episodes exceeded 50% of said population (M.S.N., 2011, p. 56). Figure 8 shows the numbers of diarrhea cases in the period 2005–2017.

Acute diarrhea is one of the most serious public health problems in Argentina because it is generally associated with unfavorable living conditions. In Argentina, hospital discharges due to intestinal infectious diseases represented 9% of all discharges in children under 5 years of age. Likewise, for this event, the greatest burden was observed in children under 2 years of age (around 75% of discharges). There is a broad consensus in highlighting the advantages that basic sanitation services provide for the health and quality of life of the population, and that these benefits are enhanced by other factors related to the infrastructure of the home,

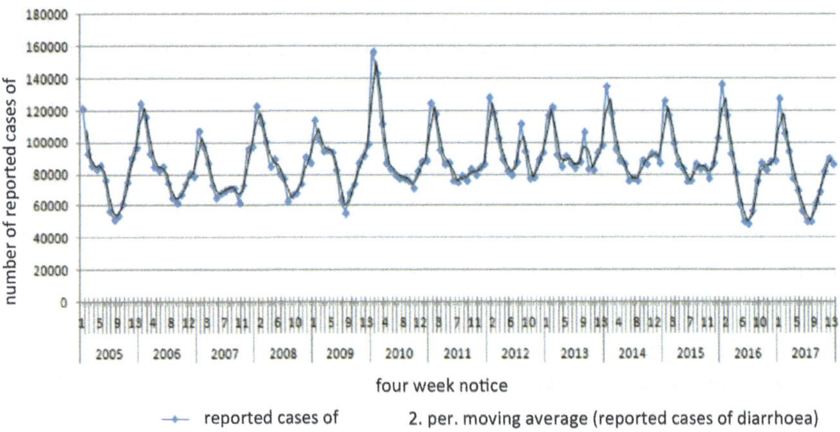

Fig. 8 Number of Diarrhea cases, Argentina (2005–2017). *Source* Surveillance area, National Directorate of Epidemiology and Analysis of the Health Situation

the socioeconomic condition of its inhabitants, and the environmental conditions of the environment that surrounds said dwelling (including both the natural and socioeconomic environment).

5 Health Mitigation and Adaptation Measures in the Face of Climate Change in Argentina

When it comes to fighting climate change to prevent the impacts it causes on the different systems of the planet, human beings apply two types of measures: mitigation and adaptation. Mitigation measures are those actions that are aimed at reducing and limiting greenhouse gas emissions, while adaptation measures are based on reducing vulnerability to the effects of climate change. Mitigation, therefore, deals with the causes of climate change, while adaptation addresses its impacts.

The Argentine economy is highly vulnerable to changes in temperature and precipitation due to its productive profile, considering that it depends on the primary agro-export sector for 19% of GDP and the manufacturing industries associated with said sector, as well as on the production of electrical energy from hydroelectricity (40%) in the slopes of the rivers originating in the Andes Mountains (Cinquantini et al., 2016, p. 23). Hence, the mitigation and adaptation measures to climate change are imperative in Argentina. The next objective is then to synthesize in Argentina some of the main public policy measures in terms of adaptation and mitigation to climate change.

5.1 Health Mitigation Measures

Climate change mitigation refers to anthropogenic intervention to reduce sources or enhance sinks of greenhouse gases. For this reason, the health sector, in addition to fulfilling its mission of protecting human health through the actions carried out in the provision of services, must understand and identify the sources of carbon due to the different activities that are carried out there without compromising the quality of care and health services provided. It is an actor that has a fundamental role in contributing to the solution of this problem, where, like other sectors of society, it must align its actions and development trajectory with the Paris Agreement in order to support the prevention of possible impacts due to climate change (Sánchez & Reyes, 2015, p. 47). The mitigation measures identified in Argentina within this line have been grouped in Table 3.

Table 3 Climate change mitigation measures in Argentina in the health sector

Redesign lighting systems, replace traditional luminaires with LED or high efficiency, automate, and implement good practices
Implementar sistemas de aire acondicionado y de refrigeración, eficientes energéticamente y libres de sustancias agotadoras de ozono, de bajo potencial de calentamiento global
Implement air conditioning and refrigeration systems that are energy efficient and free of ozone-depleting substances, with low global warming potential
Conversion of boilers or equipment that run on traditional fossil fuels to fuels such as natural gas or cleaner fuels
Promotion of the use of vehicles that use low or zero emission technologies (electric, hybrid, etc.)
Promote the installation of charging stations for electric vehicles in public hospitals
Implement telemedicine throughout the health system and report results of virtual medical examinations
Plan and optimize transport routes for supplies, biologicals, patients, and staff
Digitize the majority of documents both in the administrative area and in the provision of the health service
Ensure the correct separation of waste at the source, define which can be recovered such as glass containers for medicines, serum bags, among others
Promote high-efficiency final waste treatments such as autoclaves or physical–chemical deactivation and eliminate/reduce incineration
Treat wastewater in accordance with applicable regulations and at least install grids, sieves, and filters to separate solids before discharge into the drain
Promote water resource reuse processes

Source Sánchez and Reyes (2015)

5.2 Adaptation Measures in Health

In general, adaptation to climate change is defined as "*adjustments in natural or human systems in response to projected or actual climatic stimuli, or their effects, that can moderate damage or take advantage of their beneficial aspects*" (IPCC, 2007, p. 85). In this sense, adaptation measures aim to work on the consequences of climate change, reducing the vulnerability of each sector, and consequently reducing risk. There is currently a wide range of adaptation options in the health sector in Argentina, some of which are summarized in Table 4.

Table 4 Climate change adaptation measures in Argentina in the health sector

Strengthen the health system in the face of heat waves
Strengthen the response of the health system to cold waves
Strengthen the response of the health system and local communities to floods
Strengthen the health system to respond and communities to prevent mosquito-borne diseases
Strengthen the resilience of healthcare facilities in the face of extreme weather events
Fortalecer los sistemas de monitoreo vinculados a impactos del cambio climático sobre la salud
Prioritize health issues in the treatment of the emergency in the face of extreme events of climate change
Analyze the impact of climate change on the branches of productive activity and services in relation to occupational health
Centralize statistics on climate-related infectious diseases at the national level and prepare response programs to the appearance of infectious foci
Prevention of waterborne diseases
Implement surveillance and epidemiological control plans mainly due to the increase in temperature

Source Sánchez and Reyes (2015)

6 Conclusion

As a result of the results presented in this research, health in Argentina is severely compromised due to the probable impact that will occur with climate change. This disturbs natural ecosystems and favors ideal conditions for the spread of infections and epidemics in Argentina. It is expected that the intensification of vector diseases will have a strong impact on the country and will be more frequent and more intense in northern Argentina. On the other hand, it will affect the appearance and aggravation of respiratory and cardiovascular diseases and cancers. On the other hand, there are uncertainties. At present, the epidemiology of these diseases is unknown in the face of an increase in thermal values.

Argentina is not a historically relevant emitter of greenhouse gases, but it is a country highly vulnerable to the impacts of climate change. For this reason, in our second part of the study, emphasis was placed on both adaptation and mitigation of climate change, since it is more than likely that climatic phenomena will intensify in the future and, of course, the application of a set of public policies to face the challenges that will come. There are a large number of public policies in Argentina aimed at adaptation and mitigation to climate change. At the time of developing this document, it was possible to identify innumerable public policy initiatives that include regulatory instruments and economic instruments. However, a high degree of uncertainty persists regarding the final consequences of these public policies and their integration into a sustainable development strategy.

References

Brandt, L., Basilio, M., Introini, V., Manana, A., Ochoa, C., & Provecho, Y. (2018). *"Eliminación del paludismo en la Argentina 2018"*. Ministerio de Salud de la Nación. Buenos Aires. http://saladesituacion.salta.gov.ar/pagsala/documentos/materiales_descarga_programas_epi/direccion/paludismo/eliminaci%C3%B3n_de_paludismo_argentina_2018.pdf

Camilloni, I. A. (2018). *"Argentina y el cambio climático"*. Eudeba, Buenos Aires. https://ri.conicet.gov.ar/bitstream/handle/11336/99889/CONICET_Digital_Nro.cfeefb3b-550b-4172-bbdc-c39e534f7963_A.pdf?sequence=2

Cinquantini, M. A., Bertolino, R., Ayala, E., & Amanquez, C. (2016). *"Modelo de Inventario de Gases de Efecto Invernadero para Ciudades. La experiencia de la ciudad de Rosario, Santa Fe, Argentina"*. ANÁLISIS, N° 10—2016. https://library.fes.de/pdf-files/bueros/argentinien/12675.pdf

IPCC. (2007). Intergovernmental Panel on Climate Change. *"Contribution of working group I to the fourth assessment report of the intergovernmental panel on climate change", Climate change 2007: The physical science basis*. Cambridge University Press.

IPCC. (2013). Intergovernmental Panel on Climate Change. *"Contribution of working group I to the fifth assessment report of the intergovernmental panel on climate change", Climate change 2013: The physical science basis: Final draft underlying scientific-technical assessment.*

IPCC. (2022). Intergovernmental Panel on Climate Change. *"Sixth assessment report. Impacts, adaptation and vulnerability. Contribution to the IPCC sixth assessment report"*, 2022. https://www.ipcc.ch/report/ar6/wg2/

Madanes, N., Quintana, R. D., Kandus, P., & Camilloni, I. (2008). *"El cambio climático y sus posibles efectos en los grupos funcionales de la vegetación del delta del río Paraná (Argentina)"*. Buenos Aires, Argentina.

M.S.N. (2011). Ministerio de Salud de la Nación. *"Plan de Abordaje Integral de la Enfermedad Diarreica Aguda y Plan de Contingencia de Cólera. Guía para el equipo de salud"*. Buenos Aires. https://www.entrerios.gov.ar/msalud/wp-content/uploads/2018/05/guia-abordaje-colera.pdf

M.S.N. (2014). Ministerio de Salud de la Nación. Presidencia de la Nación. *"Perfil de país sobre cambio climático y salud"*. http://www.msal.gob.ar/determinantes/images/stories/descargas/recursos/2008_cambio_climatico_y_salud-perfil_de_pais.pdf

M.S.N. (2015). Ministerio de Salud de la Nación. *"Programa Nacional de Chagas"*. http://www.msal.gob.ar/images/stories/bes/graficos/0000000622cnt-03-guia-para-la-atencion-al-pacientecon-chagas.pdf

M.S.N. (2019). Ministerio de Salud de la Nación. Ciudad Autónoma de Buenos Aires. *"Clima y salud en la Argentina: diagnóstico de situación 2019"*. Buenos Aires. https://bancos.salud.gob.ar/recurso/clima-y-salud-en-la-argentina-diagnostico-de-situacion-2019

O.P.S. (2019). Organización Panamericana de la Salud. *"Dengue: Información general"*. Disponible en: https://www.paho.org/hq/index.php?option=com_content&view=article&id=4493:2010-informacion-general-dengue&Itemid=40232&lang=es

Sánchez, L., & Reyes, O. (2015). *"Medidas de adaptación y mitigación frente al cambio climático en América Latina y el Caribe. Una revisión general"*. Documento de Proyecto, Comisión Económica para América Latina y el Caribe (CEPAL), https://repositorio.cepal.org/bitstream/handle/11362/39781/1/S1501265_es.pdf

Smith, T. M., & Smith, R. L. (2007). *"Ecología"*, Pearson Educación, Sexta edición.

S.M.N. (2018). Servicio Meteorológico Nacional. *"Olas de Frio"*. Buenos Aires. https://www.smn.gob.ar/caracterizaci%C3%B3n-estad%C3%ADsticas-de-largo-plazo

S.M.N. (2013). Servicio Meteorológico Nacional. *"Informe especial debido a la ocurrencia de una ola de calor excepcional en Argentina durante diciembre de 2013"*. Disponible en: http://www.smn.gob.ar/serviciosclimaticos/clima/archivo/informe_temperatura_dic13.pdf

S.M.N. (2019). Servicio Meteorológico Nacional. *"Olas de Calor"*. Buenos Aires. https://www.smn.gob.ar/caracterizaci%C3%B3n-estad%C3%ADsticas-de-largo-plazo

Building Scenarios of Social and Health Vulnerability to Climate Change: A Study for Municipalities in the Mato Grosso do Sul, Brazil

Rhavena Barbosa dos Santos, Isabela de Brito Duval, Júlia Alves Menezes, Martha Mecedo de Lima Barata, and Ulisses Eugenio Cavalcanti Confalonieri

Abstract In recent years, multidisciplinary studies have sought to understand the interaction between environmental factors that modulate people's risks and susceptibilities to climate change. This study aimed to assess the socio-environmental and health vulnerability of the population in the state of Mato Grosso do Sul, Brazil, to climate change through the application of a vulnerability index. The results show sensitivity and adaptive capacity as bottlenecks found in this study. Challenging conditions include unequal access to health services and population occupation concentrated in large urban centers.

Keywords Climate change · Brazil · Vulnerability index · Public health · Socioeconomic factors · Government programs

In recent years, the debate on the impacts of climate change has become an emerging demand (Brooks et al., 2005, p. 151; Confalonieiri et al., 2014, p. 123; O'Brien et al., 2004, p. 303). Multidisciplinary studies have sought to understand the interaction between the environmental factors of a territory that modulates the risks and susceptibilities of populations facing climate changes, both in the present and the future. From this point of view, initiatives that address the concept of vulnerability in its scope stand out (Confalonieri et al., 2014, p. 124; O'Brien et al., 2004, p. 304).

According to the Intergovernmental Panel on Climate Change (IPCC) vulnerability can be defined as "the propensity or predisposition to be adversely affected" (IPCC, 2014, p. 564), comprising inherent components of exposure, sensitivity,

R. B. dos Santos (✉) · I. de Brito Duval · J. A. Menezes · U. E. C. Confalonieri
Grupo de Estudos Transdisciplinares em Saúde e Ambiente, Instituto René Rachou, Fundação Oswaldo Cruz, Belo Horizonte, Minas Gerais, Brazil
e-mail: rhavena.santos@gmail.com

M. M. de Lima Barata
Laboratório de Avaliação e Promoção da Saúde Ambiental., do Instituto Oswaldo Cruz/FIOCRUZ, Rio de Janeiro, Brazil

R. Akhtar (ed.), *Climate Change and Human Health Scenarios*, Global Perspectives on Health Geography, https://doi.org/10.1007/978-3-031-38878-1_27

and adaptive capacity. In public health, vulnerability represents a multidimensional construct comprising several biophysical, socioeconomic, and institutional factors that acts to modulate the capacity of populations or systems to cope with adverse impacts of extreme events (Phibbs et al., 2016, p. 7; WHO, 2008, p. 6). Both approaches are valuable to identify and evaluate the basal conditions that shapes different human populations predisposition to climate change (Menezes et al., 2021, p. 2).

Very common to quantify and operationalize vulnerability assessments, indicators have been presented as an efficient methodology, as they allow the comparison and synthesis of different information about the system studied (Brooks et al., 2005, p. 162; Malik et al., 2012, p. 3; Menezes et al., 2018, p. 2; O'Brien et al., 2004, p. 306; Quintão et al., 2017, p. 3). It can guide decision makers in the construction of public policies aimed at adaptation, as the factors that need to be adjusted or that favor the resilience of the evaluated systems can be disaggregated (Füssel & Klein, 2006, p. 322; Menezes et al., 2021, p. 14; Santos et al., 2019, p. 301).

In Brazil, states such as Rio de Janeiro, Minas Gerais, Espírito Santo and Amazonas already developed studies aimed at municipal mapping of the vulnerability of the human population to climate change (Barata & Confalonieri, 2014, p. 11; Menezes et al., 2018, p. 3; Quintão et al., 2017, p. 2; Santos et al., 2019, p. 301; Vommaro et al., 2020, p. 3). However, it is still necessary to expand this methodology to other territories seeking to subsidize adaptation actions. The present study aims to assess the socio-environmental and health vulnerability of the population of the state of Mato Grosso do Sul, Brazil, to climate change.

1 State of Mato Grosso do Sul

The state of Mato Grosso do Sul (MS) is located in the Midwest region of Brazil and is the sixth largest state in the country, with a total area of 357,145,532 km^2. Its political-administrative organization encompasses 79 municipalities grouped into nine Planning Regions (PR) (Campo Grande Region, Grande Dourados Region, Bolsão Region, Cone Sul Region, Pantanal Region, Leste Region, Norte Region, Sudoeste Region, Fronteira Sul Region) (Semagro, 2015, p. 16) (Fig. 1).

In 2021, the estimated population for the state was 2,839,188 inhabitants, with a population density of 6.86 in hab/km^2. The population is mostly urban, although there is a significant contingent of rural and traditional populations, mainly indigenous groups, which represent 9% of the entire indigenous population in the country (Ibge, 2021; Semagro, 2015, p. 12).

The climate is tropical, with rainy summers and dry winters. The average annual temperature varies between 26 and 23 °C (Fialho, 2014, p. 9) and rainfall varies between 1300 and 1700 mm.

The Cerrado biome, a neotropical savannah ecosystem considered a biodiversity hotspot, covers approximately 61% of the total area of the state, although portions of Amazonian vegetation and Atlantic rainforest are also observed (Ceped, 2013, p. 21;

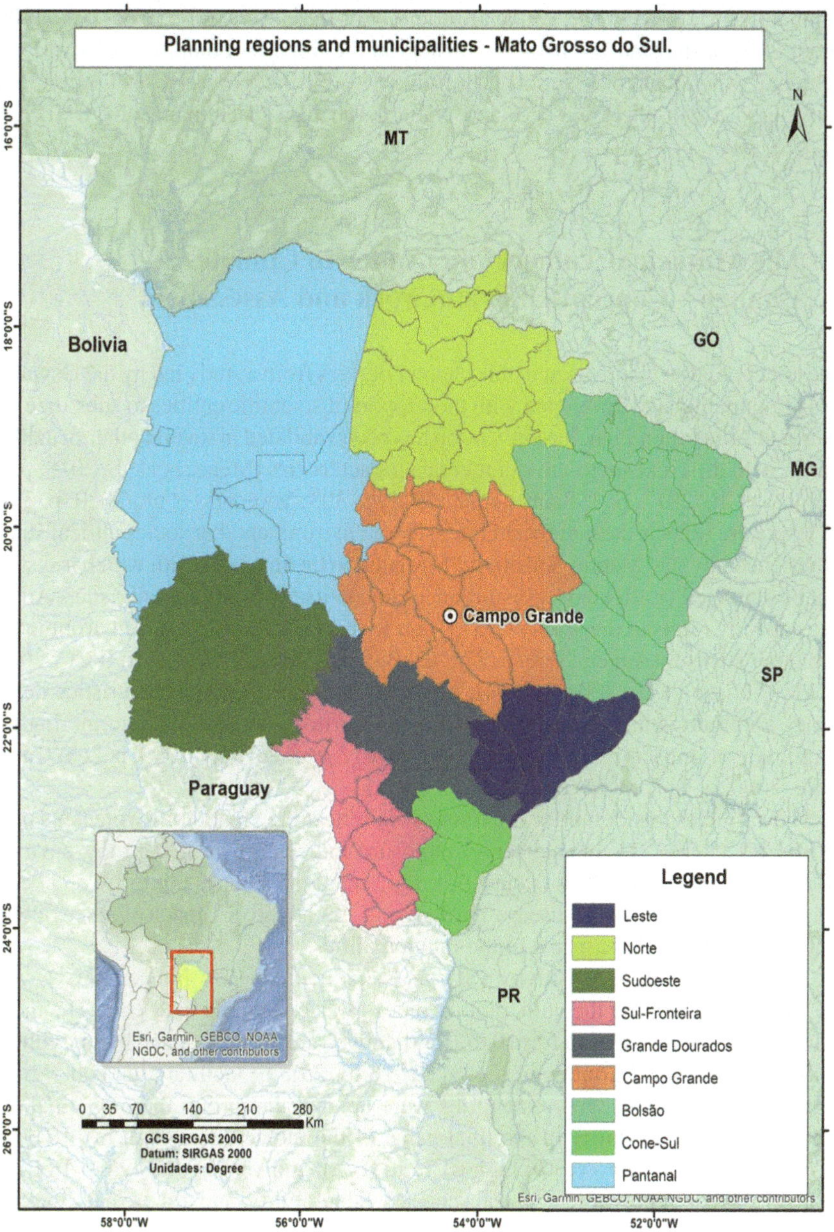

Fig. 1 Division of the state of Mato Grosso do Sul into 9 planning regions. Highlight for the location of the capital Campo Grande

Semade, 2015, p. 51). The Pantanal biome is in the northwest portion, occupying about 21% of the state's territory. Considered the largest wetland in the world, the Pantanal was declared a National Heritage Site by the Federal Constitution, and considered a Biosphere Reserve and Natural Heritage of Humanity by UNESCO (Ceped, 2013, p. 21).

2 The Municipal Vulnerability Index to Climate Change—Conceptual Framework and Assessment

The conceptual structure used in this chapter derives from a study led by the Oswaldo Cruz Foundation and developed with the support of scientific entities to subsidize the National Adaptation Plan. Its application has been validated in several other Brazilian states, with different socio-environmental characteristics (Menezes et al., 2018, p. 7; Quintão et al., 2017, p. 3; Santos et al., 2019, p. 303; Vommaro et al., 2020, p. 3).

Exposure reflects the nature and intensity of environmental or sociopolitical stress experienced by the system. Sensitivity is related to the intensity with which a system can be damaged or affected by disturbances due to its intrinsic characteristics. Adaptive capacity refers to the ability of a system to react to accommodate environmental and other stresses (Adger, 2006, p. 272; Confalonieri et al., 2018, p. 14; IPCC, 2014, p. 21; O'Brien et al., 2004, p. 304). The interaction of these characteristics determines a unique vulnerability profile for each territory/population, determining the magnitude of impacts related to climate change (Adger & Kelly, 1999, p. 265; Cutter et al., 2000, p. 717; Menezes et al., 2018, p. 4).

The variables that comprise the framework reflect the vulnerability precepts used by the IPCC (Fig. 2). In this sense, exposure indices are related to the environmental factors and disasters occurrence; the sensitivity indices comprise socioeconomic characteristics and disease burden; and the adaptive capacity indices cover institutional e political aspects of the municipalities.

The exposure (EI), sensitivity (SI), and adaptive capacity indices (ACI) were aggregated to compose the vulnerability index (VI), on a municipal scale. The index values are in a scale ranging from 0 to 1–indicative values of lower and higher vulnerability, respectively (Confalonieri et al., 2018, p. 25). To facilitate the comparison between the territories, four categories were created that group the municipalities according to the similarity of the final value of the indices. They are: low (0.00 to 0.25); moderate (0.26 to 0.50); high (0.51 to 0.75), and very high (0.76 to 1.00). A score of zero does not indicate the absence of vulnerability, and a score of 1 does not mean complete vulnerability–in relation to other municipalities in the state, a given municipality is more or less vulnerable (Confalonieri et al., 2018, p. 23).

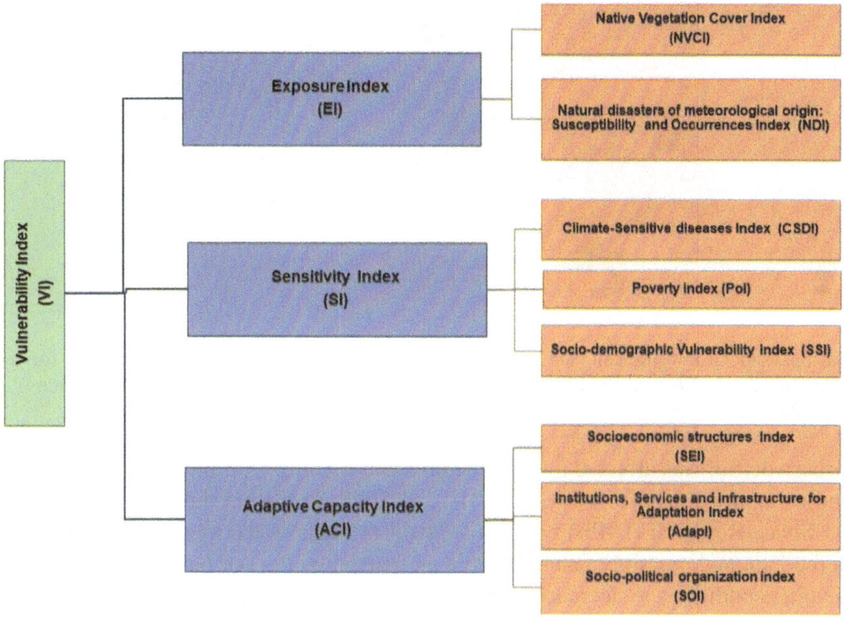

Fig. 2 Conceptual model of vulnerability to climate change

3 Identification of Local Vulnerabilities

The VI application showed high vulnerability in all Planning Regions of the state (Fig. 3a). The Cone Sul and Sul Fronteira regions stand out, with 0.61 and 0.73, respectively (Fig. 3b). SI contributed substantially to this finding, as well as the low adaptability observed by the ACI. In regions such as Grande Dourados and Bolsão, high exposure (IE) draws attention. In the next section, some findings will be highlighted for each index that composes the VI of the state of Mato Grosso do Sul.

4 The Exposure Index

The results reflect a lower vegetation cover in the eastern portion of the state (Fig. 4). There is also a great susceptibility to disasters for the most part of the state–around 55% of the territory has a value greater than 0.5 for the water stress indicator (measured by consecutive dry days). Marengo and collaborators (2021), when investigating weather patterns related to the drought that affected Pantanal between 2019 and 2020, identified a 60% reduction in precipitation compared to a normal year, which caused conditions of water stress in approximately 32% from the biome. The authors point out that 76% of the population of the states of Mato Grosso and Mato

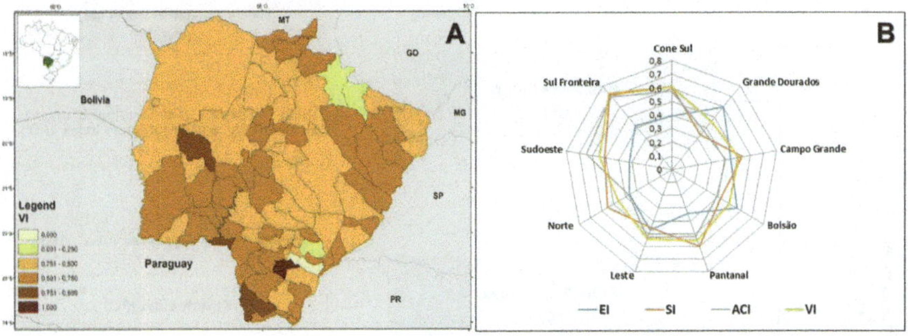

Fig. 3 Mato Grosso do Sul vulnerability index (VI). **a** Spatial distribution of the VI for the municipalities and **b** average values of the VI and its sub-components for the planning regions

Grosso do Sul was affected by extreme drought, causing multiple damages in health and economy. Also, for the years 2019 and 2020, data from the National Institute for Space Research (INPE) point to a 210% increase in the number of forest fires in the Pantanal, possibly resulting from a composite effect of land use and extreme weather conditions. A decrease in the volume of rainfall was observed, intensified by the reduction in the flow of moisture from the Amazon region, which created hot and dry climatic conditions, pushing the limits of flammability of vegetation in a context of poor management and negligent environmental regulation (Correa et al., 2022, p. 5; Garcia et al., 2021, p. 3; INPE, 2022; Libonati et al., 2020, p. 2018; Marengo et al., 2021, p. 16).

On the other hand, an increase in the frequency and intensity of extreme events such as heavy rainfall was observed for other parts of MS in early years. The Civil Defense–responsible for preventing or mitigating disasters–recorded more than a

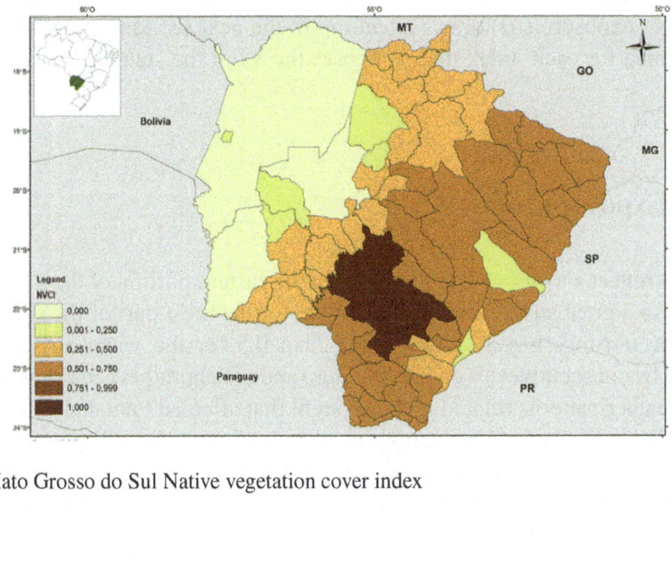

Fig. 4 Mato Grosso do Sul Native vegetation cover index

hundred occurrences caused by some weather phenomenon during the year 2017. This number had not been recorded since 1974, when the monitoring the occurrence of extreme events was initiated (www.defesacivil.ms.gov.br). It is worth mentioning that the Pantanal biome is dependent on a hydrological cycle that varies between drought, flood, and ebb events. This dynamic is directly related to geomorphological formations of the biome; occurrence of climatic events of hydrometeorological origin; and rainfall in the Upper Paraguay Basin (Marcuzzo et al., 2010, p. 172; Marengo et al., 2021, p. 2; Spacki, 2014, p. 58). All together, these findings point to the possibility that, in the near future, an increase in the frequency of dry periods will be observed, negatively interfering in the Pantanal environment and contributing to the imbalance of its ecosystem (Correa et al., 2022, p. 11; Marengo et al., 2021, p. 17).

From a public health point of view, the occurrence of the mentioned events, when associated to vulnerability conditions, can negatively impact human health, damage health infrastructures, and cause deaths, injuries, and changes in the epidemiological patterns of diseases. Regarding air pollution from forest fires, is acknowledged that exposure to smoke, atmospheric dust and aeroallergens may be associated with climate-sensitive cardiovascular and respiratory problems (IPCC, 2022). In addition, these characteristics interfere with aspects related to lack of hygiene, inadequate storage and contact with contaminated water, even in the air quality to which the population is exposed, leading to health impacts that extend over time. For the IPCC 2022, in recent years, the increase in weather events has exposed millions of people to food insecurity, reducing water security and has hampered efforts to achieve the Sustainable Development Goals (IPCC, 2022, p. 15) (Fig. 5).

Putting the traditional populations in perspective, indigenous, and Pantanal communities are sometimes directly dependent on the environment in which they live, making them particularly sensitive to environmental stress. Therefore, changes in the level of precipitation, increased temperatures, and long periods of drought can mean losses in crops and territories, leading to food insecurity and illness processes, including mental illness. In the context of Pantanal communities, the occurrence of prolonged droughts can interfere with the main source of income, which is directly

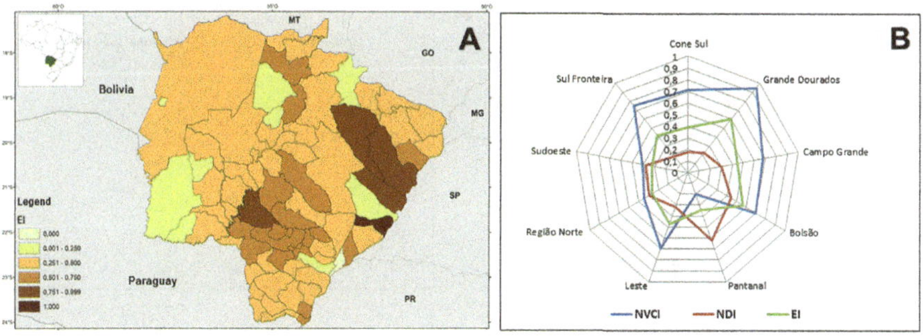

Fig. 5 Mato Grosso do Sul exposure index (EI). **a** Spatial distribution of the EI for the municipalities and **b** average values of the EI and its sub-components for the planning regions

or indirectly derived from rivers (Da Silva et al., 2014, p. 90; Spacki, 2014, p. 68). These findings are in line with the evidences from IPCC, when it states that in recent years, all over the world, sudden losses in production and access to food, aggravated by the decrease in diet diversity, have increased malnutrition in many communities, especially for Indigenous People, small food producers and low-income households, with children, the elderly and pregnant women being particularly affected (IPCC, 2022, p. 16).

5 The Sensitivity Index

The impacts of climate change on health are mediated by natural and human systems, including economic and social conditions and disturbances (IPCC, 2022). The SI encompass indicators that reflect issues known to be related to determinants of health such as basic sanitation, social and economic conditions that directly reflect on the current and future health standards of the territory (Fig. 6a). A group of municipalities with greater sensitivity was observed in a central corridor that extends from the north to the south of the state.

In this sense, the low performance in the Poverty Index (IP) (Fig. 6b) stands out. Covering socioeconomic aspects in addition to the monetary aspect, this indicator reflects the deprivation of essential elements for human well-being, which is in line with the World Organization's concept of health–"a state of complete physical, mental, and social well-being and not merely the absence of disease or infirmity." These data corroborate information from the Brazilian Institute of Geography and Statistics (IBGE), which indicates that 445,000 people in Mato Grosso do Sul (almost 16% of the state's population) lived below the poverty line, surviving on an income of up to $116 per month or $3.8 per day in 2017 (www.ibge.gov.br). However, with the recent COVID-19 pandemic, it is possible that this proportion has increased, driven by reduced income and job losses (Schmidt et al., 2021, p. 87).

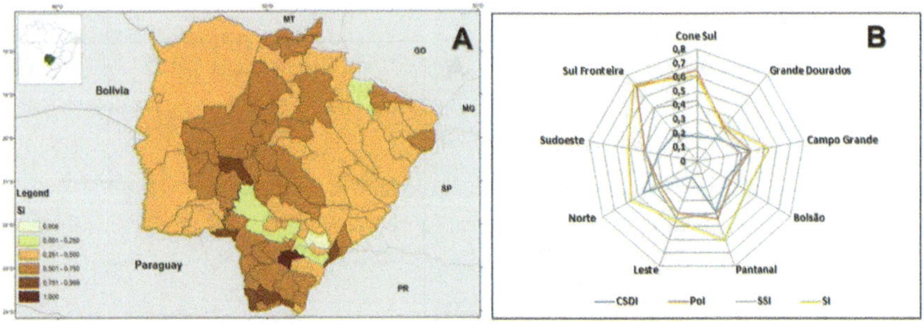

Fig. 6 Mato Grosso do Sul sensitivity index (SI). **a** Spatial distribution of the SI for the municipalities and **b** average values of the SI and its sub-components for the planning regions

The SI also covers basic sanitation conditions in locations that are known to be related to waterborne and infectious diseases. The latest IPCC report (2022) points out with a high degree of confidence that the risks of waterborne and food-borne diseases have increased regionally due to climate-sensitive aquatic pathogens, including *Vibrio* spp. In addition, there is also evidence of an increased risk of contamination by toxic substances from harmful freshwater cyanobacteria. Corrêa (2019), when studying the Brazil/Paraguay border in Mato Grosso do Sul, found that the basic sanitation conditions of people living in municipalities in the border region little changed between 2007 and 2016–those without access to treated water and sewage fell from 22.98 to 19.33%. Recently, basic sanitation issues have been discussed under the auspices of law n. 14,026, of July 15, 2020, which inaugurated a new sanitation framework based on Public–Private Partnership initiatives, a measure adopted to address the insufficiency of public resources to meet the demand (Brasil, 2020; Santos & Souza, 2022, p. 23). Even so, it is worth noting the need for reflections on this new framework that is being established, so that the services provided to society are concentrated on the less favored, known to be located in the border regions of the MS, in order to universalize and give continuity to essential services for the human well-being.

Alterations in vector-borne diseases are a well-known facet of climate change impacts and was evaluated by the climate sensitive disease index (CSDI), which considered the temporal trend of the incidence for the most prevalent diseases in the state–dengue, American tegumentary and visceral leishmaniasis, leptospirosis, deaths from diarrhea in children under 5 years of age and accidents with venomous animals (snakes, spiders, and scorpions) (IPCC, 2022). The literature shows some interactions such as the migration of farmers due to climate change, which is linked to the (re)emergence of Leishmaniasis in cities; higher temperatures causing displacement of vectors populations to colder areas; and precarious sanitation conditions favoring the permanence and proliferation of vectors (Confalonieri et al., 2017; Opas, 2021).

Most of the state had IDAC values below 0.250–a category of low sensitivity, although it is possible to highlight the Campo Grande, Pantanal and North PRs with the highest indices. However, it is possible that climate-driven factors exacerbate the current scenario of infectious diseases observed for most of the PRs, synergistically interacting with basal vulnerability conditions addressed by the other indices of the present study, causing changes in health scenarios. Reducing regional inequalities can be an important policy to increase the state's resilience and reduce impacts related to climate change.

6 Mitigation Strategies, Adaptation Measures, and Public Policies

Globally, climate change drives monitoring actions through conferences, public policies, and other actions aimed at the periodic monitoring of the situation in human systems. The same has been observed at the national level, where the Brazilian government was concerned with stimulating the reduction and management of climate risks, for example through the National Policy on Climate Change–PNMC (Law no. 12.187/2009) and the National Adaptation Plan (PNA), instruments that address the adaptation of natural, human, productive, and infrastructure systems. The actions provided for in the PNA are in line with the commitments assumed by Brazil within the scope of the Conference of the Parties on Climate Change and converge to a set of government initiatives aimed at mitigating and adapting to the effects of climate change (Brasil, 2016, p. 8–9).

However, an important cycle of Brazilian climate policy came to an end in 2020, when the deadline for meeting the targets registered in the PNMC expired. The PNMC's strengths are many, allowing the country to translate and incorporate into law the first national commitments to reduce emissions, to present a holistic view of the emissions of the entire Brazilian economy, in addition to laying the foundations for instruments of the Brazilian carbon market and of the annual emissions report (Potenza et al., 2021, p. 3). Although the country has achieved the aggregate goal of reducing emissions by 36.8% by 2020 (maximum 2068 $GtCO_2e$), Potenza et al (2021) point out that Brazil has not changed its emission trajectory–strongly influenced by deforestation–nor the type pollution profile, which made it impossible, in practice, to use the PNMC as a driving instrument for the low carbon economy. The emission level verified in 2020 was the highest since 2006, mainly driven by deforestation in the Amazon and Cerrado (INPE, 2022; Potenza et al., 2021, p. 4). Other sectors of the economy such as agriculture, energy, and industrial processes also contributed significantly to the increase observed in GHG emissions. This points to the need for greater investments regarding the implementation of public policies aimed at the development of sustainable economic activities.

Total GHG emissions in Mato Grosso do Sul amounted to around 81 $MtCO_2e$ in 2021, according to the Greenhouse Gas Emissions Estimation System–SEEG (Observatório do Clima, 2021), a significant drop compared to the beginning of the time series, in 1990 (151$MtCO_2e$). However, emissions increased compared to 2010, mainly due to the growth of the LULUCF sector–pasture-to-agricultural conversion and forest-to-pasture conversion (Brasil, 2021). In recent years, the highest percentage of emissions comes from agriculture (56%), followed by changes in land use (33%) and the energy sector (9%) (Observatório do Clima, 2021).

Regarding mitigation actions, locally, the state of Mato Grosso do Sul formalized the State Policy on Climate Change (PEMC) and the MS Carbon Neutral State Plan (Proclima) in 2021. The first provides for the conditions for the necessary adaptations to the expected impacts in the state, while the second sets the bold goal of neutralizing all GHG emissions by 2030. Focusing on the sectors that emit the most GHGs, the

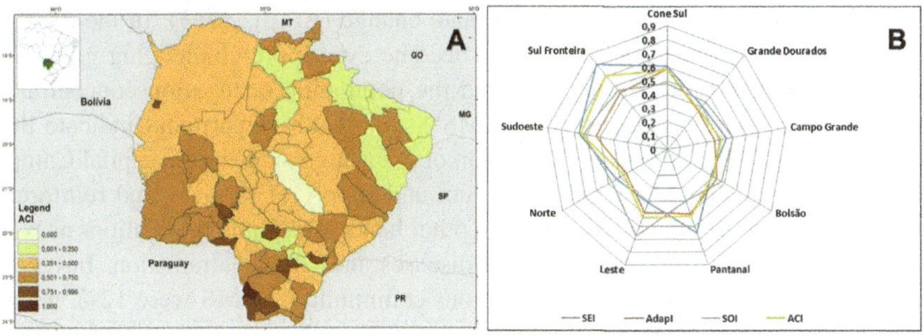

Fig. 7 Mato Grosso do Sul adaptive capacity index (ACI). **a** Spatial distribution of the ACI for the municipalities and **b** average values of the ACI and its sub-components for the planning regions

state presents initiatives for low carbon agriculture, technologies for crop-livestock-forest integration, recovery of degraded lands and Net Zero industrial enterprises. In addition, it has almost the entire energy matrix as clean.

Considering adaptation measures, the characteristics evaluated by the indices presented in the previous section–EI and SI–when associated with political-institutional conditions (ACI) may accentuate or attenuate the vulnerability patterns (VI) of Mato Grosso do Sul municipalities. The spatial distribution of ACI (Fig. 7) and its related indices demonstrated the low adaptive capacity of the state in the face of the impacts of climate change, being the Socioeconomic Structures Index–reflects conditions of employment/income, health care, and quality of education–the one with the highest values. It is also worth mentioning the municipalities close to the border with Paraguay, which have historical problems linked to low social and economic development and are also places with a social organization that, at times, may require international border articulation to reduce inequalities.

In general, the spatial distribution of the ACI and its sub-indices presents similarities with adaptive capacity indices from the Information and Analysis System on the Impacts of Climate Change (AdaptaBrasil MCTI), a national platform aiming at consolidating, integrating, and disseminating information on the impacts of climate change, providing subsidies to the competent authorities for adaptation actions in Brazil (https://sistema.adaptabrasil.mcti.gov.br). AdaptaBrasil results indicates that the adaptation conditions regarding drought hazards of the Mato Grosso do Sul in strategic sectors such as food security, water resources, and energy presents similar distribution to the ACI indices. The adaptive indices for the mentioned strategic sectors presented values ranging from low to medium categories for the most part of the municipalities, except for the energy sector, for which the municipalities showed quite adapted.

In the field of health, the Ministry of Health prepared the Health Sector Plan for Mitigation and Adaptation to Climate Change. This is structured in four axes (Health surveillance; health care; health promotion and education and health research) and aims to establish goals and strategies to contribute to mitigation measures and direct actions to adapt the processes and services of the System Unified Health System

(SUS) in the face of the impacts of climate change (Brasil, 2013). In the context of Mato Grosso do Sul, such initiatives become even more challenging given the inequity of access to health services and the population occupation concentrated in a few urban centers. Data from the MS State Health Department indicate that approximately 32% of the total population of the state resides in the capital Campo Grande, which demands a greater structure and supply of services and reinforces unequal access to health services. On the other hand, traditional populations as well as indigenous still suffer severely from diseases and child malnutrition. In recent years, child malnutrition rates in indigenous communities in MS were 12%, which represents a drop compared to previous years, but is still twice the national average (Mato Grosso do Sul, 2015).

7 Final Considerations

Sensitivity and adaptive capacity were the bottlenecks found in this study (Fig. 3), the indices showed regions susceptible to planning with low adaptive capacity.

Exposure, although individually not the biggest problem when compared to the other indices, deserves to be highlighted for assessing important conditions for maintaining the quality of life and health of the population, especially with regard to disasters of natural origin, which can directly impact the population and tend to become more and more frequent with climate change.

Finally, it should be noted that although Brazil has one of the largest and most complex health systems in the world–the Unified Health System (SUS)–the evidenced data showed great challenges experienced by the country in facing issues related to climate change, health and local vulnerabilities. The SUS proposes to offer the entire population residing in Brazil universal, integral, equitable, and free access to health. Provision of services ranging from Primary Care to the highest levels of health care, including laboratory services, surveillance vaccines, among others. However, many adaptations are necessary, from the budgetary sphere to the qualification of professionals, so that this system is able to meet the demands arising from climate change. The increase in resilience and adaptive capacity of this and other instances capable of reducing the impact of climate change on the population is directly related to the creation and implementation of public policies and government management measures designed for this purpose.

References

Adger, W. N. (2006). Vulnerability. *Global Environmental Change, 16*, 268–281.

Adger, W. N., & Kelly, P. M. (1999). Social vulnerability to climate change and the architecture of entitlements. *Mitigation and Adaptation Strategies for Global Change, 4*, 253–266.

Barata, M. M. D. L., & Confalonieri, U. E. C. (2014). Mapa da Vulnerabilidade da População do Estado do Rio de Janeiro aos Impactos das Mudanças Climáticas nas Áreas Social, Saúde e Ambiente. Rio de Janeiro, Fundação Oswaldo Cruz, 146pp.

Brasil. Lei nº 14.026, de 15 de julho de 2020. (2020). Atualiza o marco legal do saneamento básico. Published em: 16/07/2020, Ed. 135,1,Pág. 1, Órgão: Atos do Poder Legislativo.

Brasil. Ministério da Saúde. (2013). Plano Setorial da Saúde para Mitigação e Adaptação à Mudança do Clima. Brasília, 98 pp.

Brasil, Ministério do Meio Ambiente. (2016). Plano Nacional de Adaptação à Mudança do Clima: estratégias setoriais e temáticas. Brasília, Distrito Federal, Ministério do Meio Ambiente, v. 2.

Brooks, N., Adger, W. N., & Kelly, P. M. (2005). The determinants of vulnerability and adaptive capacity at the national level and the implications for adaptation. *Global Environmental Change, 15*(2), 151–163.

Ceped. Centro universitário de estudos e pesquisas sobre desastres. (2013). Atlas brasileiro de desastres naturais: 1991 a 2012: volume Mato Grosso do Sul, Santa Catarina, ed. 2, 122pp.

Confalonieri, U. E. C., Chang, M., Nahas, M. N., & Barata, M. (2018). Modelo Conceitual para avaliação municipal da vulnerabilidade humana á mudança do clima no Brasil: contribuições da FIOCRUZ ao Plano Nacional de adaptação. Brasilia, 100pp.

Confalonieri, U. E. C., & Fonseca, A. F. Q. (2013). Health security in Amazonia. Report for Global Canopy Programme and International Center for Tropical Agriculture as part of the Amazonia Security Agenda project. [s.l: s.n.].

Confalonieri, U., Lima, A. C. L., Brito, I. F., & Quintão, A. F. (2014). Social, environmental and health vulnerability to climate change in the Brazilian Northeastern Region. *Climate Change, 127*, 123–137.

Confalonieri, U., Menezes, J. A., Margonari, C. (2017). Environmental change and the emergence of infectious diseases: A regional perspective from South America. In *Examining the role of environmental change on emerging infectious diseases and pandemics* (pp. 109–137). IGI Global.

Corrêa, A. S. (2019). Políticas Públicas para a integração regional da faixa de fronteira de Mato Grosso do Sul: um estudo da "Borda Fronteira Brasil/Paraguai". 2019. Tese (Doctorate in Desenvolvimento Regional e Agronegócio)—Centro de Ciências Sociais Aplicadas, University Estadual do Oeste do Paraná, Toledo, PR, 245pp.

Correa, D. B., Alcantara, E., Libonati, R., Massi, K. G., & Park, E. (2022). Increased burned area in the Pantanal over the past two decades. *Science of the Total Environment, 835*, 155386.

Cutter, S. L., MitchelL, J. T., & Scott, M. S. (2000). Revealing the vulnerability of people and places: A case study of Georgetown County, South Carolina. *Annals of the Association of American Geographers, 90*, 713–737.

Da Silva, C. J., Albernaz-Silveira, R., & Nogueira, P. S. (2014). Perceptions on climate change of the traditional community Cuiabá Mirim, Pantanal Wetland, Mato Grosso, Brazil. *Climatic Change, 127*, 83–92.

Fialho, F. (2014). Estudos Estados Brasileiros: Mato Grosso do Sul 2000–2013. Editora Fundação Perseu Abramo,168pp.

Füssel, H. M., & Klein, R. J. T. (2006). Climate change vulnerability assessments: An evolution of conceptual thinking. *Climate Change., 75*(3), 301–329.

Garcia, L. C., Szabo, J. K., Roque, F. O., Pereira, A. M., Da Cunha, C. N., Damasceno-Junior, G. A., Morato, R. G., Tomas, W. M., Libonat, I. R., & Ribeiro, D. B. (2021). Record-breaking wildfires in the world's largest continuous tropical wetland: integrative fire management is urgently needed for both biodiversity and humans. *Journal of Environmental Management, 293*.

INPE. Instituto Nacional de Pesquisas Espaciais. Programa Queimadas. (2022). Available at: http://terrabrasilis.dpi.inpe.br/queimadas/portal/

Instituto brasileiro de geografia e estatística—Ibge. (2021). Perfil dos Municípios Brasileiros. (2021). Disponível em: http://www.ibge.gov.br/home/estatistica/economia/perfilmunic/2011/default.shtm. Available at: 25 Ago 2022.

Instituto brasileiro de geografia e estatística—Ibge. (2022). População: censo demográfico 2010—Estados. Brasília: IBGE, 2010. Disponível em: http://www.ibge.gov.br/estadosat/perfil.php?sigla=ms. Available at: 25 Ago 2022.

IPCC. (2014). Annex II: Glossary. In J. Agard, E. Schipper, J. Birkmann, M. Campos, C. Dubeux, Y. Nojiri, L. Olsson, B. Osman-Elasha, M. Pelling, M. J. Prather, M. G. Rivera-Ferre, O. C. Ruppel, A. Sallenger, K. R. Smith, S.T. Al Clair, K. J. Mach, & B. TE. Mastrandrea (Eds.), *Climate change 2014: Impacts, adaptation and vulnerability. Part B: Regional aspects. Contribution of working group II to the fifth assessment report of the intergovernmental panel on climate change.* Cambridge University Press, pp. 1757–1776.

IPCC. (2022). Summary for policymakers. In: H.-O. Pörtner, D. C. Roberts, E. S. Poloczanska, K. Mintenbeck, M. Tignor, A. Alegría, M. Craig, S. Langsdorf, S. Löschke, V. Möller, & A. Okem (Eds.), *Climate Change 2022: Impacts, adaptation, and vulnerability.* Contribution of working group II to the sixth assessment report of the intergovernmental panel on climate change. Cambridge University Press, pp. 3–33.

Libonati, R., Dacamara, C. C., Peres, L. F., Sander de Carvalho, L. A., & Garcia, L. C. (2020). Rescue Brazil's burning Pantanal wetlands. *Nature, 588*, 217–219.

Malik, S. M., Awan, H., & Khan, N. (2012). Mapping vulnerability to climate change and its repercussions on human health in Pakistan. *Global Health, 8*, 31.

Marcuzzo, F. F. N., et al. (2010). Chuvas no Pantanal brasileiro: análise histórica e tendência futura. In 3°. Simpósio de Geotecnologias no Pantanal, 2010, Cáceres/MT. Embrapa Informática agropecuária/INPE, 171–180.

Marengo, J. A., Cunha, A. P., Cuartas, L. A., Deusdará Leal, K. R., Broedel, E., Seluchi, M. E., & Bender, F. (2021). Extreme drought in the Brazilian Pantanal in 2019–2020: Characterization, causes, and impacts. *Frontiers in Water, 3.*

Mato grosso do Sul. Secretaria de Estado de Saúde. Vigilância em saúde. (2015). Available at: http://plataforma.saude.gov.br/cidades/

Menezes, J. A., Confalonieri, U. E. C., Madureira, A. P., Duval, I. B., Santos, R. B. D., & Margonari, C. (2018). Mapping human vulnerability to climate change in the Brazilian Amazon: The construction of a municipal vulnerability index. *PLoS ONE, 13*, 2.

Menezes, J. A., Madureira, A. P., Santos, R. B., Duval, I. B., Regoto, P., Margonari, C., Barata, M. M. L., & Confalonieri, U. (2021). Analyzing spatial patterns of health vulnerability to drought in the Brazilian semiarid region. *International Journal of Environmental Research and Public Health, 18*, 12.

O'Brien, K., Leichenko, R., Kelkar, U., Venema, H., Aandahl, G., Tompkins, F., Javed, A., Bhadwal, S., Barg, S., Nygaard, L., & West, J. (2004). Mapping vulnerability to multiple stressors: Climate change and globalization in India. *Global Environmental Change, 14*(4), 303–313.

Observatório do Clima. Publicações. dados. (2021). Available at: https://www.oc.eco.br/publicacoes/

Opas. Organização Pan-Americana da Saúde. (2021). Mudança do clima para profissionais da saúde: Guia de bolso, ISBN: 978-92-75-72184-1.

Phibbs, S., Kenney, C., Severinsen, C., Mitchell, J., & Hughes, R. (2016). Synergising public health concepts with the Sendai framework for disaster risk reduction: A conceptual glossary. *International Journal of Environmental Research and Public Health*, 13,12,1241.

Potenza, R. F., Quintana, G. D. O., Cardoso, A. M., Tsai, D. S., Cremer, M. D. S., Silva, F., & Azevedo, T. (2021). Análise das emissões brasileiras de Gases de Efeito Estufa e suas implicações para as metas climáticas do Brasil 1970–2020. *Sistema de Estimativas de Emissões e Remoções de Gases de Efeito Estufa (SEEG)*, 54.

Quintão, A. F., Brito, I., Oliveira, F., Madureira, A.P., & Confalonieri, U. C. (2017). Social, environmental and health vulnerability to climate change: The case of the municipalities of Minas Gerais, Brazil. *Journal of Environmental and Public Health*, 2821343

Santos, R. B., Menezes, J. A., Confalonieri, U. E. C., Madureira, A. P., Duval, I. B., Garcia, P. P., & Margonari, C. (2019). Construção e aplicação de um índice de vulnerabilidade humana à mudança do clima para o contexto brasileiro: A experiência do estado do Espírito Santo. *Saúde e Sociedade, 28*(1), 299–321.

Santos, R. F., & Souza, A. (2022). O saneamento básico em Mato Grosso do Sul: Relações fronteiriças. *Revista Eletrônica Da Associação Dos Geógrafos Brasileiros Seção Três Lagoas, 1*(35), 13–48.

Schmidt, B., Noal, D. S., Melo, B. D., Freitas, C. M., Ribeiro, F., & Passos, M. F. D. (2021). Saúde Mental e Atenção Psicossocial a Grupos Populacionais Vulneráveis por Processos de Exclusão Social na Pandemia de Covid-19, 87–98p. In G. C. Matta, S. Rego, E.P. Souto, & J. Segata (Eds.), *Os impactos sociais da Covid-19 no Brasil: populações vulnerabilizadas e respostas à pandemia.* Observatório Covid 19; 221 p.

Semade. Secretaria de estado de meio ambiente e desenvolvimento econômico. (2015). Diagnóstico Socioeconômico De Mato Grosso Do Sul. Available at: http://www.semade.ms.gov.br/wp-content/uploads/sites/20/2015/03/Diagnostico_Socioeconomico_de_MS_20151.pdf. Access: 01 de Set de 2022.

Spacki, V. (2014). Mapeamento de eventos naturais extremos e seus impactos sobre comunidades Pantaneiras, Dissertation Escola Superior de Conservação Ambiental E Sustentabilidade. Nazaré Paulista. 124p.

Vommaro, F., Menezes, J. A., & Barata, M. M. L. (2020). Contributions of municipal vulnerability map of the population of the state of Maranhão (Brazil) to the sustainable development goals. *Science of the Total Environment, 706*.

World Health Organization—WHO. (2008). Closing the gap in a generation: Health equity through action on the social determinants of health. Final report of the commission on social determinants of health. World Health Organization.

Climate Change and Forced Displacements in Brazil: The Health Context of Migrant Women

Gislene Santos

Abstract The article presents and analyzes the role of public policies in Brazil in the face of the effect of climate change, focusing on displacement and women's health. Its describes the role of the Brazilian State and the attention given to the health field of displaced women. In this context, we reflect the social and economic vulnerability of women in Brazil and the fragility the social assistance policies for this population who, as a result of the social and political dimension of disasters, lose their homes and become either homeless and/or displaced.

Keywords Natural disaster · Displacement of population · Women's health · Public policies · Brazil

1 Introduction

The relationship between climate change and migration flows is a pressing challenge in contemporary global politics. Since the late 1980s, the population—through the variable displacements of people—and nature—through the variable climate change—have become crucial problems in the field of global geopolitics. If until then, the relationship between society and the geographic environment were passive points within a territorial order specific to each nation, in the context of economic globalization the environment emerges as an international issue.

Moving beyond the political conditions that enabled the emergence of climate and migration as an issue in the global agenda, in this article we stem from a reflection in which these two variables are analyzed together. That is, we propose to discuss population displacements in direct relation with the occurrence of natural disasters. Currently, circa 38 million people in the world have been forced to leave their places of origin, and 23.7 million of these displacements, i.e., 62%, were motivated by the occurrence of some natural disaster. Such magnitude is extraordinary, given that nature is not a social and political actor able to determine the displacement of people.

G. Santos (✉)
Department of Geography, Federal University of Rio de Janeiro, UFRJ, Rio de Janeiro, Brazil
e-mail: gislenesantos@igeo.ufrj.br

R. Akhtar (ed.), *Climate Change and Human Health Scenarios*, Global Perspectives on Health Geography, https://doi.org/10.1007/978-3-031-38878-1_28

Table 1 Countries with the highest number of internally displaced people (IDPs) by disasters in 2021

Country	Number of internally displaced people
China	6,000,000
Philippines	5,700,000
India	4,700,000
Republic of the Congo	888,000
Indonesia	749,000
United States	573,000
South Sudan	506,000
Brazil	449,000
Somalia	271,000

Source GRID (2022, p. 27)

We emphasize, as our starting point, that one of the main causes of forced displacement is the unrestrained exploitation of natural resources in certain territories. Our focus will be on Brazil, which according to the GRID (2022), ranks 8th place with approximately 450,000 displaced people in 2021 (Table 1).

In Brazil, between 1991 and 2021, circa 238 million people were directly affected, with an estimate of 4065 deaths. In addition to mortality, 2766 people disappeared, whose bodies were never found, and 1,217,883 people were injured, according to data from CEPED (Atlas brasileiro de desastres naturais). Despite the sheer magnitude of the losses, Brazil still lacks solid public policies to address and mitigate the consequences of these disasters as well as to prevent them.

In this article, we reflect on this issue with particular focus on the forced displacement of women and the social assistance policies for this population who, as a result of the social and political dimension of disasters, lose their homes and become either homeless and/or displaced. As these people find themselves in an emergency homeless situation, and upon exhausting all possibilities, spatial displacement becomes their last resort and only survival strategy.

Methodologically, we relied on the database from the CEPED to record the occurrence and distribution of disasters in Brazil. For the social and economic indicators, we based ourselves on official data from the Instituto Brasileiro de Geografia e Estatística—IBGE (2021) and the Comissão Econômica para América Latina e o Caribe—CEPAL. Analytically, we corroborate the analyses by Valencio (2009) in his efforts to investigate the social and political dimensions of disasters. In the first section, we describe the territorial distribution of disasters in Brazil, in view of homeless or displaced people, and the occurrence of deaths. In the next section, we address the field of social assistance and health policies for people directly affected by disasters, with a particular focus on women. In the final considerations we discuss some of the challenges in the field of social research regarding the relationship between forced displacements, climate change, and women's health care.

2 The Territorial Distribution of Disasters in Brazil

Population displacement is a constitutive element of Brazil's territorial development. In the contemporary period, above all since the 1990s, we have witnessed a resurgence of displacements within the country, caused by profound modifications in the population's places of living. These forced displacements stem from emergency situations—caused by extreme events—which compels people to leave their places of origin. While we must analyze events such as heavy rain, floods, landslides, and drought as natural processes, we must also approach them through an intermesh between technical objects and political actions. In the hybrid between nature, technique, and politics, we must remain watchful to avoid climate determinism and/or technical discourses on climate, which have historically prevailed in Brazil's debate on migrations. This historical deterministic relationship between nature and society, in a way, continues to replicate itself in the contemporary world, as reflected in the lack of or discontinued public policies and social protection for population groups exposed to the direct effects of climate change.

This starting point is necessary insofar as nature-related displacements have historically been a staple in Brazil's ideological scenario. Between the late eighteenth century and early nineteenth century, Brazil's northeast region experienced four major droughts. The fact that drought has become one of Brazil's most pressing problems should have provided us with enough resources and social capital to face the current challenges and, above all, acknowledge that in a country characterized by deeply rooted social inequalities, the debate on the relationship between society and nature must take place within the field of politics. Thus, we must look at the climate change issue in Brazil through a critical lens.

In this article, as we set out to analyze forced displacements of people affected by natural disasters and the ensuing implications for women's health, we should further extend our attention. First, to avoid establishing correlations between natural limits and demographic control. Second, we must not underestimate the predatory use of natural resources in Brazil. Third, we must critically look at the ideological prescriptions directed toward natural disasters in Brazil. On this particular issue, disasters in Brazil fall under the auspices of civil defense agencies, which take action in cases of emergency. As Valencio (2009, p. 5) has argued, the acknowledgement of a disaster by the civil defense is, above all: "a public confirmation of a vulnerability in the State's relationship with society given the impact of a threat whose damages and losses were not satisfactorily prevented or mitigated".[1] It is important to recall that, for contractarianism theory, the justification for the State's legitimacy lies in its ability to protect man (society) from the needs of the natural world. Thus, events and natural disasters would fall more under the rubric of a given political order than a natural dimension.

[1] Original version in Portuguese: "o fenômeno de constatação pública de uma vulnerabilidade na relação do Estado com a sociedade diante o impacto de um fator de ameaça que não se conseguiu, a contento, impedir ou minorar os danos e prejuízos" (Valencio, 2009, p. 5).

Conceptual difficulties extend beyond the semantics of disasters and furthermore affect the typology of population displacement. In Brazil, we find a set of words such as *flagelados*[2] and *retirantes*,[3] historically employed to express migration flows resulting from drought. As Póvoa Neto argued (2022), the notion of migrant would only emerge in the 1970s, albeit primarily to address work-related migration than displacements caused by insufficient regional structure to address the drought crisis in the northeast. As such, men and women who displaced themselves because of the drought were classified as *flagelados* and/or *retirantes*. Amid Brazil's industrialization process, the scholarly literature erased drought-related displacements in the 1970s, and the word migrant emerged to describe work-related migratory flows. In this perspective, nature disappears from the migratory scenario insofar as an industrial and urban economic order increasingly configures the Brazilian territory. In the 1970s, circa 56% of Brazil's population lived in urban areas. The discursive field accompanied this territorial configuration with a predominant focus on work-related migrations.

Despite this new dimension of work-related mobility, many people in Brazil are still displaced, unsheltered (*desabrigados* in Portuguese), dislodged (*desalojados* in Portuguese), or *flagelados*, terms commonly employed by organizations that deal directly with disasters. We employed this typology in Fig. 1 to denote situations in which people have lost their homes and houses, i.e., losing one's goods, belongings, and home leads to displacement from one's own habitat.

One of the first revealing indicators in Fig. 1 is that the entire Brazilian territory has registered some occurrence of disasters followed by deaths and homelessness. We find an increasing incidence of disasters in the period between 1990 and 2010, with 8671 occurrences in the 1990s and 23,238 in the decade of 2000. A substantial escalation, with loss of human life and damage to human wellbeing. Regarding homelessness, with loss of housing, all 3 southern states (Rio Grande do Sul, Paraná, and Santa Catarina) stand out, as well as the southeast state of Minas Gerais, followed by the northern state of Amazonas. While the southern region registered a significant number of people who have lost their homes, the southeast region registered the highest number of deaths. In 2011, the southeast state of Rio de Janeiro alone recorded 912 deaths, 350 missing people, and 415,000 homeless people due to intense rains and landslides in the mountainous regions (according to CEPED). In the city of Nova Friburgo (Rio de Janeiro), the highest proportion of female deaths were in the age brackets of 5–9 years, 20–24 years, and 55–59 years old, according to Carmo and Anazawa (2014), encompassing losses of both female children and adults.

As we look at this territorial distribution, we find an increased number of disasters with loss of human lives over the years, above all in the southeast region which concentrates most of the Brazilian urban population. Furthermore, these situations

---

[2] Translator's note: an individual victimized by calamities or disasters (floods, droughts, earthquakes, etc.).

[3] Translator's note: An individual or group that leaves their land because of drought and poverty in search of a location that offers them better living conditions. The term was widely used in Brazil to refer to people fleeing droughts from the Northeast who migrated to major urban centers in the South-Southeast.

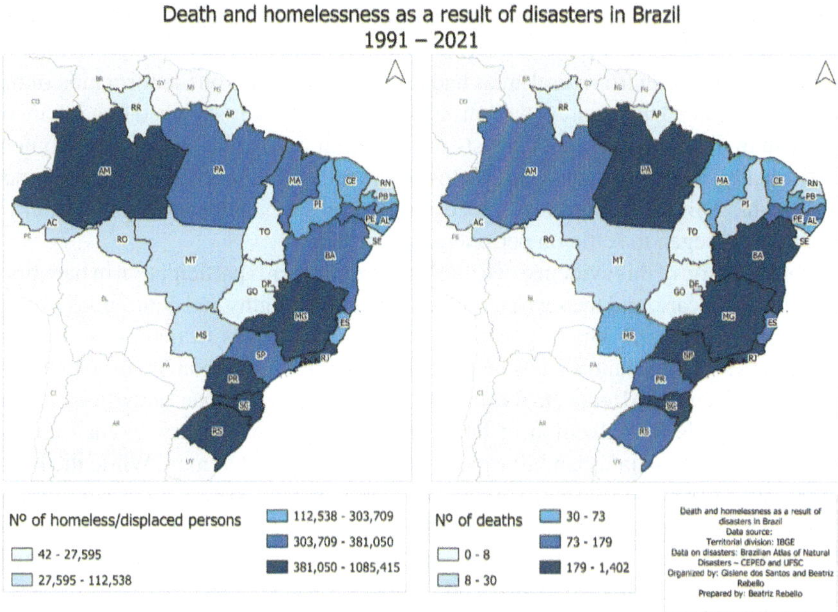

Fig. 1 Death and homelessness as a result of disasters in Brazil 1991–2021

of disaster more severely impact people inhabiting areas of elevated risk due to uncontrolled land use and land occupation.

3 The Social Conditions of Homeless Women

In Brazil, women devote more hours per week to household chores than men: circa 21 h a week while men spend 11% of their time on this activity. While the average monthly income in 2016 was around R$2245.00 for men, women earned R$1654.00 on average. The difference between men and women is also disproportionate in the weekly time devoted to part-time work: Men spend 13 h a week on average while women spend circa 30 h a week in part-time jobs. That is, they work in different places and with different employment contracts. Hence, in addition to devoting more time to domestic care activities, women perform more part-time jobs than men. In 2017, the female workload was distributed as follows: circa 6 h a week in agricultural activity; 13 h in the industry sector, and 81 h in the service sector. The greater participation of women in the service sector in contrast to agriculture work should not be understood as absence of women in agricultural activities. In fact, the contrary is true as women have been largely responsible for enhancing family farming in Brazil, with an effective participation both in agricultural production and commercialization. Women's discrepant participation in the service and agricultural sectors stems from

Brazil's high urbanization rate, as 89% of the population resided in urban areas in 2017.

Female migration from rural areas had a significant impact on the formation of the Brazilian urban population. As de Sousa Ramalho (1996) has shown, a large share of urban slum dwellers in Brazil consists of rural migrant families headed by women. Additionally, Brazil's rural exodus is closely related with a survival strategy in times of drought. This long migratory process relied on the significant participation of women who began to settle on peripheral urban areas.

Hence, on top of this evidence, we must analyze women's participation in part-time jobs, domestic care, and the service sector alongside a lengthy and continuous migratory process in Brazil which, given the enduring difficulties in managing drought emergencies in the country's northeast region, migration to urban areas emerged as a survival strategy. Moreover, the inclusion of women in urban activities did not evolve alongside formalization of labor relations. On the contrary, women have a greater participation in urban informal jobs when compared to men. While the latter represents 34.9% of informal workers, women comprise 40% of non-agricultural informal jobs. This rate is higher for black women, who make up 47.3% of the urban informal service sector. As for educational level, for women over 25 years of age, only 13.4% have completed primary education.

These indicators signal the different conditions for women in Brazil: unstable work relationships, greater time devoted to domestic care, and low educational levels. This is the national scenario that women find themselves as they strive to manage and overcome the difficulties caused by disasters and to care for a space shaped not only by risks and social inequalities, but which is ultimately unfair in the field of policies for the prevention and mitigation of natural disasters.

We must also look at the relationship between women and poverty. Although extreme poverty in Brazil is a widespread problem among men and women indiscriminately, we can say that women are the most afflicted, especially those who head households. Precariousness has an impact on multiple dimensions of social life, which, combined with situations of poverty, expose women to environmental vulnerability. According to the CEPAL, extreme poverty rates in Brazil have increased over the past 20 years for women. In 2011, the extreme poverty index for women was at 107.4, while in 2020 this index increased to 116.7.[4] This upsurge has affected women in both urban and rural areas (Table 2).

While widespread throughout the national territory, female poverty in Brazil is more pronounced in urban areas: in 2020, 24.6% of poor women in Brazil resided in rural areas while 38.5% lived in urban areas. This rate increased in the short period of one year, and in 2021, 26.1% of poor Brazilian women lived in rural areas, i.e., a 2% increase. In contrast, 42.7% of poor women lived in urban areas in 2021, an increase of 4.2%. Also noteworthy is the increase of women without income in rural areas: in 2008, women comprised 36.8% of the population without income, while in 2021

[4] The index is calculated as follows: for 100 men in extreme poverty there are 116.7 women in the same situation.

Table 2 Poverty femininity index in Brazil (2001 and 2020)

2001	National	Urban	Rural
Poverty	105.4	167.5	106.23
Extreme poverty	107.4	112.5	109.2
2020	National	Urban	Rural
Poverty	118.5	122	107.6
Extreme poverty	116.7	124.2	104.1

Source CEPALSTAT. Organized by the author

this rate was 28.5%. Despite this reduction, women continue to endure economic immobility (CEPAL).

These current indicators serve as a starting point to discuss the health of displaced women. As emphasized above, Brazil's deep-rooted social inequalities especially affect poor women, making them more susceptible and vulnerable to disasters. The map shows that the state of São Paulo (located in the southeast region) stands out both in the number of homeless people and number of deaths. Despite its higher share in the country's GDP (31.2%), the state's economic status has not translated into access to health policies to protect people vulnerable to disasters. Within the state, only the city of São Paulo maintains a city health policy for non-nationals, especially for migrant women from the neighboring country, Bolivia. Nonetheless, this was only made possible through longstanding demands from Bolivian migrant women for a maternal health policy consistent with their culture, especially regarding the use of their language for communication and in the field of sexual reproductive rights. Bolivian migrants in Brazil are largely motivated by the prospect of work, therefore their displacement does not stem from the occurrence of natural disasters. In this context, while the city of São Paulo has consolidated health policies for migrant people, it still lacks health services for women in situations of natural disasters. As noted by Valencio (2009), homeless shelters in the state of São Paulo are still precarious and offer only temporary assistance to displaced persons, leading to serious mental health symptoms such as anguish, anxiety, and distress for women who, in the context of displacements, often undertake the responsibility to maintain the family.

In 2009, the National Policy on Climate Change was founded. During its active period, from 2009 to 2019, the policy sought to incorporate a shared administrative model between different social and political fields, with the creation of several management committees. In this context, in 2015, a Social Assistance Program for disaster situations was implemented in Brazil. However, in 2019, through a set of decrees during the first year of the Bolsonaro administration (2019–2022), the Executive Health Committee was discontinued.

And other assistance initiatives, such sexual and reproductive rights policies, remain restricted to some locations and do not comprise a systematically organized national assistance policy and these actions are precarious and limited to respond to emergency situations. Furthermore, women in situations of displacement are vulnerable both in relation to care and protection of their bodies, dimensions rarely prioritized in humanitarian assistance in situations of disaster.

4 Conclusions

In the section above we described the socio-economic situation of women in Brazil. Despite longstanding internal migration patterns, in different types and forms, and the historical participation of women in this process, Brazil still lacks a national policy focused on women's migration and health. While the global geopolitical context has acknowledged the connection between climate change and population displacements as an international problem, in Brazil the population affected by disasters has not yet benefited from a solid public health policy.

While the occurrence of deaths and displacements is a widespread problem within Brazil, health care actions for displaced people remains local and focalized, and women's health care in situations of displacement has not received the attention it deserves. What we find in Brazil are precarious and emergency actions, albeit not permanent social assistance policies and long-term care strategies. This scenario is further aggravated by longstanding and deep-rooted social inequalities that directly affects women. The eradication of female poverty must be at the center of public policies, alongside actions to mitigate the social effects of disasters. The data that we presented in this work reinforce the precarious access that poor women in Brazil have to health facilities and services. We analyzed a set of variables that call attention to the social and economic plight of women in Brazil as the country experiences a process of impoverishment of women, which entails further implications in the field of health. A woman in a situation of poverty is unlikely to have the necessary conditions to care for her basic survival, a situation further aggravated by the shortage and precariousness of public facilities, which further hinder and limit women's participation in social life.

In this work we presented a description of the territorial distribution of disasters as we sought to characterize the current situation of women in Brazil. Public policies focused on mitigation are necessary, as well as guidelines and instruments to curb the unrestrained use of natural resources. An interconnection exists between the current mode of exploitation of nature and women's health. The political challenge in Brazil concerns not only the provision of emergency assistance and social care, but also changes in the productive economic foundations and in the use of nature.

Acknowledgements The author is grateful to the CNPq (National Brazilian Council for Scientific and Technological Development) for the financial contribution to this research project: 420058/2021-4.

References

Brasil. Presidência da República (PR). *Lei no 12.187, de 29 de dezembro de 2009*. Institui a Política Nacional sobre Mudança do Clima (PNMC) e dá outras providências. Brasília, 29 dez. 2009c.

Carmo, R. L., & Anazawa, T. M. (2014). Mortalidade por desastres no Brasil: o que mostram os dados. *Ciência & Saúde Coletiva, 19*(9), 3669–3681.

Centro Universitário de Estudos e Pesquisas Sobre Desastres (CEPED). (2012). *Atlas Brasileiro de Desastres Naturais: 1991 a 2010. Volume Brasil*. Florianópolis: Ed. UFSC, 2012.

Cepalstat. *Base de Datos y Publicaciones Estadísticas*. http://statistics.cepal.org/portal/databank/index.html?lang=es&indicator_id=3330&area_id=

Comissão Econômica Para América Latina e o Caribe (CEPAL). *Observatório de igualdade de gênero - América Latina e Caribe*. Disponível em. https://oig.cepal.org/pt/indicadores/indice-feminidade-da-pobreza

de Sousa Ramalho, D. (1996). Seca, Migração e Moradia: Onde fica a Mulher? Invisível? *Revista Raízes, Campina Grande, XV*(12), 31–51.

Global Report on Internal Displacement (GRID). (2022). *Children and youth in internal displacement*. IDMC, NRC. Disponível em. https://www.internal-displacement.org/sites/default/files/publications/documents/IDMC_GRID_2022_LR.pdf

Instituto Brasileiro de Geografia e Estatística (IBGE). (2022). *Estatísticas de gênero. Indicadores sociais das mulheres no Brasil*. 2021. Disponível em. https://www.ibge.gov.br/estatisticas/multidominio/genero/20163-estatisticas-de-genero-indicadores-sociais-das-mulheres-no-brasil.html?=&t=notas-tecnicas. Acesso em 08/12/2022.

Póvoa Neto, H. (2022). Planejamento regional, desenvolvimento e migrações internas na obra de Celso Furtado. In F. Fridman (Org.), *Quem planeja o Território?* (p. 571 a 596) Letra Capital.

Valencio, N. (2009). Da área de risco ao abrigo temporário: uma análise dos conflitos subjacentes a uma territorialidade precária. In N. Valencio, M. Siena, V. Marchezin, & J. Costa (Orgs.), *Sociologia dos Desastres - interfaces e perspectivas no Brasil* (pp. 34–47). RiMa Editora.

Conclusion and Suggestions

Rais Akhtar

Abstract Climate Change is the defining issue of our time and we are at a stage when developing countries an health is being impacted in both developed and developing countries. The World Health Organization has organized two international conferences in Geneva and in Paris to bring in focus health and climate issues. The issue that regions with vulnerable people and with weak health infrastructure ,and suffer from extreme weather conditions—in both developed and developing countries—will be the least able to cope without assistance to prepare and respond. There is an accumulating body of evidence showing the human health impacts of climate change. Although research on physical human effects predominates, recent studies show a significant mental health impact of climate-and-weather-related events. Awareness of climate-related mental health issues has important ramifications for the implementation of national healthcare policies.

Keyword WHO · National health care policies · Mental health · Extreme weather conditions · Hippocrates · COVID-19 pandemic

As WHO stressed that between 2030 and 2050, climate change is expected to cause approximately 250,000 additional deaths per year, from malnutrition, malaria, diarrhea, and heat stress in the developing countries as climate change affect the food and water we consume. It must be mentioned that the understanding of impacts of extreme weather events on human health and well-being dates back to Hippocrates' fifth century treatise *ON AIRS WATERS AND PLACES*. The definition of climate change induced extreme weather is when a weather event is significantly different from the average or usual weather pattern. Regions with weak health infrastructure, and vulnerable to extreme weather conditions–in both developed and developing countries–will be the least able to cope without assistance to prepare and respond. The 2003 heatwaves in western and central Europe resulted in the death of more than 72,000 people, and the 2005 Katrina hurricane, was considered the costliest disaster

R. Akhtar (✉)
Formerly Professor of Geography, University of Kashmir, Srinagar, Jammu & Kashmir, India
e-mail: raisakhtar@gmail.com

R. Akhtar (ed.), *Climate Change and Human Health Scenarios*, Global Perspectives on Health Geography, https://doi.org/10.1007/978-3-031-38878-1_29

that devastated the southern United States. It caused $191.3 billion worth of damages to the Gulf Coast. However, the disasters caused by hurricane Harvey in September 2017 proved to be even costlier with economic losses of about $190 billion. Devastating floods in Pakistan in during June–October, 2022, killed 1,739 people, and caused Rs. 3.2 trillion of damage and Rs. 3.3 trillion of economic losses. In 2022, there were 18 weather/climate disaster events in the USA with losses exceeding $1 billion each (Smith, 2023).

In light of the above, attempt has been made to edit a book on country case studies on climate change and human health scenarios encompassing both developing and developed countries.

Australia faces a difficult future, as stated by John Connell, where the natural and physical environment seemed at particular risk in 2022: the COVID-19 pandemic remained in place, floods were swamping large parts of the country, droughts simultaneously existed island, and a plague of mice were devouring what was left of rural grain production. Five meter high walls were being constructed on vulnerable northern Sydney beaches as coastal surges became more threatening. While these changes, and their outcomes in mental stress, were not all directly the result of climate change, they suggested the vulnerability of Australia to climate change and to a suite of hazards with negative effects on health, that hopefully reminded politicians and planners that multi-scalar action was crucial.

In case of Taiwan, Mei-Hui Li suggests that human activities, climate change, and COVID-19 pandemic interact in complex and multiple feedback pathways. Many human lifestyles and economic activities lead to exacerbate adverse effects of climate change worldwide. Moreover, human health and wellbeing are affected by altering environmental and social factors from direct or indirect impacts of climate change. Many human mobilities and activities are reduced or restricted to control the SARS-CoV-2 transmission globally and these strategies can also lead to change environmental conditions and social environments in different locations differently. Climate change and COVID-19 pandemic seem to alter human activities contrastingly, but effects of these two processes can overlap and produce similar outcomes in specific social groups.

Regarding the heat impact on health Oka Kazutaka reveals that Japanese government and local governments have implemented various heatstroke preventive measures to reduce heat-related health impacts such as heatstroke. However, the increase in the elderly population in Japan and rising temperatures due to climate change will further increase the health impacts. Additionally, medical crises resulting from significant health impacts would be severe in extreme heat events and health of Japanese people, mitigation measures to reduce the temperature increase are vital, and adaptation measures to reduce the risk of heatstroke and prevent medical crises.

Identify Thailand government policies and programs supportive of a climate-resilient health system, a documentary review was carried out between 2012 and 2022 to analyze policy measures that could potentially contribute to developing and sustaining a health system capable of resiliently responding to health needs. A total of 54 articles were identified. After the articles were de-duplicated, the title and abstract were checked, and 42 articles were further screened using full text. Of these, 11 met the eligibility requirements for inclusion in the final analysis. The existing policy measures were broadly categorized at the system and agency level and the level of the population. The climate change adaptation and mitigation policy measures have significant potential for health system resilience worthy of emulation as they aim to strengthen health related systems and agencies and are sensitive to healthcare needs population wide. The policies should be consistently backed by political commitment through sustained advocacy and pragmatic actionable steps to realize the objective of a climate change-resilient health system for the provision of efficient people-centered care through collaboration and partnership.

Malaysia is one of the vulnerable climate change hotspots that ostensibly witness many extreme weather events that amplify the burden of climate-sensitive diseases. Elevating rates of urbanization and population explosion in near future could magnify the implications of climate change in Malaysia, including the exacerbation of warming trends, amplification of environmental disasters and resurgence of infectious diseases. In response to this, environmental and public health institutions at all operational scales need to consciously modify their approaches to articulate evidence-based adaptation and mitigation strategies to build climate resilience across all the sectors in Malaysia. Mainstreaming behavior change into the adaptation strategies related to the extreme climate events along with infrastructure, technological, and policy advancements is another key aspect to strengthening the public health adaptive capacity and responses in Malaysia. The identification and management of constraints and barriers to climate change adaptation in anticipation of existing public health strategies are equally inevitable to address the adverse health impacts and increase the efficiency and sustainability of climate solutions in the country.

This chapter on Russia is dedicated to the experience of medico-geographical analysis of certain climate-related diseases' spread in Russia at the end of the twentieth century and the beginning of the twenty first century using the case of Tularemia and Anthrax. The role of climate change as a trigger factor causing the advancing spread of diseases has been analyzed. The potential change in ranges due to predicted climate warming was studied according to climate model INM-CM5.0. A series of maps was compiled to identify the territories prone to suitability changes for the infection foci for the period up to 2100. It was determined that regions with temperate and arctic climate may become vulnerable to the emergence of climate-related diseases in the course of environmental changes.

While still an active area of research, the increasing average temperature and decreasing atmospheric humidity due to climate change is predicted to increase the frequency and intensity of large wildfires in Canada. Wildfire smoke causes immediate respiratory distress, although there is a noted absence of research into prolonged exposure and long-term health outcomes. Further, evacuation from wildfires causes short-term hardships which leads to long-term mental health outcomes. Options to adapt to wildfires are limited and our capacity to prevent ever worsening wildfires in the future may be overwhelmed.

There is an accumulating body of evidence showing the human health impacts of climate change. Although research on physical human effects predominates, recent studies show a significant mental health impact of climate-and-weather-related events. Awareness of climate-related mental health issues has important ramifications for the implementation of national healthcare policies. This is more so in the United States (the third most populous country and a leading contributor to global greenhouse gas emission), where over one in ten adults live with severe mental illness and/or a substance use disorder (SUD). In the 2016 *Climate and Health Assessment* report, the U.S. Global Change Research Program (USGCRP) noted (with very high confidence in some instances) that exposure to weather and climate extremes increases the risk of trauma-and-stressor-related disorders, anxiety, depression and substance use, among other psychological issues. Given that mental and SUDs are the leading cause of years lived with disability globally, preventing the exacerbation of this reality by a changing climate is crucial. In this chapter, we introduce the reader to establish mental health consequences of climate change with focus on the United States.

This Chapter on Brazil presents and analyzes the role of public policies in Brazil in the face of the effect of climate change, focusing on displacement and women's health. Its describes the role of the Brazilian State and the attention given to the health field of displaced women. In this context, we reflect the social and economic vulnerability of women in Brazil and the fragility the social assistance policies for this population who, as a result of the social and political dimension of disasters, lose their homes and become either homeless and/or displaced.

Some chapters in the book have additionally discussed how the health risks from climate change are not equally shouldered by different population groups. Older people, the chronically ill, low-income households, and people living in deprived urban and coastal communities are already beginning to experience the threat of the health impacts from climate change. This book in general has also discussed how there are a range of co-benefits for human health that are associated with decarbonization and climate change mitigation action–including reducing death and diseases from air pollution, and water borne diseases.

Focus on Regional Studies

Understanding of regionally varied environmental determinants of health need to be incorporated in government health care policies of geographically diversed regions. This is vital as new environmental, climate and health issues are emerging in light of vulnerability and require rapid identification and response. Such response may assist in tackling modern environmental, climate/health challenges and deal effectively with the cross-cutting climate change health issues that suits the 2030 Agenda for Sustainable Development and strengthening health resilience to climate change impact that calls for new approach to achieve WHO global strategy to eliminate the almost one quarter of the disease burden caused by unhealthy environment.

Reference

Smith, A.B. (2023) 2022 US billion dollar weather and climate disasters in historical context. https://www.climate.gov.

Index

R. Akhtar (ed.), *Climate Change and Human Health Scenarios*, Global Perspectives on Health Geography, https://doi.org/10.1007/978-3-031-38878-1

D

E

SPRINGER NATURE

GPSR Compliance

The European Union's (EU) General Product Safety Regulation (GPSR) is a set of rules that requires consumer products to be safe and our obligations to ensure this.

If you have any concerns about our products, you can contact us on ProductSafety@springernature.com

In case Publisher is established outside the EU, the EU authorized representative is:

Springer Nature Customer Service Center GmbH
Europaplatz 3
69115 Heidelberg, Germany

The manufacturer's authorised representative in the EU is Springer Nature Customer Service Centre GmbH, Europaplatz 3, 69115 Heidelberg, Germany. If you have any concerns regarding our products, please contact ProductSafety@springernature.com

Printed and bound by CPI Group (UK) Ltd, Croydon, CR0 4YY
30/09/2026
02225322-0019